NANOCOMPOSITES AND POLYMERS WITH ANALYTICAL METHODS

Edited by **John Cuppoletti**

Nanocomposites and Polymers with Analytical Methods
Edited by John Cuppoletti

Published by InTech
Janeza Trdine 9, 51000 Rijeka, Croatia

Publishing Process Manager Romina Krebel

Technical Editor Teodora Smiljanic

Cover Designer InTech Design Team

Additional hard copies can be obtained from orders@intechopen.com

Nanocomposites and Polymers with Analytical Methods, Edited John Cuppoletti
p. cm.
ISBN 978-953-307-352-1

For EU product safety concerns:
IN TECH d.o.o., Prolaz Marije Krucifikse Kozulić 3, 51000 Rijeka, Croatia,
info@intechopen.com or visit our website at intechopen.com.

Meet the editor

Dr. John Cuppoletti is Professor of Molecular and Cellular Physiology at the University of Cincinnati (Ph.D. 1977 Univerisity of California, Davis). His research interests include pumps, carriers, and ion channels of the epithelia. In recent years, potassium and chloride channel regulation, pharmacology, and molecular engineering have become increasingly important tools with which to study the functional mechanisms of these proteins. More recently, as the X-ray crystal structures of these proteins have become available, it has become possible to combine these types of studies with computational chemistry approaches in attempts to understand how the structure relates to the function of these proteins. These transport proteins have been placed into synthetic membranes in order to obtain more detailed information regarding protein function that cannot be obtained from purely biological membranes. The natural and engineered proteins in these membranes are functional, and have led to the development of many interesting devices. Examples include drug screening devices, devices for identification of toxic bacterial products, and membranes for use in fuel cells.

Contents

Preface

This book contains chapters on nanocomposites for engineering lighter, stronger and safer materials for engineering. In addition to structural engineering materials, there are also functional materials used in biotechnology and medicine including metal nanoclusters glass (quantum dots), materials for separation of biopolymers, and materials for the repair of teeth, repair and growth of bone and cartilage. Studies of other functional materials include nanocomposites for lithium batteries. Most chapters also contain methodological and analytic approaches. The book contains chapters on materials that are in use, as well as chapters on the development and testing of materials that could be used in the future.

Human health and biotechnology can benefit from advances in nanocomposite materials. Examples in this book include materials for separations of biomolecules and quantum dots for medical imaging. There is a presentation of some very interesting work being carried out on materials which will stimulate the growth of bone (rather than simple replacement of lost structures).

Nanocomposites containing wood fibers are being studied as a green, renewable alternative to other structural materials. Use of recycled or otherwise wasted wood fibers could help slow the loss of forests.

The chapters in this book on nanocomposites are diverse. The 16 chapters comprising this book have been grouped into three main parts:

1) *New materials and analytic methods*: Research was focused on nanocomposites of polymers, ceramics, and sintered materials. Also featured is a review article on polymer nanocomposites.

2) *New materials with unique properties*: Preparation methods of metal nanoclusters (quantum dots) were given, and their properties explained in a way which makes this technology understandable to the non-expert as well as the expert. To expand on this one example, the potential use for quantum dots is wide, including semiconductors, medical imaging, solar cells and energy storage devices, improved sources of solar energy. In the area of renewable, recyclable materials there is a chapter on the uses of wood and cellulose fibers.

3) *Applications of new materials:* Studies are presented on the development of new nanocomposite materials with unique electric properties, materials for use in package insulation, and materials for better lithium batteries which show promise over previous materials. Bone repair material research being carried out shows promise in resisting degradation and recruiting, and nurturing osteoblast cells within the engineered structures which can lead to growth of new bone. This approach takes into account bone development physiology in materials design, and is highly promising.

I am pleased to have had the opportunity to work with the authors and to have served as editor of this book which expands composite materials research into so many exciting areas of development of materials, engineering, medicine and dental restoration.

The book contains a wide variety of studies from authors from all around the world. I would like to thank all the authors for their efforts in sending their best papers to the attention of audiences including students, scientists and engineers throughout the world. The world will benefit from their studies and insights. The new possibilities of the open access press bringing together such a diverse group and to disseminate widely on the web is revolutionary, and without the contributions of the group and the mechanism of InTech Open Access Publisher, this Book titled "Nanocomposites and Polymers with Analytic Methods" would not be possible.

I also wish to acknowledge the help given by InTech Open Access Publisher, in particular Ms. Romina Krebel, for her assistance, guidance, patience and support.

John Cuppoletti, Ph.D.
Department of Molecular and Cellular Physiology
University of Cincinnati
Cincinnati OH
USA

Part 1

New Materials and Analytic Methods

1

Polymer Nanocomposites: From Synthesis to Applications

S. Anandhan[1] and S. Bandyopadhyay[2]
[1]*National Institute of Technology Karnataka,*
[2]*The University of New South Wales,*
[1]*India*
[2]*Australia*

1. Introduction

Nanocomposites are composites in which at least one of the phases shows dimensions in the nanometer range (Roy et al., 1986). These are high performance materials that exhibit unusual property combinations and unique design possibilities and are thought of as the materials of the 21st century. With an estimated annual growth rate of about 25% and huge demand for engineering polymers, their potential is so promising that they are useful in several applications ranging from packaging to bio-medical. Literature survey reveals that about 18,000 publications, including papers and patents, have been published on nanocomposites in the last two decades. It has been reported that at the nanoscale (below about 100 nm), a material's property can change dramatically. With only a reduction in size and no change in the substance itself, materials can exhibit new properties such as electrical conductivity, insulating behavior, elasticity, greater strength, different color, and greater reactivity-characteristics that the very same substances do not exhibit at the micro- or macroscale. For example,

1. By the time gold crystals are just 4 nm across, the melting point drops to 700 K from its "encyclopedia value" of 1337 K (Mulvaney, 2001).
2. White crystals such as those of ZnO and TiO_2 are used as paint pigments or whitening agents, but they become increasingly colorless as the crystals shrink in size, and ZnO and TiO_2 colloids become invisible to the human eye below about 15 nm (Mulvaney, 2001).
3. Aluminum-can spontaneously combust at the nanoscale and has been used as rocket fuel.

Additionally, as dimensions reach the nanometer level, interactions at interfaces of phases become largely improved, and this is important to enhance materials properties. In this context, the surface area/volume ratio of reinforcement materials employed in the preparation of nanocomposites is crucial to the understanding of their structure–property relationships. Further, discovery of CNTs (carbon nanotubes) (Iijima, 1991) and their subsequent use to fabricate composites exhibiting some of the unique CNT related mechanical, thermal and electrical properties (Biercuk et al., 1991; Ounaies et al., 2003; Weisenberger et al., 2003) added a new and interesting dimension to this area. The possibility of spinning CNTs into composite products and textiles (Dalton et al., 2003) made further inroads for the processing and applications of CNT-containing nanomaterials. Rubber based nanocomposites are attracting considerable interest in polymer science research. Incorporation of different nanoreinforcements such as layered silicate clays, carbon nanotubes, nanofibers and silica nanoparticles into elastomers significantly enhances their mechanical, thermal, dynamic mechanical, and barrier properties along with noticeable improvements in adhesion, rheological and processing behavior (Bhowmick et al., 2010). Nowadays, nanocomposites offer new technology and business opportunities for all sectors of industry, in addition to being environmental- friendly (Choa et al., 2003). This chapter concentrates on some of the recently studied polymer nanocomposites and highlights their synthesis, properties and applications.

2. Varieties of polymer nanocomposites

Almost all types of polymers, such as thermoplastics, thermosets and elastomers have been used to make polymer nanocomposites. A range of nanoreinforcements with different shapes have been used in making polymer nanocomposites. An important parameter for characterizing the effectiveness of reinforcement is the ratio of surface area (A) of reinforcement to volume of reinforcement (V) [McCrum et al., 1996]. A plot of A/V against aspect ratio [defined as the ratio of length (l) to diameter (d)]. It can be seen from figure 1 that the predicted optimum shape for the cylindrical reinforcement to maximize A/V is

- a»1 (a fiber), and
- a«1 (a platelet).

Therefore, it can be understood that the two main classes of nanoreinforcement are fibers (e.g., carbon nanotubes) and platelets (e.g., layered silicate clays).

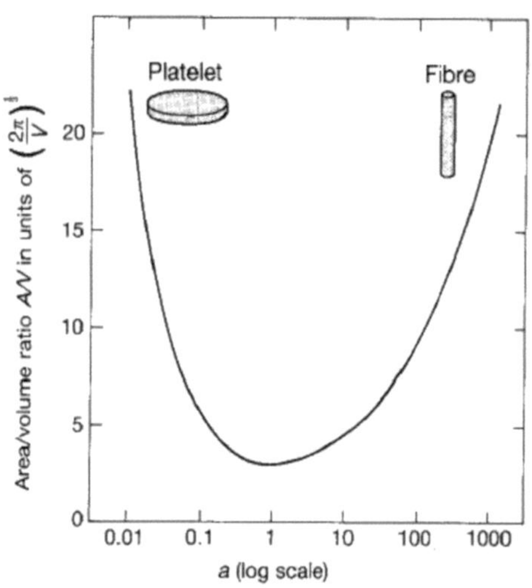

Fig. 1. Surface area to volume ratio A/V of a cylindrical particle of given volume, plotted versus aspect ratio a=l/d (from McCrum et al., 1996)

Polymer-layered inorganic platelet nanocomposites are bio-inspired materials. Mother of pearl (nacre) is a bio-nanocomposite made of: 95 percent aragonite (calcium carbonate), a brittle ceramic, and 5 percent flexible biopolymer (conchiolin). But, it is several times stronger than nylon; its toughness is almost equal to silicone. It is built like a "brick-and-mortar" structure, where millions of ceramic plates stacked on top of each other with each layer of plates glued together by thin layers of the biopolymer (Figure 2). Mixture of brittle platelets and the thin layers of elastic biopolymers make the material strong and resilient. The "brickwork" arrangement also inhibits transverse crack propagation.

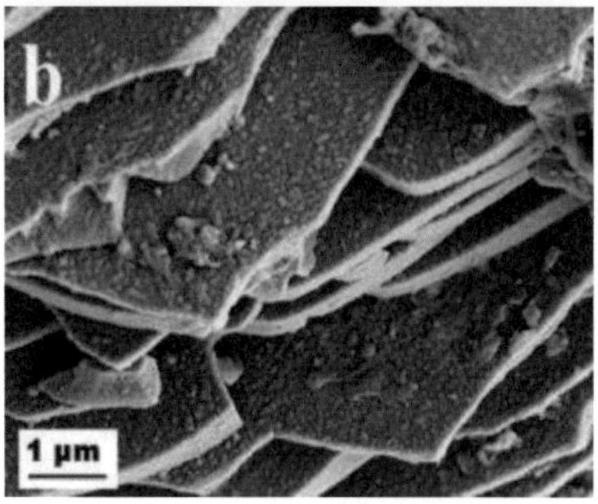

Fig. 2. Scanning electron microscope view of the fractured surface of nacre.

Table 1 presents a non-exhaustive list of possible layered host crystals that can be intercalated by a polymer. Most important one among them is layered silicate clay (figure 3). Amongst all the potential nanocomposite precursors, those based on clay and layered silicates have been more widely investigated probably because the starting clay materials are easily available and because their intercalation chemistry has been studied for a long time. Montmorillonite (MMT) is a naturally occurring 2:1 phyllosilicate, which has the same layered and crystalline structure as talc and mica but a different layer charge [Giese and Van Oss, 2002]. The MMT crystal lattice consists of 1 nm thin layers, with a central octahedral sheet of alumina fused between two external silica tetrahedral sheets (in such a way, that the oxygens from the octahedral sheet also belong to the silica tetrahedral). These layers organize themselves in a parallel fashion to form stacks with a regular van der Walls gap in between them, called interlayer or gallery (figure 4). The Na^{+} or Ca^{2+} ions residing in the interlayers could be replaced by organic cations such as alkyl ammonium ions or phosphonium ions by an ion-exchange reaction to render the hydrophilic layered silicate organophilic, so that polymer chains could be intercalated, thus causing the clay layer to be either swollen or exfoliated [Ma et al, 2004].

Fig. 3. Layers of naturally occurring phyllosilicate clay as seen under SEM.

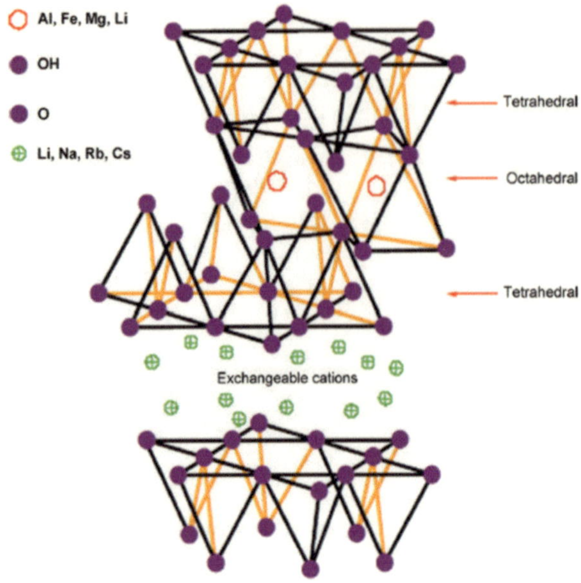

Fig. 4. The structure of a 2:1 layered silicate (from Koo, 2006).

Chemical nature	**Examples**
Element	Graphite
Metal chalcogenides	$(PbS)_{1.18}(TiS_2)_2$, MoS_2
Carbon oxide	Graphite oxide
Metal phosphate	$Zr(HPO_4)_2$
Clays and layered silicates	Montmorillonite, hectorite, saponite, fluoromica, fluorohectorite, vermiculite, kaolinite, magadiite.
Layered double hydroxides	$M_6Al_2(OH)_{16}CO_3.nH_2O$; M = Mg, Zn

Table 1. Examples of layered host crystals susceptible to intercalation by a polymer.

Carbon nanotubes are hexagonally shaped arrangements of carbon atoms that have been rolled into tubes. Carbon nanotubes are allotropes of carbon with a cylindrical nanostructure (scheme 1). Nanotubes have been constructed with length-to-diameter ratio of up to 132,000,000:1, significantly larger than any other material. The first polymer nanocomposites using CNT as reinforcement were reported in 1994 (Ajayan et al., 1994). CNTs possess high flexibility (Cooper et al., 2001), low mass density (Gao et al, 1998), and large aspect ratio (typically ca. 300-1000). CNTs have a unique combination of mechanical, electrical, and thermal properties that make nanotubes excellent candidates to substitute or complement the conventional nanoreinforcements in the fabrication of multifunctional polymer nanocomposites. Some nanotubes are stronger than steel, lighter than aluminum, and more conductive than copper. For example, theoretical and experimental results on individual single-walled CNTs (SWNT) show extremely high tensile modulus (640 GPa to 1 TPa) (Uchida and Kumar, 2005) and tensile strength (150-180 GPa) (Walt, 2004). Depending on their structural parameters, SWNT can be metallic or semiconducting, which further expands their range of applications. Because of the nearly one-dimensional electronic structures, metallic nanotubes can transport electrons over long tube lengths without significant scattering (McEuen, 1999). Chemical pretreatments with amines, silanes, or addition of dispersants improve physical disaggregation of CNTs and help in better dispersion of the same in polymer matrices (Ganguly et al., 2008). Figure 5 is a transmission electron micrograph of a portion of a nanotube showing the multiwall structure surrounding the hollow core. Figure 6 is a scanning electron micrograph showing the curvilinear structure of multi-walled nanotubes (MWNT).

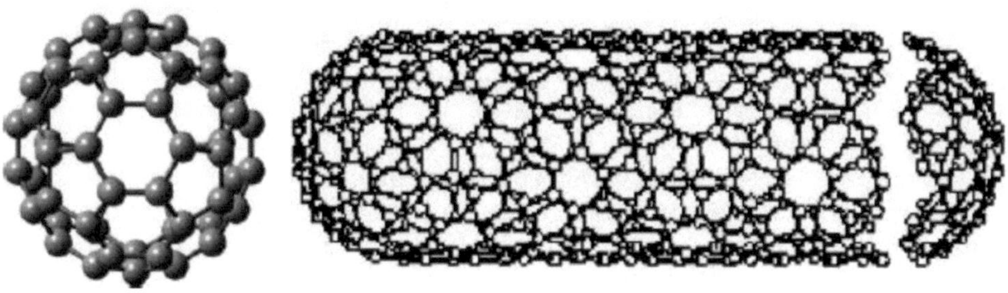

Scheme 1. Ideal structures of a fullerene and a carbon nanotube (from Wang et al., 2004).

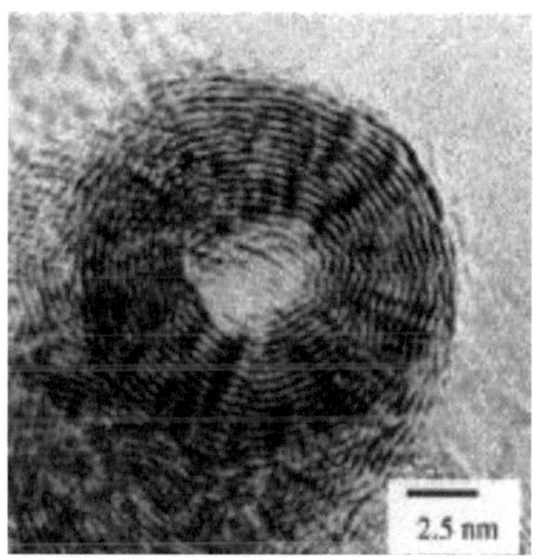

Fig. 5. TEM end-on view of a MWNT.

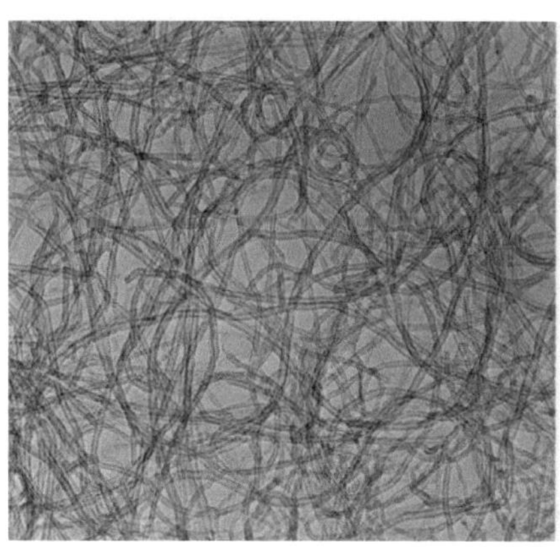

Fig. 6. Scanning electron micrograph of MWNT.

Wang et al. (2004) reviewed various methodologies for incorporating fullerene and CNT structures into polymers. The results of studies on fullerene and CNT based polymer nanocomposites show that the combination of the unique properties of CNTs and fullerenes with functional polymers will lead to novel materials with unusual mechanical, electrical, magnetic, and optical properties. Both conducting and conventional polymers can be used as matrices in the preparation of functional composites. The structures and properties of these hybrids are strongly dependent on the properties and concentrations of the active components. The combination of the unique properties of fullerenes and CNTs with polymers makes these materials potential candidates for many applications, such as data storage media, photovoltaic cells and photodiodes, optical limiting devices, photosensitive drums for printers, and so on.

Imogolite, consisting of hollow tubes with an external diameter of 2 nm and length of a few micrometers, is a naturally occurring hydrous aluminosilicate polymer found in soils of volcanic origin with a net composition $(HO)_3Al_2O_3SiOH$ (Yamamoto et al., 2005). Imogolite can also be synthesized by various routes (Suzuki and Inukai, 2010). The tubes have curved gibbsite sheets with SiOH groups on the inner surface and AlOH groups on the outer surface (scheme 2). Atomic force microscopy image of synthetic imogolite is shown in figure 7.

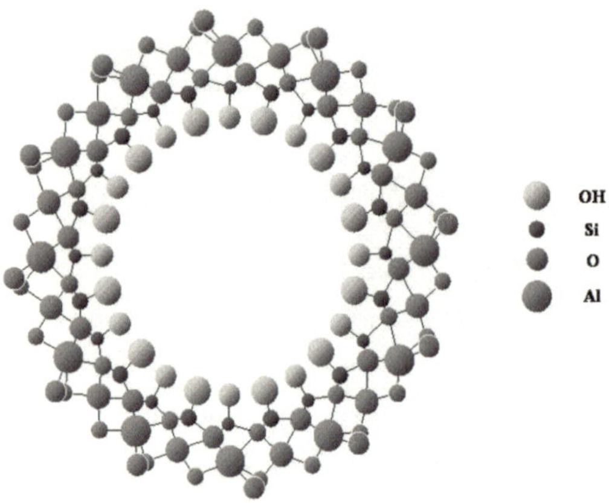

Scheme 2. Side view of the chemical structure of imogolite (from Lee et al., 2006).

Fig. 7. Atomic force microscope image of synthesized imogolite.

The structure of graphite (scheme 3) reveals that it is made of several grapheme layers those are held together by van der Waals forces (www.benbest.com, 2011) and can be exfoliated similar to layered silicate clay by chemical modification.

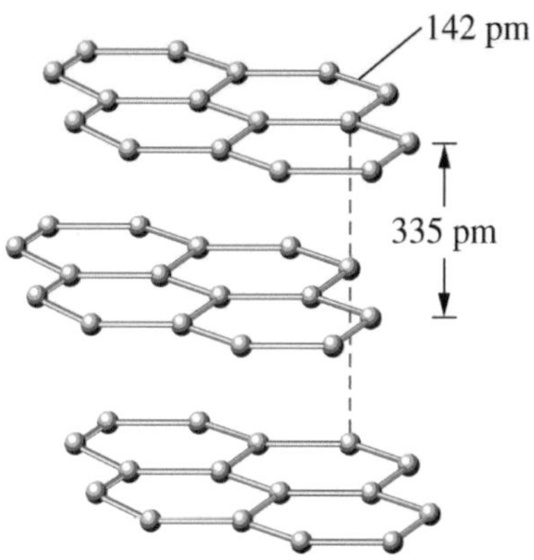

Scheme 3. 3D structure of graphene layers in graphite.

High crystallinity of graphite is disadvantageous in forming nanocomposites with polymers, as giant polymer molecules do not find spaces within the graphene sheets. This has been overcome by modifying graphite flakes with several oxidizing agents (Inagaki et al., 2004). The effective method of preparing the polymer–expanded graphite composite is by rapidly heating the pretreated (oxidized) graphite to a high temperature. The exfoliation of graphite is a process in which graphite expands by up to hundreds of times along the c-axis, resulting in a puffed-up material with a low density and high temperature resistance (Chung, 1987). The graphite thus resulted contains nanodimensional flakes (expanded graphite) providing greater surface/volume ratio for interaction with suitable polymer matrices.

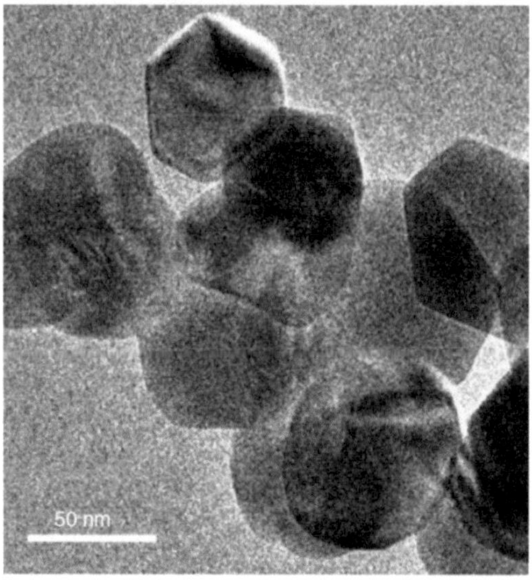

Fig. 8. Transmission electron microscopic (TEM) image of nano-zinc oxide (ZnO) (from Sahoo and Bhowmick, 2007).

Recently, several metal oxides apart from silica have been investigated and reported for rubber-based nanocomposites. Some important and commercially meaningful oxides used in rubber are ZnO, $Mg(OH)_2$, $CaCO_3$, zirconate, iron oxide, etc. When used in carboxylated nitrile rubber as curative, ZnO nanoparticles (figure 8) show excellent mechanical and dynamic mechanical properties [Sahoo and Bhowmick, 2007]. Different polymer–nano-ZnO hybrid systems based on epoxy [Liufu, 2005], poly(styrene-*co*-acrylic acid) [Ali and Iliadis, 2005], polyurethane [Zheng et al., 2005], etc. have been reported. Recent development of rubber nanocomposites by other nanoreinforcements includes piezo-rubber application by incorporating lead–zirconate by Tandon et al. [1993], Fe-containing silicone rubber by Yurkov et al. [2006], and crab shell whisker-reinforced natural rubber nanocomposites by Nair and Durfreshe [2003]. Thomas et al (2008, 2009 and 2010) extensively studied the properties of poly(ethylene-*co*-vinyl acetate) (EVA) and polystyrene nanocomposites based on nano calcium phosphate.

3. Functionalization of nanoreinforcements

On most occasions, nanoreinforcements need to be surface modified before incorporating them in to polymer matrices. This is because of the polarity difference between nanoreinforcements and majority of the polymer matrices. There are several approaches used in modifying surfaces of nanoreinforcements. The following paragraphs discuss some of the recent progresses made in the field of surface modification of nanoreinforcements.

In order to enhance the compatibility between clay and an organic polymer, one typically must ion exchange the sodium counter ion of the clay for an organophilic ion, usually an 'onium' ion, such as alkyl ammonium, sulfonium, phosphonium and imidazolium. The most common treatments that have been used to produce the organically-modified clays are ammonium salts, and this cation contains at least one long alkyl chain. The use of phosphonium salts and stibonium salts provides a marginal increase in thermal stability (Zhu et al., 2001). The formation of a nanocomposite requires the clay to be well-dispersed throughout the polymer matrix. A nanocomposite is described as intercalated if the registry between the clay layers is maintained and it is called exfoliated, or delaminated, if this registry is lost. For some properties, notably flexural modulus and permeability, it is believed that an exfoliated system shows more enhanced properties while, for fire retardancy, there is apparently no difference in behavior between exfoliated and intercalated systems. Zhang and Wilkie (2004) reported the preparation of nanocomposites of polystyrene with carbocation substituted clay. The thermal stability of this clay was greatly enhanced compared to the normal, ammonium-modified clays.

Wypych and Satyanarayana (2005) proposed a strategy to the use of synthetic and natural layered compounds, taking into account their ability to be exfoliated in the form of single layers, which can be chemically grafted with key molecules; a similar procedure can also be applied to fibrous materials. These surface-grafted molecules can carry reactive groups to be bonded to the polymer matrices. Thus adhesion between the reinforcement and the polymer matrix can be achieved.

Functionalization of MWNTs using mineral acids such as H_2SO_4 and HNO_3 is well documented in the literature. Vivekchand et al., (2007) have extensively reviewed functionalization of nanotubes and nanowires by various methodologies. For example, SWNT can be functionalized in ionic liquid using a mortar and pestle (Scheme 4) (Price et al., 2005). SWNTs can be cut into short segments by controlled oxidation by piranha (HNO_3/H_2SO_4) solutions [Ziegler, 2005].

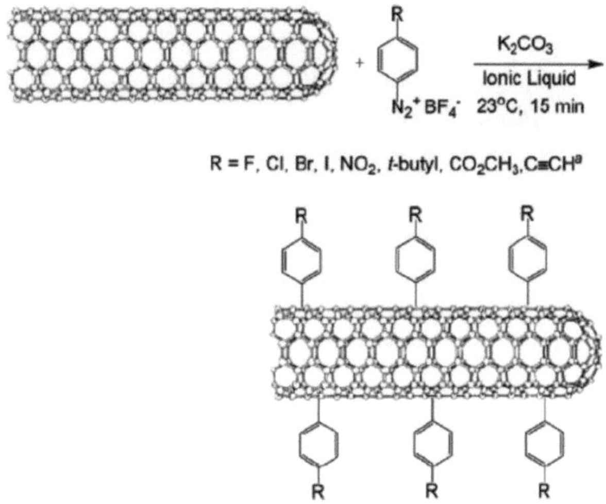

Scheme 4. Functionalization of CNT.

Metal oxide coated CNTs and inorganic nanowires can be prepared as shown in scheme 5 (Gomathi et al., 2005). The ceramic oxide coating provides improved compatibility with polar matrices as well as improved thermal stability of the nanotubes and nanowires. Natural graphite as such is not reinforcing in nature. But when modified to expanded graphite by high temperature heat treatment, this could be used as reinforcing nanoreinforcement in various polymer matrices. Recently George et al. (2007 & 2008) synthesized EVA/graphite nanocomposites using both tailor-made and commercially available expanded graphite. In both the cases, well-dispersed, fine graphite flakes were seen within the EVA matrices and this resulted in superior mechanical, dynamic mechanical, and processibility characteristics over the systems having natural graphite.

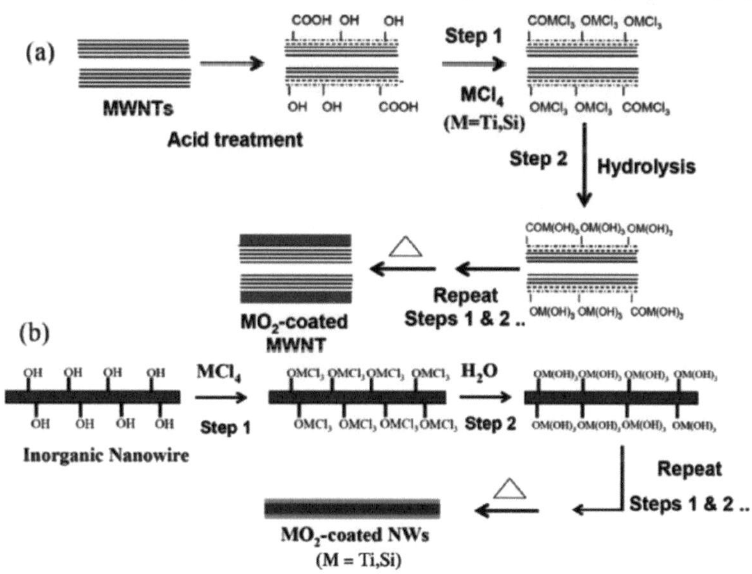

Scheme 5. The ceramic oxide coating process for (a) carbon nanotubes and (b) metal oxide nanowires (from Gomathi et al., 2005).

Imogolite nanotubes can be surface modified by 3-aminopropyltriethoxysilane (3-APS) in anhydrous toluene at 100 °C as shown in scheme 6 (Qi et al., 2008). Such surface modifications help compatibilize the inorganic nanotube imogolite with organic polymer matrices so that nanocomposites with a good interphase could be developed.

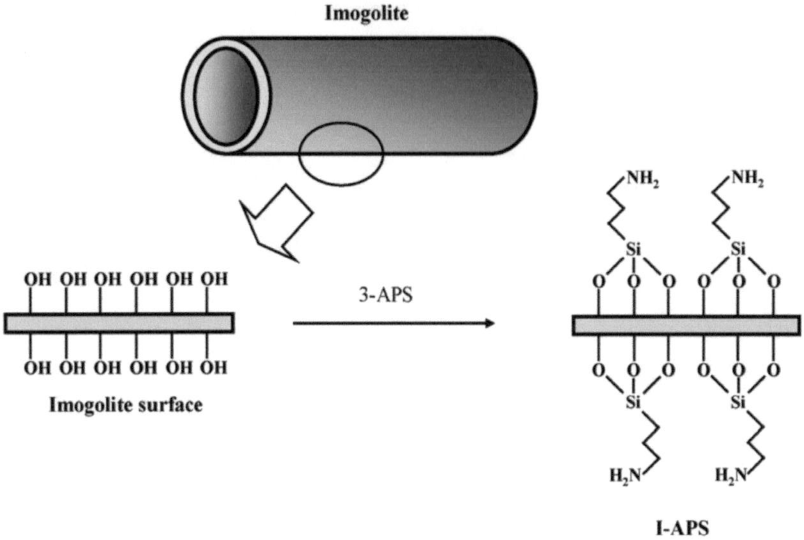

Scheme 6. Modification of imogolite surface by 3-APS (from Qi et al., 2008).

4. Synthesis of nanocomposites

Basically there are three methods to prepare nanocomposites. They are: solution casting, melt blending and in-situ polymerization. In solvent casting, a polymer, a solvent and nanoreinforcement are combined and thoroughly mixed by ultrasonication and the solvent is allowed to evaporate leaving behind the nanocomposite typically as a thin film. The solvent chosen should completely dissolve the polymer as well as disperse the nanoreinforcement. The solvent used will help in the mobility of the polymer chains which in turn helps in the intercalation of the polymer chains with the layered nanoreinforcement. In the case of melt blending, extruder or an internal mixer is used. Polymer and nano reinforcement are added in the extruder and subjected to intensive mixing for some time and nanocomposite comes out from the die. In this method, polymer mobility simply comes from thermal energy. In case of in-situ polymerization, which is widely used for PLS nanocomposite synthesis, initially the monomer and nano reinforcement are mixed. The monomer is allowed to intercalate between the silicate layers. Once the monomer is intercalated it is polymerized .The polymerization may be due to some surface modification at silicate surface or due to any functionalities present which catalyze the reaction (Alexandre and Dubois, 2000). There have been numerous reports on the synthesis of conducting polymer nanocomposites (Fang et al., 2008; Gangopadhyay and De, 2000). In an interesting work by Lee et al (2006), polypyrrole coated imogolite was synthesized by exposing $FeCl_3.6H_2O$ absorbed imogolite to pyrrole vapor under ambient conditions (scheme 7).

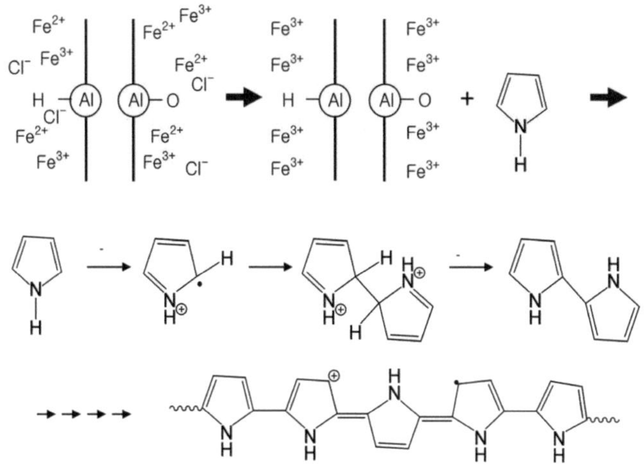

Scheme 7. Scheme to prepare polypyrrole-imogolite in a doping state with an acceptor dopant ($FeCl_3$) (from Lee et al., 2008).

5. Properties of polymer nanocomposites

5.1 Barrier properties and flame retardancy

The search for non-halogenated flame retardants has led to nanoclays, one nm thick by 1000 nm diameter. Initial research showed that the addition of as little as 5% of nano-sized clay particles could produce a 63% reduction in the flammability of nylon-6. More recent studies have shown that flame retardancy in many other polymers can be boosted by dispersing clay at the molecular level. Clays are believed to increase the barrier properties by creating a maze or 'tortuous path' (Scheme 8) that retards the progress of the gas molecules through the matrix resin (Neilson, 1967). For example, polyimide/layered silicate nanocomposites with a small fraction of O-MMT exhibited reduction in the permeability of small gases, e.g. O_2, H_2O, He, CO_2, and ethyl acetate vapors [Ray and Okamoto, 2003].

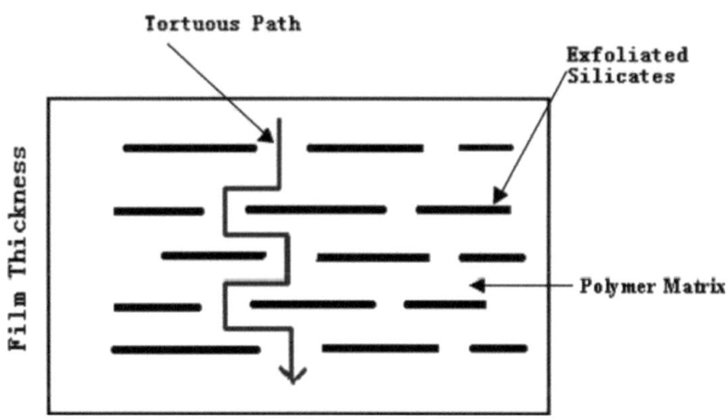

Scheme 8. Neilson's tortuous path model for barrier enhancement of nanocomposites (from Neilson, 1967).

Kashiwagi et al. [2002] studied multiwall nanotubes in polypropylene. A comparison of heat release rate curves among the three samples is shown in Figure 9. The results show that the heat release rates of the PP/MWNT nanocomposites are much lower than that of PP even though the amount of MWNTs in PP is quite small. This reduction in heat release rate is at least as much as what was previously found for clay nanocomposites in a maleic anhydride modified PP. Kashiwagi attributed the performance of the MWNT more to the presence of the small amount of iron catalyst in the MWNT. This iron may form iron oxides during combustion and iron oxides have been used as flame retardant additives. Another possibility that has been suggested is that the nanotubes function by conducting heat away from the flame zone.

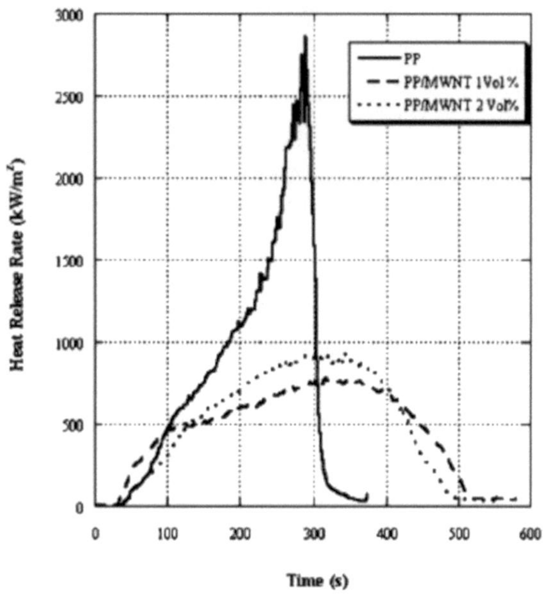

Fig. 9. Heat Release Rate for PP with CMWNT (from Kashiwagi et al., 2002).

Beyer [2002] studied the effect of adding montmorillonite nanoclays modified with a quaternary ammonium compound and carbon multiwall nanotubes in ethylene-vinyl acetate (EVA). Beyer made several observations:

1. Nanotubes did not accelerate the time to ignition like the nanoclays (the clay's acceleration was attributed to the decomposition of the quaternary ammonium compound within the nanoclay.)
2. Nanotubes are better then nanoclays at reducing the peak heat release rate at either 2.4% or 4.8% loading. The char showed much lower crack density for nanotubes. The hypothesis is that the high aspect ratio of the nanotubes formed a reinforcing composite that made the char more resistant to cracking. The reduced cracking provided better insulation to the unburned polymer surface and hence reduced the emission of volatile gasses into the flame area.
3. An equal mixture of 2.4% nanotubes and 2.4% nanoclay gave a synergistic reduction in peak heat release rate. The char showed the least amount of cracking, due to the combined reinforcement effect of both fillers. Das and Bandyopadhyay (2011) and Das et al. (2011) developed poly(methyl methacrylate) (PMMA)/MWNT nanocomposites and investigated their thermogravimetric and resistivity properties.

5.2 Optical clarity

The presence of reinforcement incorporation at nano-levels has been shown to have significant effects on the transparency and haze characteristics of films. In comparison to conventionally filled polymers, nanoclay incorporation has been shown to significantly enhance transparency and reduce haze. With butyl rubber, polyamide and acrylic rubber based clay nano composites, this effect has been shown to be due to modifications in the crystallization behavior brought about by the nanoclay particles; spherullitic domain dimensions being considerably smaller. Clays are just into thin, albeit their micro- lateral size. Thus, when single layers are dispersed in a polymer matrix the resulting nanocomposite is optically clear in the visible region (Figure 10). At the same time, there is a loss of intensity in the UV region (for $\lambda < 300$ nm), mostly due to scattering by the MMT particles. There is no marked decrease in the clarity due to nano-dispersed reinforcements compared to that of the neat-unfilled-polymer (for any relevant o-MMT loadings $\varphi \leq 9$ wt%). This is a general behavior as seen by UV/vis transmittance for thick films (3–5 mm) of polymer/MMT nanocomposites, based on PVA [poly(vinyl alcohol)] (Strawhecker and Manias, 2000), PP (Manias et al, 2001), and several epoxies. Recently, Anandhan and Arunjunairaj (2011) have obtained similar results on poly(ethylene-*co*-vinyl acetate-*co*-carbon monoxide)/O-MMT hybrid nanocomposites.

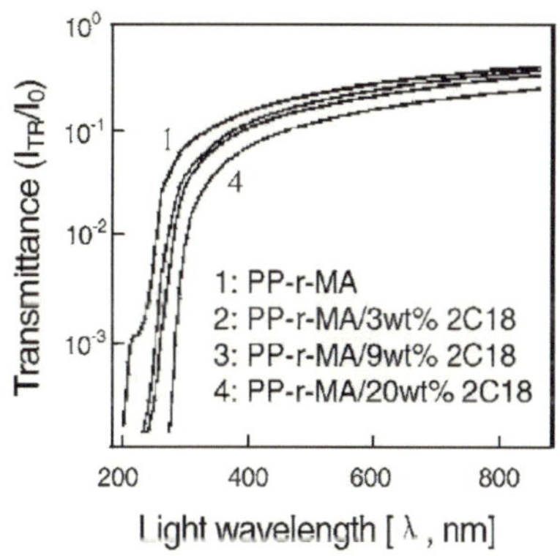

Fig. 10. UV-visible transmittance for MA-functionalized PP and its MMT nanocomposites as a function of MMT loading (from Manias et al., 2001).

Thermoplastic elastomeric polyurethane (TPU) and O-MMT nanocomposites were synthesized and characterized by Lee et al (2004). They found out that the light tranmittance of TPU/O-MMT did not reduce appreciably even after incorporation of 5 wt% clay in TPU matrix (figure 11).

Fig. 11. Visual appearance of TPU/O-MMT nanocomposite films: a) TPU, b) TPU/3wt%O-MMT, c) TPU/5 wt% O-MMT.

A plausible reason for the above observations could be that the size of nanoclay particles is less than the wavelength of visible light; hence, visible light rays are not appreciably scattered by nanoclay particles. But, visible light rays could be appreciably scattered by regions where clay particles form agglomerates. This fact could be understood by looking at scheme 9 (Wilson et al., 2002).

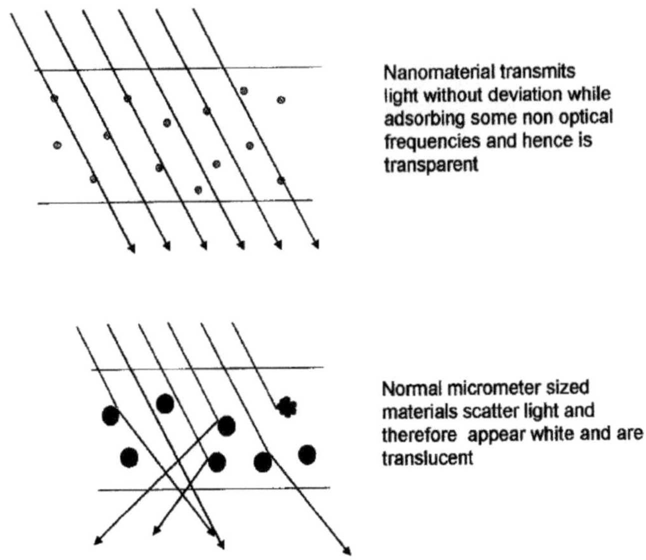

Scheme 9. Light rays passing mostly undeflected through an array of nanoparticles, so the array is transparent. When the particle size of the same material increases, light rays are scattered and the material becomes translucent (from Wilson et al., 2002).

5.3 Miscellaneous properties

Razafimahefa et al (2005) found out that a polyamide-6/clay nanocomposite fiber yarn dyes itself faster with disperse dyes than unfilled polyamide-6 yarns. Aqueous coatings of intercalated smectite clay particles in poly(N-vinyl pyrrolidone) and poly(ethylene oxide) matrices were found to have the potential to be used in inkjet media with improved receptor properties (Majumdar et al., 2003). Modification of elastomer surfaces (having low surface energy) using nanoreinforcements to increase their adhesive behavior is relatively a new approach. In a recent work, two different nanoclays such as Cloisite-15A (modified nanoclay) and Cloisite-NA (unmodified nanoclay) were incorporated into the vulcanized EPDM matrix and the adhesion strength between vulcanized EPDM to unvulcanized EPDM was investigated. It was found that Cloisite-NA showed better peel strength as compared to Cloisite-15A. A maximum of 56% improvement of peel strength was accomplished with 4 wt% Cloisite-NA concentration in EPDM. A similar improvement of adhesion is also observed with other rubbers (Bhowmick et al., 2010). For example, adhesion of thiokol rubber with aluminum was enhanced by five folds with the incorporation of 8 wt% of Cloisite-30B nanoclay (Figure 12).

PVA/Na^+-MMT nanocomposites of various compositions were studied by Strawhecker and Manias (2000). The clay layers promote a new crystalline phase different than the one of the respective neat PVA, characterized by higher melting temperature and a different crystal structure. This new crystal phase reflects on the composite materials properties. Namely, the hybrid polymer/silicate systems have mechanical, thermal, and water vapor transmission properties, which are superior to that of the neat polymer and its conventionally filled composites.

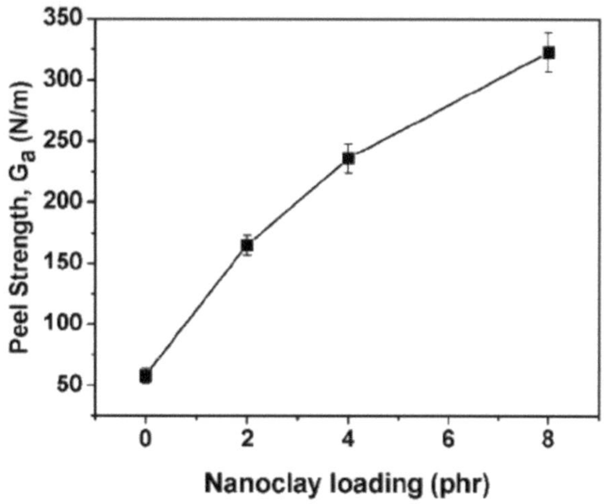

Fig. 12. Effect of concentration of Cloisite-30B nanoclay on adhesion strength of thiokol rubber against aluminum (from Bhowmick et al., 2010).

The improvement in properties of PVA could be attributed to the hydrogen bond formation between PVA and silicate layers (scheme 10). Heat distortion temperature (HDT) is the temperature at which a polymer sample deforms under a specified load. In general, improvements of HDT are reported by nanocomposite formation. Usually, a significant increase is achieved for clay contents of approximately 5 wt%, and then HDT values level off for higher clay loadings (Kojima et al., 1993; Ray et al., 2003; Nam et al., 2001).

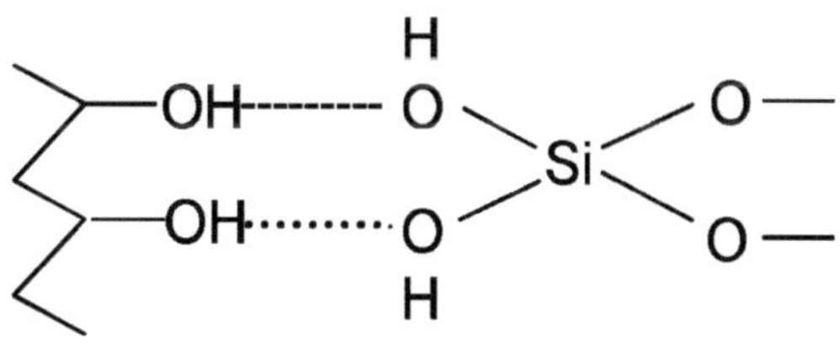

Scheme 10. Hydrogen bonded poly(vinyl alcohol/clay) organic-inorganic hybrid structure.

Atieh et al. [2005] studied the properties of natural rubber (NR)/MWCNT nanocomposites. Mechanical test results show an increase in the initial modulus for up to 12-fold in relation to neat NR. The composites containing 10–20 wt% uniformly dispersed CNT exhibited significant drop of volume resistivity [Kazumasa et al., 2005]. The effect of CNTs on the mechanical and electrical properties of ethylene-propylene-diene monomer (EPDM) rubber in comparison to that of high-abrasion furnace carbon black was studied by Ying et al. [2002]. Kar and Bhowmick [Ganguly et al., 2008] developed MgO nanoparticles and investigated their effect as cure activator for halogenated butyl rubber and obtained promising results.

Many nanocomposites based on conducting polymer matrices have been reported in the literature (Gangopadhyay and De, 2000). Lee et al (2006) studied electrical properties of polypyrrole coated imogolite and found that conductivity after modification with polypyrrole increased with polypyrrole thickness at various voltage conditions. They opined that the resultant conducting nanofibers can be utilized in electronic applications. Sengupta et al. (2010) have extensively reviewed mechanical and electrical properties of graphite and modified graphite reinforced polymer composites.

6. Applications of polymer nanocomposites

The improvements in mechanical properties of nanocomposites have resulted in major interest in numerous automotive and general/industrial applications. It includes potential for utilization as mirror housing on various types of vehicles, door handles, engine covers, and belt covers. More general applications include: packaging, fuel cell, solar cell, fuel tank, plastic containers, impellers and blades for vacuum cleaners, power tool housing, and cover for portable electronic equipment such as mobile phones and pagers.

6.1 Gas barriers for plastic bottles, packaging and sports goods

Hybrids made of poly(dimethyl siloxane) rubber and nanosilica generated in-situ by hydrolysis of tetraethyl orthosilicate can be specifically shaped, giving objects such as golf balls (figure 13) (Hu and Mackenzie, 1992; Chung et al, 1990; Mackenzie et al, 1992).

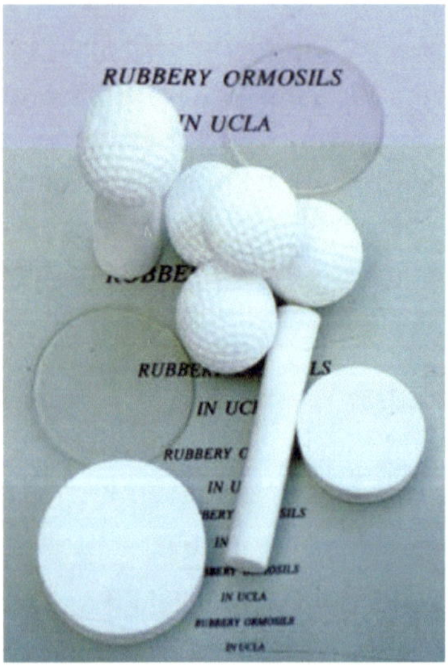

Fig. 13. Rubbery hybrids with different shapes (from Sanchez et al., 2005).

A number of polymer nanocomposites based on polymers, such as butyl rubber, styrene butadiene rubber, ethylene propylene diene monomer rubber, ethylene vinyl acetate copolymer, ethylene-octene copolymer, have been used commercially for barrier applications. These polymers can act as excellent barriers for many gases such as CO_2, O_2, N_2, and chemicals such as toluene, HNO_3, H_2SO_4, HCl, etc. Due to excellent solvent barrier properties PNCs have been utilized in chemical protective and surgical gloves in order to protect against chemical warfare agents and for avoiding contamination from medicine.

PNCs also have been widely used in food packaging and plastic containers, both flexible and rigid. Specific examples include packaging for processed meat, cheese, cereals and dairy products, printer cartridge seals, medical container seals for blood collection tubes, stoppers for medical containers and blood bags, baby pacifiers and drinking water bottles.

The incorporation of layered silicate clay reduces diffusion rate of small molecules, such as water and oxygen, through polymer films. Clay based polymer nanocomposites have been used in plastic bottle manufacturing industries for improving barrier, mechanical properties and self life of the product. To improve the bottle's properties and seek an alternative solution, plastic bottles have been tried in the beer package industry. The shelf-life of a clay based nanocomposite plastic beer bottle is more than 6 months. The first plastic beer bottle based on clay has been introduced by Honeywell. Nanocomposites have been used for beer bottle manufacture to solve many problems, such as the beer colloids instability, including biological and non-biological aspects, oxygen permeation and bad taste due to light exposure (Ke and Stroeve, 2005). A study on nitrile rubber/organoclay nanocomposites (in 10 wt % organoclay) reveals that the relative vapor permeability for water and methanol was reduced up to 85% and 42%, respectively, compared to the neat polymer.

One of the recently commercialized sports goods is double core Wilson tennis ball (figure 14). The clay nanocomposite coating of the Wilson tennis balls maintains the internal pressure for an extended period of time. The core is coated by a butyl rubber-clay (vermiculite) nanocomposite that acts as a gas barrier, doubling its shelf life. By polymer nanocomposites, more flexible coatings with gas permeability 30–300 times lower than butyl rubber has been produced. These coatings have been shown to be undamaged by strains up to 20%. This double core new tennis balls using this coating retain air longer, and it able to bounce twice as long as ordinary balls (improvement in air retention). It is anticipated this technology can be extended to the rubber industry and be incorporated into soccer balls or in automobile or bicycle tires [Koo, 2006]. Another important application in sports good is basketball shoe pouch manufactured by Triton Inc. from EVA/clay nanocomposites. Figure 15 shows clay nanocomposite pouch filled with helium inserts that fits into basketball shoe. The ultimate property of the pouch is that it can give good resilience for basket ball player while jumping; meantime, it exhibits excellent gas barrier properties. The Converse system developed this process, but, Triton System has manufactured it first [Koo, 2006].

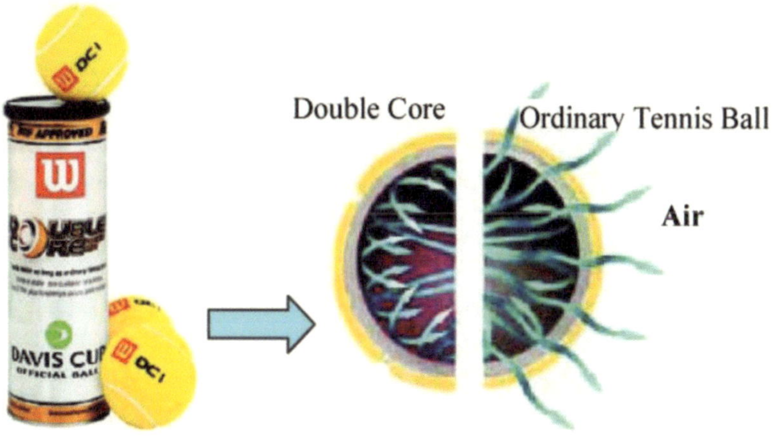

Fig. 14. The core of this Wilson tennis ball is covered by a polymer-clay nanocomposite coating (from Sanchez et al., 2005).

On the other hand tire industry is another very important application field for rubber/clay nanocomposites for improving barrier and mechanical properties of tires and tubes. The excellent air retention properties of butyl/halo butyl and chemically modified rubbers with incorporated clays are very well known in the tire industries as these nanocomposites are widely used in the inner tube and inner liner in automotive tires for improving many properties such as tensile, tear resistance, rolling resistance, dimensional stability, improved flame resistance and enhanced air retention.

6.2 Energy storage systems and sensors

Fuel cells act as electrochemical devices, which convert chemical energy of carbon, hydrogen and oxygen directly and efficiently into useful electrical energy with heat and water as the only byproducts. Due to incorporation of nano materials their efficiency increases considerably. In fuel cells, proton exchange membrane's role is to allow proton transport from the anode to the cathode, to be an electron nonconductive material and to act as a gas separating barrier (Prater, 1990). Typical membranes are made of organic polymers containing acidic functions such as carboxylic, sulfonic or phosphonic groups which dissociate when solvated with water, allowing H_3O^+ (hydrated proton) transport. Therefore, the membrane performance is related to the ionic group amount and to the hydration rate. Moreover, the membrane needs to be chemically (highly acidic medium) and thermally (from 80 to 140 °C) stable. Membranes made of silylated-sulfonated poly(ether ether ketone) (Premchand et al., 2008), poly(benzimidazole)/sulfonated silica nanoparticle nanocomposite (Suryani and Liu, 2009), fluorinated polybenzimidazole/silica nanoparticle composites (Chuang et al., 2007) were used as proton conducting membranes for fuel cell applications and were reported to be superior in terms of proton conductivity than pristine polymer membranes.

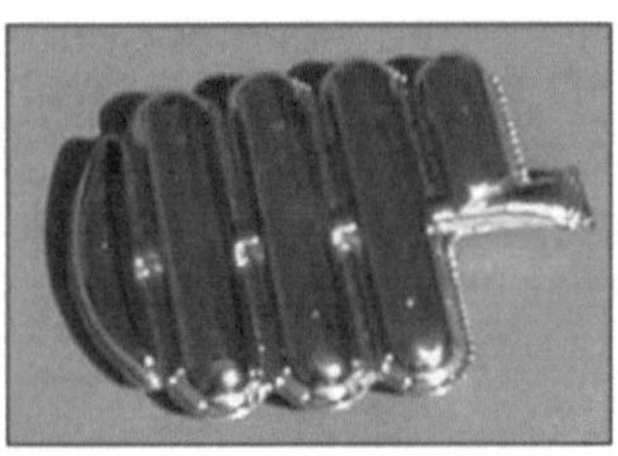

Nanocomposite pouch filled with helium

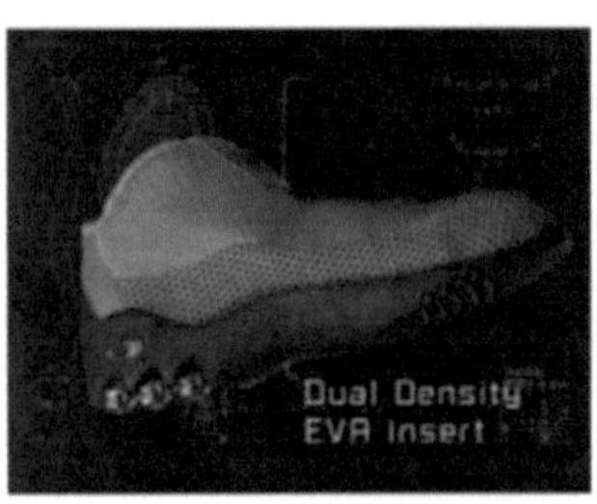

Insert that fits into the shoe

Fig. 15. Converse basketball shoe pouch by Triton Systems Inc (from Koo, 2006).

Sulphonated polystyrene ethylene butylene polystyrene/montmorillonite nanocomposites can be used as proton exchange membranes [www.azom.com, 2010] due to their superior proton exchange capacity. Several polymers are being used in fuel cell applications, such as hyperbranched polymer with a hydroxyl group at the periphery, cross linked sulphonated poly(ether ether ketone), sulfonated polybenzimidazole copolymer, phosphoric acid doped polybenzimidazole, sulfonated polyarylenethioethersulfone, sulfonated polybenzimidazole, etc. Clay incorporated thermoset polymer fuel cells exhibit higher proton conductivity, better ion exchange capacity and rate of conductivity is high even at higher humidity and improved mechanical properties [Bai and Ho, 2008]. Anis et al. (2008) developed PVA/nano heteropolyacid nanocomposites that could be used in fuel cell electrolyte membranes.

Another important storage application of clay based nanocomposite is solar cells. PLS nanocomposites based solar cells have the potential to become one of the leading technologies in conversion of sunlight to electrical energy. Because of their ease of processing from solution, fast and low cost mass production of devices is possible in a roll-to-roll printing fashion. They exhibit high efficiency, enhanced light absorption and possess excellent barrier cum mechanical strength [Yang and Loos, 2007]. Even though solar cells are used outdoors, one need not worry about damage from any mechanical actions or UV rays for clay based thermoset nanocomposites. Various polymers have been used for solar cell manufacturing, such as poly(phenylene-vinylene), polythiopene, polyfluorene, polyaniline, polypyrrole, etc. PLS nanocomposites have been widely used in sensor technology too. Many novel polymers, such as polyamidoamine dendrimer, silicone, hyper branched polymers, polyacrylamide, etc. have been used for sensing gas, atmospheric moisture, detection of solvent leaking in pipe line application, etc. Clay incorporated elastomers are being developed as sensors to detect fatigue, impact and large strain for aerospace applications.

6.3 Optical glass and membranes

Clay incorporated polymers have been shown, when employed to coat transparent materials, to enhance both toughness and hardness of these materials without sacrificing light transmission characteristics. An ability to resist high velocity impact combined with substantially improved abrasion resistance of PLS nanocomposites was demonstrated by Triton Systems. Owing to this reason and improved optical properties it has been widely commercialized in contact lens and optical glass applications (Haraguchi, 2011).

Polymer/clay nanocomposites can also be used to fabricate various types of membranes, such as solvent filters, filters for bacteria and virus (www.nanowerk.com), solid electrolytes for fuel cells, membrane for gas separation, etc (Cong et al., 2007). Organic membrane technology is used to design membranes for separating synthetic gas from natural gas. The transformation of natural gas into liquid materials requires design of inorganic membrane catalysts. Lot of research has been done on membranes based on thermoset clay nanocomposites. Polymers, such as polyimide, polybenzimidazole, cross-linked sulfonated poly(ether ether ketone),and polyacrylate, have been widely used for membrane applications. The creation of nanopores in nanocomposites by the use of nanoreinforcement is the main mechanism in gas, solvent separation and transportation. The nanopores decide the sensitivity and quality of the membrane (www.nanowerk.com).

6.4 Products with low flammability

The ability of nanoclay incorporation to reduce the flammability of polymeric materials is incredible. Gilman demonstrated the extent to which flammability behavior could be restricted in polymers like polypropylene with as little as 2% nanoclay loading. The improvement of flame resistance by incorporation of clay been commercialized in various applications including cable wire jacket, car seats (Hard Foams), packaging films, textile cloths, surface coatings for many steel products, paints; one of the higher end applications is rocket ablative materials core manufacturing (Koo, 2006). Na^+-MMT in nitrile rubber nanocomposites showed higher flame retardancy with good oil resistance as compared to a control sample [Wang et al., 2009]. Figure 16 shows nanocomposite power cable developed by Kabelwerk Eupen.

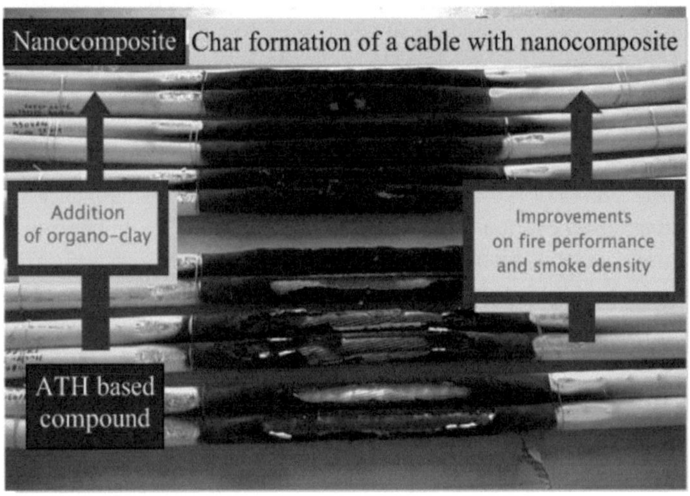

Fig. 16. Nanocomposite power cable by Kabelwerk Eupen (from Koo, 2006).

Polyurethane/clay based nanocomposites also exhibit superior flame retardancy and they have already been commercialized in automobile seats manufacturing (hard foam). Borden Chemicals' SC-1008, a resole phenolic, was selected as the resin for manufacturing rocket ablative material with MMT. The nano scale distribution of silicate layers leads to a uniform char layer that enhances the ablative performance. The formation of this char was only minutely influenced by the type of organic modification on the silicate surface of specific interactions between the polymer and the aluminosilicate surface, such as end-tethering of a fraction of the polymer chains through ionic interaction to the layer surface. The same formulation has also been used for simulated solid rocket motor casing. It exhibited excellent flame retardancy. The ablative test was performed at around 2200°C, and the results showed that the motor has not burned.

6.5 Electronics and automobile sectors

Application of thermoset/clay nanocomposites for electronics and automobile sectors is another big milestone. The ability of nanoclay incorporation to reduce solvent transmission through polymers such as specialty elastomers, polyimides, poly urethane, etc. has been demonstrated. A study reveals the significant reductions in fuel transmission through polyamide-6/66 by incorporation of nanoclay reinforcement. The major driving force for the usage of PNCs by tire companies is reduction of weight and processing costs. Clay is one of the naturally available materials and also it possesses very low density; it facilitates reduction in weight. Clay incorporated tires exhibit excellent mechanical properties as compared to ordinary tires as well as improved gas barrier performance for tubeless tire applications. Generally styrene butadiene and natural rubber based clay nanocomposites are mostly preferred for automobile tire manufacturing and butyl rubber is for tubes. It is due to improved abrasion resistance and thermal properties of the tires combined with their longer life (Koo, 2006).

The addition of clay into conducting polymers such as polypyrrole, polythiophene polyaniline, etc. sensibly increases conductivity and impedance effect with improved mechanical properties such as elongation, impact, scratch resistance and results in longer life of the product. These PNCs find many applications in solar cells, wind mills, electronic circuit boards, battery manufacturing, micro chips, transistor, etc. Nanocomposites consisting of spatially confined liquid crystals are of great interest due to the prospects of their application in optoelectronic devices, photonic crystals, depolarizers, scattering displays, information storage and recording devices such as compact disk, universal serial bus storage, and windows with adjustable transparency [Merhari, 2009].

6.6 Coatings

Coatings are important for modifying properties of surfaces. Several strategies have been tried by researchers for improving surface properties of products. One of the well versed developments is nanoclay based polymer coatings. Plenty of work has been done in this area [Jin et al, 2010; Turri et al., 2010]. Nanoclay incorporated thermoset polymer nano coatings exhibit superior properties such as superhydrophobicity, improved wettability, excellent resistance for chemicals, corrosion resistance, improved weather resistance, better abrasion resistance, improved barrier properties and resistance to impact, scratch, etc. Coating thickness depends on process parameters such as dipping time, temperature, nature of surfactant, purity of nanomaterial, etc. Turri et al. [2010] developed nano structured coatings based on epoxy/clay on steel substrates. The scratch strength of epoxy/clay nano structured coating was two times greater than the pristine epoxy coating. These kinds of coatings are best candidates for various applications, such as construction, thermal barriers for aerospace applications, automobile, and pipe line coatings for marine applications, sometimes for decorative purpose, textiles, etc. Clay and nano silver incorporated thermoset polymer nano coatings could be used to improve antibacterial properties and it has been widely used in medical sectors. At present these kinds of hybrid coatings are one of the promising ones for improving shelf life and antibacterial properties in medical field.

7. Conclusion

Among many highly hyped technological products, polymer nanocomposites are one of those, which have lived up to the expectation. Polymer nanocomposites exhibit superior properties, such as mechanical, barrier, optical, etc. as compared to micro- or macro-composites. Owing to this, polymer nanocomposites have shown ubiquitous presence in various fields of application. Polymer nanocomposites for various applications could be synthesized by proper selection of matrix, nanoreinforcement, synthesis method and surface modification of either the reinforcement or polymer (if required). Many products based on polymer nanocomposites have been commercialized. This review has tried to highlight various types of nanoreinforcements and their surface modification procedures, some unique properties of nanocomposites, various technological applications of polymer nanocomposites with some specific examples of commercialized products. Though it is not a comprehensive one, this review could give a basic idea about polymer nanocomposites to a beginner.

8. References

M. Alexandre, and P. Dubois, Materials Science and Engineering, 28 (2000) 1-63.

Ali, H.A. and Iliadis, A.A. Thin Solid Films 471, 154, 2005.

S. Anandhan, and M.Arunjunairaj, unpublished results (2011).

Anis. A, Banthia A. K., and Bandyopadhyay. S., Synthesis & Characterization of PVA/STA Composite Polymer Electrolyte Membranes for Fuel Cell Application; ASM International Journal of Materials Engineering and Performance, 2008, 17, 5, 772 - 779.

Anis. A, Banthia A.K., and Bandyopadhyay. S., Synthesis/characterization of PVA copolymer/ phosphomolybdic acid based crosslinked composite Polymer Electrolyte Membranes, J Power Sources 2008, 179, 69–80.

Atieh, M.A., Girun, N., Ahmadun, F.R., Guan, C.T., Mahdi, El. S., and Baik, D.R. J. Nanotech. 1, 1, 2005.

H. Bai, W. Ho, New sulfonated polybenzimidazole (SPBI) copolymer-based proton-exchange membranes for fuel cells. Journal of the Taiwan Institute of Chemical Engineers 2008; 40: 260-267.

G. Beyer, Fire and Materials, 26, (2002), 291-293.

A.K.Bhowmick, M.Bhattacharya, S.Mitra, P.K.Maji, A.Choudhury, J. J.George, G.C.Basak, K.D.Kumar, Kautschuk Gummi Kunststoffe, 62, 192, 2010.

Biercuk MJ, Llaguno MC, Radosvljevicm, HJ. Carbon nanotube composites for thermal management. *Applied Physics Letters*. 2002; 80(15):2767-2769.

Choa YH, Yang JK, Kim BH, Jeong YK, Lee JS, Nakayama T, Preparation and characterization of metal: ceramic nanoporous nanocomposite powders. *Journal of Magnetism and Magnetic Materials*. 2003; 266(1-2):12-19.

Y. J. Chung, S. Ting and J. D. Mackenzie, in Better Ceramics Through Chemistry IV, vol. 180, ed. B. J. J. Zelinski, C. J. Brinker, D. E. Clark and D. R. Ulrich, Materials Research Society, Pittsburg, PA, 1990, p. 981.

S. Chuang, S. L. Hsu, Y. Liu, Journal of Membrane Science 305 (2007) 353–363.

Chung, D.D.L. J. Mater. Sci. 22, 4190, 1987.

H. Cong, M. Radosz, B. Francis Towler and Y. Shen, Separation and Purification Technology,55, 2007, 281-291.

Cooper, C. A.; Young, R. J.; Halsall, M. *Composites, Part A* 2001, *32A*, 401-411.

Dalton AB, Coolins S, Munoz E, Razal JM, Ebron VH, Ferraris JP, Super-tough carbon-nanotube fibres: these extraordinary composite fibres can be woven into electronic textiles. *Nature*. 2003; 423(6941):703-703.

Das M., Ray D., Bandyopadhyay S., Banerjee S., Bandyopadhyay N. R, Basumallik A; Thermogravimetric and Resistivity Study of Ex situ and in situ Poly(Methyl Methacrylate)/Carboxylic Acid Group Functionalized Multiwall Carbon Nanotubes Composites. Journal of Applied Polymer Science, 2011, Volume 120, Issue 5, pages 2954–2961, DOI 10.1002/app.33320.

Das. M and Bandyopadhyay.S., Chapter 11: Polymethyl Methacrylate–Carbon Nanotube Composites, Page # 335-360: Carbon Nanotube Based Nanocomposites: Recent Development Editors: Kamal K. Kar and Alma Hodzic ISBN: 978-981-08-3712-9 Year: 2011 Publisher: Research Publishing, Singapore.

Fang FF, Choi HJ, Joo J. J Nanosci Nanotechnol. 2008, 8:1559-81.

R. Gangopadhyay and A. De, Chem. Mater., 2000, 12, 608-622.

A. Ganguly, J. J. George, S. Kar, A. Bandyopadhyay, A. K. Bhowmick, 'Rubber Nanocomposites based on miscellaneous nanoreinforcements', in 'Current topics in elastomers research', A. K. Bhowmick, Ed., 1st ed., CRC press, Boca Raton, 2008.

Gao, G.; Cagin, T.; Goddard, W. A., *Nanotechnology* 1998, *9*, 184-191.

George, J.J., Bandyopadhyay, A., and Bhowmick, A.K. J. Appl. Polym. Sci. 2007, doi 10.1002/APP.25067.

George, J.J. and Bhowmick, A.K. J. Mater. Sci. 43, 702, 2008.

Giese RF, Van Oss CJ (2002) Colloid and Surface Properties of Clays and RelatedMinerals, Marcel Dekker, New York.

A. Gomathi, S. R. C. Vivekchand, A. Govindaraj, C. N. R. Rao, Adv. Mater., 2005, 17, 2757–2561.

K. Haraguchi, Polymer Journal (2011) 43, 223–241.

http://www.azom.com/details.asp?ArticleID=921 (accessed on 22 September, 2010).

http://www.benbest.com/cryonics/lessons.html (accessed on 18March, 2011).

http://www.nanowerk.com/spotlight/spotid=377.php (accessed on 30 March, 2011).

Y. Hu and J. D. Mackenzie, Mater. Res. Soc. Symp. Proc., 1992, 271, 681.

Iijima S. Helical microtubes of graphitic carbon. *Nature.* 1991; 354(6348):56-58.

Inagaki, M., Tashiro, R., Washino, Y., and Toyoda, M. J. Phys. Chem. Solids 65, 133, 2004.

H. Jin, J. J. Wie, S. C. Kim, Effect of Organoclays on the Properties of Polyurethane/Clay Nanocomposite Coatings, Journal of Applied Polymer Science 2010;117: 2090-2100.

Kamigaito O. What can be improved by nanometer composites? Journal of Japan Society of Powder Metallurgy. 1991; 38:315-321.

T. Kashiwagi, et.al., *Macromol. Rapid Commun.*, 23, (2002) 761-765.

Kazumasa, Y., Haruumi, H., Tetsuro, O., and Hokoku, K. Fukuoka-ken Kogyo Gijutsu Senta, 15, 35, 2005.

Ke YC, Stroeve P. Background on Polymer-Layered Silicate and Silica Nanocomposites, Polymer-Layered Silicate and Silica Nanocomposites. 1st Edition, Elsevier, Amsterdam, 2005.

Kojima Y, Usuki A, Kawasumi M, Okada A, Kurauchi T, Kamigaito O. J Polym Sci Polym Chem 1993;31:983–986.

J. H. Koo, Polymer Nanocomposites - processing, characterization and applications, McGraw-Hill, New York, 2006, 235-261.

H. S. Lee, M. Y. Choi, S. Anandhan, D. H. Baik, S. W. Seo, 'Microphase structure and physical properties of polyurethane/organoclay nanocomposites', ACS PMSE preprints, 91, 638 (2004).

Y. Lee, B. Kim, W.Yi, A.Takahara, D.Sohn, Bull. Korean Chem. Soc. 2006, Vol. 27, 1815-1818.

Liufu, S., Xiao, H., and Li, Y.P., Polym. Degrad. Stab. 87, 103, 2005.
Ma, X, Lu H, Liang G, Yan H (2004) J. Appl. Polym. Sci., 93:608-614.
D. Majumdar, T. N. Blanton, D.W. Schwark, Applied Clay Science, 23 (2003) 265- 273.
E. Manias, A. Touny, L. Wu, K. Strawhecker, B. Lu and T. C. Chung, Chem. Mater. 2001, 13, 3516-3523
J. D. Mackenzie, Y. J. Chung and Y. Hu, J. Non-Cryst. Solids, 1992, 147&148, 271.
N.G.McCrum, C.P.Buckley, and C.B.Bucknall, Principles of Polymer Engineering, Oxford Science, New York, 1996.
McEuen, P. L.; Bockrath, M.; Cobden, D. H.; Yoon, Y.-G.; Louie, S. G. *Phys. Rev. Lett.* 1999, *83*, 5098-5101.
L. Merhari, Hybrid Nanocomposites for Nanotechnology: Electronic, Optical,Magnetic and Biomedical Applications, Springer, Berlin, 2009, pp.231-287.
P. Mulvaney, MRS bulletin, 1, December 2001.
Nam PH, Maiti P, Okamoto M, Kotaka T, Nakayama T, Takada M, Polym Eng Sci 2001;42:1907–1918.
Nair, K.G. and Durfreshe, A. Biomacromolecules 4, 666, 2003.
Nielsen L. J Macromol Sci Chem 1967;A1(5): 929–42.
Ounaies Z, Park C, Wise KE, Siochi EJ, Harrison JS. Electrical properties of single wall carbon nanotube reinforced polyimide composites. *Composites Science and Technology*. 2003; 63(11):1637-1646.
K. Prater, J. Power Sources, 1990, 29, 239.
Y.D. Premchand, M.L. Di Vona, and P. Knauth, 'Proton-Conducting Nanocomposites and Hybrid
Polymers', in 'Nanocomposites: ionic conducting materials and structural spectroscopies', P.Knooth and J.Schoonman, Eds, Springer, Berlin, 2008.
B. K. Price, J. L. Hudson, J. M. Tour, J. Am. Chem. Soc., 2005, 127, 14867–14870.
X.Qi, H. Yoon, S.H. Lee, J.Yoon, S.J.Kim, Journal of Industrial and Engineering Chemistry 14 (2008) 136–141.
Ray, S. S.; Okamato, M. Prog Polym Sci 2003, 28, 1539.
Ray SS, Yamada K, Okamoto M, Ueda K. Polymer, 2003;44:857–866.
L. Razafimahefa, S. Chlebicki, I. Vroman, E. Devaux, Dyes and Pigments 66 (2005) 55-60.
Roy R, Roy RA, Roy DM, Alternative perspectives on "quasi-crystallinity": non-uniformity and nanocomposites. Materials Letters. 1986; 4(8-9):323-328.
R. Sengupta, Bhattacharya.M., Bandyopadhyay.S., and Bhowmick. A.K., A review on the mechanical and electrical properties of graphite and modified graphite reinforced polymer composites;: Progress in polymer science doi:10.1016/j.progpolymsci.2010.11.003 Volume 36, Issue 5, May 2011, Pages 638-670.
C. Sanchez, B. Julian, P. Belleville, M. Popall, J. Mater. Chem., 2005, Vol. 15, 3559–3592.
Sahoo, S. and Bhowmick, A.K. J. Appl. Polym. Sci. 106, 3077, 2007.
Southern Clay Products, Gonzales, TX. (http://www.scprod.com/gm.html)
K. Strawhecker and E. Manias, *Chem. Mater.* 12 (2000)2943.
Suryani, Y. Liu, Journal of Membrane Science 332, 2009, 121-128
K. E. Strawhecker and E. Manias, Chem. Mater. 2000, 12, 2943-2949.

M. Suzuki and K. Inukai, 'Synthesis and Applications of Imogolite Nanotubes' in 'Inorganic and Metallic Nanotubular Materials: Recent Technologies and Applications', T. Kijima, Ed., Springer, Berlin, 2010.

Tandon, R.P., Choubey, D.R., Singh, R., and Soni, N.C. J. Mat. Sci. Lett. 12, 1182, 1993.

Thomas SP, S Jose, Abi S Aprem, S Thomas, W Yang and Sri Bandyopadhyay, Poly(ethylene-co-vinyl acetate)/Calcium Phosphate Nanocomposites: Thermo Mechanical and Gas Permeability Measurements, POLYMER COMPOSITES—-2010, DOI 10.1002/pc.20887

Thomas Selvin P, Seno Jose, Sabu Thomas and Sri Bandyopadhyay : Mechanical & Thermal Properties of Poly(ethylene-co-vinyl acetate) Calcium Phosphate Nanocomposites, J Phys Chem C 2009, 113, 1, 97-104.

Thomas SP, S.Thomas, R.Abraham, S.Bandyopadhyay-Polystyrene/calcium phosphate nanocomposites: Contact angle studies based on water and methylene iodide eXPRESS Polymer Lett. 2008, 2, 7, 528–538.

Thomas Selvin P, Sabu Thomas, Sri Bandyopadhyay, Andreas Wurm and Christoph Schick : Polystyrene/Calcium Phosphate Nanocomposites: Dynamic Mechanical and Differential Scanning Calorimetric Studies : Composites Science and Technology 2008, 68, 15-16,3220-3229.

Thomas Selvin Palakattukunnel, Sabu Thomas, P. A. Sreekumar, Sri Bandyopadhyay : Poly(ethylene-co-vinyl acetate)/calcium phosphate nanocomposites: contact angle, diffusion and gas permeability studies J Polym Res DOI 10.1007/s10965-010-9530-1

S.Turri, L. Torlaj, F. Piccinini, M. Levi, Abrasion and Nanoscratch in Nanostructured Epoxy Coatings, Journal of Applied Polymer Science, 2010, 118, 1720-1727.

Uchida, T.; Kumar, S. *J. Appl. Polym. Sci.* 2005, *98*, 985-989.

S.R.C. Vivekchand, A. Govindaraj, and C.N.R. Rao, 'Nanotubes and Nanowires: Recent Developments', in 'Nanomaterials Chemistry', C.N.R. Rao, A. Müller, and A.K. Cheetham, Eds, WILEY-VCH Verlag, Weinheim, 2007.

D.H. Walt, A. *MRS Bull.* 2004, *29*, 281-285.

C. Wang, Z. Guo, S. Fu, W. Wu, D. Zhu, Prog. Polym. Sci. 29 (2004) 1079–1141.

Wang .Q, Zhang. X, Qiao. J, Exfoliated sodium-montmorillonite in nitrile butadiene rubber nanocomposites with good properties, Chinese Science Bulletin, 2009; 54: 877-879.

Weisenberger MC, Grulke EA, Jacques D, Ramtell T, Andrews R. Enhanced mechanical properties of polyacrylonitrile: multiwall carbon nanotube composite fibers. *Journal of Nanoscience and Nanotechnology*. 2003; 3(6):535-539.

M. Wilson, K.Kannangara, G.Smith, M. Simmons, and B. Raguse, Nanotechnology: basic science and emerging technologies, CRC press, Boca Raton, 2002.

F. Wypych, K. G. Satyanarayana, Journal of Colloid and Interface Science 285 (2005) 532–543.

Yamamoto, K.; Otsuka, H.; Wada, S.-I.; Sohn, D.; Takahara, A. Soft Matter, 2005, 1, 373.

X. Yang, and J. Loos, Macromolecules, 2007, 40 (5), pp 1353–1362.

Ying, Y., Yan, Z., Li-sha, Z., and Zhi-jie, J. Tanxingti 12, 1, 2002.

Yurkov, G.Y., Astaf'ev, D.A., Nikitin, L.N., Koksharov, Y.A., Kataeva, N.A., Shtykova, E.V., Dembo, K.A., Volkov, V.V., Khokhlov, A.R., and Gubin, S.P. Inorg. Mat. 42, 496, 2006.

Zheng, J., Ozisik, R., and Siegel, R.W., Polymer 46, 10873, 2005.
Zhu J, Morgan AB, Lamelas FJ, Wilkie CA. Chem Mater 2001;13:3774–80.
K. J. Ziegler, Z. Gu, H. Peng, E. L. Flor, R. H. Hauge, R. E. Smalley, J. Am. Chem. Soc., 2005, 127, 1541–1547.

2

PVA / Montmorillonite Nanocomposites: Development and Properties

Andreas A. Sapalidis, Fotios K. Katsaros and Nick K. Kanellopoulos
N.C.S.R. Demokritos, Institute of Physical Chemistry, Greece

1. Introduction

Polymers are hardly used in their pure form in applications; they are often filled with additives. The modification of polymers, through the incorporation of additives in the micrometer range, e.g. calcium carbonate, glass beads and talc, yields, with few exceptions, multiphase systems containing the additive embedded in a continuous polymeric matrix. The additives can either improve their process ability (lubricants, antioxidants and stabilizers), or modify the modulus and the strength (carbon black, silica beads and fibers, clay, mica), the appearance (pigments and surfactants), the conductivity (carbon black and carbon nanotubes), the transport properties, the moisture content, the flammability or simply reduce their cost (Alexandre & Dubois, 2000), (Fischer, 2003), (Lagaly, 1999), (Giannelis, 1996).

On the other hand, polymer nanocomposites represent a class of material alternative to conventional filled polymers. In this type of materials, nanofillers (having at least one dimension in nanoscale range) are finely dispersed in polymer matrix offering remarkable improvement in performance properties of the polymer, including high moduli (LeBaron et al., 1999), (Giannelis et al., 1999), (Vaia et al., 1999) increased strength (Giannelis, 1998) and heat resistance (Gilman, 1999), improved gas barriers properties (Xu et al., 2001), (Sinha-Ray et al., 2002) and fire redundancy (Gilman et al., 2000), (Morgan, 2006) etc.

There are four major parameters affecting the final properties of the nanocomposites, namely:

i. the filler aspect ratio,
ii. the filler dispersion,
iii. the filler alignment and orientation and
iv. the polymer-polymer and polymer-filler interfacial interactions.

The control and the optimization of the above-mentioned structural factors are expected to enable the development of nanocomposites with predefined and superior properties. Interfaces can greatly affect the properties of a composite. The large surface area of the filler can be either advantageous or disadvantageous, depending on the type of property. For example, it may result in a composite of low strength (in spite of the possibly high strength within a single unit of the nanofiller, e.g., within a single nanofiber) due to the mechanical weakness of the interface. It may also result in a composite of high electrical resistivity due to the electrical resistance associated with the interface.

Numerous potential nanosized fillers, with high aspect ratio, have been utilized for the preparation of high efficient nanocomposites. Amongst them, clay and layered silicates have been more widely investigated, probably due to their availability and their low cost. Furthermore, their well-studied intercalation chemistry offers additional advantages for their application in nanocomposite technology.

Polymer nanocomposites are two-phase materials, in which the polymers are reinforced by nanoscale fillers. One of the most extensively used filler material is the smectite class of aluminum silicate clays, of which the most common representative is montmorillonite (MMT). MMT has been employed in many Polymer/Layered Silicates (PLS) nanocomposite systems because it has a potentially high-aspect ratio and high-surface area that could lead to materials, which could possibly exhibit great property enhancements. In addition, it is environmentally friendly, naturally occurring, and readily available in large quantities.

However, since the layered silicates in their pristine state are hydrophilic, the nanolayers are not easily dispersed in most polymers and tend to form agglomerates. Dispersion of the inorganic platelets into discrete monolayers is further hindered by the intrinsic incompatibility of hydrophilic layered silicates and hydrophobic engineering plastics. Therefore, in most cases, layered silicates must initially be organically modified to produce well-organized nanocomposites. On the other hand, MMT is capable of forming stable suspensions in water, while its hydrophilic character also promotes the dispersion of these inorganic crystalline layers in water soluble polymers, such as poly(vinyl alcohol), poly(ethylene oxide) etc.

Poly(vinyl alcohol) (PVA) is a water soluble polymer extensively used in paper coating, textile sizing, and flexible water soluble packaging films. However, the insufficient mechanical properties and the poor resistance in solvents have limited its applicability in industrial processes. Development of PVA-based nano-composites has been an emerging method to improve PVA's properties. Thus, PVA/layered silicate nanocomposite materials with improved mechanical, thermal, and permeability properties may offer a viable alternative for these applications to heat treatments (that may cause polymer degradation) or conventionally filled PVA materials, which are optically opaque.

This chapter highlights the major developments in the preparation of PVA/MMT nanocomposites during the last decades. Furthermore, the authors, based on polymer-polymer and polymer-clay interactions, describe an effective way for the preparation of well-dispersed nanocomposites with improved properties. The obtained nanocomposites were characterised by a variety of techniques, including XRD, TEM, AFM, DCS, TGA, mechanical strength, oxygen permeability and water sorption. The developed films, due to the favorable polymer-particle interactions, revealed excellent dispersion of the clay particles in the polymer matrix and improved properties.

2. Structure of polymer and layered silicates

2.1 Structure of MMT

The commonly used layered silicates for the preparation of polymer/layered silicate (PLS) nanocomposites belong to the same general family of 2:1 layered- or phyllosilicates (Alexandre & Dubois, 2000). Their unit lamellar crystal is composed of two crystal sheets of silica tetrahedron combined with one crystal sheet of alumina octahedron between them (Ray & Bousmina, 2006). The layer thickness is about 1 nm, while the lateral dimensions of these layers may range from a few nanometers to several microns or several hundreds of microns, depending on the particular layered silicate, the source of the clay and the method of preparation. Therefore, the aspect ratio i.e. the length to thickness ratio of these layers is particularly high, reaching values even greater than 1000 (McNally et al., 2003), (Pavlidou & Papaspyrides, 2008).

The layers of the silicates can easily be self-organized to form stacks with a regular van der Walls gap in between them, called the interlayer or the gallery. Isomorphic substitution within the layers (for example, Al^{+3} replaced by Mg^{+2} or by Fe^{+2}, or Mg^{+2} replaced by Li^{+}) generates negative charges that are counterbalanced by alkali or alkaline earth cations - mainly sodium and calcium ions, existing hydrated in the interlayer. As the forces that hold the stacks together are relatively weak, small molecules including water, solvents, and monomer as well as polymer, can enter into these galleries, causing the lattice to expand (Giannelis, 1996), (Nguyen & Baird, 2006). A characteristic value of *d*-spacing is reported for each type of layered silicate. The *d*-spacing is the repeat unit in the crystalline structure including the 1 nm thick platelet and the spacing in between the platelet sheets (Brindley, 1984).

The layered silicates characterized by a moderate surface charge known as the cation exchange capacity (CEC), which is generally expressed as mequiv/100 g. The charge is not locally constant, but varies from layer to layer, and must be considered as an average value over the whole crystal. In the case of tetrahedrally-substituted layered silicates, the negative charge is located on the surface of silicate layers, and hence, the polymer matrices can interact more readily with these than with octahedrally-substituted material (Ray & Bousmina, 2006). The advantages of the layered silicates that are generally considered for PLS-nanocomposites preparation are mainly related to their ability to form exfoliated structures (individual platelets) and to the straightforward and effective functionlisation of their surface. The exfoliated/intercalated structures provide high aspect ratios and high-surface areas, leading to nanocomposites with improved final properties. On the other hand, the surface chemistry of the layers can be adjusted through ion exchange reactions with organic and inorganic cations.

Montmorillonite, which is the best-known member of a group of clay minerals, called "smectites" or "smectite clays", is the most widely used layered silicate for the preparation of nanocomposites. Montmorillonite has a 2:1 structure, which allows sharing of the oxygen between Al and Si. Additionally, there are hydroxyl groups at the edges of each clay platelet. The structure of montmorillonite is shown in Figure 1, while its general formula is (Pinnavaia & Beall, 2000):

$$(Na,Ca)_{0.33}(Al,Mg)_2(Si_4O_{10})(OH)_2 \cdot nH_2O \tag{1}$$

Montmorillonite is a hydrophilic, inorganic material, which usually contains hydrated Na+ or K^+ ions (Krishnamoorti et al., 1996). Therefore, in this pristine state, layered silicates are only miscible with hydrophilic polymers, such as poly(ethylene oxide) (PEO) (Aranda & Ruiz-Hitzky, 1992) or poly(vinyl alcohol) (PVA) (Strawhecker & Manias, 2000), (Yu et al., 2003). To render layered silicates miscible with hydrophobic polymer matrices, one must convert the normally hydrophilic silicate surface to an organophilic one, making the intercalation of polymer chains into inorganic galleries more effective.

Generally, this can be done by ion-exchange reactions with cationic surfactants including primary, secondary, tertiary, and quaternary alkylammonium or alkylphosphonium cations. This, results in expansion between the clay galleries, due to the larger molecules inserted between the layers. The reaction also changes the clay from hydrophilic to hydrophobic, making it more compatible with the organic matrix. Additionally, the alkylammonium or alkylphosphonium cations can provide functional groups that can react with the polymer matrix, or in some cases initiate the polymerization of monomers to improve the adhesion between the inorganic and the polymer matrix (Messersmith & Giannelis, 1995), (Gilman et al., 2000).

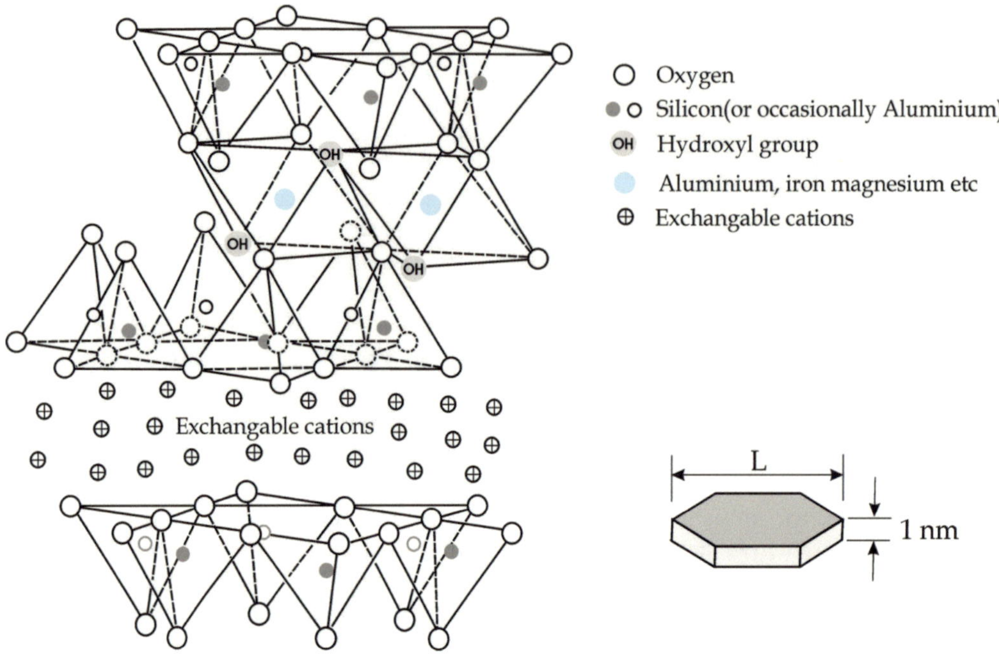

Fig. 1. Structure of 2:1 layered silicate showing two tetrahedral sheets of silicon oxide fused to an octahedral sheet of aluminum hydroxide and Platelet structure

2.2 Polyvinyl alcohol (PVA)

PVA is the most widely produced water soluble polymer today. It has been available since 1924, when its synthesis via the saponification of poly-(vinyl acetate) was first described by Herrmann and Haehnel (Herrmann & Haehnel, 1924). PVA is currently produced from the parent homopolymer poly(vinyl acetate) (PVAc) via hydrolysis (methanolysis) (Hay & Lyon, 1967). Its application has mainly been focused on fibre industry. Recently, it has received much attention in non-fibre applications specifically in pharmaceutical, biomedical and biochemical applications, due to its attractive characteristics, such as biocompatibility, biodegradability, and water-solubility. PVA has been also used for membranes (Chuang et al., 2000), drug delivery system (Brazel & Peppas, 1999), and artificial biomedical devices (Kobayashi et al., 2003).

However, the undesirable properties of PVA, including the poor solvent resistance and anti-ageing behaviour, the insufficient strength and the low heat stability, have restricted its further applications. Thus, the preparation of PVA-based conventional (Sapalidis et al., 2007) as well as nano-composites (Chang et al., 2003), (Yu et al., 2003), (Podsiadlo et al., 2007) in order to improve the mechanical, the thermal, and the gas barrier properties have attracted several research studies.

2.3 Nanocomposite structure

In general, the degree of dispersion of the clay platelets into the polymer matrix determines the structure and the final properties of the nanocomposites. Depending on the nature of the components used (layered silicate, organic cation and polymer matrix) and the method of preparation, three main types of composites may be obtained, when layered clay is associated with a polymer (Figure 2):

a. *Non mixing composites:* In this case, the interactions between the clay particles and the polymer are very weak and thus, the polymers cannot enter into clay galleries. A phase separated composite is formed with relatively poor mechanical properties (Alexandre & Dubois 2000).
b. *Intercalated structures*: in which the polymeric chains are intercalated between the silicate layers resulting in a well ordered multilayer morphology, built up with alternating polymeric and inorganic layers. Intercalation results a separation of about 2-3 nm between the layers, which is independent of the clay to polymer ratio (Yeh & Chang, 2008). The properties of this type of nanocomposites resemble those of ceramic materials.
c. *Exfoliated or delaminated structures*: in which the clay layers are well separated from one another and individually dispersed in the continuous polymer matrix (Giannelis, 1996). In this case, the distance between two platelets may be in the range of 5-10 nm or even more (Dennis et al., 2001). In such systems, the polymer–clay interactions are maximised leading to significant changes in mechanical and physical properties. It is generally accepted that exfoliated systems exhibit better mechanical properties than intercalated ones (Varlot et al., 2001), (Chin et al., 2001).

2.4 Techniques used for the characterization of nanocomposites

Since the properties of the nanocomposites are defined by the dispersion of the inorganic nanoplatelets into the polymer, one needs to know the degree of exfoliation / intercalation of a particular sample and compare it to other samples. A number of methods have been reported in the literature for this purpose (Krishnamoorti et al., 1996), (Morgan & Gilman, 2003), (Van der Hart et al., 2001). Generally, the state of dispersion and exfoliation of

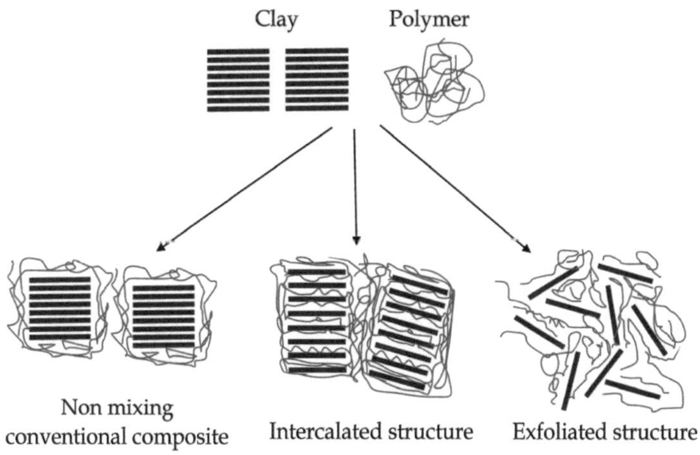

Fig. 2. Schematic representation of the three types of composite structures depending on polymer -clay interactions

nanoparticles has typically been studied using X-ray diffraction (XRD) analysis and transmission electron microscopy (TEM) observations. Due to its easiness and availability, XRD is most commonly used to probe the nanocomposite structure (Giannelis et al., 1999), (Biswas & Ray, 2001), (Ray & Okamoto, 2003). The nanocomposite structure, namely, intercalated or exfoliated, may be identified by monitoring the position, shape, and intensity of the basal reflections, from the distributed silicate layers. For example, in an exfoliated nanocomposite, the extensive layer separation associated with the delamination of the original silicate layers in the polymer matrix results in the eventual disappearance of any coherent X-ray diffraction from the distributed silicate layers. On the other hand, for intercalated nanocomposites, the finite layer expansion associated with the polymer intercalation results in the appearance of a new basal reflection corresponding to the larger gallery height. Additionally, in some cases, XRD is employed to study the kinetics of the polymer melt intercalation (Vaia et al., 1996).

Although XRD offers a convenient method to determine the interlayer spacing of the silicate layers in the original layered silicates and in the intercalated nanocomposites (within 1-4 nm), little can be said about the spatial distribution of the silicate layers or the structural homogeneity of the nanocomposites. In addition, some layered silicates do not exhibit well-defined basal reflections, consisting that way difficult the determination of the intensity pattern and the shape of the relative peaks (Zanetti et al., 2000).

Therefore, conclusions concerning the mechanism of nanocomposites formation and their structure based solely on XRD patterns are only tentative. On the other hand, TEM allows a qualitative understanding of the internal structure, spatial distribution and dispersion of the nanoparticles within the polymer matrix, and views of the defect structure through direct visualization. However, special care must be exercised to guarantee a representative cross section of the sample.

Both techniques are essential tools for evaluating nanocomposite structure (Morgan & Gilman, 2003). However, when layer spacing exceeds 6-7 nm in intercalated nanocomposites or when the layers become relatively disordered in exfoliated nanocomposites, XRD features weaken to the point of not being useful.

To this end, numerous advanced techniques can be employed for the characterization of the nanocomposites. Recent simultaneous small angle X-ray scattering (SAXS) and XRD studies yielded quantitative characterization of nanostructure and crystallite structure in some nanocomposites (Bafna et al., 2003). In addition, neutron membrane diffraction technique can reveal information about both the dispersion and the orientation of the inorganic layers in the polymer matrix (Katsaros et al., 2009).

In contrast to the conventional electron microscopy, Atomic Force Microscopy (AFM) represents an effective alternative method to study the dispersion of the nanofillers without any limitations regarding the sample preparation, the contrast and the resolution. Atomic force microscopy, although increasingly being used in polymer structure characterization (Magonov & Reneker, 1997), (Magonov, 2000), surprisingly has not been widely used to investigate the layered silicate nanocomposite systems (Clement et al., 2001), (Yalcin & Cakmak, 2004), (Maiti & Bhowmick, 2006). The development of this technique has helped to image surface topography on nanoscale. Thus, apart from the investigation of clay dispersion AFM can provide information about surface topography of the samples. In addition, AFM can also measure fundamental properties of sample surfaces, e.g., local adhesive or elastic properties on nanoscale (Malwitz et al., 2004), (McNally et al., 2003), (Jiang et al., 2005).

2.5 PVA/MMT nanocomposites

Poly(vinyl alcohol) - clay nanocomposite materials have widely been studied in the past. Greenland (Greenland, 1963) reported the first fabrication of PVA/MMT composites by a solvent casting method, using water as a co-solvent. Ogata and coworkers (Ogata et al. 1997) applied the same technique for the production of PVA/MMT composites. Furthermore, Strawhecker and Manias (Strawhecker & Manias, 2000) have also developed PVA/MMT nanocomposites films with improved properties, by means of solvent casting method using low viscosity, fully hydrolyzed atactic PVA. The obtained nanocomposite films (clay content up to 5 wt %) exhibited both intercalated and exfoliated regions.

Recently, Chang and coworkers (Chang et al., 2003) reported the preparation of PVA-based nanocomposites with three different types of clays, namely pristine MMT and two different types of organically modified MMT (Dodecylamine and 12- aminolauric modified montmorillonite). They applied the same solvent casting method, but used also N,N-dimethylacetamide (DMAc) as co-solvent in addition to water. XRD patterns and TEM observations of their nanocomposites revealed the formation of exfoliated nanocomposites when pristine clays were used for the fabrication of nanocomposites. On the other hand, intercalated nanocomposites were produced with organo- modified clays. This implies that the hydrophilic character of clay promotes the effective dispersion of the inorganic crystalline layers in water-soluble polymers.

Based on in-situ intercalative polymerisation, Yu et al. (Yu et al., 2003) reported the synthesis of a series of PVA/MMT nanocomposites, using AIBN as initiator. In the first step of their preparation, the vinyl acetate monomers were intercalated into the organically modified clay galleries. Free radical polymerisation was followed in the subsequent step. The final PVA/MMT nanocomposites were obtained after hydrolysis of the vinyl acetate groups. The structural characterization of the developed materials confirmed the formation of mixed intercalated/exfoliated regions.

More recently a highly ordered Poly(vinyl alcohol) / montmorillonite nanocomposites were produced by the layer by layer (LBL) process (Podsiadlo et al., 2007). The LBL process is based on sequential adsorption of nanometer-thick monolayers of oppositely charged compounds (such as polyelectrolytes, charged nanoparticles, and biological macromolecules) to form a multilayered structure with nanometer-level control over the architecture. The obtained nanocomposites exhibited superior mechanical properties without significant reduced in optical transparency.

At equilibrium, the nanocomposite structure predicted from thermodynamics corresponds to an intercalated periodic nanocomposite with *d*-spacing around 1.8 nm, which is expected to be independent of the polymer-to-silicate ratio (Lee et al., 1998). However, thermodynamics can only predict the equilibrium structure. On the other hand, the nanocomposite structure is mainly determined by kinetic factors. In water solutions, PVA and MMT layers remain in colloidal suspension. During slowly drying, the silicate layers remain distributed and embedded in the polymer gel. Additional drying removes all of the solvent, and although the thermodynamics would predict the MMT layers to re-aggregate, the slow polymer dynamics entrap some of the layers and keep them separated. Therefore, the platelets remain dispersed in the polymer matrix. Obviously, the kinetic constraints imposed by the polymer become less important as the polymer-to-silicate fraction decreases, and consequently, for higher amounts of MMT, intercalated structures are formed. For these periodic structures, the variation of the d-spacing with wt% of MMT reflects the different polymer-silicate weight ratios, and upon increasing the amount of MMT the intercalated d-spacing converges to the equilibrium separation of 1.8 nm (Ray & Bousmina, 2006)

The structures obtained during the addition of clay particles in a polymer solution are determined by the polymer-polymer and polymer-clay interactions (Lim et al. 2001). These interactions enable the chains to receive certain conformations and to build structures around the inorganic layers. Depending on the nature of the adhesion (strong or weak), these arrangements lead to the formation of larger assemblies like structured intercalated regions. Theoretical and experimental results demonstrate that the adhesive role of a polar polymer between hydrophilic clay layers, the so called "glue effect", tends to strongly prohibit complete dissociation of the layered structure of the clay, resulting in an ordered intercalated state (Lee et al., 2006).

In pure PVA, the hydrogen bonds are the dominant interaction responsible for both structure and molecular dynamics. As a consequence of these bonding interactions, water is capable to destroy inter- and intra- chain hydroxyl bonds, affecting PVA's crystalline regions and acting as a plasticizer, increasing the free-volume of the amorphous phase (Hodge et al. 1996). On the other hand, in PVA/MMT nanocomposites, the interactions of the surface of the clay with the polymer can be attributed to the formation of hydrogen bonds between the PVA hydroxyl groups and the negatively charged clay surface (Grunlan et al., 2004), (Hernández et al., 2008).

Thus, in previous studies, the authors trying to maximise the polymer-clay interactions, utilized mainly fully hydrolysed PVA. In this chapter, targeting to weaker polymer-polymer interactions, a PVA matrix with 88% hydrolysis grade, was used for the preparation of nanocomposite films. Furthermore, due to the presence of metal cations in the clay lattice, increased interactions between clay platelets and acetoxy groups of the PVA are expected (Stathi et al. 2009), boosting the intercalation /exfoliation of the inorganic layers. To this end, a series of PVA/ MMT clay composites were prepared by solvent film casting method using water as solvent. The obtained nanocomposites were characterised by a variety of techniques including XRD, TEM, AFM, DCS, TGA, mechanical strength, oxygen and water permeability and water sorption. The developed films, due to the favorable polymer-particle interactions, revealed excellent dispersion of the clay particles in the polymer matrix and improved properties.

3. Synthesis of PVA/MMT nanocomposites

3.1 Materials and methods

Low viscosity, partially hydrolyzed atactic Poly(vinyl alcohol) Mowiol® 5-88 (Average molecular weight: 37000 g/mol - Sigma) was used for the preparation of the nanocomposites. Na-MMT (cationic exchange capacity of 80 meq/100g) was supplied by S&B Industrial Minerals S.A. The purity (in montmorillonite) of raw clay used in this study was about 76%.

The developed nanocomposite films were characterised by a variety of microscopic and macroscopic techniques. The dispersion of the clay platelets was investigated by XRD, AFM and TEM. XRD patterns were recorded on a Siemens XD-500 diffractometer using CuKα1 radiation source, while Electron Transmission images were obtained using a Jeol JEM 2011 TEM. In addition, the Surface morphology was examined by a Digital Instruments Nanoscope III atomic force microscope (AFM), using tapping mode.

The oxygen permeability experiments at variant %RH were performed on a PBI Dansensor OPT-5000 instrument, according to ASTM F2622-08 method. The samples, prior to their testing, were conditioned in oven at a temperature of 50 ± 2 °C for 48 h (ASTM D 618 – Procedure B).

The mechanical strength of the films was measured by means of a Thumler GmbH Tensile Tester Model (cell load 250N -PA 6110 Nordic Transducer), using specimens of 3 cm width and 6 cm length. Prior the testing, the samples were pre-equilibrated at constant relative humidity.
A Modulated DSC Model 2920 TA Instruments was used to measure the thermal properties of the films, while a SETARAM SETSYS Evolution 18 Analyser (RT-1750°C) was applied for the Thermogravimetric analysis of the samples. Additionally, the optical clarity of the materials was studied by UV-Vis spectroscopy (Cary 100 Varian Inc. UV-Vis spectrophotometer). The average samples' thickness used for optical measurements was 50±0.02 μm.
Finally, the sorption isotherm experiments were carried out on a homemade gravimetric system, equipped with a CI Electronics Ltd® microbalance and a SS gas/vapour dosing unit. The mass changes, the vapor pressure and the cell temperature were continuously recorded by means of LabVIEW® software.

3.2 Preparation of the PVA/MMT films

A PVA solution in water (10 wt%) was produced (6 hours mixing at 90°C) and used as stock solution. The samples were prepared by mixing MMT water suspensions with the polymer solutions in quantities that gave 5, 10 and 20 wt% clay loading on the final films, namely PVA/MMT05, PVA/MMT10 and PVA/MMT0 respectively. Initially, the mixture of polymer and MMT (100ml) was stirred for half hour at 80°C and then sonicated for an extra half hour. The suspension (10 to 20 ml regarding the final film thickness) was poured in square (12x12 cm) polystyrene Petri dishes and left to dry slowly at 25°C for about 15 days. The average thickness of the samples used for permeability and mechanical tests was 0.1 ± 0.02 mm

4. Results and discussion

4.1 Morphology of developed composites

4.1.1 Microscopic techniques

Transmission Electron Microscopy (TEM) was used to for the preliminary characterization of the nanocomposites formed, with the emphasis on the dispersion of inorganic layers in the polymer matrix. Typical TEM images are shown in Figure 3 (a), (b), (c) for the 20 wt % MMT sample (PVA/MMT20). TEM observations reveal the existence of silicate layers in the exfoliated state. Furthermore, some larger intercalated tactoids could also be identified.
In addition to TEM, Atomic force microscopy (AFM) was also applied for the determination of the clay dispersion in PVA matrix. Figure 3(d) shows the corresponding AFM image of the PVA/MMT20 film, in tapping mode. The presence of disc shaped objects of about 50-140 nm can be attributed to the clay particles, while the linear parts (of about 20 nm) may be related to the polymer chains. In addition to the organization, a specific orientation of clay platelets is clearly obvious, in a wide area of the sample (about 2 μm^2). Although these organized regions may not be representative of the whole sample, reveal that the increased polymer-clay interactions lead to formation of exfoliated and intercalated structures. The obtained results are in good agreement with TEM observations. Thus, AFM can be a good alternative to electron microscopy for the analysis of nanocomposites, without any limitations regarding the contrast and the resolution.

4.1.2 XRD patterns

Wide angle X-ray diffraction is probably the most indicative technique to investigate the interactions between the polymer and inorganic layers. The XRD patterns of MMT powder, pure PVA film and nanocomposite films prepared in this study are shown in Figure 4. The peak at $2\theta = 7^{o}$ can be attributed to the basal spacing (~12.6 Å) of the MMT, while the other peak (around 29°) can be related to clay impurities. In addition, the peak at $2\theta = 19.4^{o}$ (d = 4.57 Å) be present in pure polymer, can be attributed to crystal reflections of PVA.

On the other hand, the observed increase in the basal spacing of PVA/MMT20 sample (around 27Å) reveals the formation of intercalated structures, due to the polymer chains in the clay galleries. This expansion of the clay's basal spacing is due to favorable interactions between the clay's surface and the polymer groups. It must be noted that intercalated regions are obtained at lower clay concentration than previous studies have reported (Strawhecker & Manias, 2000).

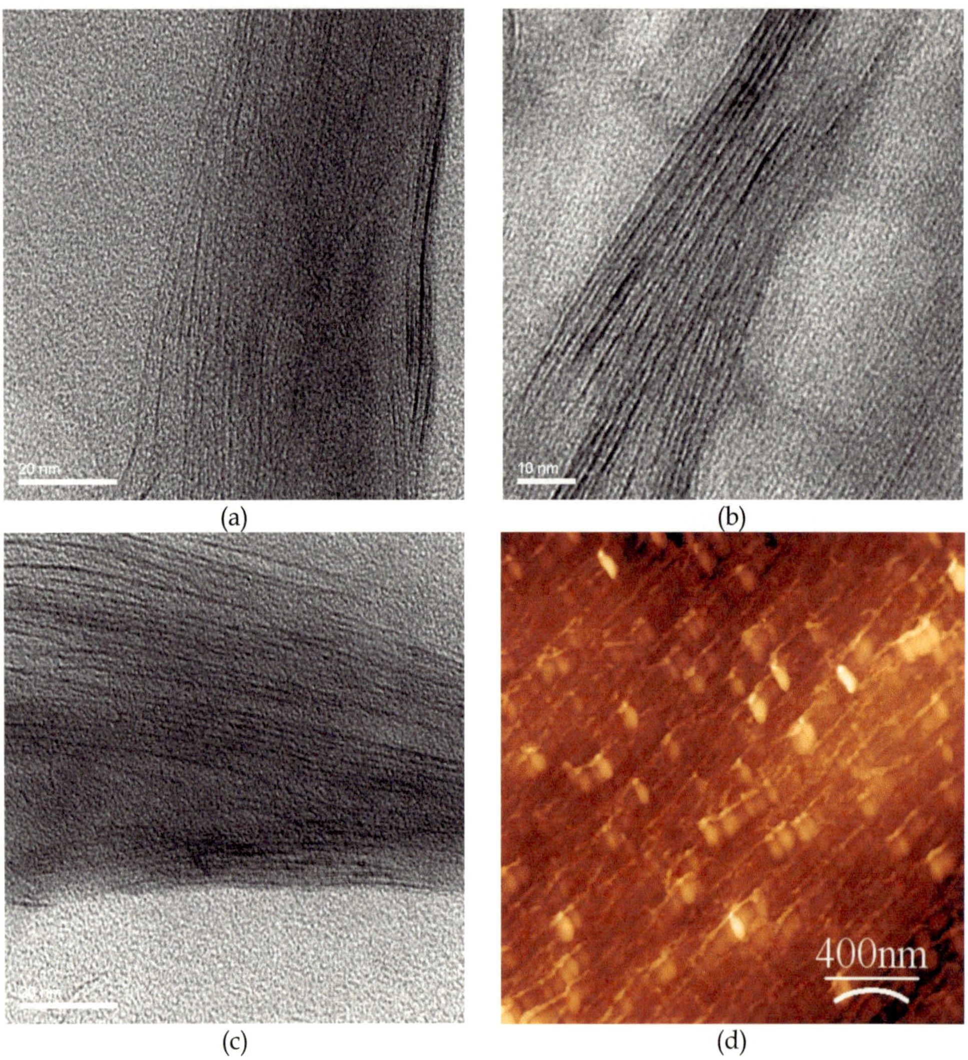

Fig. 3. TEM micrographs (a), (b), (c) and AFM photo (d) of of PVA/MMT20 film

In the case of the PVA/MMT05 and PVA/MMT10 samples the peak corresponding to the basal distance is not apparent. Furthermore, the presence of a shoulder at clay's basal spacing reveals that a small part of the clay formed aggregates (Figure 4). Moreover, the increased background of these patterns suggests the existence of exfoliated inorganic layers in extended regions of the nanocomposite. These results are in agreement with the results obtained from the microscopic techniques.
In order to evaluate the orientation of the clay platelets, in regard to the film's surface, the following test was performed: the sample was measured in two forms, namely film and powder. The first time, the sample was left intact as a film and it was placed on top of the sample holder. The second time, the sample was fragmented into as small as possible pieces and measured again. The obtained patterns are shown in Figure 5. It is obvious that the peak corresponding to the basal spacing of the clay is less pronounced in the case of the powder sample, indicating that there is a specific orientation of clay particles, parallel to the surface of the nanocomposite film.

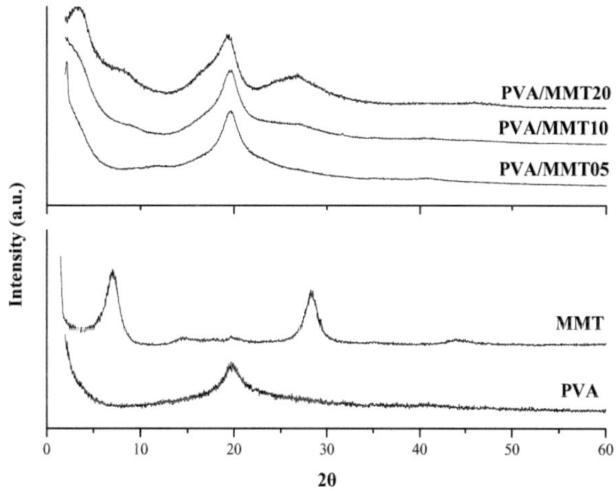

Fig. 4. XRD patterns of pure Montorillonite, PVA and Nanocomposite films

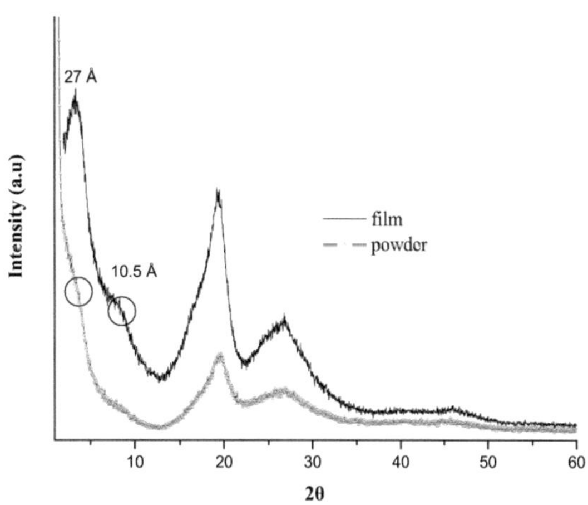

Fig. 5. XRD patterns of PVA/MMT20 sample powder and in film form

4.2 Gas permeability

Gas permeability is a very crucial property for various applications, including medical and industrial, which also apply to everyday materials, such as those used in packaging. The permeation of molecules through a non-porous polymeric membrane is determined by the capability of the molecules to be sorbed on the membrane's surface and body, to diffuse within and finally to be desorbed on the other side of the membrane.

The overall process is the product of the following coefficients:

a. Solubility (S) – the partition coefficient of sorbed, in regard to total molecules,
b. Diffusivity (D) - the rate of transport of molecules through the polymeric matrix.

The molecules' rate of transport through a polymer, as a result of the combined effects of diffusion and solubility, can be expressed by permeability coefficient (P_e), which is related to D and S by the expression:

$$P_e = D \times S \tag{2}$$

In this case, P_e incorporates both kinetic and thermodynamic properties of the polymer-permeant system.

The mass transport mechanism of gasses permeating a nano-platelet reinforced polymer is considered to be similar to that in a semicrystalline polymer. In most theoretical studies the nanocomposite is considered to consist of a permeable phase (polymer matrix), in which impermeable layers are dispersed (Cussler et al., 1998). The permeability of a nanocomposite system is mainly influenced by the following factors:

a. the volume fraction of the clay particles,
b. their orientation in relation to diffusion direction, and
c. the aspect ratio of the platelets.

It is generally accepted that the transport mechanism within the polymer matrix follows Fick's law, and that the matrix maintains the same properties and characteristics as the neat polymer. Thus, the diffusivity of the homogeneous matrix material D_0 is not influenced by the presence of particles. While this is likely a good assumption in many situations, certain exceptions are present in the literature, in which the anisotropic fillers in semicrystalline polymer matrices can influence the extent and morphology of crystalline regions, and thereby modify the matrix diffusivity.

Based on these assumptions, the tortuous path that is formed due to the presence of impermeable particles is analogous to the reduction of the diffusion coefficient (Figure 6). The tortuosity effect is related to the degree of the dispersion as well as to the aspect ratio of the layers and their orientation. Additionally, a decrease of the solubility is expected in the nanocomposite due to the reduced polymer matrix volume. Since the clay content in nanocomposites is usually small, the contribution of latter factor in overall permeability is less important.

Several models have been developed in order to predict the mass transfer through nanocomposites. A simple permeability model for a regular arrangement of platelets has been proposed by Nielsen (Nielsen, 1967):

$$\frac{P_e^{comp}}{P_e^{matrix}} = \frac{1-\varphi}{1+a/2\varphi} \tag{3}$$

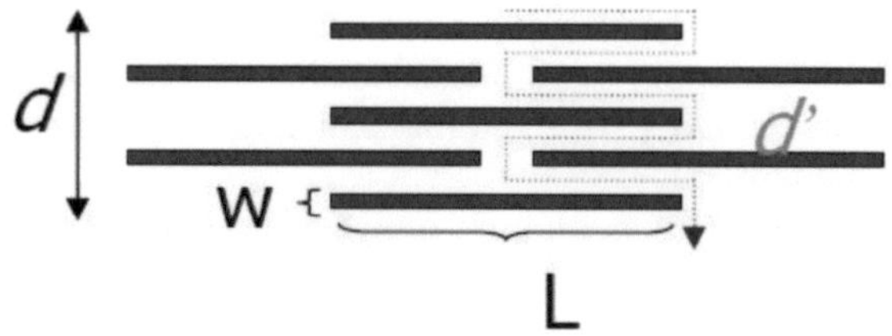

Fig. 6. Tortuous path according to Nielsen model

where P_e^{comp} and P_e^{matrix}, are the permeability coefficients for the composite and pure polymer film respectively, while

$$a = L/W \tag{4}$$

is the aspect ratio of the nanoplatelets and φ, the volume fraction of the clay into the composite. Although its simplicity, Nielsen model has been widely accepted and fits experimental observations well in several cases, especially for small volume fractions (below 10%) (Sun et al., 2008), (Wang et al., 2005)

Since the Nielsen model assumes more or less a perfect arrangement of the clay particles several other permeability models were based on the assumption of non-perfect alignment of platelets. A novel equation (Bharadwaj, 2001) introduces the parameter S, which is connected to the apparent mean angle, θ, of the platelet's long end and of the diffusion front by the following expression:

$$S = \frac{1}{2}(3\cos^2\theta - 1) \tag{5}$$

The reduction in the permeability of the nanocomposite can then be described by following equation:

$$\frac{P_e^{comp}}{P_e^{matrix}} = \frac{1-\varphi}{1+\frac{L}{2W}\varphi\left(\frac{2}{3}\right)\left(S+\frac{1}{2}\right)} \tag{6}$$

In the case that the platelets are randomly orientated (S=0), tortuosity decreases and diffusion is enhanced more, than in the case of vertically to the diffusion oriented platelets (S=1 reduces to Nielsen model). It should be noted that both the abovementioned models are based on the fact that diffusion is the dominant mechanism that determines the overall permeability.

Figure 7 presents the oxygen permeability at 23°C at different relative humidity values. A significant improvement of the gas barrier properties is observed for all studied samples. Thus, the permeability of the PVA/MMT20 at 50% RH decreases to almost 10% of the corresponding value of the pure PVA film. On the other hand, the permeability of the nanocomposites increases at higher %RH, although the obtained values are still much lower than the permeability of the pure PVA, at the same relative humidity. In this case, the water acts as plastisiser, facilitating the transport of oxygen molecules through the film.

The application of Nielsen model to the experimental data resulted in an aspect ratio (α) of ~70 in the case of 50% RH measurements. The calculated aspect ratio value is similar to that derived from AFM measurements. This implies that larger portion of the clay layers are delaminated within the polymer matrix.
At higher %RH, the predicted from Nielsen model aspect ratios decrease (to 20, 10 and finally to 2 for 65%, 75% and 85% RH respectively), revealing that the barrier effect of clays becomes less significant, as relative humidity increased. In these cases, the flow properties are mainly determined by the transport of the gas molecules though the swelled polymer.
It has also been reported that the presence of the inorganic platelets can affect the free volume of the polymeric phase, especially when there is low affinity between polymer and clay. In such a case, the permeability is determined by the counterbalance effects of reduction due to impermeable clay layers and the increase due to enhanced free volume (Incarnato et al., 2003).
In general, strong polymer-clay interactions can affect crystallinity, molecular orientation, and packing of the molecules near the nanoplatelets, leading to an enhancement in gas barrier properties, while poor adhesion between nanoplatelets and matrix usually results in opposite effect (Chaiko & Leyva, 2005), (Osman et al., 2003).

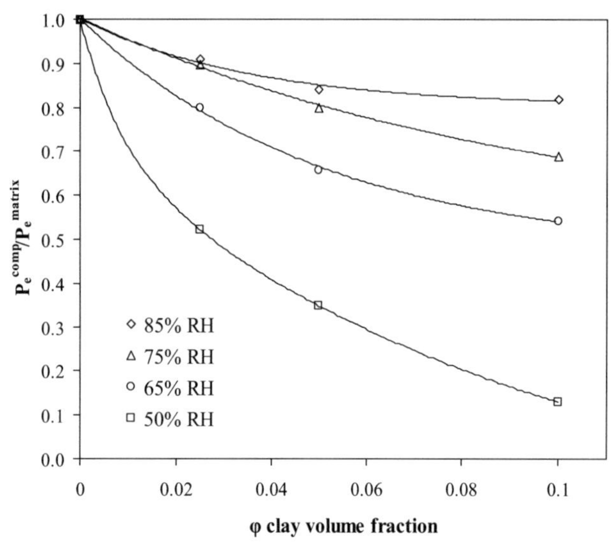

Fig. 7. Oxygen permeability of composite films presented in relation to pure PVA values

4.3 Mechanical properties

The clearest advantage of nanocomposites against conventional composite materials is the great improvement in their tensile properties, by small nanofiller addition. This result is caused mainly by the high surface area of the filler, which interact in the nanoscale with the matrix. The mechanical properties of polymers are influenced by many factors such as exposure to solvents, temperature, aging etc. Thus, it is of great importance to determine the mechanical properties under standard conditions.
In the case of the produced PVA/MMT nanocomposites, the measurements were performed at ambient humidity (~50% RH) and 23 °C, while the samples were pre-equilibrated at 45 and 70 %RH prior their testing. Figure 8 demonstrates the measurements for the Young's modulus and the elongation of the samples for 45% and 70% RH. In ambient humidity

conditions (45 %RH), Young's modulus value increases along with clay content increment, resulting to a maximum of 2.6 times for the case of PVA/MMT20 (in comparison with pure PVA). The great improvement in the Young modulus in PVA/MMT nanocomposites can be attributed to the strong interaction between matrix and silicate layers via formation of hydrogen bonds, due to the strong hydrophilicity of the clay edges (Sengwa, 2009).
At elevated RH (70%), the influence of the clay is much greater. Therefore, for the PVA/MMT05 sample the increase is more than 9 times and for PVA/MMT20 almost 193 times, respectively. The difference in the values, obtained between the two different relative humidity conditions, is due to the water's strong plasticization ability for PVA. Thus, while water can not influence the strength of clay particle, affects greatly the mechanical properties of PVA. Therefore, the effect of MMT on mechanical properties of the nanocomposites is more pronounce at higher %RH.

4.4 Thermal properties

An additional important aspect of nanocomposites is their increased heat resistance. The thermogravimetric curves of the samples are shown in Figure 9. The major weight losses are observed in the range of 200-500 °C, which corresponds to the structural decomposition of the PVA. Moreover, the nanocomposites exhibit significant less weight loss, in regard to pure PVA. Thus, for the PVA/MMT20 sample the weight loss at 400 °C is around 40% less of that for the neat polymer. It is also clear that the onset degradation temperature of the nanocomposites is slightly increased by the incorporation of the clay.

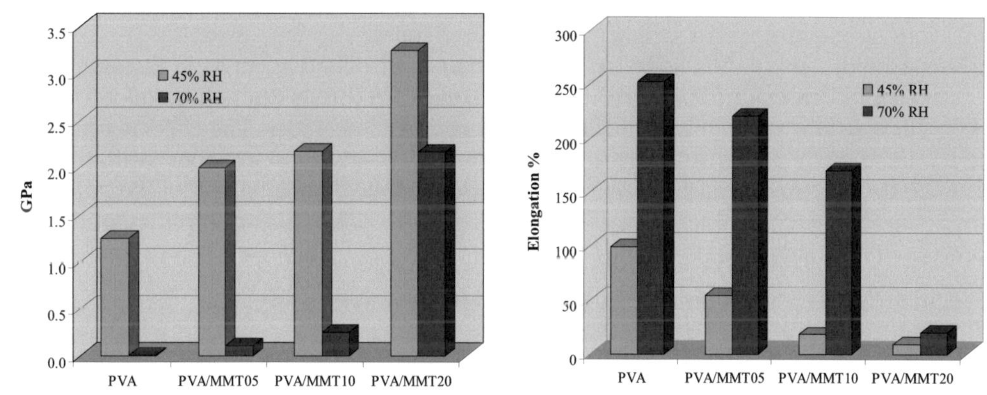

Fig. 8. Mechanical properties of developed nanocomposites (Left: Young Modulus, Right: Elongation values)

The enhancement in the thermal stability of the nanocomposites is due to the presence of clay nanolayers, which act as barriers to maximize the heat insulation and to minimize the permeability of volatile degradation products through the material (Chang et al., 2003).
Furthermore, DSC experiments reveal a decrease in crystallinity of the polymer with increasing clay content. Additionally, a second endothermic peak around 210 °C is observed (Figure 9 b), which can be attributed to the existence of a new crystal phase, induced by the presence of the clays. The enthalpy of fusion of this peak is linearly related to the clay content, suggesting that the inorganic layers are well dispersed, either intercalated or exfoliated, in the polymer. These results are in good agreement with the results obtained from the study of similar polymer-clay systems (Strawhecker & Manias, 2000).

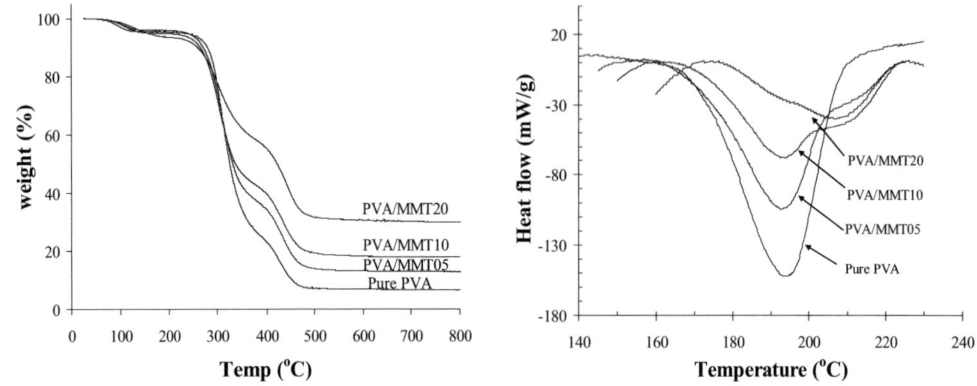

(a) (b)

Fig. 9. TGA (a) and DSC (b) curves of PVA/MMT naonocomposites

4.5 Optical properties

In general, the optical clarity of the nanocomposites can be related to the dispersion of the inorganic paltelets into polymer matrix: well dispersion in nanoscale will lead to exfoliated composites with high optical clarity. Reinforcing agents, in micrometer scale, usually scatter the light and thus reduce the light transmittance and the optical clarity of the composites. The same effect is observed when clay aggregates are formed. Due to their size (200-800 nm), clay aggregates cause strong scattering and/or absorption, resulting in very low transmission of the UV -Vis light. On the other hand, efficient dispersion in nanoscale, due to increased polymer- particle interfacial interactions, eliminates scattering and allows the preparation strong yet transparent films, coatings and membranes. The UV-Vis spectra of the prepared nanocomposites are shown in Figure 10. Contrary to previous studies (Yu et al., 2003) the prepared nanocomposite films retain the optical transparency of the pure PVA film in the visible region, even at high clay loading (up to 20% wt). This implies that the clay platelets are well dispersed in the PVA matrix.

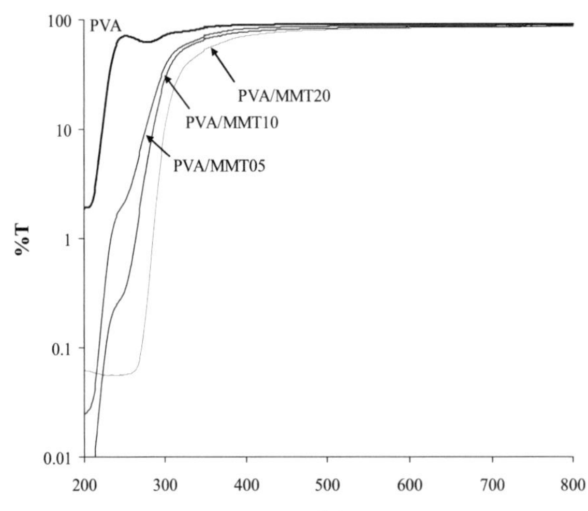

Fig. 10. UV-Vis spectra of developed nanocomposites films

4.6 Water sorption

The relationship between the concentration of water in a material at constant temperature and the water relative pressure (p/p_0) or water activity (a_w) is described by the water sorption isotherm.

The sorption isotherm can provide useful information concerning the sorption mode and the interactions involved in the sorption process. Several theoretical models have already been used to describe sigmoidal type isotherms. Models available in the literature to describe moisture sorption isotherm can be divided into several categories (Al-Muhtaseb et al., 2004): (a) models based on a monolayer adsorption (BET model), (b) models based on a multi-layer and condensed film (GAB model), (c) semi-empirical (Ferro-Fontan, Henderson and Halsey models) and (d) empirical models (Smith and Oswin models). In general, adsorption isotherms can be used for the determination of both the diffusion coefficient and the solubility and, therefore, the permeability of the films using Equation 2.

Sorption isotherm curves are obtained by plotting, the concentration of sorbed water (q^{Ads} in mg water/g sample) versus the water relative pressure (p/p_0). The sorption isotherms of pure PVA, MMT and PVA/MMT20 film are presented in Figure 11.

The shape of the nanocomposites water sorption isotherms is similar to that of the reference matrices, showing that the sorption mechanism is mainly governed by the same mechanism as in polymer matrix. In general, a sigmoidal shape corresponding to type IV sorption mode in the classification of Brunauer–Emmett–Teller is clearly observed for both films (Brunauer et al., 1940). This curve shape is typical for many hydrophilic materials (Gocho et al., 2000), (Masclaux et al., 2010).

At low water concentrations ($p/p0<0.4$) the behavior of nanocomposite film is almost identical to that of pure polymer. At higher relative pressures, an increase in both adsorption capacities is observed due to the enhanced molecular mobility of the polymer chains in the presence of water, which acts as plasticiser. Moreover, the sorbed amount for pure PVA film is much greater than the corresponding value of nanocomposite film.

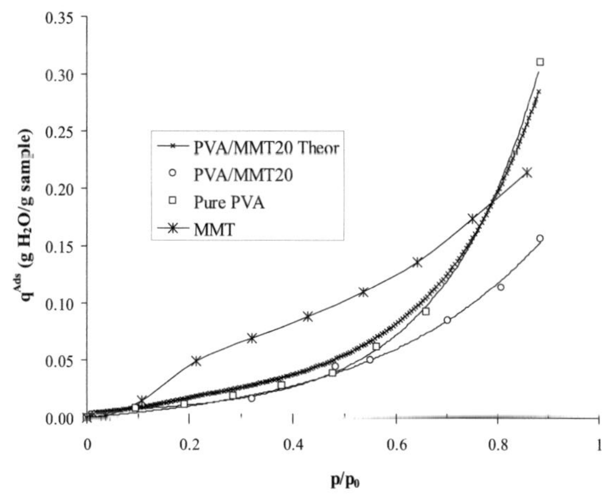

Fig. 11. Experimental water adsorption isotherms of pure PVA and PVA/MMT20 films and theoretical isotherm of PVA/MMT20 assuming no interactions between clay layers and polymer.

It should be noted that in the case of PVA/MMT20 the water uptake is lower than the calculated capacity, considering the contribution of each component, namely PVA and MMT, by means of an additive law, depending on their weight percentage (Figure 11). It could then be concluded that some sorption sites in the polymer are not available for water sorption in the composite, probably due to their involvement in the formation of polymer/nanoclay interactions. This hypothesis is in good agreement with the results obtained by the other techniques, used in this study. Therefore, water sorption isotherms can be considered as a valuable technique for the characterization of nanocomposites, providing useful information not only about macroscopic characteristics of the materials such as solubility and the diffusivity coefficients, but also about the interfacial nanoscale phenomena occurring between polymer and inorganic layers.

5. Conclusions

A series of PVA/Montmorillonite nanocomposites were prepared by effective dispersion the inorganic platelets into PVA matrix, via solvent casting technique. The developed nanocomposites were studied by a variety of microscopic and macroscopic techniques. Morphological studies using TEM, proved the excellent dispersion of the clay particles into the polymer matrix and the formation of exfoliated and intercalated structures. These results were in agreement with the obtained XRD patterns. On the other hand, AFM revealed, apart from the organization, the specific orientation of the clay particles, parallel to the surface of the nanocomposite. Furthermore, the orientation of the MMT layers was also identified by combined diffraction experiments in film and in powder form.

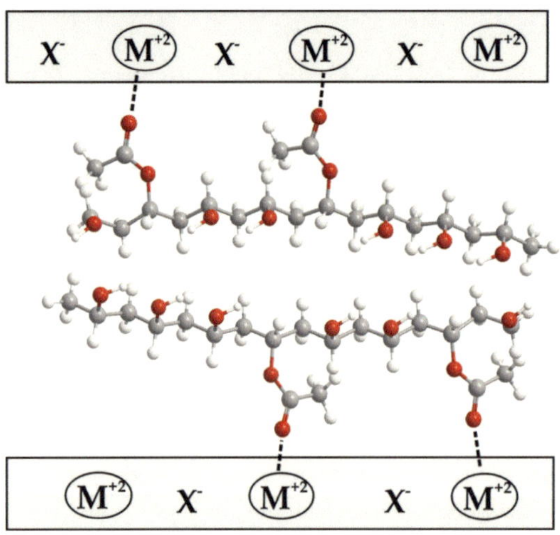

Fig. 12. Enhanced interactions between of partially hydrolised PVA and clay particles

The developed films, due to the favorable polymer-particle interactions, revealed excellent dispersion of the clay particles in the polymer matrix and improved mechanical strength, increased heat resistance and advanced gas barrier properties, retaining their optical transparency even at high clay loadings (20 wt %).

The application of the theoretical permeability model (Nielsen approximation) to the experimental data enabled the calculation of the theoretical aspect ratio of the clay (~70), which was in good agreement with the values obtained from the literature. This implied that the clay platelets were well-dispersed in the PVA matrix. Similar results were also obtained from water isotherm technique. In this case, the deviation of the experimental isotherm from the theoretical curve, calculated by the volume fraction additive law, revealed that a portion of the adsorption sites were used for the establishment of polymer-clay interactions.

Generally, the final properties of a nanocomposite are determined by the competitive effects of polymer-polymer and polymer-clay interactions. In PVA/MMT nanocomposites, the polymer -clay mixing can be settled by the interactions between the functional groups of PVA and the negative charge on clay surface. In our case, the acetoxy groups of the partially hydrolised PVA together with the presence of metal cations in the clay lattice, lead to the formation of strong polymer - clay interactions (Figure 12), promoting the dispersion of the inorganic layers into the polymer matrix. These interactions enabled the formation of intercalated and /or exfoliated structures, enhancing the overall material's performance.

6. References

Alexandre, M. & Dubois, P. (2000). Polymer-layered silicate nanocomposites: preparation, properties and uses of a new class of materials. *Materials Science and Engineering*, 28, 1-63

Al-Muhtaseb, A.H.; McMinn, W.A.M. & Magee, T.R.A. (2004). Water sorption isotherms of starch powders Part 1: mathematical description of experimental data. *Journal of Food Engineering*, 61, 297–307

Aranda, P. & Ruiz-Hitzky E. (1992). Poly(ethylene oxide)-silicate intercalation materials.' *Chemisty of Materials*, 4, 1395-403

Bafna, A.; Beaucage, G.; Mirabella, F. & Mehta, S. (2003). 3D Hierarchical orientation in polymer-clay nanocomposite films. *Polymer*, 44, 1103-15

Bharadwaj, R. (2001). Modeling the Barrier Properties of Polymer-Layered Silicate Nanocomposites. *Macromolecules*, 34, 9189-9192

Biswas, M. & Ray, S.S. (2001). Recent progress in synthesis and evaluation of polymer montmorillonite nanocomposites. *Advances in Polymer Science*, 155, 167-221

Brazel, C.S. & Peppas, N.A. (1999). Mechanisms of solute and drug transport in relaxing, swellable, hydrophilic glassy polymers. *Polymer*, 40, 3383-98

Brindley, G.W. (1984). In Crystal Structures and their X-Ray Identification, Mineralogical Society No.5, Brindley, G.W. & Brown, G., Eds, London, 169-172

Brunauer, S.; Deming, L.S. & Teller, E. (1940). On a theory of the van der Waals adsorption of gases. *Journal of the American Chemical Society*, 62, 1723–1732

Chaiko, D.J.; Leyva, A.A.(2005). Thermal transitions and barrier properties of olefinic nanocomposites. *Chemistry of Materials*, 17, 13–19

Chang, J.H.; Jang, T.G.; Ihn, K.J. & Sur G.S. (2003). Poly(vinyl alcohol) nanocomposites with different clays: pristine clays and organoclays. *Journal of Applied Polymer Science*, 90, 3204-14

Chin, I.-J.; Thurn-Albrecht, T.; Kim, H.-C.; Russell, T.P. & Wang, J. (2001). On exfoliation of montmorillonite in epoxy. *Polymer*, 42, 5947–52

Choudalakis, G. & Gotsis, A.D. (2009). Permeability of polymer/clay nanocomposites: A review. *European polymer Journal*, 45, 967-984

Chuang, W.Y.; Young, T.H.; Chiu, W.Y. & Lin, C.Y. (2000). The effect of polymeric additives on the structure and permeability of poly(vinyl alcohol) asymmetric membranes. *Polymer*, 41, 5633-41

Clement, F.; Lapra, A.; Bokobza, L.; Monnerie, L. & P. Menez. (2001). Atomic force microscopy investigation of filled elastomers and comparison with transmission electron microscopy - application to silica-filled silicon elastomers. *Polymer*, 4, 6259–6270

Cussler, E.L.; Hughes, S.E.; Ward III, W.J. & Aris, R. (1988). Barrier Membranes. *Journal of Membranes Science*, 38, 161-174

Dennis, H.R.; Hunter, D.L.; Chang, D.; Kim, S.; White, J.L.; Cho, J.W. & Paul D.R. (2001). Effect of melt processing conditions on the extent of exfoliation in organoclay-based nanocomposites. *Polymer*, 42, 9513–22

Fischer, H. (2003). Polymer nanocomposites: from fundamental research to specific applications. *Materials Science and Engineering: C*, 23, 763–72.

Giannelis, E.P. (1996). Polymer Layered Silicate Nanocomposites. *Advanced Materials*, 8, 29-35

Giannelis, E.P. (1998). Polymer-Layered Silicate Nanocomposites: Synthesis, Properties and Applications. *Applied Organometallic Chemistry*, 12, 675-680

Giannelis, E.P.; Krishnamoorti, R. & Manias, E. (1999). Polymer-Silicate Nanocomposites: Model Systems for Confined Polymers and Polymer Brushes. *Advances in Polymer Science*, 138, 107-147

Gilman, J.W. (1999). Flammability and thermal stability studies of polymer layered-silicate (clay) nanocomposites. *Applied Clay Science*, 15, 31-49

Gilman, J.W.; Jackson, C.L.; Morgan, A.B.; Harris, R.; Manias, E.; Giannelis, E.P.; Wuthenow, M.; Hilton, D. & Phillips, H. (2000). Flammability Properties of Polymer−Layered-Silicate Nanocomposites. Polypropylene and Polystyrene Nanocomposites. *Chemistry of Materials*, 12, 1866-73

Gocho, H.; Shimizu, H.; Tanioka, A.; Chou, T.J. & Nakajima, T. (2000). Effect of polymer chain end on sorption isotherm of water by chitosan. *Carbohydrate Polymers*, 41, 87–90.

Greenland, D.J. (1963). Adsorption of poly(vinyl alcohols) by montmorillonite. *Journal of Colloid and Interface Science*, 18, 647-64

Grunlan, J.C.; Grigorian, A.; Hamilton, C.B. & Mehrabi, A.R. (2004). Effect of clay concentration on the oxygen permeability and optical properties of a modified poly(vinyl alcohol. *Journal of Applied Polymer Science*, 93, 1102- 109

Hay, J.M. & Lyon, D. (1967). Vinyl alcohol: a stable gas phase species. *Nature*, 216, 790–1.

Hernández, M.C.; Suárez, N.; Martínez, L.A.; Feijoo, J.L.; Mónaco, S.L. & Salazar, N. (2008). Effects of nanoscale dispersion in the dielectric properties of poly(vinyl alcohol)-bentonite nanocomposites. *Physical Review E*, 77, 051801

Herrmann, W.O. & Haehnel, W. (1924). *Germany Patents*, No. 480866

Hodge, R.M.; Bastow, T.J.; Edward, G.H.; Simon, G.P. & Hill A.J. (1996). Free Volume and the Mechanism of Plasticization in Water-Swollen Poly(vinyl alcohol). *Macromolecules*, 29, 8137-43

Incarnato, L.; Scarfato, P.; Russo, G.M.; Di Maio, L; Iannelli, P.; Acierno, D. (2003). Preparation and characterization of new melt compounded copolyamide nanocomposites. *Polymer* , 44, 4625–4634.

Jiang, T.; Wang, Y-H.; Yeh, J-T. & Fan, Z-Q. (2005). Study on solvent permeation resistance properties of nylon6/clay nanocomposite. European Polymer Journal, 41, 459-466

Katsaros, F.K.; Steriotis, Th.A.; Sapalidis, A.A. & Favvas, E.P. (2009). Neutron diffraction studies of polymer/Clay nanocomposites, in *BENSC Experimental Reports 2008*, A. Rodig, A. Brandt, H.A. Graf (Eds), page 212, Berlin

Kobayashi, M.; Toguchida, J. & Oka, M. (2003). Preliminary study of polyvinyl alcohol-hydrogel (PVA-H) artificial meniscus. *Biomaterials*, 24, 639-47.

Krishnamoorti, R.; Vaia, R.A. & Giannelis, E.P. (1996). Structure and dynamics of polymer-layered silicate nanocomposites. *Chemistry of Materials*, 8, 1728-1734

Lagaly, G. (1999). Introduction: from clay mineral-polymer interactions to clay mineral-polymer nanocomposites. *Applied Clay Science*, 15, 1-9

Lan, T.; Kaviratna, P.D.; Pinnavaia, T.J. (1994). On the nature of polyimide clay hybrid composites. *Chemistry of Materials*, 6, 573–575

LeBaron, P.C.; Wang, Z. & Pinnavaia, T.J. (1999). Polymer-layered silicate nanocomposites: an overview. *Applied Clay Science*, 15, 11–29

Lee, J.Y., Baljon, A.R.C.; Loring, R.F. & Panagiopoulos, A.Z. (1998). Simulation of polymer melt intercalation in layered nanocomposites., *Journal of Chemical Physics*, 109, 10321-30

Lee, S.S.; Hur, M.H.; Yang, H.; Lim, S. & Kim, J. (2006). Effect of interfacial attraction on intercalation in polymer/clay nanocomposites. *Journal of Applied Polymer Science*, 101, 2749-2753

Lim, Y.T. & Park, O.O. (2001). Phase morphology and rheological behavior of polymer/layered silicate nanocomposites. *Rheologica Acta*, 40, 220-229

Magonov, S.N. (2000). AFM analysis of polymers In: Encyclopedia of analytical chemistry, R.A. Meyers, (Ed), Wiley, Chichester, pp. 7432–7491, ISBN: 9783527293131

Magonov, S.N. & Reneker, D. (1997). Characterization of polymer surfaces with atomic force microscopy. *Annual Review of Materials Science*, 27, 175-222

Maiti, M. & Bhowmick, A.K. (2006). New insights into rubbereclay nanocomposites by AFM imaging. *Polymer*, 47, 6156-6166

Malwitz, M.M.; Dundigalla, A.; Ferreiro, V.; Butler,P.D.; Henk, M.C. & Schmidt, G. (2004). Layered structures of shear-oriented and multilayered PEO/silicate nanocomposite films. *Physical Chemistry Chemical Physics*, 6, 2977-2982

Masclaux, C.; Gouanvé, F. & Espuche, E. (2010). Experimental and modelling studies of transport in starch nanocomposite films as affected by relative humidity. *Journal of Membrane Science*, 363, 221–231

McNally, T.; Murphy, W.R.; Lew, C.Y.; Turner, R.J., & Brennan, G.P. (2003). Polyamide- 12 layered silicate nanocomposites by melt compounding. *Polymer*, 44, 2761–72

Messersmith, P.B. &, Giannelis E.P. (1995). Synthesis and barrier properties of poly(ε-caprolactone)-layered silicate nanocomposites. *Journal of Polymer Science Part A: Polymer Chemistry*, 33, 1047-57

Morgan, A.B. & Gilman, J.W. (2003). Characterization of poly-layered silicate (clay) nanocomposites by transmission electron microscopy and X-ray diffraction: a comparative study. *Journal of Applied Polymer Science*, 87, 1329-38

Morgan, A.B. (2006). Flame retarded polymer layered silicate nanocomposites: a review of commercial and open literature system. *Polymers for Advanced Technologies*, 17, 206-17

Nguyen, Q.T. & Baird, D.G. (2006). Preparation of Polymer–Clay Nanocomposites and Their Properties. *Advances in Polymer Technology*, 25, 270–285

Nielsen, L.E. (1967). Models for the Permeability of Filled Polymer Systems. *Journal of Macromolecular Science, Part A: Pure and Applied Chemistry*, 1, 929-941

Ogata, N.; Kawakage, S. &, Ogihara, T. (1997). Poly(vinyl alcohol)-clay and poly(ethylene oxide)-clay blend prepared using water as solvent. *Journal of Applied Polymer Science*, 66, 573-81

Osman, M.A.; Mittal, V.; Morbidelli, M.; Suter, U.W. (2003). Polyurethane adhesive nanocomposites as gas permeation barrier. *Macromolecules*, 36, 9851–9858

Osman, M.A.; Mittal, V.; Morbidelli, M.; Suter, U.W.(2004). Epoxy-layered silicate nanocomposites and their gas permeation properties. *Macromolecules*, 37, 7250–7257

Pavlidou, S & Papaspyrides, C.D. (2008). A review on polymer–layered silicate nanocomposites. *Progress in Polymer Science*, 33, 1119–1198

Pinnavaia, T. & Beall, G. (2000), Polymer-Clay Nanocomposites. Wiley: New York.

Podsiadlo, P.; Kaushik, A.K.; Arruda, E.M.; Waas, A.M.; Shim, B.S.; Xu, J.; Nandivada, H.; Pumplin, B.G.; Lahann, J.; Ramamoorthy, A. & Kotov, N.A. (2007). Ultrastrong and Stiff Layered Polymer Nanocomposites. *Science*, 318, 80-83

Ray S.S. & Okamoto, M. (2003). Polymer/layered silicate nanocomposites: a review from preparation to processing. *Progress in Polymer Science*, 28, 1539-641

Ray, S.S. & Bousmina, M. (2006). Biodegradable polymer/layered silicate nanocomposites, In: *Polymer nanocomposite*, Y-W. Mai & Z-Z. Yu (Ed.), 57-129, Woodhead Publishing Limited and CRC Press LLC, ISBN-13: 978-1-85573-969-7

Sapalidis, A.A.; Katsaros, F.K.; Romanos, G.E.; Kakizis, N.K. & Kanellopoulos, N.K. (2007). Preparation and characterization of novel poly-(vinyl alcohol)–Zostera flakes composites for packaging applications. *Composites Part B: Engineering*, 38, 398-404

Sengwa, R.J.; Choudhary, S.; Sankhla, S. (2009). Dielectric spectroscopy of hydrophilic polymers-montmorillonite clay nanocomposite aqueous colloidal suspension. *Colloids and Surfaces A: Physicochemical and Engineering Aspects*, 336, 79–87

Sinha-Ray, S.; Yamada, K.; Okamoto, M. & Ueda, K. (2002). Polylactide-Layered Silicate Nanocomposite: A Novel Biodegradable Material. *Nano Letters*, 2, 1093-1096

Stathi, P.; Papadas, I.T.; Enotiadis, A.; Gengler, R.Y.N.; Gournis, D.; Rudolf, P. & Deligiannakis, Y. (2009). Effects of Acetate on Cation Exchange Capacity of a Zn-Containing Montmorillonite: Physicochemical Significance and Metal Uptake. *Langmuir*, 25, 6825–6833

Strawhecker, K.E. & Manias, E. (2000). Structure and Properties of Poly(vinyl alcohol)/Na+ Montmorillonite Nanocomposites. *Chemistry of Materials*, 12, 2943-2949

Sun, L.; Boo, W-J.; Clearfield, A.; Sue, H-J.; Pham, H.Q. (2008). Barrier properties of model epoxy nanocomposites. *Journal of Membrane Science*, 318, 129–136

Vaia, R.A.; Jant, K.D.; Kramer, E.J. & Giannelis, E.P. (1996). Microstructural evaluation of melt-intercalated polymer-organically modified layered silicate nano- composites. *Chemistry of Materials*, 8, 2628-35

Vaia, R.A.; Price, G.; Ruth, P.N.; Nguyen, H.T. & Lichtenhan, J. (1999). Polymer/layered silicate nanocomposites as high performance ablative materials. *Applied Clay Science*, 15, 67-92

Van der Hart, D.L.; Asano, A. & Gilman, J.W. (2001). Solid-state NMR investigation of paramagnetic nylon-6 clay nanocomposites. 2. Measurement of clay dispersion, crystal stratification, and stability of organic modifiers. *Chemistry of Materials*, 13, 3796-3809

Varlot, K.; Reynaud, E.; Kloppfer, M.H.; Vigier. G. & Varlet, J. (2001). Clay-reinforced polyamide: preferential orientation of the montmorillonite sheets and the polyamide crystalline lamellae. *Journal of Polymer Science Part B: Polymer Physics*, 39, 1360–70.

Wang, Y.; Zhang, H.; Wu, Y.; Yang, J.; Zhang, L. (2005). Preparation and properties of natural rubber/rectorite nanocomposites. European Polymer Journal ,41, 2776–2783.

Xu, R.; Manias, E.; Snyder A.J. & Runt, J. (2001). New Biomedical Poly(urethane urea)−Layered Silicate Nanocomposites. *Macromolecules*, 34, 337-9

Yalcin, B. & Cakmak, M. (2004). The role of plasticizer on the exfoliation and dispersion and fracture behavior of clay particles in PVC matrix: a comprehensive morphological study. *Polymer*, 45, 6623-6638

Yeh, J.-M. & Chang, K.-C. (2008). Polymer/layered silicate nanocomposite anticorrosive coatings. *Journal of Industrial and Engineering Chemistry*, 14, 275–291

Yu, Y.H.; Lin, C.Y.; Yeh, J.M. & Li, W.H. (2003). Preparation and properties of poly(vinyl alcohol)-clay nanocomposite material. *Polymer*, 44, 3553-60.

Zanetti, M.; Lomakin, S. & Camino, G. (2000). Polymer layered silicate nanocomposites. *Macromolecular Materials and Engineering*, 279, 1-9

3

Silver and Gold Nanoparticles on Sol-Gel TiO_2, ZrO_2, SiO_2 Surfaces: Optical Spectra, Photocatalytic Activity, Bactericide Properties

Anna Eremenko[1], Natalia Smirnova[1], Iurii Gnatiuk[1], Oksana Linnik[1], Nadezhda Vityuk[1], Iuliia Mukha[1] and Aleksander Korduban[2]
[1]*Chuiko Institute of Surface Chemistry of National Academy of Sciences of Ukraine,*
[2]*Kurdyumov Institute of Metallophysics of National Academy of Sciences of Ukraine,*
Ukraine

1. Introduction

Development of new nanomaterials with metal nanoparticles (Ag, Au, Cu, Rh, Pd, *etc.*) deposited on oxide surfaces, embedded within pores or encapsulated in its matrices have gained much attention in material science because of expanding applications of such composites in optics, medical diagnostics, analytical chemistry, catalysis, photocatalysis *etc.* The most widely used catalyst is titanium dioxide. Titania effective properties could be improved by mixing with other oxides (ZrO_2, SiO_2, ZnO) that act as additives to control structure-sorption, optical and electronic properties. Incorporation of metal nanoparticles into a solid matrix of titania enhances their quantum efficiency (Kim et al, 2001; Alberius, 2002). Semiconductor-metal composite nanoparticles have been shown to facilitate charge rectification in the semiconductor nanostructures that is beneficial for maximizing the efficiency of photocatalytic reactions (Kamat, 2003, Subramanian et al, 2001). The metal NPs, being adsorbed or incorporated into titania matrix, modify the interface and/or alter the pathways with which photogenerated charge carriers undergo recombination or surface reactions. Metal nanoparticles embedded in dielectric matrixes are promising composite materials for optical applications as systems with enhanced third-order electronic susceptibility $\chi^{(3)}$ [SiO_2-ZrO_2-Ag]. Aggregation and reshaping of metal nanoparticles and other processes occurring at the nanometal/porous matrix interfaces on preparation or post-reaction treatments, as well as mutual influence of their electronic structures, physical or chemical interactions of phases, affect many characteristics of nanocomposites (Chan et al, 2004; Epifani et al, 2000; Gonella et al, 1999; He J. et al, 2002; He C. et al, 2002; Liz-Marzan et al, 1996; Kelly et al, 2003; Kreibig and Vollmer, 1995; Shacham et al, 2004; Shter et al, 2007; Song et al, 2005).

There are hundreds publications devoted to the fabrication techniques of nanosized titania-based photocatalyst and the properties description of obtained materials. The methods of synthesis including sol-gel process with thermoinduced (Antonelli & Ying, 1995), photo- or chemical reduction of metal ions deposited on the TiO_2 surface (Smirnova et al, 1992), as well as combination of ion-exchange and reduction process (Gnatyuk, 2005) were proposed for preparation of films containing small particles of metals and semiconductor (Kim et al, 2001; Alberius, 2002; Antonelli & Ying, 1995; Smirnova et al, 1992; Gnatyuk et al, 2005).

The sol-gel techniques are most effective and popular on preparation of metal/oxide or metal/organics nanocomposites (Chan et al, 2004; Epifani et al, 2000; Gonella et al, 1999; He J. et al, 2002; He C. et al, 2002; Liz-Marzan et al, 1996; Kelly et al, 2003; Kreibig and Vollmer, 1995; Shacham et al, 2004; Shter et al, 2007; Song et al, 2005; Traversa et al, 2001). However it is not analyzed in the literature the interrelation of the synthetic conditions and physico-chemical properties of obtained materials, and hence the recommendations for the preparation of effective stable photocatalyst based on titania films and modified with small amount of noble metal nanoparticles (NPs).

In this work we present improved photo- thermochemical production of TiO_2, ZrO_2, ZnO and SiO_2,- metal functional films, as well as binary and ternary oxide composites via template assisted sol-gel method and doped with Ag and Au nanoparticles (NPs). Metal NPs have been synthesized by photochemical, chemical or thermal reduction of appropriate ions and embedded into oxide matrices via adsorption from their colloids or by the direct thermoreduction of metal ion/oxide composition with attempt to enhance their photocatalytic (M/TiO_2-based composites) and bactericide (M/SiO_2 composites) activity. Investigations of optical properties, XPS, surface morphology, electronic structure and photocatalytic activity have been performed.

2. Experimental part

All reagents were used as received. Template sol-gel method was applied for preparation of mesoporous silica, titania and zirconia films at glass and silicon substrates. Detailed procedure for the films synthesis with embedded noble metal nanoparticles is described in (Krylova et al, 2009). To form Ag and Au nanoparticles embedded within the oxide films, an appropriate amounts of $AgNO_3$ or $HAuCl_4$ were added to the precursor sols. Concentration of the Ag^+ ions was varied from 1 to 30 at.% and Au^{3+} ions from 1 to 7 at.% compared to molar concentration of alkoxides.

For film deposition onto glass or silicon wafers, dip-coating technique was utilized. After deposition of the film, gelation and gel ripening, it was dried in air at room temperature for 2 h (dried samples). Then the dried films were sintered in a furnace at a heating rate $\beta = 2$ °C/min to 250 °C, and at $\beta = 0.25$ °C/min from 250 to 350 °C. Template burns out at these temperatures and this process should be carefully carried out for keeping the ordered porous structure of the oxide film/Ag, Au nanoparticles. Then temperature was elevated to 500, 550 and 600 °C at $\beta = 3$ °C/min and the systems were kept at a certain temperature for 3 h.

Zinc ions modified titanium dioxide mesoporous films were synthesized by sol-gel method. The film covering on the clean glass substrate was performed by means of dip-coating techniques. The generation of the gold particles in the film structure was performed by different procedures. The first two methods were grounded on the admixing of (3 mol %) tetrachloroauric ions during titanium and zinc sol formation. Particularly, the first one consists of the drying of every layer at 60 °C for 30 min and UV light action for 10 min. The slow heating to 500 °C for 6 hours was done (TiO_2/ZnO/Au_{ph-t}, colored in pink). The second way was to treat every layer at 200 °C for 10 min with final irradiation by UV light for 30 seconds and annealing at 500° C for 2 h (TiO_2/ZnO/Au_t, colored in blue). The third method is based on the adsorption process. Previously synthesized and calcined (500° C) Zn^{2+}/TiO_2 film was dip-immersed in tetrachloroauric acid solution adjusted to pH 4 for 30 sec. These films were dried and exposed to UV light for 60 sec. The films were coded as TiO_2/ZnO/Au_{ads} film (colored in dark-blue). Hence, Au^{3+} ions were reduced 1) in the sol of zinc and titanium metalorganic species; 2) in the sol where crystallization of TiO_2 could take place; and 3) on the surface of anatase doped with zinc ions. A source of UV light was 1000 W middle - pressure mercury lamp. The absorption spectra of the films were recorded by Lambda 35 UV-Vis spectrophotometer (PerkinElmer) in the range of 200-1000 nm. It must be noted that the duration of film irradiation was experimentally established, no change in the SPR band intensity was observed after longer irradiation exposure and the reproducibility of the film synthesis was achieved for all three procedures.

The solutions of tetracycline hydrochloride (Aldrich) were prepared by the dissolving of an appropriate amount of the antibiotics in freshly distilled water. The film was immersed in 40 ml of $2{\cdot}10^{-5}$ mol/L (9.6 mg/L) TC solution until complete adsorption in the dark occurred and then irradiated by 1000 W middle-pressure mercury lamp. The reaction temperature was kept constant (25 °C) during irradiation. The change of the absorption spectra was recorded by Lambda 35 UV-vis spectrophotometer (PerkinElmer). The reaction rate was estimated as the pseudo-first order and calculated as a change in the TC absorption intensity at λ=357 mn. (Linnik et al, 2009).

SEM images were registered on SEM LEO-1530. EDS spectra were obtained using Brucker AXS X-ray detector.

XPS spectra were registered on ES-2402 with PHOIBOS-100_SPECS using Mg Kα line of 200 W Mg X-ray tube as a radiation source at 1253,6 eV. Vacuum in a camera maintained at $2{\cdot}10^{-7}$ Pa. The film's size covered on Si wafer was 10x10 mm. The XPS signals were fitted using Gaussian-Newton method.

Photooxidation of Rhodamine B dye aqueous solutions (1x10 -5 mol/l) in the presence of a film with a mass of about 1 mg, was performed in a quartz reactor with water-cooling under vigorous stirring at pH = 6-7. The UV light was provided by a 1000 W low-pressure mercury lamp (λ = 254 nm, P = $2.91\mathrm{x}10^{-7}$ quants/s). The changes of the Rhodamine B concentration were monitored by absorption measurements at 554 nm.

Optical spectra (transmission mode) of the films were recorded using a Lambda UV-Vis (Perkin Elmer) spectrometer.

3. Optical properties of sol-gel films modified by metal nanoparticles

Optical properties of metal nanoparticles containing sol-gel derived films are strongly dependent on the synthesis history of the samples.

Two different routes of synthesis of SiO_2, TiO_2 and ZrO_2 and mixed oxide films modified with Ag and Au nanoparticles that can be described as one-step thermal reduction of metal ions in the inorganic matrix during its sintering and a procedure that includes two stages, namely photoreduction of the metal ions with subsequent thermal treatment of composite films at elevated temperatures have been studied. Both proposed synthesis strategies led us to the formation of metal nanoparticles containing inorganic films but with different optical and surface properties.

3.1 SiO_2, TiO_2 and ZrO_2 films with embedded Ag and Au nanopartiles obtained by thermal - induced reduction

Thermal-induced formation of the metal nanoparticles in the mesoporous silica, titania, zirconia and mixed oxide matrixes during their sintering at temperatures higher than 500 °C is a result of electron attachment to Ag^+ and Au^{3+} ions from organic substances of the sol-gel films or their decomposition products. Inorganic SiO_2, TiO_2 and ZrO_2 films with uniformly distributed Ag and Au nanoparticles and intensive coloration, characteristic for surface plasmon absorption of silver and gold nanoparticles have been obtained when the films doped with different amounts of metal ions were heat treated at elevated temperatures.

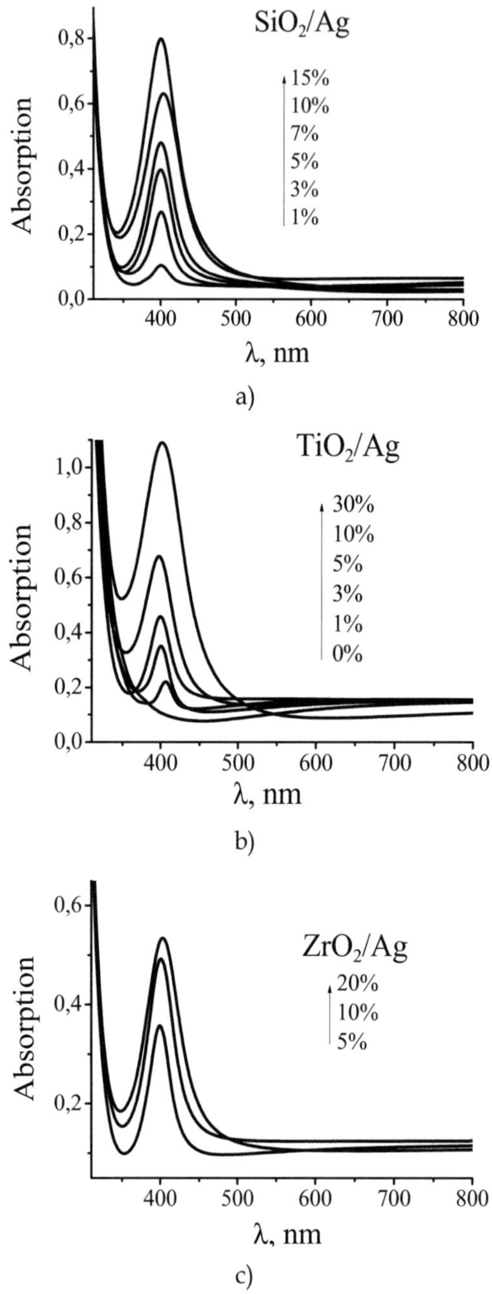

Fig. 1. Absorbance spectra of films sintered at 500 °C (2 hrs) with increasing Ag^+ content in the initial sol for a) SiO_2/Ag; b) TiO_2/Ag and c) ZrO_2/Ag.

In the absorption spectra of silica, titania and zirconia films doped with silver ions and sintered at 500 °C (Fig. 1), an intensive, symmetric absorption band at around 400 nm appeared due to the silver nanoparticles formation induced by thermal reduction of metal ions as it was described before. The uniform distribution of Ag nanoparticles within the surface and near-surface layers of SiO_2 and TiO_2 films was confirmed by SEM images of the corresponding samples (Fig. 2). The mean particles size of Ag nanoparticles varies depending on the host matrix and is equal to 5-9 nm in the case of silica and 3-5 nm for titania films. We have attributed this change of the particles size of the formed nanoparticles to the differences in crystallinity of the matrixes. Obviously, amorphous structure of silica at 500 °C favors continuous movement of silver ions and formed nanoparticles that leads to the larger metal nanoparticles in the final structures.

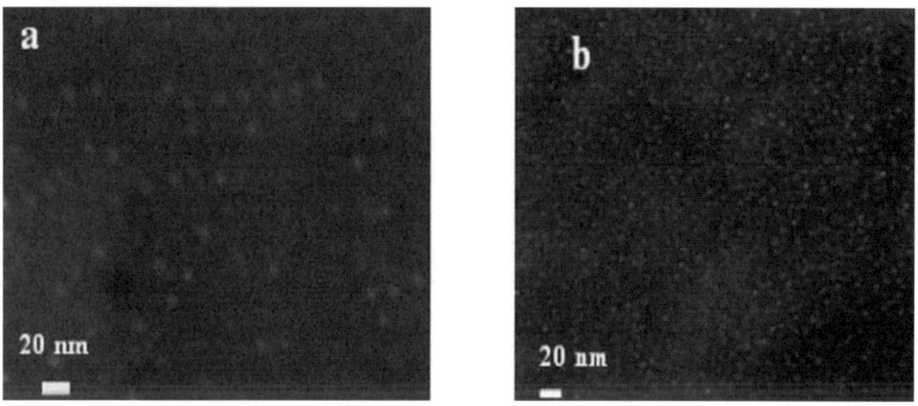

Fig. 2. SEM images of mesoporous films sintered at 500 °C: a) SiO_2/Ag 10% and b) TiO_2/Ag 10%.

Elevating of the sintering temperatures of the inorganic matrixes with embedded Ag nanoparticles leads to certain spectral changes in the absorption spectra of the films: shape of the surface plasmon band, position and intensity. These changes depend also on the initial content of the silver ions in the films. Thermal transformations of the absorption spectra of silver nanoparticles containing films depending on the concentration of silver dopant and heat treatment conditions are illustrated in the Fig. 3 for the case of TiO_2 as the host matrix. At low silver content (0.3-5 mol.%) increase of the sintering temperatures up to 550-600 °C was accompanied with the decrease of the surface plasmon band intensity and the appearance of new long-wave band at about 470-520 nm for the films with 5 mol.% of silver and heat treated at 600 °C.

This effect was attributed to the evaporation of nanosized silver from the outer surface of the films and formation of larger silver particles in the latter case. When SiO_2, TiO_2 and ZrO_2 films were doped with 10 mol.% of silver, elevating of the heat treatment temperatures leads to the increasing of the integral absorbance of the films in the spectral range of absorption of silver nanoparticles, and again, to the appearance of the long-wave band, mentioned before. For the films with further increased silver content up to 30 mol.% only steady increasing of intensity of silver nanoparticles absorption could be noticed. We have concluded that continuous reduction of silver ions takes place compensating evaporation loses of the nanosized silver, when the films with bigger amount of silver ions are subjected to the elevated sintering temperatures, leading to the increasing of integral absorbance of the films. The formation of the long-wave band in the absorption spectra of the silver nanoparticles containing inorganic films might correspond to the appearance of nonspherical silver particles (Kelly et al, 2003; Link et al, 2001).

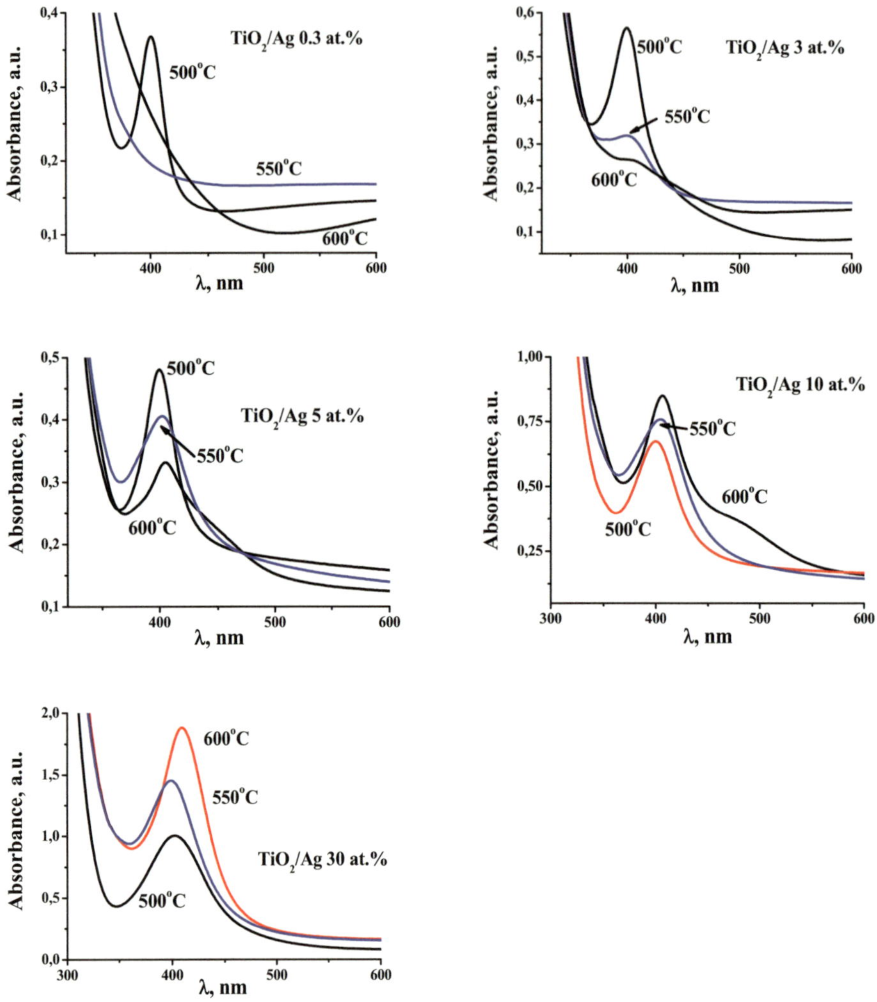

Fig. 3. Evolution of the absorbance of TiO_2/Ag films depending on the silver content and heat treatment temperature.

Thermal-induced formation of gold nanoparticles in the inorganic matrixes starts at lower temperatures due to the differences in thermal stability of the metal ions sources. Chloroauric acid used in the films as Au^{3+} source, is less thermally stable in comparison with silver nitrate. The scheme of thermal transformations of $HAuCl_4$ is presented below:

$$HAuCl_4 \xrightarrow{120°C} AuCl_3 \xrightarrow{185–200°C} AuCl \xrightarrow{290°C} Au$$

Heat treatment of the SiO_2/Au 1% film at temperatures more than 400°C was accompanied with the formation of gold nanoparticles within the silica matrix. The wide surface plasmon band at 535 nm characteristic for Au nanoparticles appeared in the absorption spectrum of the composite film (Fig. 4).

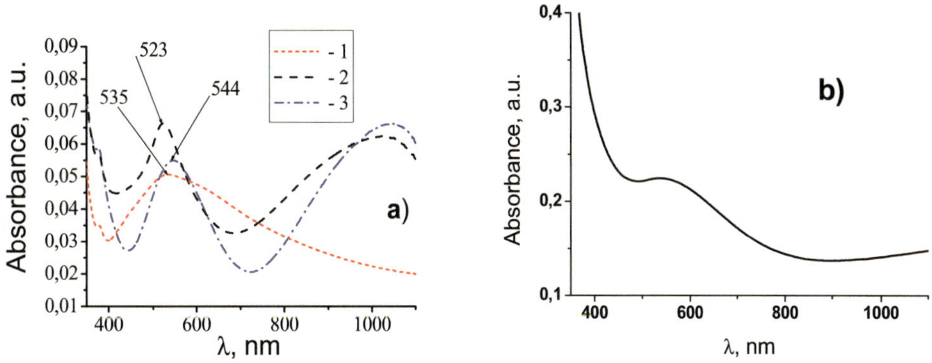

Fig. 4. Absorbance spectra of a) SiO_2/Au 1% film sintered at different temperatures (3 hrs) 1) 400 °C; 2) 450°C; 3) 500 °C and b) TiO2/Au 1% film sintered at 500 °C (2 hrs).

Elevation of sintering temperature of the films up to 450°C caused some blue shift and increase of intensity of the main absorption band of Au nanoparticles (523 nm) with spherical shape and formation of the additional long-wave band at approximately 1000 nm, characteristic for the absorption of the trigonal prism shape gold nanoparticles in accordance with (Huang et al, 2004). Thermal treatment of the SiO_2/Au 1% film at 500°C leads to decrease of intensity and red shift up to 544 nm of the short-wave absorption band of gold nanoparticles with increase of intensity of the long-wave absorption band. We attribute these spectral changes to the growth of spherical and prismatic gold nanoparticles as the result of thermal facilitated movement of the particles.

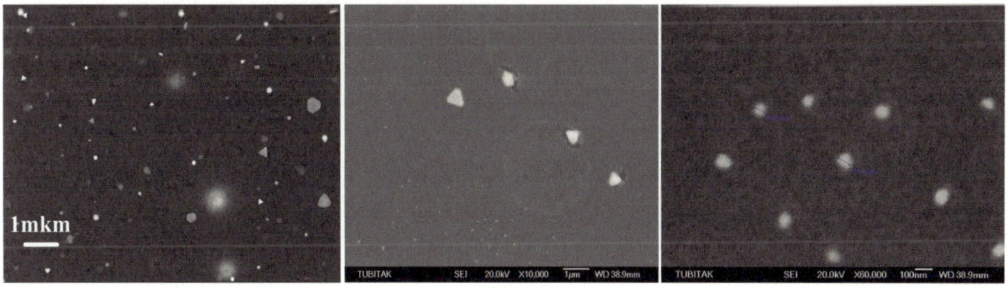

a) b) c)

Fig. 5. SEM images of the mesoporous films with embedded gold nanoparticles a) SiO_2/Au 1% and b, c) TiO_2/Au 1%.

In the SEM images of surface of silica and titania films modified with gold nanoparticles (Fig. 5) we have observed formation of large gold nanoparticles of different shape which conform with the conclusions of analysis of optical spectra of the films. Bimodal particles size distribution was observed for thermally reduced Au nanoparticles in the analyzed host matrixes with mean particles size in the range of 90-110 nm and 230 nm. Formation of the large trigonal prisms with the size up to 400-500 nm predominantly on the film surface confirms the assumption of the growth mechanism of the particles proposed earlier for silver nanoparticles in SiO_2, TiO_2 and ZrO_2 matrixes (Krylova et al, 2009).

The hydrophilic-hydrophobic interactions between hydrophilic oxide surface (host matrix) and hydrophobic surface of metal nanoparticles will govern metal nanoparticles onto the outer surface of the host. On the other hand, thermal-induced reduction of the metal ions and charged clusters by volatile organics formed during template and organic ligands decomposition/burning out facilitates transfer of reduced atoms, clusters by the gaseous flow onto the outer surface of the films. The other factor - steric hindrance in confined space of the pores, which hinder nanoparticles growth - is much weaker at the outer surface, where the forces of mutual ions/atoms/nanoparticles interactions are smaller than within pores. In other words, particles have more freedom for their diffusion, clusterization and agglomeration (Krylova et al, 2009).

3.2 TiO_2 and TiO_2/ZnO films modified with Ag nanopartiles by photoreduction of metal ions and subsequent thermal treatment

For the photoreduction of metal ions on the surface of inorganic matrix with the aim of formation of metal nanoparticles on it, titanium dioxide was chosen as appropriate one due to its photocatalytic activity. It was observed, that when TiO_2 mesoporous films have been UV irradiated after adsorption of silver ions from the solutions of silver nitrate or silver ammonia complex, photodeposition of nanosized silver islands takes place. This photodeposition process is widely used in literature for the production of novel catalysts, functional surfaces, optical elements, etc. (Sclafani et al, 1997; Subramanian et al, 2001).

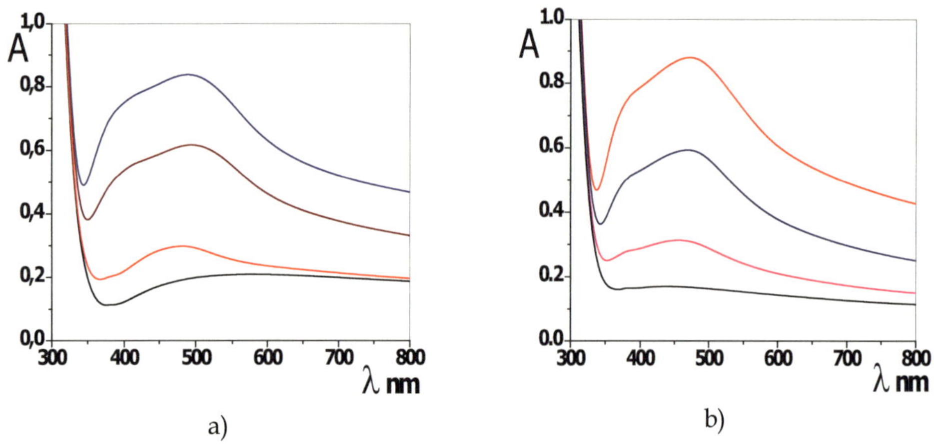

Fig. 6. Evolution of absorption spectra of TiO_2 films with photoreduced silver nanoparticles depending on the accumulation of silver NPs during irradiation time when: $AgNO_3$ (a), and $[Ag(NH_3)_2]NO_3$ (b) are used as silver ion sources.

In the absorption spectra of the TiO_2/Ag films with photodeposited silver particles broad complex absorption surface plasmon band of nanosized silver with maxima at 390 and 490 nm was detected. This band corresponds to the fromation of silver particles with broad particles size distribution (Epifani et al, 2000; Lance et al, 2003). In the case of photodeposition of silver nanoparticles from silver ammonia complex $[Ag(NH_3)_2]NO_3$ the surface plasmon band of silver has more distinct shape with maxima at 390 and 460 nm. The intensity of the band in both cases increases with increasing of the deposited silver content. In addition some maximum displacement towards longer wavelength, characteristic for the agglomeration of the silver particles can be detected.

As the shape and position of the formed silver nanoparticles absorption depends on the silver source used, to control the dispersion of silver particles on the surface of photoactive films, we have decided to incorporate amphoteric zinc oxide into TiO_2 matrix, which sites after leaching in basic solutions (ammonia), may play a role of ion exchange sites for silver complex adsorption and further silver nanoparticles nucleation upon photoreduction and subsequent thermal treatment, as was previously proposed for Mg^{2+} modified wet TiO_2 films by He et al (He J. et al, 2003).

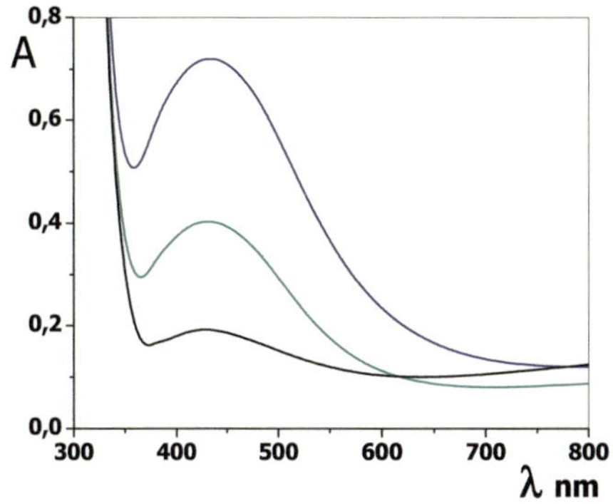

Fig. 7. Evolution of absorption spectra of TiO2/1% ZnO films with photoreduced silver nanoparticles depending on the accumulation ofAg NPs during irradiation when [Ag(NH3)2]NO3 was used as silver ion source.

It was observed, that in the absorption spectra of the UV irradiated TiO_2/ZnO mesoporous films with adsorbed silver ions after ion exchange in silver ammonia complex solutions broad surface plasmon band of nanosized silver with maximum at 440 nm was formed, whereas the band at 390 nm was not observed contrary to the results when silver nitrate was applied. It indicates more uniform silver particles size distribution.

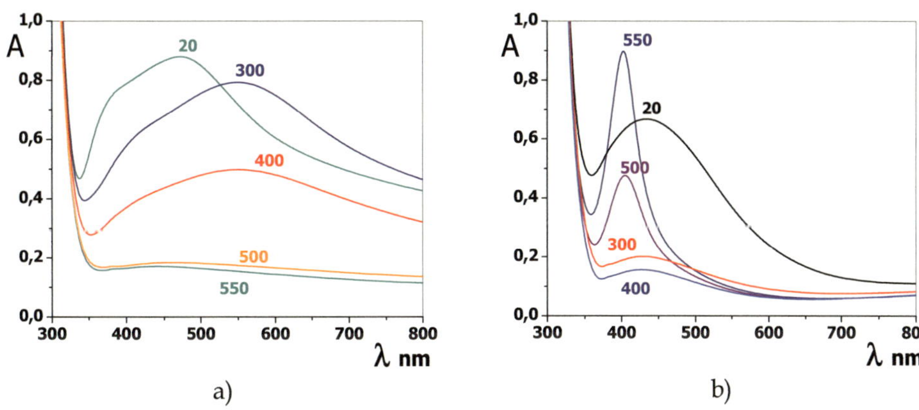

Fig. 8. Absorption spectra of TiO_2 (a) and TiO_2/1% ZnO (b) films with photoreduced silver nanoparticles after thermal treatment at different temperatures (indicated near the spectral bands).

Subsequent thermal treatment of the TiO_2 and TiO_2/ZnO films with photodeposited silver caused significant changes in the absorption spectra of the composite films. Red shift and decrease of intensity of the surface plasmon band of nanosized silver was observed for the TiO_2/Ag films upon heating up to 400 °C with further complete disappearance of the absorption typical for Ag nanoparticles after film sintering at 500-550 °C, that can be attributed to the formation of extremely small silver particles on the film surface and/or partial dissolution of the silver nanodrops in the TiO_2 crystalline matrix (He J. et al, 2003; Krylova et al, 2009).
Different absorption spectra evolution was observed for the TiO_2/ZnO films with photodeposited silver. Surface plasmon band of silver particles disappears at all upon film heating up to 300-400 °C whereas after increasing of the sintering temperature of the composite up to 500-550 °C distinct intensive absorption band of spherical silver nanoparticles at 400 nm appears in the absorption spectrum of the film.

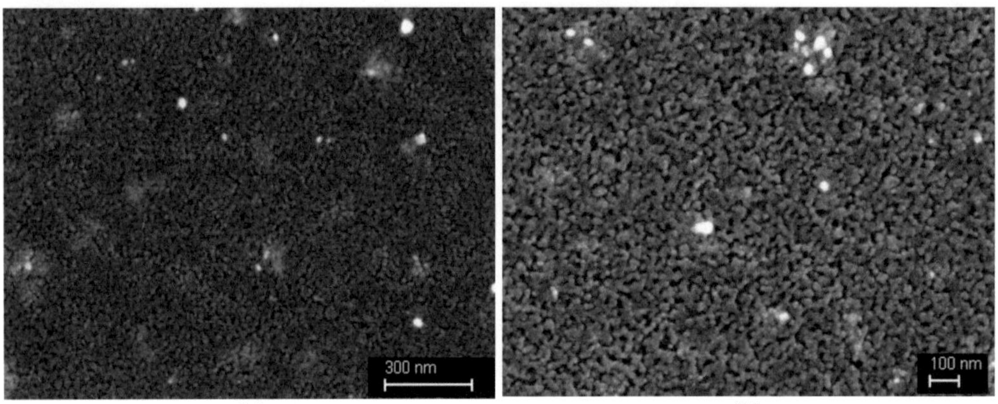

(a) (b)

Fig. 9. SEM images of mesoporous TiO_2 (a) and TiO_2/1% ZnO (b) films with photoreduced silver after thermal treatment at 550°C.

We believe that during ion exchange taking place at the zinc oxide rich sites of the TiO_2 matrix when the latter is being immersed into silver ammonia complex, the sites within the TiO_2 matrix for further nucleation of silver nanoparticles have been appeared.
Uniform distribution of silver nanoparticles over mesoporous TiO_2 and TiO_2/ZnO surfaces with the size of 4-20 nm and 50-100 nm after the cycles of photoreduction and thermal treatment was confirmed by SEM microscopy analysis of films (Fig. 9).

3.3 TiO_2 and TiO_2/ZnO films modified with Au nanopartiles produced by photo- and thermal treatment

The procedures of metal particles synthesis influence on the maximum position of SPR band of Au NPs (Fig.10). For the films obtained by thermo- and photo-thermo procedures, the gold particles are spherical whereas adsorption brings to the nanorods with low aspect ratio as it is indicated in (Rodrı´guez-Ferna´ndez et al, 2005; Eustis & El-Sayed, 2005; Eustis & El-Sayed, 2006). It must be noted that the more intensive and well-defined SPR maxima were registered for the films containing zinc ions indicating the higher dispersion of metal crystallization centers near zinc ions in titania matrix preventing the NP's aggregation in the clusters (Manujlov et al, 2008).

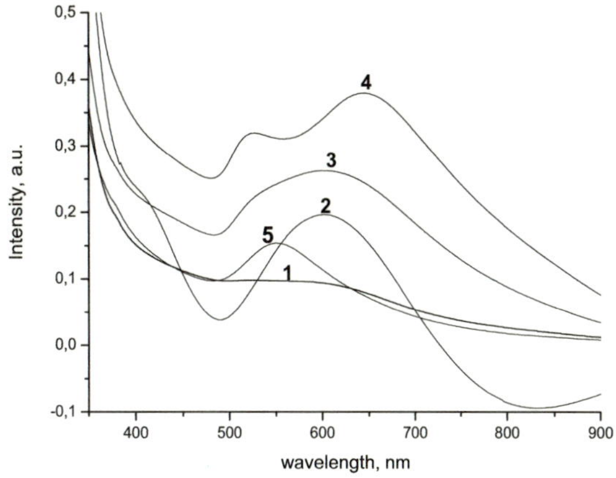

Fig. 10. Absorption spectra of gold-doped films: 1. TiO_2/Au_t; 2. $TiO_2/ZnO/Au_t$ (λ_{max}= 600 nm); 3. TiO_2/Au_{ads} (λ_{max}= 600 nm); 4. $TiO_2/ZnO/Au_{ads}$ (λ_{max}= 518 and 645 nm); 5. $TiO_2/ZnO/Au_{tph-t}$ (λ_{max}= 545 nm).

The elemental maps of Ti, O, and Zn showed the homogeneous distribution of titanium, zinc and oxygen on the surface of $TiO_2/ZnO/Au$ films (not shown here). SEM images and corresponding Au maps (Fig. 11) depict the different sizes and homogeneous distribution of gold particles with the size in the range of 25-180 nm (dominate 25-40 nm) for the $TiO_2/ZnO/Au_t$ and 1-35 nm (dominate 3-10 nm) for $TiO_2/ZnO/Au_{ph-t}$ films (the particle size distribution is not shown). The gold particles in $TiO_2/ZnO/Au_{ads}$ are 25-250 nm (the major fraction is 70-110).

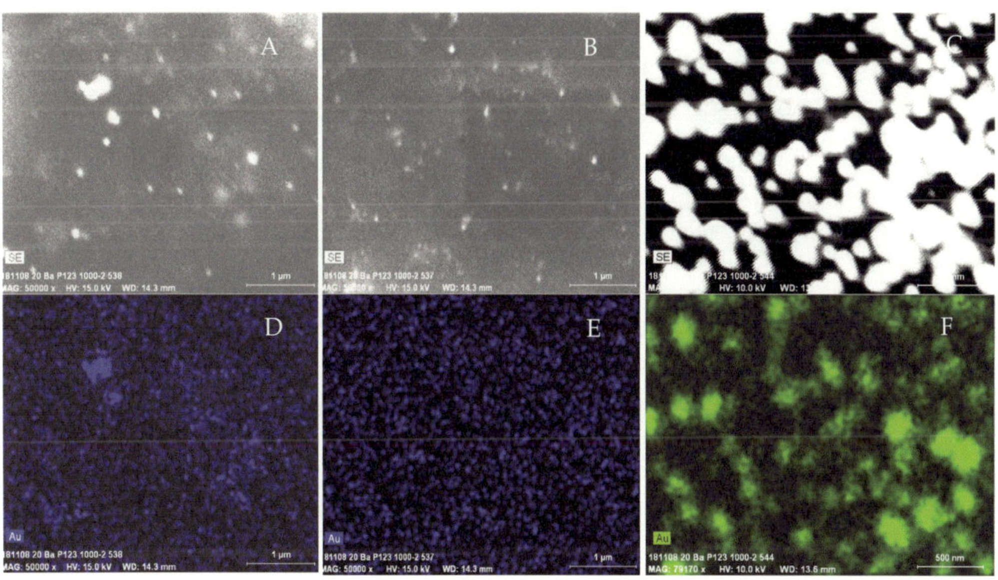

Fig. 11. SEM images and Au map of $TiO_2/ZnO/Au$ films: A, D- Au_t, B, E- Au_{ph-t}; C, F- Au_{ads}.

X-ray analyzed EDS spectra testify the zinc ions on the surface of the $TiO_2/ZnO/Au_t$ and $TiO_2/ZnO/Au_{ph-t}$ films contrary to Zn^{2+}/TiO_2 and $TiO_2/ZnO/Au_{ads}$ (not shown here). The enrichment of the film surface by zinc ions is suggested to be due to the formation of complexes between zinc acetate and tetrachlorautic ions during sol ageing resulting in the localization of zinc ions near photoformed gold particles on the surface.

3.4 $TiO_2/ZrO_2/SiO_2$ films with embedded Ag and Au nanopartiles by thermal induced reduction

From the point of view of photocatalytic applications it was interesting to synthesize triple mixed oxides, which surface would have new active surface sites active in catalytic processes. In addition, modification of the last with metal nanoparticles could have positive influence on the charge separation in the semiconductor during its excitation by UV light. To fulfill these requirements we have synthesized by sol-gel method mixed $TiO_2/ZrO_2/SiO_2$ films modified with Ag and Au nanoparticles through thermal reduction of noble metal ions in the mixed matrix during its sintering.

No absorption characteristic for nanosized silver particles was detected in the absorbance spectra of $TiO_2/ZrO_2/SiO_2$/Ag films with different percentage of silver after their sintering at 500°C. Similar picture was already described for TiO_2/Ag films with photodeposited silver particles after heat treatment at temperatures above 500°C. We believe that this means rather formation of the tiny silver particles on the surface of nanocomposite films with oxide shell than the absence of the nanosilver. SEM microscopy of surface of 21%TiO_2/9%ZrO_2/70%SiO_2/Ag film (Fig. 12 a) proves the formation of flake-like particles in the subsurface/surface region of the film with the size of about 30 nm, most probably, corresponding to the silver particles covered with thick oxide shell. This will be further confirmed by the XPS investigations of the surface of the composites.

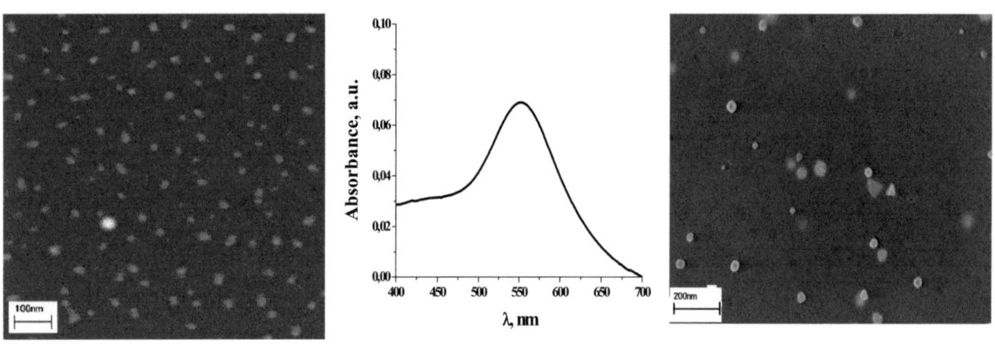

a) b) c)

Fig. 12. SEM images of surface (a, c) and absorbance spectrum (b) of $TiO_2/ZrO_2/SiO_2$/Ag (5%) (a) and $TiO_2/ZrO_2/SiO_2$/Au(3,4%) (b, c) films after sintering at 500 °C.

In absorption spectrum of $TiO_2/ZrO_2/SiO_2$/Au film (Fig. 12 b) the surface plasmon band of gold nanoparticles is observed at 550 nm. The surface of $TiO_2/ZrO_2/SiO_2$/Au films is covered with gold nanoparticles of different shapes - triangular pyramids and spheres that can be seen from SEM image (Fig. 12 c). Big amount of voids on the films surface, with the shape corresponding to the one of metal nanoparticles formed, confirms heat induced movement of reduced metal clusters/nanoparticles as a result of their agglomeration and/or leaving of the surface into gas phase.

Optical absorption spectra of $TiO_2/ZrO_2/SiO_2$/Au films with different content of gold introduced into the sol for films are presented in the Fig. 13. The position of SPR band shifts consistently to longer wavelengths region from 549 to 554 nm with increasing Au^{3+} concentration from 1 to 7 mol.% in the films. According to data available in the literature

(Epifani et al, 2000; Selvan et al, 1998; Sung-Suh et al, 2004) it means an increase of the size of Au nanoparticles after thermal treatment of the films.

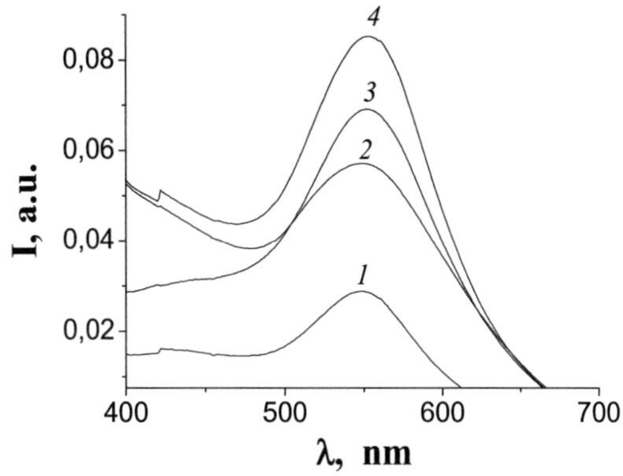

Fig. 13. Optical absorption spectra of $TiO_2/ZrO_2/SiO_2$ films with various percentages of gold: *1* – 1 % Au; 2 – 3,4 % Au; 3 – 5 % Au; 4 – 7 % Au.

In summary, synthesis strategy employed for the preparation of composite materials determines final localization of noble metal nanoparticles. Single step thermal reduction of noble metal ions during sintering of sol-gel films leads to the distribution of metal nanoparticles within the bulk of the films as well as partial localization of the particles on the films surface. Dual step modification of the films surface by photoreduction of silver ions and subsequent composites thermal treatment results in uniform modification of surface rather than bulk of the films by Ag nanoparticles with narrow particles sizes distribution. The last is being determined by the leaching of ZnO from the surface of composite films with subsequent localization of tiny Ag nanoparticles on these sites.
Photoreduction of tetrachloroauric ions in the "wet" films leads to the formation of the homogeneous distributed gold nanoparticles with the sizes to 10 nm where the subsequent thermal treatment did not cause to its aggregation. The mechanism of thermal reduction is differed from the former case that is the reason of enlargement of gold particles due to the easier diffusion of gold (I) ions formed on the first stage of thermal treatment (scheme). Photoreduction of adsorbed $AuCl_4^-$ ions on the Zn^{2+}/TiO_2 film surface creates rod-shaped nanosized gold particles.

4. XPS investigations

4.1 SiO_2/Ag and SiO_2/Au

Electronic structure of the composite films has been investigated by means of X-Ray photoelectron spectroscopy. The main contribution to the Ag*3d*-line of the SiO_2/Ag film and to the Au4f-line of the SiO_2/Au film have the components with the binding energy (E_B) Ag$3d_{5/2}$ = 370.84 eV (Fig. 14 a) and E_B Au$4f_{7/2}$ = 85.61 eV (Fig. 14 b), correspondingly. The position of these lines is slightly shifted towards higher energies comparing to the ones reported for the gold and silver foils [Kamat et al, 2002].
As we reported previously [Suzer/Jashan article], these shifts might be explained by the chemical interactions between metal nanoparticles and oxide matrix. By applying external voltage bias during XPS analysis of the SiO_2/Ag-Au sol-gel films linear parallel shift of the Si2p, Au4f, Ag3d and O1s signals was observed revealing chemical interaction between metal nanoparticles and silicon oxide matrix.

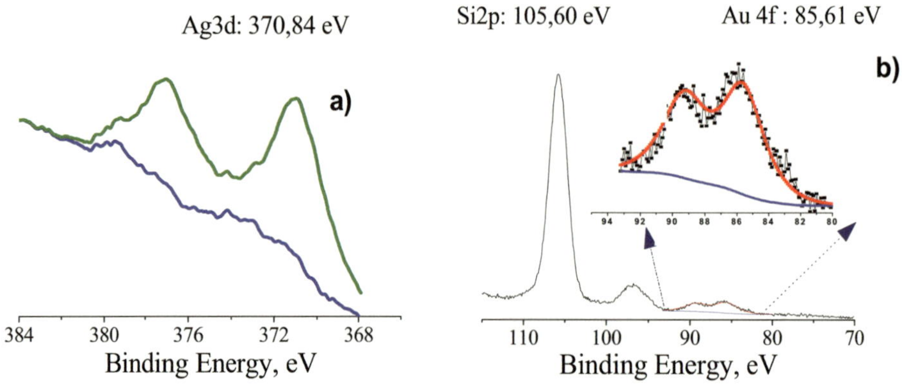

Fig. 14. XPS-spectra of Ag3d a) and Si2p and A4f b) -levels for SiO_2/ Ag and SiO_2/ Au mesoporous films.

4.2 TiO_2/Ag and TiO_2/ZnO/Ag films

In the XPS spectrum of the TiO_2 film with silver nanoparticles produced by thermal induced silver reduction (Fig. 15) we have observed that the Ag*3d*-line is formed by the contribution of three silver states with E_B Ag$3d5_{/2}$ = 367.7 eV (silver oxide), 368.4 eV (metallic silver) and at 368.8 eV (charge transfer state).

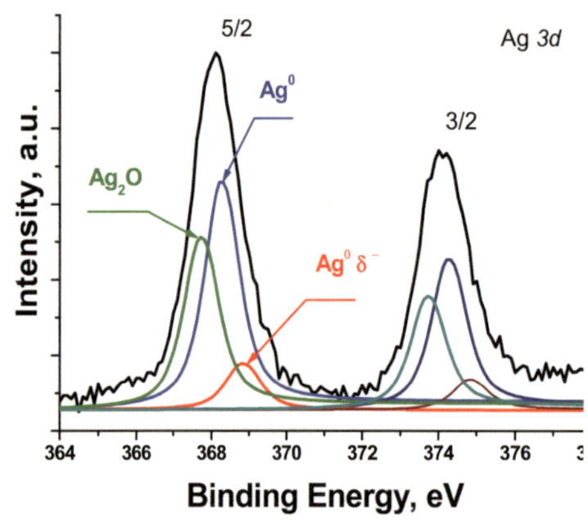

Fig. 15. XPS-spectra of Ag3d–levels for TiO_2/ Ag film heat treated at 500°C.

Silver is a metal that has anomalous properties in E_B shifts when being oxidized, i.e. the Ag*3d* peaks shift to lower E_B values (Weaver & Hoflund, 1994). Usually, positive E_B shifts in the metal core-level peaks are observed when metal is oxidized that is explained by considering the electronegativity differences between metal atom and cation. Factors such as lattice potential, work function changes, and extra-atomic relaxation energy leads to negative E_B shift in the case of Ag and some Cd compounds (Xin et al, 2005).

Thus, it can be concluded that in the process of thermal reduction of silver ions while TiO_2 inorganic matrix formation we obtain silver nanoparticles embedded into inorganic matrix covered with silver oxide shell. On the other hand, when silver nanoparticles containing nanocomposites are prepared through photoreduction and subsequent thermal treatment, different states of silver, contributing to the Ag*3d*-line in the XPS spectra, have been observed in comparison to the samples described before.

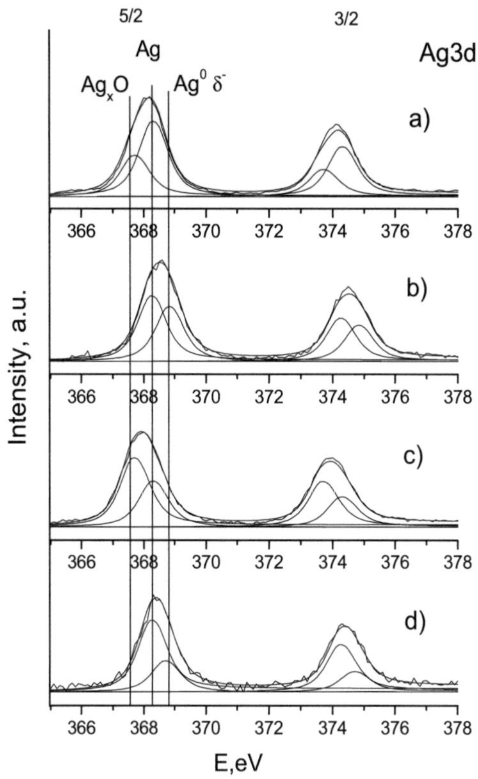

Fig. 16. XPS-spectra of Ag3d-levels for as-prepared samples with photoreduced silver particles (a, c), and after thermal treatment at 500°C (b, d) of TiO_2/Ag (a, b) and TiO_2/ZnO/Ag (c, d) films.

Ag$3d_{5/2}$ component for as-prepared TiO_2/Ag and TiO_2/ZnO/Ag films stands at 368.2 and 368.0 eV respectively (Fig. 16, a, c). Thermal treatment results in the peak narrowing and their shifts toward higher binding energy by 0.4eV. The peak decomposition reveals the presence of Ag in metallic state peaked at 368.35 eV and Ag_2O with peak at E_B = 367.7 eV. The values are in good agreement with those (368.22 eV) reported in (Krylova et al, 2006; Gregg & Sing, 1982). These results indicate that the silver nanoparticles formed on TiO_2 under given experimental conditions (UV irradiation, ambient atmosphere, room temperature) are chemically very reactive and were easily oxidized with Ag_2O shell formation. Authors (Nguyen, 1999) reported that the grows of silver oxide overlayer up to 6 nm on Ag^o – TiO_2 interface is a function of plasma exposure time at room temperature. Higher intensity of oxide peak for TiO_2/ZnO/Ag film as compared to TiO_2/Ag supports our assumption about more homogeneous distribution of smaller Ag nanoparticles on this surface. Tendency to oxidation might be increased significantly with decrease of particle size and increase of portion of surface atoms exposed to interface.

Annealing at 500°C results in the complete decomposition of silver oxide, no peaks are observed at low E_B side near 367.7 eV for TiO_2/Ag as for TiO_2/ZnO/Ag samples (Fig. 16, b, d). For the last one Ag_2O decomposition leads to Ag^o peak intensity growth (Fig 16, d) that coincides with narrow SPR band appearance in the absorption spectra (Fig 8 b).

For both samples two components were found to form $Ag3d_{5/2}$ peak: one of them at 368.4 eV corresponds to metallic silver and the other one that has binding energy higher by 0.4 eV (368.8 eV) than that for Ag^o. Observed shift towards higher E_B after thermal treatment is similar to reported for Ag nanoparticles in SiO_2, SiN_x and TiO_2 thin films (Gun`ko & Mikhalovsky, 2004). This effect was observed also for Pt (Crepaldi et al, 2003) indicating charge transfer from semiconductor matrix to the metal.
XPS data confirm our suggestion that Ag^o is still presented on the TiO_2/Ag film, but the disappearance of SPR band in the TiO_2/Ag spectra could be caused by the formation of very small Ag particles on the TiO_2 surface or by partial "dissolving" of certain critical size silver nanodrops in the crystalline matrix as it was described elsewhere (Shacham et al, 2004; Shter et al, 2007). Escape of the metal nanoparticles from TiO_2/1%ZnO/Ag film after 500°C treatment leads to the more homogeneous particle size distribution through the film profile because of more intensive evaporation of silver droplets from the outer surface of the films occurs. The smaller particles that manifested in the intensive SPR peak in the absorption spectra were formed in restricted media inside the film pores, where Zn^{2+} ions were replaced by Ag^+ one and converted to Ag^o as a result of photoreduction. Similar results are reported for temperature dependence of Ag nanoparticles distribution through the depth profile of Ag-TiO_2 sol-gel films (Crepaldi et al, 2003).

4.3 TiO_2/ZnO/Au films

Nonsymmetrical Ti (2p) peaks registered in the spectra of all samples were deconvoluted as the sum of 458.9 and 458.5 eV peaks corresponded to Ti-O-Ti and Ti-O-Zn bonds (not shown here). No XPS peak of Zn (2p) is obtained for $Au_{ads}/Zn^{2+}/TiO_2$. The peak attributed to the formation of Ti-O-Zn (E_{BE}=1022.5 eV) is predominated over the Zn-O-Zn one (E_{BE}=1021.7 eV) for $TiO_2/ZnO/Au_t$ and $TiO_2/ZnO/Au_{ph\text{-}t}$ in comparison with TiO_2/ZnO (Fig. 16). The XPS results and X-ray analyzed EDS spectra clearly show that gold NP's are responsible for the acceleration of anatase crystallization and formation of $Zn_2Ti_3O_8$ phase detected by XRD analysis (Smirnova et all, 2010).

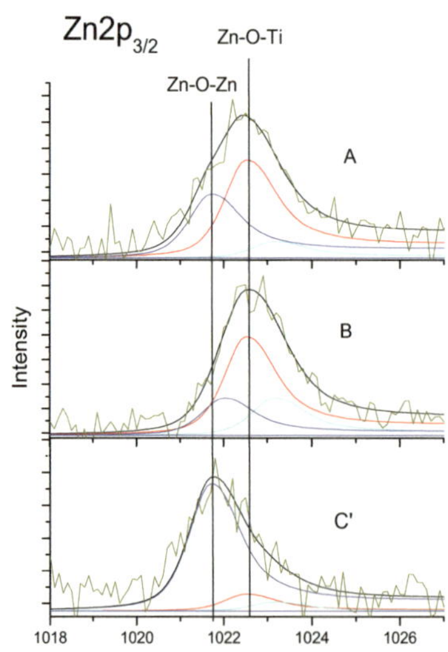

Fig. 17. XPS spectra of $Au_{ph\text{-}t}/Zn^{2+}/TiO_2$ (A), $Au_t/Zn^{2+}/TiO_2$ (B) and Zn^{2+}/TiO_2 (C').

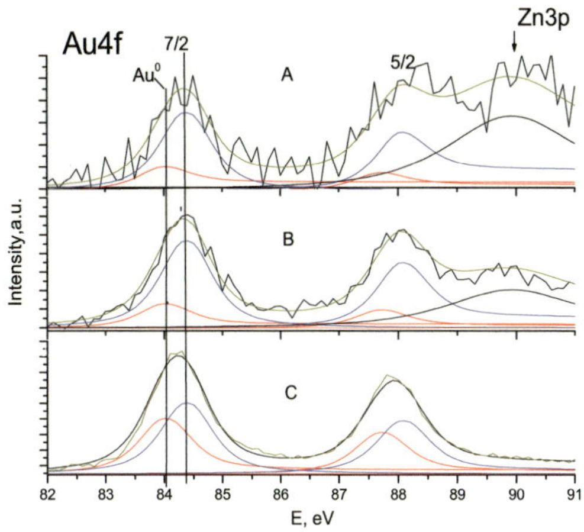

Fig. 18. XPS spectra of $Au_{ph\text{-}t}/Zn^{2+}/TiO_2$ (A), $Au_t/Zn^{2+}/TiO_2$ (B) and $Au_{ads}/Zn^{2+}/TiO_2$ (C).

Binding energy (BE) of Au $4f_{7/2}$ peaks (84.3 eV) is deconvoluted in doublet for the gold containing films. BE signals at 84.0 and 84.4 eV can be related to the bulk of metallic gold and gold clusters, respectively, as reported (Ozkaraoglu et all, 2007). The intensity of signal attributed to the clusters predominates the intensity of bulk gold for $TiO_2/ZnO/Au_{ph\text{-}t}$ and $TiO_2/ZnO/Au_t$ samples (Fig. 17 A and B) contrary to $TiO_2/ZnO/Au_{ads}$ (Fig. 18). It is concluded that the correlation between the gold sizes and contribution of Au clusters on the surface is achieved.

4.4 $TiO_2/ZrO_2/SiO_2$ films with embedded Ag and Au

The XPS spectra of silver and gold nanoparticles in ternary oxide films are presented in Fig. 19 and Fig. 20 respectively. In silver Ag 3d region the fitted spectra consist of two main peaks - Ag $3d_{5/2}$ and Ag $3d_{3/2}$ doublet. Ag $3d_{5/2}$ peak position at 367.99 eV (Fig. 18, thick line) is in good agreement with 368.196 eV reported in (Nirmalya et all, 2007), 368.22 eV in (Matsuoka et all, 1997) and 367.98 eV in (Zhang et all, 2000) for metallic silver.

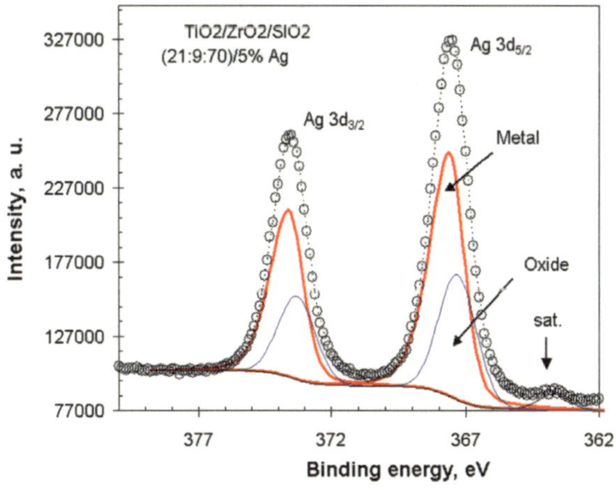

Fig. 19. Fitting procedure for Ag 3d spectra for mixed oxides film modified with silver nanoparticles $TiO_2/ZrO_2/SiO_2$ (21:9:70)/ 5%Ag: circles – experimental data, dashed line – fitted curve; thick line - metallic silver, thin line – silver oxide.

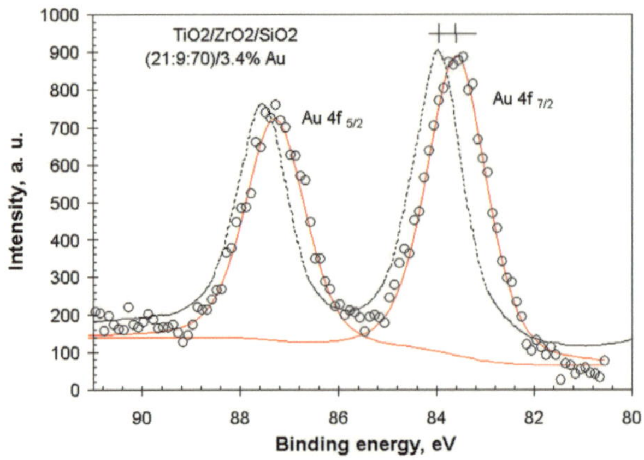

Fig. 20. Fitted Au 4f spectra of $TiO_2/ZrO_2/SiO_2$ (21:9:70)/3.4%Au film and gold reference spectra: circles - experimental data, thin lines – fitted curve; dashed line - literature data (Matsuoka et all, 1997).

Fitting procedure results for gold Au 4f region are presented in Fig. 20. In this figure the literature data reported in (Matsuoka et all, 1997) are compared with the experimental data. Peaks corresponding to gold oxide were not detected. Small shift of Au $4f_{7/2}$peak by 0.2 eV towards lower binding energy values was found.

Au 4f doublet peaks were detected by XPS at sufficient intensity, peaks position and shape (Fig. 20) coincide with literature data (Matsuoka et all, 1997) nevertheless considerable disagreement between Au molar and surface atomic concentrations was found (3,4 mol.% and 0,4 at.%). This disagreement can be related to the differences in Au concentration over the film profile.

Hence, XPS measurements confirm that gold and silver generated nanoparticles are mostly present in metallic state. The silver state in the composite films strongly depends on the chosen synthesis strategy. Oxide layer of silver nanoparticles is being destroyed by thermal treatment (500 °C) for the films prepared by photoreduction with further thermal treatment whereas for the films prepared through single step thermal reduction silver particles are covered by oxide shell. The growth of gold clusters fraction and Zn-O-Ti bond formation were registered in XPS spectra for the materials with smaller particle sizes prepared by photo- or thermoreduction.

5. Photocatalytic properties

Among studied oxide composite materials modified with metal nanoparticles titanium dioxide itself or its mixtures with other oxides: SiO_2, ZrO_2, ZnO demonstrate good photocatalytic properties under illumination with UV or visible light. As was already mentioned before, modification of photocatalysts with metal nanoparticles either embedded within the film or deposited onto their surface strongly increases photoactivity of the photoactive catalytic materials through the improvement of charge separation processes (Sclafani et al, 1997; Submanian, 2001).

Photocatalytic activity of the titania based materials studied in present investigation has been widely examined in different environmentally important processes.

5.1 TiO_2/Ag films

The photocatalytic activity of mesoporous TiO_2/Ag films with thermally reduced Ag nanoparticles was tested using the process of xantene dye Rhodamine B (RB) degradation in aqueous solutions. Under UV irradiation (λ = 254 nm) the absorption peaks corresponding to Rhodamine B diminished and finally disappeared, indicating degradation of the dye. No new absorption bands were observed, in contrast to Sung-Suh et al, 2004, where significant (-50 nm) shift of the long wave absorption band of Rhodamine B was reported indicating de-ethylation process. So we can suggest that Rhodamine B photooxidation in our experiments proceeds through the degradation of chromophore chains in the solution. To compare the photoactivity of TiO_2 and TiO_2/Ag films with Ag contents from 1 to 10 at.%, the rates of the dye photodegradation process were calculated from a pseudo first-order reaction rate approximation for equal reaction conditions (see Table 1).

Sample	UV light		Visible light	
	k, min^{-1}	$\tau_{0,5}$, h	k, min^{-1}	$\tau_{0,5}$, h
Blank experiment	1.6×10^{-3}	7.1	$2.5\cdot10^{-6}$	76.7
TiO_2	3.5×10^{-3}	3.3	$1.3\cdot10^{-5}$	15.3
TiO_2/Ag 1% (at.)	6.0×10^{-3}	1.9	-	-
TiO_2/Ag 3% (at .)	6.2×10^{-3}	1.9	$1.3\cdot10^{-5}$	14.7
TiO_2/Ag 5% (at.)	6.8×10^{-3}	1.7	$4.2\cdot10^{-5}$	4.6
TiO_2/Ag 10% (at.)	4.3×10^{-3}	2.7	$4.9\cdot10^{-5}$	3.9

Table 1. The rate constants of photodegradation of RB in the presence of TiO2/Ag films.

As it is showed in the table, a maximal efficiency (two times higher in comparison to pure TiO_2 film) was observed for TiO_2/Ag films with an Ag content of 5 at.%. Further increase of the dopant content leads to a decrease of TiO_2/Ag photocatalytic activity. According to Refs. 4. 5. and 13 (Manujlov et all, 2008) doping of the semiconductor with low concentrations of noble metal nanoparticles is advantageous to promote charge separation processes but with increasing metal concentrations the processes of hole trapping by negatively charged metal nanoparticles become predominant leading to depressing of photocatalysis.
In the case of visible light excitation of TiO_2/Ag with different silver nanoparticles content it was observed steady increase of the photocatalytic activity of films with increasing of silver nanoparticles loading in the samples (see Table). Obviously, this difference of the photocatalytic activity of the TiO_2/Ag film dependence on the silver nanoparticles content in the case of excitation with UV or visible light is related to the different mechanisms of the dye molecule degradation (Sung-Suh et al, 2004).

5.2 TiO_2/ZrO_2/SiO_2 films with embedded Ag and Au nanoparticles

Photocatalytic degradation of Rhodamine B dye in the presence of ternary TiO_2/ZrO_2/SiO_2 films modified with noble metal nanoparticles under UV-light irradiation proceeds according to the same scheme as for the TiO_2/Ag and ZnO/TiO_2/Ag films, i.e. without/with minor impact of de-ethylation procecess but mainly through the dye chromophore degradation steps. As it was expected, ternary oxide systems with embedded noble metal nanoparticles have higher photocatalytic activity than that for unmodified TiO_2/ZrO_2/SiO_2 films. The rate constants are presented in the Table 2.

Sample	Rate constants, $k*10^{-3}$ min^{-1}
$TiO_2/ZrO_2/SiO_2$	1.4
$TiO_2/ZrO_2/SiO_2$/5 %Ag	2.0
$TiO_2/ZrO_2/SiO_2$/3,4 %Au	6.3

Table 2. The rate constants of photodegradation of RB in the presence of catalysts.

Comparing the photocatalytic activity of the films of ternary system, modified with gold and silver nanoparticles, it was found that in the presence of films containing silver, the rate constant of decomposition of the Rhodamine B is lower in comparison with Au modified films, when almost four-fold increase of photocatalytic activity in comparison with bare $TiO_2/ZrO_2/SiO_2$ film was observed. As the gold containing photocatalyst has the most promising photoactivity, Au nanoparticles concentration influence on the photocatalytic activity of $TiO_2/ZrO_2/SiO_2$ films was studied. We have found that there exists an optimum gold nanoparticles content (5 mol.%) in the ternary oxide system when the photocatalytic activity of the films increases by almost one order of magnitude comparing to the one for unmodified films. Further increasing of gold concentration cases deterioration of photocatalytic activity of the samples. The rate constants of the process of photocatalytic degradation of Rhodamine B dye in the presence of $TiO_2/ZrO_2/SiO_2$/Au films with different Au content are summarized in the Table 3.

$TiO_2/ZrO_2/SiO_2$/Au, mol.%	Rate constants, $k*10^{-3}$ min^{-1}
1	6.0
3,4	6.3
5	10.2
7	5.3

Table 3. The rate constants of photodegradation of RB in the presence of $TiO_2/ZrO_2/SiO_2$/Au films with different gold content.

The observed dependence of the $TiO_2/ZrO_2/SiO_2$ films photocatalytic activity on the gold concentration has been attributed to the differences in size, quantity and homogeneous distribution of gold nanoparticles on the catalyst surface. As was proved by the AFM investigations of surface of TiO_2/ZrO_2/SiO2/Au films with different gold content (Vityuk et all, 2007), the surface coverage with gold nanoparticles continuously increases with increasing gold concentration in the films. Most probably, on the surface of the $TiO_2/ZrO_2/SiO_2$/Au films with low surface coverage with gold nanoparticles (1 and 3.4 mol.%), the recombination processes are more efficient than that in the films containing 5 mol.% of gold. Further increase in concentration of gold (more than 5 mol.%) is accompanied with aggregation of nanoparticles, which in turn leads to a screening of surface from irradiation and prevents direct contact of dye molecules with the surface of catalyst.

5.3 Photocatalytic activity of Ag (Au) ZnO/TiO_2 coatings in tetracycline hydrochloride (TC) degradation

The biological active compounds as antibiotics, hormones, preservatives and anesthetics have been identified in the aquatic environment and soil as the result of the extensive development of pharmaceutical industry and the wide consumption by human. Antibiotics are widely used against microbial infection in medicine and veterinary as well as the feed additives increasing the growth and prevent pathogens in animal farms and fisheries raising

the concerns over proliferation of antibiotic-resistant bacteria. Tetracycline is one of the most frequently prescribed groups of antibiotics. Residues of TC and their metabolites were detected in eggs, meat and animals based on different exposure methods (Ruyck et all, 1999; Zurhelle et all, 2000). Their accumulation in human organism can produce arthropathy, nephropathy, central nervous system alterations, spermatogenesis anomalies, possible mutagenesis and photosensitivity in human beings. Due to their antibacterial nature, antibiotic contaminated waters cannot be effectively eliminated by traditional biological methods (Kummerer et all, 2000). Tetracycline presents a class of the compound that is sensitive to light and classified as a phototoxic drug. Thus, TC residues in culture pond and on the surface of soil follow through the formation of the TC derivatives that could be even more dangerous for public health.
Destruction of the chemical bonds of the complex organic structure such as antibiotics tetracycline (Fig. 21) signified the unpredicted type of the products and thus the effect of reaction conditions on the reaction mechanism. TC molecule can behave as a cation, a neutral/zwitterion, an anion or a dianion depending on pH of the systems. Ring A absorbs only in the 250-300 nm area, whereas the BCD ring chromophore contributes to both 250-300 and 325-400 nm absorption bands as described in (Schneider et all, 2003).

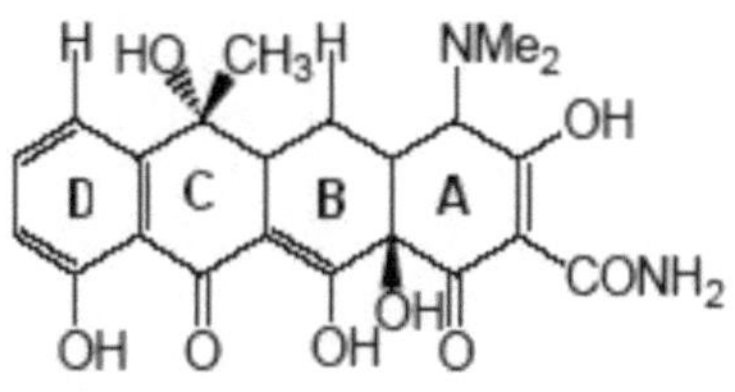

Fig. 21. The structural formula of TC.

Tetracycline presents a class of the compound that is sensitive to light and classified as a phototoxic drug. The photodecomposition products of TC, such as dedimethylaminotetracycline, anhydrotetracycline, lumitetracycline, oxytetracycline, quinone form, and fourteen others were detected at different reaction conditions (Davis et all, 1979; Moore et all, 1983; Sanniez et all, 1980). Furthermore, it is identified seven products of the TC photolysis under conditions similar to the natural ones (Oka et all, 1989).
The preliminary estimation of the photocatalytic efficiency can be performed by detailed monitoring of the TC absorption spectra during irradiation. The shifts of the maxima at 275 and 357 nm and the appearance of absorption at 400-500 nm are attributed to the formation of tetracycline derivates (Paola et all, 2004).
However, the steady and simultaneous fall of these two maxima without absorption in visible are registered at $TiO_2/ZnO/Au_t$ and $TiO_2/ZnO/Au_{ph-t}$ films leading to the suggestion of tetracycline degradation rather then oxidation. (Linnik et all, 2009). As seen from Fig. 22, irradiating the tetracycline at $Ag_t/ZnO/TiO_2$ brings to the less effective degradation than at $Au_t/ZnO/TiO_2$. In the presence of $TiO_2/ZnO/Ag$ films synthesized either thermo or adsorption methods, the TC transformation accompanied by the formation of oxidized products. Antibiotics adsorption on $TiO_2/ZnO/Au_{ph-t}$ film is reached to 58 % contrary to the adsorption inability of $TiO_2/ZnO/Au_{ads}$ and $TiO_2/ZnO/Au_t$ films. Comparing the absorption spectra of the film before the contact and after adsorption of TC the new absorption band at 420 nm is appeared (Fig. 23 A). Irradiation for 90 min brings to the gradual intensity decrease in the absorption spectra of the film as well as TC solution. Hence, the TC degradation takes place on the surface of $TiO_2/ZnO/Au_{ph-t}$ film through the adsorption-desorption equilibrium.

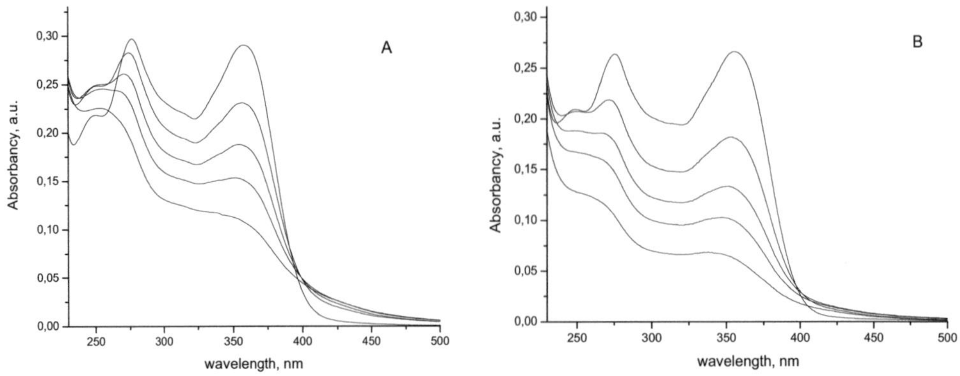

Fig. 22. Evolution of absorption spectra of tetracycline solution over TiO2/ZnO/Agt (A) and TiO_2/ZnO/Au_t (B) during 90 min irradiation.

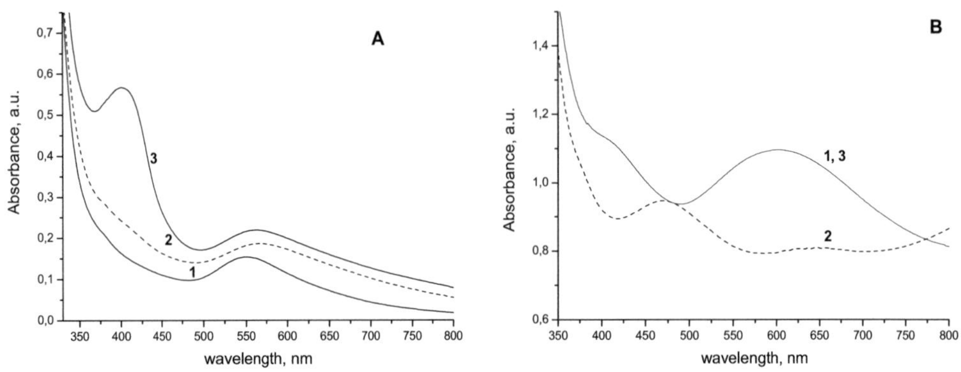

Fig. 23. Absorption spectra of $Au_{ph\text{-}t}$/ZnO/TiO_2 (A) and Au_t/ZnO/TiO_2 (B) films: 1-before contact; 2-after 90 min of irradiation; 3- after adsorption in the dark.

The shift of SPR band after 90 min irradiation in TC solution is observed in the absorption spectrum of the TiO_2/ZnO/Au_t film (Fig. 23 B). The initial view of the spectrum is achieved by keeping of the film in the dark and air with for 24 hours. The blue shift of SPR is supposed to be due to the oxidation of gold by the hole and/or hydroxyl radical and/or other radical generated during photocatalytic degradation of TC molecules accompanying the depletion of gold metal particles. The formed gold ions might be incorporated on the interface of Au particle and TiO_2. In the dark, the catalytic reduction of these ions by metal gold could followed as describe in (Gachard et all, 1998).

Photolysis of TC molecules (FW=444) resulted in the formation of anhydrotetracycline (FW=428), 4-epi-anhydrotetracycline (FW=426), oxytetracycline (FW=460), lumitetracycline (FW=399), dedimethylamino TC (FW=400), quinoid TC (FW=416) as the main products. The surface of the films is monitored using matrix free Laser desorption/ionization mass spectrometry (LDI) with the experimental parameters registered in the same manner as the liquid phases. The $Au_{ph\text{-}t}$/ZnO/TiO_2 and Au_t/ZnO/TiO_2 films as well as the solutions of TC before and during 90 min irradiation were compare for the level of TC degradation by MALDI measurements. Formation of aforesaid products is observed neither in the liquid phase nor on the film (Linnik et all, 2009). It is shown that the destruction of TC molecules after 90 min irradiation is more efficient on the surface of $TiO_2/ZnO/Au_{ph-t}$ resulting in the appearance of the adsorbed product molecules with the less m/z values. The desorbed ion yield fragments of TC intermediates from the surface of $TiO_2/ZnO/Au_{ph-t}$ film after 20, 60, and 90 min irradiation are depicted in the scheme. TC molecule is adsorbed by amide group of ring A that leads to its destruction after 20 min irradiation. The adsorbed cyclic compounds are still present on the surface of the film after 60 min irradiation. Two main desorbed ion yield fragments (the highest percent values) show the mineralization of TC molecule.

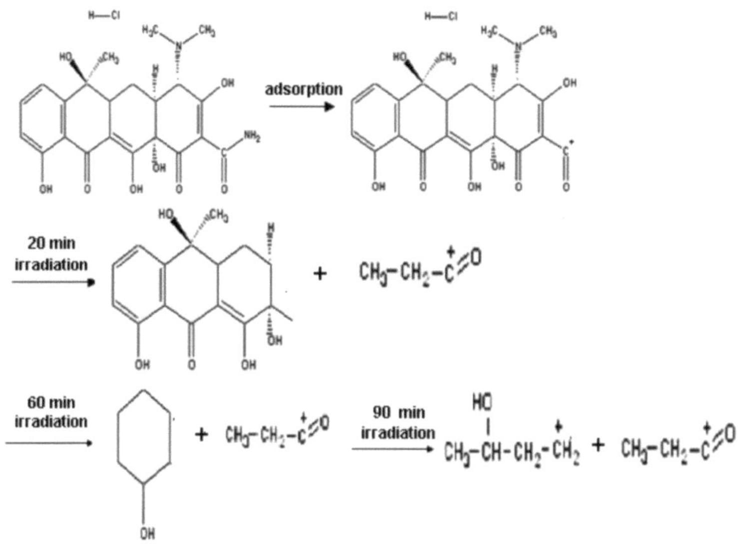

Scheme of the intermediates formation during the photocatalytic destruction of TC at $TiO_2/ZnO/Au_{ph-t}$ film.

Thus, the efficiency of photocatalytic TC destruction is affected by the size of gold nanoparticles where the mineralization of TC molecule is achieved on the surface of $TiO_2/ZnO/Au_{ph-t}$ film. Contrary to the gold-containing films, the oxidation of TC is observed over $TiO_2/ZnO/Ag$ composites.

6. Bactericide properties of stabilized silver and gold nanostructures on the surface of disperse silica

The nanoparticles of Ag and Au are used for the delivery of drugs (Jain, 2005; Paciotti et all, 2004; West & Halas, 2003), the treatment of wounds, the decontamination of water, and as bactericidal agents (Baker et all, 2005; Elchiguuerra et all, 2005; Pal et all, 2007; Shrivastava et all, 2007; Weir et all, 2008). Many bacteria are resistant to antibiotics, and it is therefore necessary to look for new bactericidal materials. Silver nanoparticles have a broad spectrum of antibacterial activity and are nontoxic for humans at low concentrations (Jin & Zhao, 2009). In large amounts, however, their toxicity is high, and the biocompatibility is substantially lower than for gold nanoparticles (Browning et all, 2009; Huang et all, 2008; Nallathambly et all, 2008; Song et all, 2010). The effectiveness of the nanoparticles can be intensified by depositing them on a support with a highly developed surface. In this respect highly dispersed silica (HDS) is an ideal material. An important problem in the synthesis of nanoparticles is prevention or retardation of their aggregation and also oxidation at the stage of formation of the silver nanoclusters. In contrast to their state in the films, on the surface of dispersed silica the silver nanoparticles are unstable.

Therefore, in view of the diversity of potential applications of dispersed silica containing Ag nanoparticles on the surface, particularly as bactericidal agents, the stabilization of the nanoparticles of silver and their protection against oxidation and the influence of the surroundings represent an urgent task from the theoretical and applied view point. The principle of the stabilization of Ag nanoparticles with polymers or surfactants was applied successfully to the synthesis of composite systems based on highly dispersed silica (HDS) containing stable nanoparticles of silver with PVP and/or SDS as stabilizers (Mukha et all, 2009, 2010).

Samples of Ag/SiO_2 were synthesized by the adsorption of previously prepared colloidal solutions of nanosized silver

on the surface of dispersed silica. Nanosized silver in colloidal solution was obtained by chemical reduction from silver nitrate in the presence of sodium tetrahydroborate and a binary stabilizer - SDS and PVP. The obtained silver colloids were adsorbed on the surface of HDS. The concentration of Ag on the HDS amounted to 0.05%. The use of a binary stabilizer raises the stability of the silver nanoparticles. The interaction of the SDS and PVP consists of a combination of two processes: the formation of a complex as a result of hydrophobic interaction between the hydrocarbon unit of the SDS and the methylene groups of the PVP and electrostatic interaction between the head groups of the SDS and the partial charges of the nitrogen and oxygen of the pyrrolidol ring].

To explain the inhibitor effects of silver on bacteria it was suggested that silver reacted with proteins by combining the thiol (-SH) groups, leading to the bacteria inactivation (Traversa et all, 2001). In this work we examined the antibacterial activity of Ag/SiO_2 and Au/SiO_2 suspensions and Ag (Au) colloids. Ag/SiO_2 powders were tested after treating at 85-90°C, Au/SiO_2 - after 500°C. The probes have been diluted in distilled water with the concentration of 0,0016% wt. of metal NPs and 3,13 % wt. of HDS in suspension. Colloids were tested with the concentration of 0,0016 % wt. of metal NPs.

$AgNO_3$ solution has been used in the control experiments in the equal concentration as Ag NPs. Binary stabilizer PVP/SDS and $NaBH_4$ have been studied in the control with the same amount as in the colloids. Initial HDS has been tested also at the concentration 3,13% wt. in presence of the stabilizers and $NaBH_4$.

The results of antimicrobial activity of metal NPs in colloids and suspensions against E.coli, S.aureus и C.albicans are presented in the Table 4.

Essential reduction value for bacteria E.coli (5 lg) and fungi C.albicans (4 lg) in colloids achieved after 1 hour of exposure of microbial cells with Ag NPs. Staphylococcus were more Ag NPs-resistant, particularly 4,35 lg reduction achieved only after 4 hour. Decrease of Ag concentration allowed revealing mentioned phenomenon. C.albicans bacteria were the most sensitive among of studied objects.

The control $AgNO_3$ solution didn't show antimicrobial action. The same result also was revealed for PVP/SDS and $NaBH_4$ mixture. Thus presented experimental data indicate a high antimicrobial activity of silver colloids to all microorganisms.

Embedding of Ag NPs on SiO_2 surface slightly decrease activity of Ag/SiO_2 suspension. The exposure time increases and changes in interaction character of Ag NPs with the microbial cells appear. The contact time for 4 lg reduction achievement for C.albicans remained the same as in colloid (1 h). At the same time Ag/SiO_2-resistance of E.coli rised. The reduction value was only 3,58 lg after 4 hour contact time. On the contrary, S.aureus was more sensitive then in colloid. But generally antimicrobial activity of Ag NPs/SiO_2 complex remained high. The nanosized gold at the same concentration range doesn't show antibacterial action on indicated microorganisms.

Samples and concentrations	Exposure time, h	test-strains, lgR*		
		E. coli	*S. aureus*	*C. albicans*
Ag NPs (0.0016 % wt) colloid	1	> 5.22	< 1.57	> 4.27
	2	> 5.22	2.07	> 4.27
	4	> 5.22	4.35	> 4.27
	24	> 5.22	> 5.24	> 4.27
initial microorganisms amount, lgN_0		7.37	7.39	6.42
Ag NPs (0.0016 % wt)/SiO2 (3.13 % wt)	1	< 1.39	3.89	> 4.54
	2	1.48	5.14	> 4.54
	4	3.58	> 5.17	> 4.54
	24	> 5.06	> 5.17	> 4.54
SiO2 (3.13 % wt) (control)	1	< 1.39	< 1.5	< 0.87
	2	< 1.39	< 1.5	< 0.87
	4	< 1.39	< 1.5	< 0.87
	24	< 1.39	< 1.5	< 0.87
initial microorganisms amount, lgN_0		7.21	7.32	6.69

Table 4. The antimicrobial activity of Ag NPs in colloids and suspensions.

7. Summary

In summary, synthesis strategy employed for the preparation of composite materials determines final localization of noble metal nanoparticles. Single step thermal reduction of noble metal ions during sintering of sol-gel films leads to the distribution of metal nanoparticles within the bulk of the films as well as partial localization of the particles on the films surface. In addition, intensive plasmon resonance absorption of metal nanoparticles can be obtained. Dual step modification of the films surface by photoreduction of silver ions and subsequent composites thermal treatment results in uniform modification of surface rather than bulk of the films by Ag nanoparticles with narrow particles sizes distribution. The last is being determined by the leaching of ZnO from the surface of composite films with subsequent localization of tiny Ag nanoparticles on these sites.

Such nanocomposites modified with metal nanoparticles are beneficial toward improving the efficiency of the photocatalytic oxidation – mineralization processes.

Photoreduction of tetrachloroauric ions in the "wet" films leads to the formation of the homogeneous distributed gold nanoparticles with the sizes to 10 nm where the subsequent thermal treatment did not cause to its aggregation. The mechanism of thermal reduction is differs from the former case that is the reason of enlargement of gold particles due to the easier diffusion of gold (I) ions formed on the first stage of thermal treatment.

The efficiency of photocatalytic Tetracycline destruction is affected by the size of gold nanoparticles where the mineralization of TC molecule is achieved on the surface of $TiO_2/ZnO/Au_{ph-t}$ film. Contrary to the gold-containing films, the oxidation of TC is observed over $TiO_2/ZnO/Ag$ composites.

The fixation of the metal nanoclusters on silica surface occurs due to a) interaction between the functional groups of stabilizer shell of NPs and OH-groups of silica and b) location of Me NPs within secondary pores of HDS globules. Obtained composite systems containing strongly bounded and homogeneously distributed on silica surface Ag NPs serve as potential wide-spectrum antimicrobial materials for medical and pharmaceutical application.

Thus, using photo-thermoreduction of appropriate ions within sol-gel oxide semiconductor/dielectric/metal composite films indicate simple and convenient way to produce improved photocatalysts sensitive to the visible, self-cleaning coatings, effective antimicrobial medium etc.

8. Acknowledgment

A. E., N. P. and Yu. M. thank the Governmental Target Scientific and Technical Center of Ukraine for financial support of part 6 of this work (Project N 5.16.1.7)

9. References

Alberius, P., Frindell, K., Hayward, R., Kramer, E., Stucky, G., Chmelka, B. (2002). General predictive synthesis of cubic, hexagonal and lamellar silica and titania mesostructured thin films. *J. Chem. Mater.*, Vol. 14, pp. 3284 - 3294.

Antonelli, D.M., Ying, J.Y. (1995). Synthesis of hexagonally packed mesoporous TiO_2 by modified sol-gel method. *Angew. Chem., Int. Ed. Engl.*, Vol. 34, pp. 2014 - 2017.

Arabatis, I.M., Stergiopolulos, T., Bernard, M.C. et al. (2003). Silver-modified titanium dioxide films for efficient photodegradation of methylorange. *Appl Catal B,* Vol.42, pp.187-201.

Baker, C., Pradhan, A., Pakstis, L., Pochan, D.J., Shah, S.I. (2005). Synthesis and antibacterial properties of silver nanoparticles. *J. Nanosci. Nanotechnol.*, Vol.5, pp.244-249.

Browning, L., Lee, K., Huang, T. (2009) Random walk of single gold nanoparticles in zebrafish embryos leading to stochastic toxic effects on embryonic developments. Nanoscale, Vol.1, pp.138-152.

Chaki, N.K., Tsunoyama, H., Negishi, U., Sakurai, H., Tsukuda T. (2007). Effect of Ag-Doping on the Catalytic Activity of Polymer-Stabilized Au Clusters in Aerobic Oxidation of Alcohol. *J. Phys. Chem. C.*, Vol.111, No.13, pp.4885 - 4888.

Chan, Y., Zimmer, J.P., Stroh, M., Steckel, J.S., Jain, R.K., Bawendi, M.G. (2004). Incorporation of Luminescent Nanocrystals into Monodisperse Core–Shell Silica Microspheres. *Adv. Mater.* Vol. 23-24, pp.2092-2097.

Crepaldi, E.L., Soler-Illia, G.J., Grosso, D., Cagnol, F., Ribot, F., Sanchez, C. (2003). Controlled Formation of Highly Organized Mesoporous Titania Thin Films: From Mesostructured Hybrids to Mesoporous Nanoanatase TiO_2. *J. Am. Chem. Soc.*, Vol. 125, pp.9770-9786.

Elchiguuerra, L.J., Burt, L.J., Morones, R.J., Camacho-Bragado, A., Gao, X., Lara H. H., Yacaman, M.J. (2005). Interaction of silver nanoparticles with HIV-1. *J. Nanobiotechnol.*, Vol. 3, pp.1-10.

Epifani, M., Giannini, C., Tapfer, L., Vasanelli, L. (2000). Sol-gel synthesis and characterization of Ag and Au nanoparticles in SiO_2, TiO_2, and ZrO_2 thin films. *J. American Ceram. Society.*, Vol. 83, No.10, pp. 2385 - 2393.

Eustis, S. & El-Sayed, M. A. (2005). The Aspect Ratio Dependence of the Enhanced Fluorescence Intensity of Gold Nanorods: Experimental and Simulation Study. *J. Phys. Chem. B*, Vol.109, No.34, pp.16350-16356.

Eustis, S. & El-Sayed, M. A., (2006). Determination of the Aspect Ratio Statistical Distribution of Gold Nanorods in Solution from a Theoretical Fit of the Observed Inhomogeneously Broadened Longitudinal Plasmon Resonance Absorption Spectrum. *J. Appl. Phys*, Vol. 100, No. 4, Article ID 044324.

Gachard, E., Remita, H., Khatouri, J., Keita, B., Nadjo, L., Belloni, J. (1998). Radiation-induced and chemical formation of gold clusters. *New J. Chem.*, Vol. 22, pp.1257-1265.

Gnatiuk, Yu., Smirnova, N., Eremenko, A., Oranska, O., Chuiko, O. (2005). New nanostructured mesoporous film materials based on ZrO_2/TiO_2 for photooxidized destruction of 2,4- dinitroanilin. *Reports of NAS of Ukraine,* Vol. 8, pp. 143 - 148.

Gnatyuk, Yu., Manuilov, E., Smirnova, N. et al. (2006). Sol-gel produced mesoporous Ag/TiO_2 coatings effective in rhodamine B photooxidation. *Springer,* In: Kassing R et al (ed) NATO Science Series. II. Mathematics, Physics and Chemistry. Functional properties of Nanostructured Materials. Vol.223, pp.485-490, TheNetherlands.

Gnatyuk, Yu., Smirnova, N., Eremenko, A. et al. (2005). Design and photocatalytic activity of mesoporous TiO_2/ZrO_2 thin films. *Ads Sci & Technol.* Vol.23, pp.497–508.

Gonella, F., Mattei, G., Mazzoldi, P., Battaglin, G., Quaranta, A., De, G., Montecchi, M. (1999).Structural and optical properties of silver-doped zirconia and mixed zirconia-silica matrices obtained by sol-gel processing. *J.Chem. Mater.*, Vol. 11, pp. 814 - 821.

Gregg, S.J. & Sing, K.S. (1982). *Adsorption, surface area and porosity,* second ed. Academic Press,.ISBN 0123009561, London.

Gun'ko, V.M. & Mikhalovsky, S.V. (2004). Evaluation of slitlike porosity of carbon adsorbents. Carbon, Vol.42, pp. 843- 849.

He, C., Yu, Y., Hu, X., Larbot, A. (2002). Influence of silver doping on photocatalytic activity of titania films. *Appl. Surf.Sci.,* Vol. 200, pp. 239-247.

He, J., Ichinose, I., Kunitake, T., Nakao, A. (2003). In situ synthesis of noble metal nanoparticles in ultrathin TiO_2-gel films by combination of ion-exchange and reduction processes. *Langmuir,* Vol. 18, pp.10005-100010.

Hirakawa, T., Kamat, P.V. (2004). Electron Storage and Surface Plasmon Modulation in $Ag@TiO_2$ Clusters. *Langmuir,* Vol. 20, pp.5645-5647.

Huang, T., Nallathamby, P. & Xu, X. (2008). Photostable single-molecule nanoparticle optical biosensors for real-time sensing of single cytokine molecules and their binding reactions. *J. Am. Chem. Soc.,* Vol.130, pp.17095-17105.

Huang, W., Qian, W. & El-Sayed, M. (2004). Coherent Vibrational Oscillation in Gold Prismatic Monolayer Periodic Nanoparticle Arrays. *Nano Letters*, Vol. 4, pp. 1741-1747.

Huang, X., El-Sayed, I. H., Qian, W., El-Sayed, M. A. (2006). Cancer Cell Imaging and Photothermal Therapy in the Near-Infrared Region by Using Gold Nanorods. *J. Am. Chem. Soc.*, Vol.128, pp.2115-2120.

Jain K. K. (2005). Nanotechnology-based drug delivery for cancer. *Technol. Cancer Res. Treat.*, Vol. 4, pp. 407-416.

Jain, P. K., Lee, K. S., El-Sayed, I. H., El-Sayed, M. A. (2006). Calculated absorption and scattering properties of gold nanoparticles of different size, shape, and composition: applications in biological imaging and biomedicine, *J.Phys. Chem. B*, Vol.110, No. 14, pp.7238-7248.

Jin, Y., Zhao, X. (2009). Cytotoxicity of Photoactive Nanoparticles, in Safety of Nanoparticles. *Springer*, edited by Tomas J. Webster, pp.19-31, ISBN978-0-387-78607-0, New York, USA.

Kamat, P. (2003). Influence of Metal/Metal Ion Concentration on the Photocatalytic Activity of TiO_2-Au Composite Nanoparticles. *Langmuir*, Vol.19, pp.469-474 469.

Kamat, P. V. (1993). Photochemistry on Nonreactive and Reactive (Semiconductor) Surfaces. *J.Chem. Rev.*, Vol. 93, pp. 207 - 300.

Kamat, P.V. (1997). Composite Semiconductor nanoclusters. In: Kamat PV, Meisel D (ed) Elsevier Science. Semiconductor Nanoclusters - Physical, Chemical and Catalytic Aspects. Amsterdam

Kamat, P.V., Flumiani, M., Dawson, A. (2002). Metal-metal and metal-semiconductor composite nanoclusters. *Coll. Surf. A: Physicochem. Eng. Aspects*, Vol. 202, pp.269-279.

Kelly, K. L., Coronado,E., Zhao, L.L., Schatz, G.C. (2003). The Optical Properties of Metal Nanoparticles:

Kim, J., Song, K.C., Foncillas, S., Pratsinis, S.E. (2001). Dopants for synthesis of stable bimodally porouse titania. *J. Europ. Ceram. Society*, Vol. 21, pp. 2863 - 2872.

Kreibig, U., Vollmer, M. (1995). Optical properties of metal clusters, *Springer*, ISBN10 0387578366, 3540578366, Berlin.

Krylova, G.V., Gnatyuk, Yu.I., Smirnova, N.P., Eremenko, A.M., Gunko, V.M. (2009). Ag nanoparticles deposited onto silica, nitania and zirconia mesoporous films synthesized by sol-gel template method. *J. Sol-Gel Sci. Technol*, Vol. 5, No.2, pp. 216-228.

Kummerer, K., Al-Ahmed, A., Mersch-Sundermann, V. (2000). Biodegradation of some antibiotics, elimination of the genotoxicity and affection of wastewater bacteria in a simple test. *Chemosphere*, Vol.40, pp.701-710.

Lance, K.K., Coronado, E., Zhao, L., Schatz, G.C. (2003). The optical properties of metal nanoparticles; the influence of size, shape and dielectric environment. *J. Phys. Chem. B*, Vol. 107, pp. 668-677.

Liao, S., Donggen, Y., Yu, D. et al. (2004). Preparation and characterisation of ZnO/TiO_2, $SO_4^{2-}/ZnO/TiO_2$ photocatalyst and their photocatalysis. *J Photochem Photobiol A*, Vol.168, pp.7–13.

Link, S. & El-Sayed, M. A. (2001). Spectroscopic determination of the melting energy of a gold nanorod. *J Chem. Phys,* Vol.114, pp. 2362-2369.

Linnik, O., Manuilov, E., Snegir, S., Smirnova, N., Eremenko, A. (2009). Photocatalytic degradation of tetracycline hydrochloride in aqueous solution at ambient conditions stimulated by the gold containing zinc-titanium oxide films. *J. Adv. Oxid. Technol.,* Vol.2, No.12, pp.551-560.

Liz-Marzan, L.M., Giersig, M., Mulvaney, P. (1996). Synthesis of Nanosized Gold–Silica Core–Shell Particles. Langmuir, Vol.12, pp. 4329-4335, doi: 10.1021/la 9601871.

Lopez, T., Gomez, R., Sanchez, E., Tzompantzi, F. (2001). Photocatalytic ativity in the 2,4-Dinitroaniline decomposition over TiO_2 sol-gel derived catalysts. J. *Sol-Gel Science and Technol.,* Vol. 22, pp. 99 - 107.

Manujlov, E., Gnatyuk, Yu., Smirnova, N., Eremenko, A., Vorobets, V., Kolbasov, G., Guobiene, A., Tamulevicius, S. (2008). Mesoporous TiO_2 and TiO_2/ZnO/Ag films: sol-gel synthesis, photoelectrochemical and photocatalytic properties. *NATO Science for Peace and Security Series C: Environmental Security,* "Sol-Gel Methods for Materials Processing. Focusing on Materials for Pollution Control, Water Purification, and Soil Remediation". (Eds: P. Innocenzi, Yu.Zub and V. Kessler), pp. 427-434.

Marci, G., Augugliano, V., López-Muñoz, M.J. et al. (2001). Preparation, characterization and photocatalytic activity of polycrystalline ZnO/TiO_2 systems. I. Surface and bulk characterization. *J Phys Chem B,* Vol.105, pp.1026-1040.

Matsuoka, J., Naruse, R., Nasu, H., Kamiya, K. (1997). Preparation of gold microcrystal-doped oxide optical coatings through adsorption of tetrachloroaurate ions on gel films. *J. of Non-cryst. sol.,* Vol.218, pp.151 - 155.

Mukha I., Eremenko, A. M., Korchak, G. & Mikhienkova, A. (2010). Antibacterial Action and Physicochemical Properties of Stabilized Silver and Gold Nanostructures on the Surface of Disperse Silica. *J. Water Resource Protection,* Vol.2, pp.131-136.

Mukha, Yu., Eremenko, A. M. & Smirnova, N. P. (2009).Formation, physical - chemical and bactericide properties of stabilized silver nanostructures on the surface of disperse silica, *Chem., Phys.& Technol. Surf.,* Vol.15, pp. 255-266.

Nallathamby, P., Lee, K., Xu, X. (2008). Design of stable and uniform single nanoparticle photonics for *in vivo* dynamics imaging of nanoenvironments of zebrafish embryonic fluids. *ACS Nano,* Vol.2, pp.1371-1380.

Nguyen, V. &, Do, D.D. (1999). A New Method for the Characterization Of Porous Materials. Langmuir, Vol.15, pp. 3608- 3615.

Oka, H., Ikai, Y., Kawamura, N., Yamada, M., Harada, K., Ito, S., Suzuki, M. (1989). Photodecomposition products of tetracycline in aqueous solution. *J. Agric. Food Chem.,* Vol.37, pp.226-231.

Ozkaraoglu, E., Tunc, I. & Suzer, S. (2007). X-ray induced reduction of Au and Pt ions on silicon substrates. *Surf. Coat. Technol.,* Vol.201, pp.8202-8204.

Paciotti, G.F., Myer, L., Weinreich, D., Goia, D., Pavel, N., McLaughlin, R.E., Tamarkin, L., (2004). Colloidal Gold: A Novel Nanoparticle Vector for Tumor Directed Drug Delivery. *Drug Delivery,* Vol. 11, pp. 169-183.

Pal, S., Tak, Y. K., Song, J. M. (2007). Does the Antibacterial Activity of Silver Nanoparticles Depend on the Shape of the Nanoparticle? A Study of the Gram-Negative Bacterium Escherichia coli. *Appl. Environ. Microbiol.*, Vol.73, pp.1712-1720.

Paola, A., Addamo, M., Augugliaro, A., Garcia-Lopez, E., Loddo, V., Marci, G., Palmisano, L. (2004). Photolytic and TiO_2-Assisted Photodegradation of Aqueous Solutions of Tetracycline. *Fresenius Environ. bull.* Vol.13, pp.1275-1280.

Portales, H., Saviot, L., Duval, E., Fujii, M., Hayashi, S., Fatti, N., Vallee, F. (2001). Resonant raman scattering by breathing modes of metal nanoparticles. *J. Chem. Phys.*, Vol. 115, pp.3444 - 3447.

Ptashko, T., Smirnova, N., Eremenko, A., Oranska, E., Huang, W. (2007). Synthesis and photocatalytic properties of mesoporous TiO_2/ZnO films with improved hydrophylisity. *Ads Sci & Technol*, Vol.25, pp.35-43.

Rodrı´guez-Ferna´ndez, J., Pe´rez-Juste, J., Mulvaney, P., Liz-Marza´n, L.M.(2005). Spatially-Directed Oxidation of Gold Nanoparticles by Au(III)-CTAB Complexes. *J. Phys. Chem. B*, Vol. 109, No. 30, pp.14257-14261.

Ruyck, H., Ridder, H., Renterghem, R., Wambeke, F. (1999). Validation of HPLC method of analysis of tetracycline residues in eggs and broiler meat and its application to a feeding trial. *Food Addit. Contam.*, Vol.16, No.2, pp.47-56.

Sanniez, W.H.K., Pilpel, N. (1980). Photodecomposition of sulfonamides and tetracyclines. *J.Pharm. Sci.,* Vol.69, pp.5-8.

Schneider, S., Schmitt, M. O., Brehm, G., Reiher, M., Matousek, P., Towrie, M. (2003). Fluorescence kinetics of aqueous solutions of Tetracycline and its complexes with Mg2+ and Ca2+. Photochem. Photobiol. Sci., Vol.2, pp.1107-1117.

Sclafani, A., Mozzanega, M.N. & Herrmann, J.M. (1997). Influence of silver deposits on the photocatalytic activity of titania. *J. Catal*, Vol. 168, pp.117 - 120.

Selvan, S.T., Ono, Y., Nogami, M. (1998). Polymer-protected gold clusters in silica glass. *J. Materials Letters.*, Vol. 37, pp.156 - 161.

Shacham, R., Mandler, D., Avnir, D. (2004). Electrochemically Induced Sol–Gel Deposition of Zirconia Thin Films (pages 1936–1943). J.Chem A Europ., Vol.10, pp.1936-1943.

Shi, Z.M. & Lin, L.N. (2009). Influence of La^{3+}/Ce^{3+} doping on phase transformation and crystal growth in TiO_2-15%ZnO gels. *J Non Cryst Solids*, Vol.355, pp.213-220.

Shrivastava, S., Bera, T., Roy, A., Singh, G., Ramachandrarao, P., Dash, D. (2007). Characterization of enhanced antibacterial effects of novel silver nanoparticles. *Nanotechnology,* Vol.18, pp.1-9, doi: 225103.

Shter, G.E., Behar-Levy, H., Gelman, V., Grader and, G.S., Avnir, D. (2007). Organically Doped Metals—A New Approach to Metal Catalysis: Enhanced Ag-Catalyzed Oxidation of Methanol. *Adv Function Mater.*, Vol.17, pp.913–918).

Smirnova, N., Eremenko, A., Bykovskaya, L., Kulikov, S., Chuiko, A. (1992). Fluorescence spectra of adsorbed heteroaromatic molecules at selective laser excitation *J.Molec. St.,* Vol. 266, pp. 3417 - 422.

Smirnova, N., Vorobets, V., Linnik, O., Manuilov, E., Kolbasov, G. (2010). Photoelectrochemical and photocatalytic properties of mesoporous TiO_2 films modified with silver and gold nanoparticles. *Surf. Interface Anal.*, Vol.6-7, No.42, pp.1205-1208.

Song, J., Atay, T., Shi, S., Urabe, H., Nurmikko, A. (2005). Large Enhancement of Fluorescence Efficiency from CdSe/ZnS Quantum Dots Induced by Resonant Coupling to Spatially Controlled Surface Plasmons. Nano Lett., Vol.5, pp.1557-1561.

Song, Y., Nallathamby, P.D., Huang, T., Elsayed-Ali, H.E., Xu, X.H. (2010). Correlation and Characterization of 3D Morphological Dependent Localized Surface Plasmon Resonance Spectra of Single Silver Nanoparticles Using Dark-field Optical Microscopy and Spectroscopy and AFM *J. Phys. Chem. C.*, Vol.**114**, No. 1, pp. 74-81.

Subramanian, V., Wolf, E., Kamat, P. (2001). Semiconductor-Metal Composite Nanostructures. To What Extend Do Metal Nanoparticles Improve the Photocatalytic Activity of TiO_2 films? *J Phys Chem B*, Vol.105, p.11439-11448.

Sung-Suh, H., Choi, J., Hah, H., Koo, S., Bae, Y. (2004). Comparison of Ag deposition effects on the photocatalytic activity of nanoparticulate TiO_2 under visible and UV light irradiation. *J. Photochem. Photobiol. A.*, Vol. 163, pp. 37 - 44.

The Influence of Size, Shape, and Dielectric Environment. Phys.Chem.B, Vol.107, No.3, pp 668–677.

Traversa, E., Vona, M.L., Nunziante, P., Licoccia, S. (2001). Photoelectrochemical Properties of Sol-Gel Processed Ag-TiO_2 Nanocomposite Thin Films. J Sol-Gel Sci Technol., Vol.22, pp.115-123.

Vityuk, N.V., Petrik, I.S., Eremenko, A. M., Smirnova, N. P., Gorbik, P.P. (2007). Synthesis, optical and photocatalytic properties of $TiO_2/ZrO_2/SiO_2$ films modified the gold nanoparticles. *Chem. Phys. and Tech. Surf.*, Vol. 13, pp.145-151.

Vorobets, V., Manujlov, E., Smirnova, N. et al. (2008). Electro- and photocatalytic properties of electrodes based on mesoporous TiO_2-ZnO-Ag films. Chem.Phys. & Technol.Surf., Vol.14, pp.382-390.

Wang, C.T. & Lin, J.C. (2008). Surface nature of nanoparticle zinc-titanium oxide aerogel catalysts. *Appl Surf Sci*, Vol.254, pp.4500 4507.

Weaver, J.F. & Hoflund, G.B. (1994). Surface characterization study of the thermal decomposition of AgO. *J Phys Chem* Vol.98, pp.8519-8524.

Weir, E., Lawlor, A., Whelan, A., Regan, F. (2008). The use of nanoparticles in anti-microbial materials and their characterization. Analyst, Vol.133, No.7, pp.835-845.

West, J. L., Halas, N. J., (2003).Engineered Nanomaterials for Biophotonics Applications: Improving Sensing, Imaging, and Therapeutics. Ann. Rev. Biomed. Eng., Vol.5, pp.285-292.

Xin, B.F., Jing, L.Q., Ren, Z.Y. et al. (2005). Effects of simultaneously doped and deposited Ag on the photocatalyic activity and the surface states of TiO_2. *J Phys Chem B*, Vol.109, pp.2805-2809.

Yashan, G. R., Krylova, G. V. & Eremenko, A. M. (2008).Bactericide properties of gold and silver nanoparticles in solution and on high disperse silica surface, *Chem., Phys.& Technol. Surf.*, Vol. 14, pp.524- 533.

Zhang, W.F., He, Y.L., Zhang, M.S., Yin, Z., Chen, Q. (2000). Raman scattering study on anatase TiO_2 nanocrystals. *J. Phys.D: Appl.Phys.*, Vol.33, pp. 912 - 916.

Zorn, M.E., Tompkins, D.T., Zelter, W.A. et al. (2000). Catalytic and photocatalytic oxidation of ethylene on titania-based thin films. *Environ Sci Technol*, Vol.34, pp.5206–5210.

Zurhelle, G., Muller-Seitz, E., Petz, M. (2000). Automated residue analysis of tetracyclines and their metabolites in whole egg, egg white, egg yolk and hen's plasma utilizing a modified ASTED system. *J. Chromatogr. B: Biomed. Sci. Appl.*, Vol.739, No.1, pp.191-203.

4

New Composite Materials Modified with Nano-Layers of Functionalized Polymers for Bioanalysis and Medical Diagnostics

Dmitry Kapustin, Anna Prostyakova, Yana Bryk, Elena Yagudaeva and Vitaly Zubov
Shemyakin-Ovchinnikov Institute of Bioorganic Chemistry, Russian Academy of Sciences, Russian Federation

1. Introduction

The essence of modern biotechnology, in particular, chemical, biological and medical biotechnology, has been changed drastically, first due to development of recombinant DNA technology. Thus, so called molecular biotechnology has been arisen. This new discipline is based on the integrated use of methods borrowed from molecular and cell biology, microbiology, genetics, biochemistry and chemical engineering. These methods together with classical microbiological and immunological methods are widely used, for instance for laboratory diagnostics of infection diseases. The new fine technologies now used not only in scientific laboratories but also in routine medical practice, are connected with development of artificial multiple copying of DNA, i.e. the method known as polymerase chain reaction (PCR). However, along with many advantages of PCR-diagnostics there is a serious problem relating to an optimal choice of the techniques for the sample preparation. At present, there is no single universal approach for DNA isolation from different sources (e.g. from bacterial lysate or blood). To solve this specific aim the special adsorbents should be used depending on physical-chemical and sorption properties of the target compound.

Thus, the success of biomedical biotechnology is to a large extent determined by availability of new effective materials and more universal approaches for biopolymer isolation and purification. As we will show below an optimal approach is based on the use of solid (porous) matrices coated with specific polymers providing the selective sorption of biopolymers. Special attention should be paid not only to retaining of functional activity of isolated substances, but also to elimination of labor- and time-consuming protocols as well as to development of scalable and automatable techniques.

Therefore, the reasons of necessity in new multipurpose matrices are evident. Such matrices should not induce irreversible denaturation of biopolymers and, at the same time, they should provide effective separation of the components contained in the complicated mixture (such as clinical sample).

The development of biocompatible composites modified with fluoropolymer- and polyaniline-based modifiers is in the scope of this work. Both the polytetrafluoroethylene (PTFE) and the polyaniline (PANI) were shown to be effective for one-step DNA isolation from the complex mixtures containing DNA, RNA and proteins (as well as low-molecular substances) (Kapustin et al., 2003, 2006). A number of different methods were described, which relates to the synthesis, characterization of the properties and practical use of the adsorbents coated with PTFE and PANI. However, original techniques of manufacture of such materials did not find a wide application due to technological difficulties connected with their synthesis or purification. For example, in a case of PTFE the difficulty is determined by need of irradiation source, in a case of PANI the main complication is concerned with formation of insoluble polymer particles, which are not retained with the adsorbent surface, and therefore this overlaborates the procedure of a final product washing out (Kapustin et al., 2006).

Usually manufacturing of the chromatographic composite adsorbents is based on a number of consequtive physical-chemical transformations. Each step of such process can result in significant change of the morphology and sorption properties of the resulting product. For example, the synthesis of composite adsorbent based on porous silica can be accompanied with the change of sizes and shape of pores of the matrices due to partial dissolving of silica. The surface concentration of silanol groups and their steriometric configuration can be also changed. Such processes are not always be predicted synonymously even using modern analytical techniques. Therefore, uncertainty of the adsorbent synthesis is contained previously in the properties of the starting materials.

Moreother, the method of the composite adsorbent preparation itself can comprise a certain extent of ambiguousness. Sometimes it is impossible to find the direct connection between the structure of the obtained adsorbent and the conditions of its synthesis. Therefore, selection of optimal conditions (even within the framework of determined method) requires the additional investigations. In particular, the most of inventions comprises namely such selection as «know-how». The characteristics describing the composites with complicated structure, which can not be exactly determined using modern analytical and identification methods, should include their qualitative composition (the atoms of determined elements), the results of physical-chemical tests, detailed protocol of the synthesis procedure, and the information relating to specific chromatographic properties of the described material.

Currently, a vide variety of approaches using the composite adsorbents finds an application in practice. These approaches can use organic and inorganic matrices, low-molecular and polymer modifiers, they can be routine or sometimes they require to use sophisticated equipment, and so on. Nevertheless, the procedures of isolation and purification of biopolymers are based on four main physical-chemical processes of the substances separation, as it presented in the table 1.

Generally speaking, these methods are based on different solubility of nucleic acids, proteins and polysaccharides (precipitation of nucleic acids with organic solvents, chromatography of biopolymers, batch-processes, in particular adsorption of biopolymers on magnetic particles, etc). However, most of them are multistage, time- and labor consuming because they are based on the concept of retaining nucleic acids from the biological samples with the adsorbent as the first step followed by elution from the chromatographic material (column, membrane, etc) on the second step. Variety of practical tasks (separation of different classes of biologically active substances including biopolymers, isolation of the determined class of biopolymers from the complex initial mixture, separation and purification of individual substances of similar chemical nature, and so on) determines a large choice of biocompatible composite materials, however such materials are difficulty considered as universal ones. In the context of this review, the universal adsorbent should be compatible with different classes of biopolymers contained in the initial sample, i.e. it should provide both the separation of specific biopolymers keeping their initial structure and properties (avoiding an irreversible denaturation) and the high selectivity providing the isolation of needed fraction from the complicated sample. It is evidently, that the specific sorption properties of the used composites are determined by the chemical structure and the sorption properties of the polymer modifier. Therefore, the development of new "universal" adsorbents should be started from the selection of the corresponding modifier providing the formation of uniform stable coating on the matrices of different nature. The modifier should be characterized with high selectivity under bioseparation conditions.

Principal of action	Corresponding methods
	1. Extraction
Separation of the mixtures containing liquid or solid components due to selective dissolving of individual components in special solvents. Target substance should be better soluble in a solvent (extractant) than in a medium.	Phenol-chlorophorm extraction of proteins and nucleic acids. Alkali extraction of plasmid DNA using the glass beads. Liquid-liquid extraction (partitioning).
	2. Precipitation
Precipitation due to heating, cooling, concentration, or dilution. Precipitation in the presence in the presence of special added substances.	Precipitation of proteins via ultracentrifugation. Preparative ultracentrifugation of DNA under cesium chloride density gradient. Re-precipitation of proteins and nucleic acids in alcohols. Precipitation of proteins with ammonium sulphate, dextran sulphate, etc.
	3. Adsorption
Particular case of extraction. Concentration of target substance or admixture occurs at the interface between adsorbent and liquid medium.	Purification of DNA using magnetic particles. Purification of the proteins and peptides via "batch-method" using magnetic beads. Purification of DNA using the porous silica.
	4. Chromatography
Methods elaborated to separate the mixtures of substances or particles, based on the difference between the rates of their migrating in a system of immiscible phases, which move relatively to each other.	Corresponding examples include large number of applications.

Table 1. The main physical-chemical processes of the substances separation.

It is desirable that the use of such modifier could provide the one-step isolation/separation protocol. Actually, few materials are known, which could be effective for the separation of nucleic acids and proteins utilizing the simpler separation concept (as compared to the traditional multistep protocols) i.e. not nucleic acids but proteins and other components are retained with the chromatographic material (Saburov et al., 1991, Kapustin et al., 1998). However, the polymers showing that kind of the separation behavior i.e. polythetrafluoroethylene (PTFE) and polyaniline (PANI) do possess neither mechanical nor processing properties allowing to manufacture "good" porous adsorbents from those materials. Therefore the practical realization of the specific adsorption behaviour of PTFE and PANI can be achieved using composite matrices modified with nano-layers of the corresponding polymers.

Indeed, the composite adsorbents modified with PTFE and PANI were shown to be effective for one-step purification of nucleic acids from proteins. However, the techniques originally used for synthesis of PTFE- and PANI-modified composite adsorbents despite a certain similarity of their chromatographic properties were quite different, complicated and difficult for scaling up. In the case of PTFE-containing material it was γ-ray induced post-polymerization of tetrafluoroethylene (TFE) in adsorption layers at low temperatures (Kapustin et al., 2006). In the case of PANI-containing material it was oxidative precipitation polymerization of aniline in aqueous phase followed by deposition of PANI nano-particles onto the carrier surface. Furthermore, the original adsorbents prepared by the above-described techniques were not always effective for obtaining purified samples of nucleic acids from some "complicated" biological samples, i.e. whole blood, plant tissues, etc, containing in parallel with proteins the powerful PCR inhibitors such as hem or chlorophylls, or for separation of DNA and RNA, double-strained DNA (dsDNA) from single-strained DNA (ssDNA), etc. Therefore, further development of the adsorbent composition, its surface structure and manufacture technology is required.
This chapter describes the advance in development of the composite sorption materials modified with fluoropolymers and polyanilines for one-step separation and purification of biopolymers (such as nucleic acids and proteins), and, to some extent, reflects the chronology of elaboration of the specific approaches in this field. The main milestones relating to development of alternative techniques (in respect to traditional ways of the synthesis) of fluoropolymer-and PANI-modified adsorbents, the examples of their practical application in analytical procedures and medical diagnostics are discussed in the corresponding subsections.

2. Silica-based adsorbents as effective systems for isolation and purification of biopolymers (traditional approaches in bioseparation)

One of the most commonly used solid supports in liquid chromatography is unmodified silica (Snyder et al 2010). There are many approaches based on use of unmodified silica for the purification of nucleic acids from biological samples (Rother et al. 2011). The mechanism providing this type of bioseparation is based on the specific interaction between silica's surface and nucleic acids, in particular, in the presence of chaotropic salts at high concentration and high pH values. Silica is weakly acidic under physiological conditions, and its surface is represented mostly by Si–O- rather than Si–OH groups . Therefore, in the presence of very high chaotrope salt concentrations, cations in solution can form relatively stable layer of counter ions around the negatively charged silica surface and change their charge into positive. As a rule, the chaotropic agents disrupt the hydration shells of biomolecules in aqueous solution and denature them. An exception is DNA, which is remarkably stable against chaotrope denaturation. The phosphate groups of nucleic acids are strongly acidic, therefore, nucleic acids bind effectively to the positively charged silica's surface, but other molecules are left in solution. The DNA–silica complex is then washed in a salt solution or an alcohol/water mixture to remove weakly bound "impurities", then DNA can be eluted with low salt concentration buffer or water.
Analogous multistep mechanism of nucleic acid isolation/purification is a basis of the most modern techniques (it is true for both unmodified silica or silica modified with polymers). Nevertheless, the silica itself is characterized by relatively low stability under high acidic or alkali conditions, high level of non-specific sorption, and requires specific conditions

providing the optimal surface concentration of required functional groups, so additional chemical modification is needed. A great number of effective adsorbents were obtained as result of silica modification with low-molecular substances. However, the properties and application features of such materials are not in the scope of the present review; however, the majority of them are characterized with the same disadvantages as it was observed in a case of unmodified silica application.

In early 1970's a great practical interest had been arisen to polymer-modified silica as adsorbents for chromatography of proteins and peptides, enzymes, nucleic acids and viruses. Such interest was determined by the advantages provided with rigid-skeletal materials as compared to traditional soft organic resins and gels. Wide use of cross-linked dextranes, agarose and polyacrylamide, which were developed in early 1960's, was limited due to their low mechanical stability. The letter development of highly cross-linked resins partially overcames the mentioned disadvantages, but the cardinal way to solve this problem is the development of composite polymer-coated materials.

Known methods of producing of composite adsorbents can be nominally divided into two large groups. The first one includes the methods based on adsorption or chemosorption of polymers from their solutions onto the silica's surface. The second group involves the processes of radical or ionic polymerization of monomers in a presence of (mostly silica) carrier.

Physical adsorption of polymers on the carrier surface is the simplest way to obtain the composite adsorbents. It can be realized via direct sorption of polymers, which are retained on the surface due to multipoint contact between partially positively charged macromolecules with cation-exchange surface of silica by means of hydrogen or ionic bonds. In the case of chemical adsorption the silica surface can be preliminary treated with the suitable agents (such as functionalized silanes, different additional bifunctional reagents, etc) to activate the surface and provide the stable polymer coating after the contact of the polymer modifier with activated carrier.

The examples of the corresponding modifiers include a large variety of polymers and oligomers. The most illustrative examples of modifiers, which are suitable for physical adsorption of macromolecules on the silica support include dextrans or starch (as well as its derivatives), poly-N-vinylpyrrolidone (and its copolymers with wide range of functionalized comonomers), poly-N-vinyltriazol, copolymers of poly-4-vinylpyridine with styrene, polyvinyl alcohol, polyethyleneimine, polyallylamine, polyethylene glycol (and its derivatives), olygomeric phases based on N-substituted amides of carbonic acids, polybutadyene, and many others.

Methods of producing the adsorbents based on graft polymerization usually include less number of steps as compared to methods based on adsorption of polymers. At the same time, determination of the optimal conditions of such process has significant importance, so the correlation between the synthesis conditions and the properties of the final product is less evident as compared to adsorption-based techniques. The examples of the corresponding methods include polymerization of wide number of vinyl monomers in the presence of carrier and bifunctional cross-linking agents, or using different modes of the carrier surface activation (such as ionizing irradiation, μ- and γ-irradiation, and so on).

The use of polymer modifiers allows to obtaining of the stable and effective composite materials combining the required structure and mechanical properties of the carrier with specific sorption properties of the modifier. However, the majority of the developed materials provide multistep mechanism of the biopolymers separation/purification, which is similar, in principle, to the process where unmodified silica support is used. Therefore, to elaborate more efficient variants of one-step separation of different classes of biopolymers (primary, nucleic acids and proteins), the specific types of polymer modifiers are required.

3. Perfluoropolymers as effective modifiers for the preparation of composite adsorbents for one-step separation of biopolymers

The requirements to the properties of an "ideal adsorbent" for separation of biomolecules or particles, comprise a few controversial features, which can hardly be achieved in a single material. Those include insolubility, permeability to macromolecules, high rigidity, well-defined porosity independently of solvent, large specific surface area, low non-specific adsorptivity, physico-chemical and biological stability, and facile derivatization. Therefore, it should come as not surprise that investigators have attracted their attention to the fluorine-containing polymers. Fluoropolymers are known have a complex of unique physical-chemical and sorption properties due to the presence of fluorine atoms in the macromolecules. The simplest example of the perfluorinated polymer is the polytetrafluoroethylene (PTFE). Its polymer chain forms almost ideal cylinder with the outer shell of fluorine atoms. The surface formed with such macromolecules is characterized with extremely low non-specific sorption that is really important property, in particular, for effective use of fluoro-containing materials in bioseparation of nucleic acids. PTFE as adsorbent was first described by Hjerten (Hjerten et al., 1978) and was tested on examples of purification of proteins and tRNA. Recently William (William et al., 1986) presented a new fluorocarbon adsorbent for reversed-phase HPLC of peptides and proteins. However, low specific area and capacity (1.1 mg tRNA/g) and relatively low limits of working pressure do not allow to the use of this adsorbent for preparative chromatography. Moreover, because of its insolubility in common solvents, infusibility, and softness, the manufacture of rigid mesoporous matrices from PTFE as a final product is impossible. This task is to be accomplished via creation of defect-free polymer layers on a solid porous matrix by means of direct (graft) polymerization of TFE in adsorption layers. The polymerization is initiated both by active centres generated on the surface of starting materials during irradiation, like, for example, in work of Turkin, Zubov, Saburov et al., 1988 and with the use of chemical initiators (e.g. fluoroolefin ozonides). The polymerization of TFE on the surface of inorganic materials has been described in the literature (Ivanov, Saburov & Zubov 1992). As shown by preliminary experiments, during vapour- or liquid-phase monomer feeding, the grafted polymer does not form a continuous layer and occurs as single globular entities on the surface of an inorganic support. An effective method for the preparation of grafted polymers is the use of the post-effect, i.e., grafting due to active centres stabilized on a solid support. To find optimal synthesis conditions for the desired morphology of the composite obtained by means of this method, several modes of the process were examined. Controlled porous glass (CPG) with an average pore diameter of 50 nm and an average particle size of 100–150 μm was used as a support in this case. The modes of irradiation, temperate regime, and conditions of the monomer introduction into reaction system were varied in this study. It turned out that the most effective mode providing the required morphology of the composites includes irradiation of the matrix with the pre-adsorbed TFE at -196°C followed by slow heating of the support with TFE preliminarily sorbed up to room temperature, and post-polymerization during warming the system outside the irradiation zone ("joint radiolysis" at -196°C). From mercury porosimetry data, it is possible to determine the thickness of the polymer coating, which varies from 1–3 to 5–10 nm depending on the amount of TFE introduced. These values correspond to those calculated from the amount of the grafted polymer and the total pore surface area of the matrix. In addition, the surface of the synthesized adsorbents is poorly wetted by both polar and nonpolar liquids, i.e., become fully non-wettable.

The bioseparation behaviour of thus obtained PTFE-modified CPG is illustrated by a typical chromatogram of a mixture containing the *E. coli* plasmid pBR 322, RNA, and associated proteins (Ivanov et al, 1992). It is seen that DNA elutes from the column with the first fraction in the excluded volume, while RNA is weakly retained but likewise elutes in the isocratic regime. However, the total protein fraction is completely eluted from the column only with an aqueous organic eluent. The resulting mixture of proteins can be separated in the reversed-phase HPLC mode. The mechanism of retention of individual proteins on fluoropolymer adsorbents resembles that on classical C18 phases. However, the elution volume in this case is much smaller. Consequently, fluoropolymer composite adsorbents are effective in separation of both nucleic acids and proteins. A high selectivity of fluorinated polymers is their general property, which is provided by a very low surface energy of the relevant coatings. This advantage has been realized using the cartridge variant of bioseparation, which is described in details below.

Thus, the use of perfluoropolymers as modifier of composite adsorbents provides more sufficient, from the practical point of view, one-step scheme of biopolymers separation.

4. From perfluoropolymers to partially fluorinated polymer modifiers. An effective alternative to the traditional techniques

Interactions observed between the fluoropolymers and biopolymers have been attracted the attention of scientists for many years. As an example, the article of G. Xindu and P.W. Carr (Xindu & Carr, 1983) can be mentioned. Authors described the adsorption aspects of fluorocontaining substances (including perfluoroalcyl-containing protein) at the fluorocarbon surface. The main conclusion was that the nature of the interaction between fluorinated proteins with fluorocarbon surface is not known yet. Latter works were also devoted to the study of the interaction between different fluorinated phases and low-molecular and polymer substances. These studies were carried out using not only perfluoro-containing materials, but also the chromatographic systems containing insignificant part of hydrogen atoms substituted by the fluorine. Therefore, the further development of bioseparation methods based on the use of fluorocontaining adsorbents was not unexpected. Early works in this field were devoted to study of functionalized particles of polychlortrifluoroethylene (PTCFE), which is chemically stable, hydrophobic and mechanically stable material. The treatment of PTCFE with metal-organic substances (such as n-butyllithium, aryllythium, alkylmagnesium, arylmagnesium) resulted in the preparation of alkyl-containing materials with low content of chlorine. n-buthyl-PTCFE adsorbent was shown to be effective for separation of aromatic compounds via reverse-phase mechanism of chromatography. Phenyl-PTCFE adsorbents (having substituted aromatic rings) were effective in separation of some drug agents via both the reverse-phase and the ion-exchange mechanisms. At the same time the adsorbent modified with trialkyl (C_8-C_{10}) methylammonium or quaternary ammonium substance RPC-5 was suitable to separate biological substances such as mRNA, nucleotides, nucleosides, DNA-fragments. Unfortunately, this adsorbent was not stable enough for the repeated use. Berendsen

(Berendsen et al., 1986) obtained good results using silica modified with (heptadecafluorodecyl)dimethylsilane (HFD). Such adsorbents were shown to be effective for separation of fluorocontaining low-molecular organic substances. These materials retained nonpolar compounds weaker as compared to hydrocarbon-containing adsorbents. Therefore, it was suggested to desorb proteins from the fluorocontaining surface reducing the content of organic component in eluent (i.e. under non-denaturating conditions), and thus, this approach justified the use of HFD-modified adsorbents.

In early 1990's, Saburov, Zubov et al (Saburov et al., 1991) have developed the composite adsorbents based on porous silica modified with trifluorostyrene. Polymer modifier was obtained by radical copolymerization of trifluorostyrene with methylvinyltrietoxisilane. In the first case the suspension of the support particles was prepared in the solution of copolymer in absolute toluene followed by boiling, filtration and washing with hot toluene. End-capping of the residual silanols was carried out with hexamethyldisilane. The alternative way of the modification was based on the treatment of the carrier particles with γ-source followed by contact with vapour of trifluorostyrene (Saburov al., 1988). Dimer of trifluorostyrene was removed by washing with toluene. The both obtained adsorbents had the similar adsorptive and chromatographic properties and can be characterized as intermediate materials if compared to silica modified with TFE and silica modified with polystyrene. In both cases the obtained materials demonstrated the high hydrolytic stability. The materials were used as a basis to prepare a wide range of ion-change adsorbents with different surface capacity (from 95 to 300 μmol/g), and they were successfully used for separation of insuline and desaminoinsuline as well as nucleotidephosphates.

Another approach for producing fluoropolymer-coated adsorbents was suggested by Kapustin in 1998 (Kapustin et al., 1998), for synthesis of the silica adsorbent modified with partially fluorinated polybutadyene. This approach is based on the *in situ* fluorination of thin films of oligomers or polymers preliminary immobilized on the carrier surface. The fact, that unsaturated hydrocarbons can be fluorinated using such agents as xenon difluoride, is of special interest, because introduction of the fluorine atoms into the polymer coating provides modification of the surface not disturbing the volume properties of the coating. Standard method of direct fluorination suggests the mixture of gaseous fluoride with nitrogen. However, the fluoride is toxic, very corrosive that requires safety measures during all operations. The use of alternative fluorinating agent simplifies the process of modification.

The first step of the process is deposition of thin uniform layer of the non-fluorinated precursor on the porous silica matrix. It is achieved by impregnation of the pre-evacuated matrix with the precursor solution in volatile solvent. After evaporation of the solvent the material was fluorinated. A typical example of the precursors are oligobutadienes of MM 5000 - 10000. It was shown earlier (Zubov et al., 1998), that XeF_2 effectively interacts with the surface of polybutadiene and modifies it. Fluorination under oxygen-free conditions excludes the appearance of polar groups that can significantly increase the non-specific sorption. Fluorination of the polymers immobilized on the silica surface accompanied with their chemical structuring (cross-linking) due to the recombination of macroradicals and interaction between radicals with multiple bonds. The result is formation of cross-linked fluorinated polymer coatings (Kapustin et al., 1998). By-products (first of all, hydrofluoric acid and the products of its interaction with silica) and XeF_2 absorbed on pores surface, then can be removed under vacuum at the higher temperature.

The main feature of the suggested method is obtaining of fluorinated polymer phase directly during the process of the adsorbent producing in contrast with the techniques using fluoromonomers. The fluoropolymer coating is additionally setted due to formation of intermolecular bonds.

Obtained adsorbents were used for isolation of bacterial plasmid DNA (from *E.coli*, JM 109) and DNA from nuclear-containing human cells using the cartridges containing about 140 mg of adsorbent. Such procedure was shown to be very simple, especially in comparison with earlier elaborated multi-step methods (such as centrifugation under density gradient after an alkali treatment of cell lysates or subsequent treatment of the lysate with phenol and chlorophorm). The observed results of DNA isolation from the model mixture containing DNA, RNA and proteins were the similar as compared to use of PTFE-coated adsorbent, i. e. DNA was eluted at first, then RNA was eluted using the same buffer solution, and the proteins were desorbed from the material using 50% methanol. The ratio of intensivities at λ=260 nm and λ=280 nm measured for the collected DNA-containing elates was approximately 2:1, and it confirmed the high extent of DNA purification from the proteins.

The relatively simple protocols for DNA isolation from a human blood were elaborated. These methods included either column chromatography of the sample obtained after isolation and lysis of the nuclei-containing cells from a blood, or very simple "batch-method" based on the sorption of the components in the mixing volume followed by centrifugation and collection of the supernatants.

In the first case DNA, but not RNA, was eluted from the column with TE-buffer in the excluded volume, then the retained components were desorbed under linear gradient of iso-propanol. In the second case the "batch-procedure" was repeated twice using the same sample, and the content of DNA was measured spectroscopically in the both collected supernatants. The fist aliquot of the collected supernatants contained about 60% of purified DNA, and the second one contained about 40% of DNA.

The most effective application of thus prepared adsorbent was achieved when one-step isolation of DNA from the bacterial lysates was carried out (like it has been demonstrated using PTFE-coated adsorbents). However, in spite of evident advantage of this synthetic method in comparison with irradiation technique, it seems to be more suitable for modification of non-porous supports, first due to necessity in laborious washing procedure of the porous adsorbent to get rid of by-products of fluorination. Nevertheless, it was demonstrated, that preliminary immobilized oligomeric or polymer phases on the carrier's surface can be chemically modified to reach the required extent of fluorination.

Materials coated with partially fluorinated polymers have additional advantage, i.e. they in addition can be chemically modified to improve hydrophilicity of the surface or to introduce specific functional groups. Therefore, fluoropolymer-modified silica can be successfully used not only in bioseparation. Matrices coated with fluoropolymers are considered as perspective candidates for immobilization of bioligands. In this case the presence of specific functional groups on the fluorinated surface is required. Direct synthesis of fluoropolymers containing specific functional groups is not a simple task. Sometimes, more efficient way to obtain the functionalized matrices is formation of thin solid fluorocontaining coating on the carrier surface by casting of preliminary prepared fluoropolymer (e.g. copolymer of TFE with functionilized fluorocontaining monomers) from its solution in perfluoro-containing solvent (mixture).

An example of preparation of the support for solid-phase synthesis of oligonucleotides is given below. This material is based on porous silica coated with the copolymer *poly-*

ethylene-*co*-{methyl-3-(1-(difluoro((trifluoroethenyl)oxy)methyl)-1,2,2,2-tetrafluoroethoxy)-2,2,3,3-tetrafluoropropanate} (EVE) followed by specific chemical modification. The corresponding method of synthesis includes such steps as incubation of the support particles in the solution of polymer followed by ultrasonic treatment, evaporation of the solvent and treatment with nucleoside. This method was elaborated by authors of the present chapter in 2007 (unpublished results).
Chemical structure of EVE allows to carrying out the different polymer analogous reactions due to presence of active esther groups as it shown on Fig. 1.

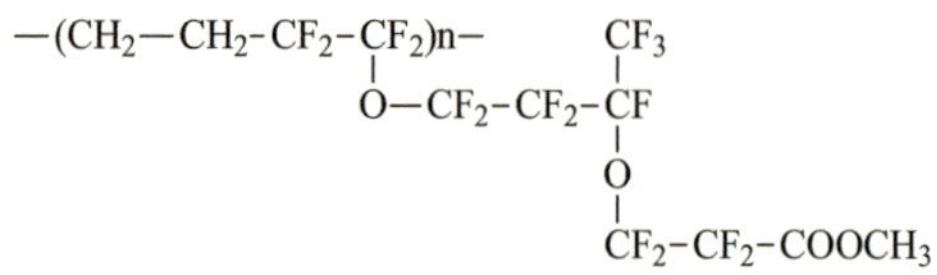

Fig. 1. The chemical structure of *poly*-ethylene-*co*-{methyl-3-(1-(difluoro ((trifluoroethenyl) oxy)methyl)-1,2,2,2-tetrafluoroethoxy)-2,2,3,3-tetrafluoropropanate} (EVE).

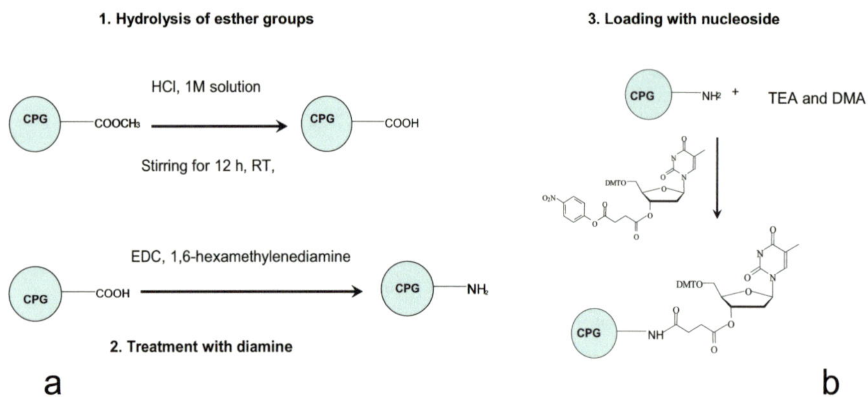

Fig. 2. Synthesis of fluorinated support based on CPG modified with EVE for solid-phase synthesis of oligonucleotides.

The scheme of the chemical modification of side groups of immobilized fluoropolymer (EVE) is presented on Fig. 2a. Thus modified CPG was treated with Nucleoside (DMT-dT-SNPE from Proligo GmbH, Germany) in accordance with the scheme represented on Fig. 2b. Surface concentration of nucleoside residues depends on degree of amination and can be controlled. Stability of the nucleoside-loaded CPG was studied using the standard Trytil-assay. Nucleoside-loaded EVE-coated CPG was stored for 9 months. The concentration of immobilized nucleosides was found about 9.25 μmol/g. This value corresponds to analogue value (9.33 μmol/g) which was measured for fresh prepared sample. Thus, stable support for solid phase oligonucleotides synthesis was developed and characterized.
Abovementioned data confirmed that fluoropolymer-coated porous silica composites can be effectively used for one-step isolation of nucleic acids, purification of other biopolymers and for immobilization of bioligands. However, these important properties are also demonstrated by some other polymers i.e. polyanilines. The composite materials modified with polyanilines via traditional presipitative polymerization methods and by alternative ways are described below.

5. Polyaniline as alternative effective modifier for producing the composites for one-step bioseparation

The unique physicochemical properties of PANI explain the wide use of PANI-based materials in various areas of science and technology. Polyaniline is characterized by a high chemical stability and is insoluble in most of common solvents. The specific features of the structure of the PANI macromolecule make PANI-containing materials effective candidates for use in biomedical applications, in particular, for separation of complex mixtures of biopolymers (Kapustin et al., 2003). An addition, reversible changes of PANI properties depending on pH of medium, make it possible to regulate the sorption properties of "smart" PANI coatings.

5.1 Oxidation precipitative polymerization of aniline as traditional way of preparation PANI-coated adsorbents

Polyaniline-modified adsorbents behave in a similar to fluoropolymer-coated adsorbents manner as regards the separation of mixtures containing nucleic acids and proteins. However, the use of PANI-containing adsorbents offers an impressive advantage in separation of mixtures of proteins, since it does not require the addition of an organic component to the mobile phase. A typical example of the pH-gradient separation of a model protein mixture is given in (Kapustin et al., 2006). The proteins eluted from the column in the order of ascending pI values of the individual components. The mechanism of retention and desorption of proteins on PANI coatings is complex and seems to be determined by both ionic interactions of the sorbate charged groups with ionic sorption sites on the adsorbent surface and the ability of a PANI coating to reversibly bind protein molecules via hydrogen bonding. The following examples illustare the similarity of the sorption properties of PTFE- and PANI-containing adsorbents based on TrisoporTM-500 CPG. Samples of 100–200 µl of a mixture containing DNA, RNA and proteins were applied into cartridges packed with adsorbents. The cartridges were incubated for 3 min at room temperature and centrifuged for 1 min at 240 g. Then, the collected eluates (80–100 µl) were analyzed. The results of electrophoresis and the spectroscopy data showed that the eluates obtained using both types of the adsorbents contained DNA cleaned of protein up to 90%.

It turned out that PANI coatings effectively bind metal-containing compounds, such as heme and chlorophylls (powerful PCR inhibitors). Figure 3 presents the results of electrophoresis in 1% agarose gel of DNA specimens obtained via passing samples of whole blood and leaves of *Nicotiana Tabacum L.* after preliminary treatment with concentrated lysing solutions, through cartridges containing 140 mg of the adsorbent each. The eluates were additionally subjected to spectrophotometric analysis. It was shown that the purified DNA samples are practically free of heme or chlorophylls (judging from the absence of absorption at characteristic wavelengths) and, as such, are quite suitable for PCR analysis.

In spite of fact that the PANI-coated adsorbents are effective for one-step bioseparation of nucleic acids there are a number of difficulties, firstly connected with the complicated procedure of the final product purification from unbound and loosely bound polymer particles. Therefore, development of new approaches for the synthesis of PANI-coated composites is important practical task.

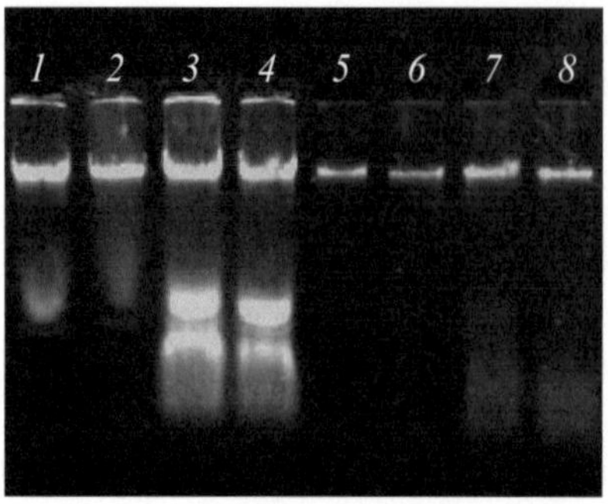

Fig. 3. Results of electrophoresis in 1% agarose gel of DNA samples obtained via passing the lysates of cow whole blood and *Nicotiana Tabacum L.* leaves through cartridges packed with the PANI-containing adsorbent: (*1, 2*) blood lysate, (*3, 4*) tobacco leaf lysate, (*5, 6*) eluates containing blood DNA, and (*7, 8*) eluate containing plant DNA.

5.2 Matrix polymerization of aniline is an effective alternative to precipitative process

Polyaniline is prepared via chemical or electrochemical (Stejskal., 2002) oxidation of aniline. The chemical polymerization of aniline in an acidic medium in the presence of a solid surface frequently leads to the formation of thin polymer coatings (Sapurina & Stejskal, 2008). The mechanism of oxidative polymerization is complex and is still under debate despite a large number of publications on the issue. However, it is generally adopted that the first step of the reaction is the protonation of the monomer. Then, the protonated form of aniline is oxidized yielding the radical ion, followed by the formation of dimers, oligomers, and, finally, a solid insoluble polymer. The characteristic features of the process are the presence of an induction period, the autocatalytic character, and usually observed slowing down of the reaction before the complete depletion of the monomer. Direct evidence for autocatalysis by the polymer is that the addition of a portion of PANI to reaction system immediately after mixing of monomer and oxidant solutions changes the shape of the polymerization kinetic curve. The induction period shortens, depending on the mass of added PANI (Kapustin et al., 2003), and the monomer conversion curve demonstrates a second step. The presence of silica carrier has practically no effect on the kinetic pattern of aniline polymerization as compared to polymerization in the absence of the support. In other words, the silica surface has no catalytic effect on the polymerization of aniline. As a result, aniline polymerization produces not only a film on the silica surface but also proceeds in the bulk of the reaction mixture. For example, in the case of aniline polymerization in the presence of CPG, up to 40% of the monomer is consumed for the formation of a suspension of PANI particles.

The necessity of aniline protonation at the first step of oxidative polymerization supposes that insoluble solid polyacids (acid cation exchangers) should be used as matrices for the reaction. Therefore, silica modified with polymers having sulfo groups on the surface can be used as support. To prepare such supports a number of approaches were developed. In particular, silica was modified with PTFE on which polystyrene (PS) crosslinked with divinylbenzene as grafted by means of radiation post-copolymerization. Then, the PS-containing silica was sulfonated. In the case of aniline polymerization in the presence of surface-sulfonated supports, the induction period almost disappeared as compared to the polymerization in the absence of the support. No formation of PANI particles in the reaction volume was detected. Since the morphology of the original support is retained after modification and the polymeric coating efficiently shields the silica surface, the obtained composite was proposed to be used for DNA isolation from plant lysates, i.e. from the complex mixtures containing DNA, RNA, proteins, peptides, polysaccharides, surfactants, chlorophylls, and low-molecular compounds. Lysates of tobacco leaves (*Nicotiana tabacum L.*) were applied onto cartridges with the PANI-containing adsorbents based on both polysulphostyrene-coated silica (CPG-PSS) and unmodified CPG. When the lysates are passed through the adsorbents, DNA is eluted in the excluded volume, whereas proteins, chlorophylls, and a considerable part of low molecular substances are retained. The degree of purification of DNA from proteins and chlorophylls was estimated using electrophoresis and spectrophotometry. The plant lysates were completely decolorized upon passing through cartridges with CPG-PSS-PANI (the absorbance of the initial lysate at 416 nm was 0.19, whereas the absorbance of the eluate obtained using CPG-PSS-PANI was 0.04). At the same time, the eluates obtained in the case of CPG-PANI had a pale light-green colour (absorbance 0.10). The ability of the adsorbent containing the residual sulfo groups to bind chlorophylls is explained, most likely, by the interaction of the porphyrin fragments of chlorophylls with the sulfo groups on the adsorbent surface. The degree of purification of DNA isolated from the plant tissue lysates using CPG-PSS-PANI enables their direct use for amplification by the PCR method. The corresponding amplicons are shown on Fig. 4.

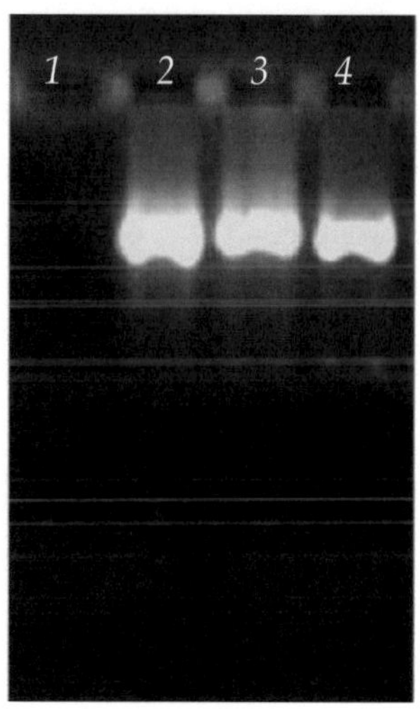

Fig. 4. Electrophoregram of the DNA amplicons (*2*—*4*) obtained by the PCR with the eluates containing DNA from the *N. tabacum* L. tobacco leaves. The eluates were obtained using the CPG—PSS—PANI sorbent (*1* is negative control, H_2O).

The obtained coating is in fact the surface polycomplex of PANI and sulfonated PS. In other words, it is the self-doped derivative of PANI, i.e. the polymer structure containing SO_3H functional groups. Matrix polymerization offers an efficient strategy to control both the kinetics and mechanism of the polymerization and the structure of resulting polymers. Thus, polymerization yields polycomplexes formed between the original matrix and the formed polymer. In particular, matrix polymerization of PANI was assumed by Yagydaeva, Kapustin, Prostyakova and Zubov in 2009 as a basis for preparing composite biosorbents, in which the products of the matrix oxidative polymerization of aniline on aromatic poly(sulfonic acids) were used as modifiers of porous silica. Of special interest is the use of various poly(sulfonic acids) (PSAs) as dopants (Yagudaeva et al., 2007). Some systems remain phase-homogeneous since PANI develops directly on the water-soluble matrix (PSA). Previous experiments showed that when the synthesis of PANI is performed in the presence of the aromatic polyacid poly(*p,p*'-(2,2'-disulfonic acid)diphenyleneisophthlamide (*iso*-PSA), the local concentration of the protonated aniline in the vicinity of a polyacid macromolecule increases, and turns out to be higher than that in solution. Thus, polymerization of aniline in the presence of PSA should proceed without formation of a suspension of PANI particles in the bulk of the reaction mixture. This concerns the polymerization of aniline both in the solution of PSA and in the presence of PSA-modified silica. Besides, the use of the preformed PSA-PANI stable polycomplexes as polymer modifiers of the support surface makes it possible to simplify the preparative technique of the composite adsorbents manufacturing. Furthermore, owing to the presence of sulfo groups in a polycomplex, the adsorbent surface possesses additional functionality. Hence, the sorption behaviour of the materials can be controlled. In particular, the matrix synthesis of PANI was performed in the presence of *iso*- and *tere*-PSAs.

The sorption behaviour of PANI-containing adsorbents prepared via polymerization of aniline on PSA preimmobilized on the surface of the silaminated silica, as well as via immobilization of the preformed *iso*-PSA–PANI complexes on the surface of silaminated silica, was compared to the corresponding characteristics of CPG-PANI and PSS-PANI. Model mixtures containing DNA, RNA and proteins (BSA) were applied on cartridges packed with the prepared adsorbents (nearly 100 mg). Elution was performed at various pH values or various polarities of eluents (during desorption of protein components of the mixture). Eluates were analyzed by electrophoresis and spectrophotometry. The yield of DNA in the case of CPG-PSS-PANI and CPG—*iso*-PSA-PANI (prepared by the first method) was above 80%, and RNA was absent in both cases. DNA elutes in the excluded volume, while RNA and protein retained and the degree of protein sorption with the CPG-*iso*-PSA-PANI is higher than the degree of protein sorption with the CPG-PSS-PANI. On the contrary, the adsorbent prepared in the presence of *tere*-PSA-PANI irreversibly retained almost all nucleic acids. Thus, in terms of the sorption behavior, PANI, which was formed on the surface of *tere*-PSA, differs from that formed on the surface *of iso*-PSA or on the surface of the unmodified silica. Therefore, *tere*-PSA was found to be not suitable for the synthesis of PANI-containing adsorbents. The character of sorption of nucleic acids with the adsorbents containing *iso*-PSA is nearly the same as that of previously prepared PANI adsorbents containing sulfo groups chemically bonded to the surface. The efficiency of DNA isolation depends on the nature of the matrix polymer and increases with an increase of the content of sulfo groups.

In the case of *iso*-PSA-PANI simultaneous presence of additional sorption sites (i.e. residual sulfonic acid groups) in the obtained coatings allow to additional possibilities for bioanalysis, e.g. separation of ss- and dsDNA depending on pH of medium, as it shown on Fig. 5. To study this effect the authors of the present chapter used the model system containing dsDNA isolated from *Agrobacterium tumifaciens C58* and ssDNA obtained after reverse transcription (using reverse trancriptase RNAse H+) of RNA isolated from Tobacco Mosaic Virus followed by hydrolysis of RNA at pH > 8.0. Aliquotes of thus prepared mixture were applied onto the cartridges packed with CPG-*iso*-PSA-PANI, then collected eluates were used in PCR analysis (using the corresponding specific primers). It is seen, that increase of pH value from 8 up to 9 results in increasing of the yield both ss- and dsDNA in the collected eluates. At the same time, ssDNA/dsDNA ratio (equals ~ 1:3) in these eluates is constant.

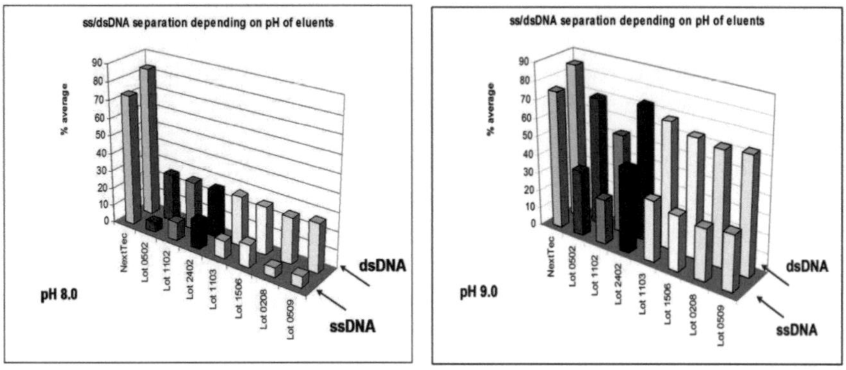

Fig. 5. Separation of ssDNA and dsDNA depending on pH of medium using *iso*-PSA-PANI-coated adsorbent.

It could be said in conclusion, that the use of preformed *iso*-PSA–PANI complexes for obtaining PANI-containing adsorbents appreciably simplifies the technique and, hence, scaling up of the manufacture of the said materials.

5.3 Unique properties of PANI coatings open new possibilities for bioseparation and bioanalysis

Introduction of additional functional groups into fluorocontaining macromolecules results in a widening the fields of their applications, similarly, introduction of additional functionality to PANI coatings opens the new possibilities of their use in bioanalysis. An example of such effective approach was demonstrated in the patent WO 2011/0014308 A1 (Vaczine-Schlosser et al., 2011). This invention relates to a process for so called 'laser desorption ionization mass spectrometry' using a polymer coating as UV absorption medium onto which the sample probe of interest (containing proteins or peptides) is deposited. The corresponding polymer coatings, in particular, were obtained by copolymerization of aniline with substituted anilines such as 3-amino-benzoic acid (poly-ANI-*co*-3-ABA).

Mass spectrometry (MS) is a widely used analytical method for determining the molecular mass and structure of various compounds. It involves transfer of the sample molecules into the gas phase and ionization of the molecules. Molecular ions are separated by using electric or magnetic fields in high vacuum according to their mass-to-charge (m/z) ratios. During the last decades, MS has proven to be an outstanding technique for accurate and sensitive analysis of biopolymers, like proteins and peptides. With the introduction of soft ionization techniques such as electro spray ionization (ESI) and matrix-assisted laser desorption ionization (MALDI) or surface enhanced laser desorption ionization (SELDI), it became possible to transfer into the gas phase and ionize these non-volatile, large, and thermally labile molecules without their deep degradation.

However, the latter technique requires use of so called "matrices", i.e. energy absorbing substances (usually aromatic acids like α-cyano-4-hydroxy cinnamic acid) mixed with an analyte. The matrices not only make possible volatilize the analyte, but produce considerable background noise in the lower molecular part of the mass-spectrum. Therefore, there is a strong need for a process that is not limited by strong background signals, or even such process, which makes the addition of a matrix compound superfluous. The said invention provides a process comprising depositing of a sample probe on a polymer surface comprising a UV absorbing aromatic monomer unit followed by irradiating the sample probe and/or the surface with a UV laser beam thereby effecting an ionization and/or desorption of the sample molecule, and determining the mass of the ionized sample molecule.

A main idea of the said invention is to replace a matrix substance by a surface of polymeric material deposited on the sample chip. The inventors found that this can be achieved by the use of a polymer having UV absorbing aromatic units, which may be selected from, in particular, aniline or an aniline derivative. The UV radiation may be absorbed by the polymer and the absorbed energy is transferred to the sample molecules. During the laser irradiation step, the surface may be heated. However, the polymers of this invention show a good thermal and oxidative stability even under such condition.

The ability of aniline-containing coatings to bind a variety of proteins was studied using Spectral Phase Interference. Investigation of the kinetics and effectiveness of pH-dependent binding of proteins and peptides to the polymer surface was studied using thin glass slides (90 - 120 µm) modified with PANI coatings. It was shown that a reversible sorption of different proteins on the PANI-modified glass surface could be achieved depending on the pH. Cytrochrome C showed a reversible adsorption to the surface at a pH of 7.2 and was desorbed at a pH of 2. It was furthermore shown that cytochrome C (M 12 000), casein (M 20 000), myoglobine (M 17 800), IgG (M 125 000) and poly-L-lysine (M 150 000) could be bound to the surface of the PANI-coated silicon-strip at pH of 7.2. Calf thymus DNA and poly-uridine acid potassium salt were not bound to the surface under the same conditions.

Thus, silicon strips coated with aniline-based polymer films are suitable to retain specific biological entities and function as chromatographic supports as well.

To visualize protein retention on the modified PANI-ABA Si-strips, model proteins with different pI were labelled with luminescent (emission at 546 and 581) semiconductor (CdSe)ZnS nanocrystals. Protein solution droplets were put onto the modified surfaces (10 µl of solution, 0.5 mg protein/ml) and incubated for 5 min at ambient temperature. Excess was removed and sorption was optically visualized. The results are presented on figure 6 and in table 2. The results show that PANI-ABA surfaces can be used to separate proteins depending on their pI values. Separation of acid and alkali protein/peptides can be carried out directly on the silicon strip surface using tris-HCl buffer or another appropriate solution.

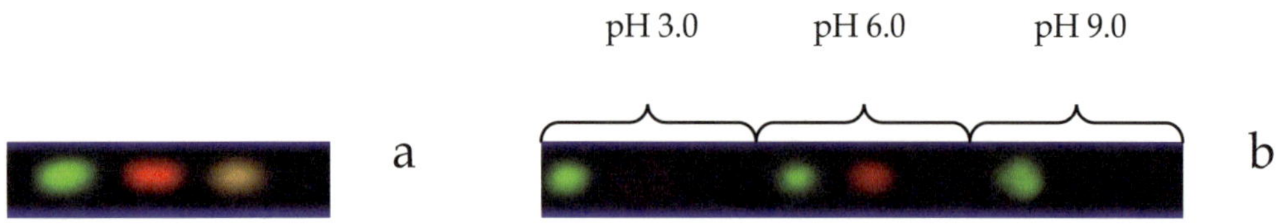

Fig. 6. Visualization of the proteins having different pI values and labelled with semiconductor nanocrystals under UV-source. The individual proteins before washing with the buffer solution *(a)* and retention of cytochrome C, myoglobine and pepsin on silicon-strips coated with *poly*-ANI-*co*-3-ABA by precipitation polymerization *(b)*. The correspondence between the spot colour and individual proteins is shown in the table 2.

Protein	pI	Colour	Before washing	pH 3.0	pH 6.0	pH 9.0
Cytochrome C	10.3	Green	+	+	+	+
Myoglobine	6.9	Red	+	-	+	-
Pepsine	2.8	Yellow	+	-	-	-

Table 2. The correspondence of the spot colour to individual proteins with different pI and results of the retention of cytochrome C, myoglobine and pepsin on silicon-strips coated by precipitation polymerization. (+) indicates good observation of UV excitation, and (-) indicates no observation of UV excitation.

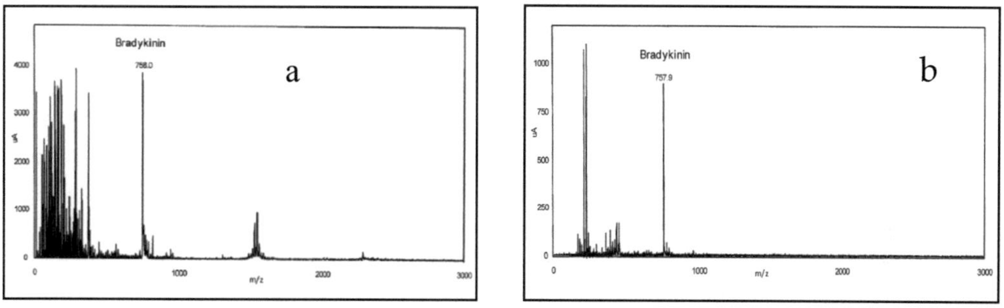

Fig. 7. MS spectra of peptide standards on a silicon strip modified with PANI-PABA with *(a)* and without an addition of CHCA as matrix *(b)*.

PANI-PABA coated Si-strips were analyzed with and without matrix addition, in order to compare the efficiency of conventional sample preparation techniques for analysis of compounds in the low mass region. In the experiments, the two most common matrix types were used: a-cyano-cinnamic acid (CHCA) and sinapinic acid (SPA) (Bio-Rad). 1 mL matrix solution was added to the surface and dried. The matrix addition step was repeated. Spectra were acquired under the same experimental conditions. The results show that without addition of extra matrix, no background peaks can be observed from the PANI-PABA coated arrays. However, addition of the standard matrix causes the appearance of background ions in the low mass range with high intensity, which hinders the analysis of low molecular weight compounds. Figure 7 shows MS spectra of peptide standards on a silicon strip modified with PANI-PABA (with a molar ratio of aniline : 3-ABA of 3:1). The results illustrate that the polymer coating indeed exhibits inherent MALDI matrix activity. The arrays shows good matrix activity with the formation of intensive protonated molecular ions ([M+H]+). No background peaks are observed in the spectra.
Thus, new effective analytical system based on ability of PANI coatings to separate the proteins and absorb UV energy was developed.

6. New approach of the support surface activation based on ozone-induced polymerization

The described above polymer modifiers (i.e. fluorinated polymers and polyanilines) are effective for one-step separation of nucleic acids. At the same time they are quite different not only by the chemical structure, but also by the mechanisms of their formation, although in both cases chemical deposition results in stable uniform nano-thick polymer layers on the matrix surface. Here we describe an approach to manufacture such composites using similar chemical steps.

For that purpose, an 'universal' activated solid support providing formation of such layers is required. Ozone is known to chemically activate different solid materials due to formation of various peroxide groups and/or ozonides on the surface (Robin, 2004). This is true for most organic matrices, however, silica does not react with ozone itself, but usually it contains different admixtures, in particular, iron oxide (in average, from 0.5 up to 3.5% depending on silica preparation technology). Such admixtures after ozone treatment, in principal, can form different peroxide ion- or radical-containing chemical structures, which can initiate polymerization of different monomers in thin layers. Last phenomenon was not studied in details until now.

Indeed, Prostyakova, Kapustin and Zubov (2011) have demonstrated, that after ozone treatment of silica samples followed by removing of unabsorbed ozone and other gaseous products, iodometric titration confirms the formation of peroxide species. The amount of those species is negligible in "pure" silica materials but increases for the samples with considerable amount of admixtures (presumably iron oxides). Thus, the ozone treated silica can be tested as heterogeneous initiator allowing modification of its surface by direct polymerization of various monomers via different polymerization mechanisms, such as radical polymerization of TFE (utilization of the forming surface radicals) and oxidative polymerization of aniline (utilization of surface peroxides). Indeed, incubation of the ozonated silica in contact with TFE at room temperature results in polymerization of this gaseous monomer at the surface of the carrier. In order to improve wettability and separation properties of the obtained materials, silica was additionally modified with copolymers of TFE with hexafluoroethylene (HFP) (to stabilize the radicals on the ends of propagated PTFE chains (Zubov et al., 2007), and with allylamine (AA) or allyl alhogol (AAl) (to introduce the additional functional groups). The comonomers were introduced either simultaneously as a mixture of monomers in the vapor phase or in a post-polymerization mode by introduction of the second monomer after the polymerization of TFE on the first stage was completed. In the last cases considerable concentrations of amino or hydroxylic gropus were found in the composites. TFE, AA and AAl can be polymerized only via radical mechanism. Therefore, ozonated silica can be considered as effective heterogeneous radical initiator. In fact, it also causes polymerization the other radically polymerizable monomers such as styrene, acrylamide, etc. The result of this polymerization is a porous silica-polymer composite with the variable polymer content up to ~ 25%. It should be noted, that morphological characteristics of the prepared adsorbents were similar as compared to the PTFE-coated CPG after irradiation induced post-copolymerization of TFE (Kapustin et al., 1998).

It could be presumed that the heteroperoxides immobilized at the silica surface can induce not only radical polymerization process, but also initiate oxidative precipitate aniline polymerization, thus giving rise to an adsorbent modified with surface-bound polyaniline. Indeed, when ozonized silica was incubated with acidic aqueous solution of aniline not containing an oxidizer, PANI was formed only at the surface of the matrix. It can be presumed that peroxide centres serve as heterogeneous oxidizers and polymerization sites, and the following growth of PANI macromolecules is continued in adsorption layers of the carrier surface. Uniformity of PANI deposition at the surface of the porous substrate was confirmed by mercury porosimetry and the chemical (alkali) stability test. The measured thickness of the PANI coating was about 8.5 nm.

The separation properties of the materials prepared by means of developed ozone-initiation technique were demonstrated on the examples of nucleic acids isolation from different sources. Modern methodology of separation of biological mixtures is based on use of compact spin-cartridges, as contrasted with traditional column chromatography methods.

Developed fluoropolymer-coated adsorbents were tested as effective materials for one-step isolation and purification of dsDNA and RNA from the bacterial lysates of *Agrobacterium tumefaciens C58* and model mixtures containing fragmented dsDNA or ssDNA (synthetic oligonucleotide). The materials coated with copolymers of TFE with other comonomers were also tested for separation of NA depending on their secondary structure, where bacterial RNA or single strained short oligonucleotides were retained to a different extent depending on the chemical structure of the immobilized polymer and the charge of the adsorbent surface. The retention of ssDNA and RNA increases in a row PTFE-HFP - PTFE-AAl - PTFE-AA. At the same time dsDNA yield increases in comparison with the silica modified with PTFE-AA, when adsorbent having surface hydroxylic groups is used.

To estimate the extent of DNA purification provided with developed fluoropolymer coated adsorbents the eluates collected after the isolation of DNA from bacterial lysates of *A. tumifaciens C58* culture were tested by real-time PCR analysis. The DNA yield after using of the adsorbent coated with PTFE-HFP (by irradiation-induced technique) was assumed as 100%. To compare efficiency of PCR fragment amplification the cross point (Cp) determination method was used (i.e. the determination of Cp as the cross-point of the maximum of the second derivative of amplification curve with the base line. The obtained results show that the yield of purified DNA in the collected eluates are in average by 2-fold higher after use of PTFE-HFP-coated adsorbents in comparison with the materials modified with copolymers of TFE with AA or AAl. However, all those types of PTFE-containing adsorbents provide the isolation of purified DNA, which is suitable for direct PCR diagnostics.

For medical point of view, the most important adventure provided by the developed technique is the clinical diagnostics of the human pathogens using PANI-coated adsorbents obtained via ozone-initiation technique. Such a possibility was demonstrated, in particular, on the examples of the human DNA (iC gene) and viral DNA (T4 phage) isolation (and PCR-detection) from whole human blood and the model mixtures containing blood and viral T4 phage particles (viral particles were added to the blood sample before DNA isolation). For this purpose the different protocols of the sample lysis were used. In particular, the most effective protocol included thermal lysis of the blood samples. The protocols based on the use of developed PANI-coated adsorbent (comprising only 8 manipulations) were compared to the standard sample preparation procedures (comprising 20 manipulations) using the commercial kits Probe-GS DNA Isolating Kit (DNA Technology Inc., Russia) and the mini-columns from NextTec GmbH (Germany). The purified DNA were used in PCR. The concentrations of amplified PCR fragments were compared to the concentration of PCR fragment, which was obtained after isolation of DNA using the Probe-GS DNA Isolating Kit. The determined average values of Cp and relative quantity of DNA are presented in the table 3a (for amplification of human iC gene) and in the table 3b (for amplification of viral DNA).

Mode of the sample preparation	An average Cp		StD, %		Concentration, %		StD, %	
	a	b	a	b	a	b	a	b
Initial lysate	-	-	-	-	-	-	-	-
"Sample GS" kit (DNA-Technology, Russia)	27.0	22.4	0.2	0.3	1.0	1.0	3.7	4.9
NextTec spin-column (NextTec GmbH, Germany)	26.3	18.8	0.6	0.4	1.5	11.6	11.3	5.4
PANI-coated adsorbent (based on ozone treated silica)								
Thermal lysis	26.0	18.4	0.0	0.4	2.0	15.3	0.03	4.8
One-step enzyme lysis	26.0	19.0	0.3	0.1	1.9	9.8	5.1	1.0
Two-step enzyme lysis	26.4	19.0	0.3	0.1	1.4	9.9	5.6	1.4

Table 3. Viral *(a)* and human *(b)* DNA amplification in PCR.

Indeed, the use of prepared PANI-coated adsorbent results in noticeable decrease of an average Cp value in PCR detection of both the human DNA and the viral DNA in the blood samples, in particular, providing by 8-fold higher yield of the human iC gene and by 10-fold higher yield of viral DNA.

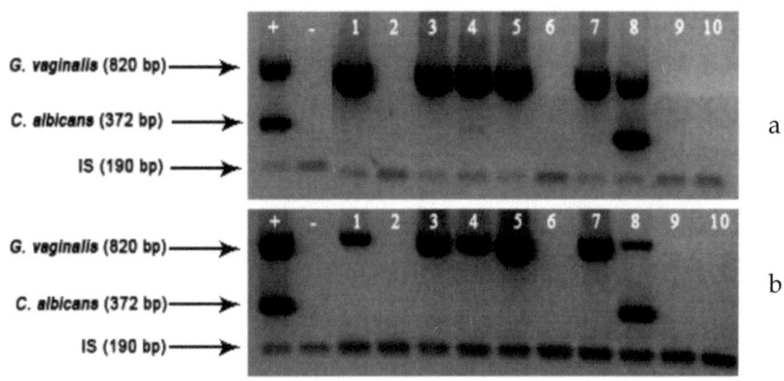

Fig. 8. PCR detection of *Gardnerella vaginalis* and *Candida albicans* in the randomly selected urinogenitalic smears from 10 patients. The results of the electrophoresis in 2% agarose gel. Isolation of DNA was carried out using the standard procedure developed by Diatom™DNA Prep 100 Kit (IsoGen Inc., Russia) (a) and the cartridges packed with the developed PANI-coated adsorbent (b). The arrows show the location of PCR fragments relating to *G. vaginalis* (820 bp), *C. albicans* (372 bp) and internal standard (190 bp).

Another example of diagnostic use of the developed composites is given by the comparison of effectiveness of the sample preparation for practical clinical PCR diagnostics when using the developed PANI-coated adsorbents (protocol I) and the standard Diatom™DNA Prep 100 Kit (IsoGen Inc., Russia) (protocol II). Protocol II includes effective enzymatic lysis of the sample, but it is relatively more labour- and time-consuming procedure as compared to the protocol I. Urinogenital smears were taken from 10 randomly selected patients and treated in accordance with the mentioned protocols I and II. Simultaneous detection of two groups of microorganisms including *Gardnerella vaginalis* and *Candida albicans*, and *Chlamydia trachomatis, Ureaplasma urealyticum, Mycoplasma hominis,* and *Mycoplasma genitalium* was carried out. The collected eluates were assessed by electrophoresis in 2% agarose gel. Figures 8 and 9 show the obtained results. Although the yield of DNA fragments after use of the protocol I was somewhat lower comparably with use of the protocol II (due to dilution of the sample by some 6 times in the course of the sample preparation), the results of PCR tests were identical. Moreother, the purity of isolated DNA was higher in the first case (it was confirmed by the more effective amplification of the internal standard).

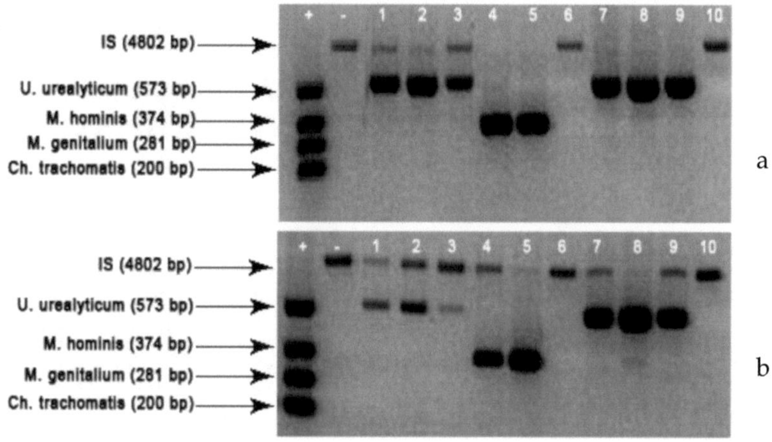

Fig. 9. PCR detection of *Chlamydia trachomatis, Ureaplasma urealyticum, Mycoplasma hominis,* and *Mycoplasma genitalium* in the randomly selected urinogenitalic smears from 10 patients (using the same samples as in the Figure 8). The results of the electrophoresis in 2% agarose gel. Isolation of DNA was carried out using the standard procedure developed by Diatom™DNA Prep 100 Kit (IsoGen, Inc, Russia) (row a) and the cartridges packed with the developed PANI-coated adsorbent (row b). The arrows show the location of PCR fragments relating to internal standard (4802 bp), *U. urealyticum* (573 bp), *M. hominis* (374 bp), *M. genitalium* (281 bp), and *Ch. trachomatis* (200 bp).

Thus, the developed facile and scalable ozone-induced technique of manufacture of biocompatible composite adsorbents is a perspective alternative as compared to known irradiation post-polymerization and oxidative polymerization techniques. Silica carriers treated with ozone can be used as effective heterogeneous initiators for polymerization of different monomers, which can be polymerized by radical or oxidative mechanisms.

7. Summary

Investigation of the functional properties of the adsorbents based on activated silica pursued several goals. Firstly, it was the comparison of the properties of the materials obtained by irradiation, oxidative and ozone-induced techniques. Secondly, it was a demonstration of the possibility to introduce the additional functional groups into polymer layer under relatively mild conditions. The last advantage enables to control the sorption properties of the adsorbents as well as their hydrophilicity. Thirdly, it was the demonstration of the specific properties provided by immobilized PANI phase, such as retention of PCR inhibitors from the blood containing samples and effective isolation of pathogenic DNA from the clinical samples.

The table 4 gives a comparison of the main properties of the described fluoropolymer- and PANI-coated materials. Analysis of the represented data allows to conclusion that fluoropolymers as well as polyanilines can be considered as multipurpose biocampatible polymer modifiers for producing the sorption materials for bioseparation and bioanalysis.

Purpose	**Material (the polymer modifier)**				
	Perfluoropolymers (PTFE and etc)	*Partially fluorinated polymers*	*PANI*	*Aniline-containing polymers (polycomplexes and copolymers)*	*Combination of fluoropolymer with PANI*
Main approaches					
One-step separation of proteins and nucleic acids	+	+	+	+	+
One-step isolation of nucleic acids from the complicated mixtures	+	+	+	+	+
Specific approaches					
Separation of the proteins	+	-	+	+	+
Immobilization of bioligands	-	+	-	+	+
Bioanalytical applications	-	-	+	+	+
One-step separation of ss- and ds- nucleic acids	-	-	-	+	+

Table 4. Fields of application of fluoropolymer- and PANI-coated adsorbents.

The use of fluoropolymers and polyanilines as modifiers of the composite adsorbents provides their effectiveness in such main applications as separation of different classes of biopolymers (like nucleic acids and proteins) by one-step procedure. Moreover, the further complication of the isolation procedure and/or structure of the polymer modifier results in realization of more specific approaches such as separation of the mixtures containing the biopolymers of the same class (e.g. proteins and peptides, ssDNA and dsDNA, or DNA and RNA) as well as their use as supports in bioanalytical application. In last cases the polymer coating immobilized on the carrier surface acts as a whole supramolecular structure which reversibly changes its properties depending on the conditions of environment.

Thus, the review of such "interdisciplinary" branch of the macromolecular chemistry like a development of multifunctional and multipurpose composite biocompatible materials illustrates the tendency to move from "chemical design" to "adaptive chemistry" , like it has been analogously formulated by Nobel laureate Jean-Marie Lehn in his works devoted to self-organizing structures in biomaterial science.

8. Acknowledgements

The work in part was financially supported by the EU Sixth Framework Program (*"Novel and Improved Nanomaterials, Chemistries and Apparatus for Nano-Biotechnology"* NACBO, EU ref. NMP4-CT-500804, and *"Development of new and cost effective methods for non-invasive diagnosis of human pathogens"* DIAGNOSIS, EU ref. LSHB-CT-037212).

9. References

Berensden K.A., Pikaart K. A., de Galan L. & Olieman C. (1986). (Heptafluorodecyl)dimethylsilyl bonded phase for reversed-phase liquid chromatography. *Analitical Chemistry.* Vol. 52. p.1990 - 1993

Hjerten S. & Hellman U. (1978) Chromatographic desalting, deproteinization and concentration of nucleic acids on columns of polytetrafluoroethylene. *Journal of Chromatography A* Vol. 159, p. 47-55q

Ivanov A. E, Saburov V. V. & Zubov V. P. (1992) Polymer-Coated Adsorbents for the Separation of Biopolymers and Particles . *Advances in Polymer Science* Vol. 104.p. 136-175

Kapustin D.V., Saburov V.V., Zavada L.L., Evstratov A.V., Barsamyan G.B., Zubov V.P. Composite fluorine-containig adsorbents for separation and purification of biopolymers. *//Russian Journal of Bioorganic chemistry - 1998 - 24. p. 770 - 777*

Kapustin D.V. Yagudaeva E.Yu., Zavada L.L., Zhigis L.S., Zubov V.P., Yaroshevskaya E.M., Plobner L., Leiser R-M. & Brem G. A (2003) Composite Polyaniline-Containing Silica Sorbent for DNA Isolation. *Russian Journal of Bioorganic chemistry* Vol. 29, No. 3. p.310 - 315.

Kapustin D. V., Yagudaeva E. Yu. & Zubov V. P. (2006) New Polymer-Coated Materials for One-Step Separation of Nucleic Acids. In: *Frontiers in DNA Research.* C.R. Wood (ed) p.113-136, Nova Science Publishers, Inc. ISBN 1-59454-925-7, New York

Kapustin D.V., Prostyakova A.I., Ryazantsev D.Yu. & Zubov V.P. (2011) Novel composite matrices modified with nanolayers of polymers as perspective materials for separation of biomolecules and bioanalysis. *Nanomedicine.* Vol. 6, No. 2, p. 241-255.

Robin J.J. (2004) Overview of the Use of Ozone in the Synthesis of New Polymers and the Modification of Polymers. *Advances in Polymer Science*, Vol. 167, p. 235-245.

Rother D., Sen T., East D. & Bruce I.J. (2011) Silicon, silica and its surface patterning/activation with alkoxy- and amino-silanes for nanomedical applications. *Nanomedicine.* Vol. 6, No. 2, p. 281-300

Saburov V.V., Muidinov M.R., Guryanov S.A., Kataev A.D., Turkin S.I. & Zubov V.P. (1991) Perfluoropolymer-containing silicon sorbents and their application in reverse-phase chromatography of biologically-active substance. *Russian Journal of Physical Chemistry A.* Vol. 65, No.10, p. 2692-2698

Snyder L.R., Kirkland J.J. & Dolan J.W. (2010) *Introduction to Modern Liquid Chromatography (3rd Edition).* John Wiley & Sons, Hoboken. ISBN: 0470167548 NJ, USA

Saburov V.V., Zubov V.P., Turkin S.I., Kisilev E.M., Dmitriev K.N. & Tsar'kova M.S. (1988) Methods of preparing modified macroporous silica for chromatography of biopolymers. *Patent RU №2080905*

Sapurina I. & Stejskal J. Review. (2008) The mechanism of the oxidative polymerization of aniline and the formation of supramolecular polyaniline structures. *Polymer Inernational.* Vol. 57. p. 1295–1325

Stejskal J. Polyaniline. Preparation of a conducting polymer (IUPAC Technical Report) // *Pure Appl. Chem. - 2002 - 74 - 5. p. 857–867*

Vogelstein B. & Gillespie D. (1979) Preparative and analytical purification of DNA from agarose. *Proc. Natl Acad. Sci. USA.* Vol. 76. No. 2. p. 615–619.

William D.W. & Kabra P.M. (1990) Extended life for blood serum analysis columns using dual zone chromatographic materials. *Anal. Chem.* Vol. 62. p.807-810

Xindu G. & Carr P.W. (1983) Use of fluorinated bonded phases in reversed-phase high-performance liquid chromatography of proteins. *J. Chromatogr.* Vol. 269. p.96−102

Yagudaeva E. Yu., Bukina Ya. A.,Prostyakova A. I.,Zubov V. P.,Tverskoy V. A. & Kapustin, D. V. (2009) Oxidative polymerization of aniline on the surface of silica in the presence of poly(sulfonic acids) as a method of preparing efficient biosorbents. *Polymer science series A.* Vol. 51. No. 6. p. 675-682 ISSN: 0965-545X

Yagudaeva E. Yu., Muidinov M. R.,Kapustin D. V. & Zubov, V. P. (2007) Oxidative polymerization of aniline on the surface of insoluble solid poly(sulfo acids) as a method for the preparation of efficient biosorbents. *Russian chemical bulletin.* Vol. 56. No. 6. p. 1166-1173. ISSN: 1066-5285

Zubov V. P., Kapustin D. V., Generalova A. N., Yagudaeva E. Yu., Vikhrov A. A., Sizova S. V. & Muidinov M. R. (2007) Modification of Solids with Polymer Nanolayers as a Process for Manufacture of Novel Biomaterials. *Polymer Science, Ser. A,* Vol. 49, No. 12, pp. 1247–1264. ISSN: 0965-545X

Vaczine-Shlosser G., Ribbing C., Bachman P.K., Zubov V.P. & Kapustin D.V. (2011) Surface coating for laser desorbtion ionization mass spectrometry of molecules. *Patent WO 2011004308 (A1)*

5

Polymer Nanocomposite Materials Based on Carbon Nanotubes

Adam J. Proud, Rabin Bissessur and Douglas C. Dahn
University of Prince Edward Island, Canada

1. Introduction

In the past two decades, mobile devices have decreased significantly in size and yet their capabilities and storage capacities continue to grow dramatically. Not unexpectedly, the energy demands of these devices are rather substantial, which has led to a huge increase in the level of research into batteries. In recent years, millions of dollars of research funding have been directed towards the development of more efficient battery systems, with a large focus on lithium ion batteries. These batteries are among the most popular for devices with high energy demands due to their high energy capacities. However, despite the impressive performance of lithium ion batteries to date, there is still significant room for improvement.

Lithium ion batteries have progressed significantly since they were first developed in the early 1970s. These early systems consisted of lithium metal anodes combined with titanium disulfide cathodes; however, they demonstrated limited cell potentials and these chalcogenidic cathodes were soon replaced by layered oxide systems (Whittingham, 1976). Many such layered oxides were studied by the Goodenough group (Thackeray et al., 1983; Mizushima et al., 1980) with great success which led to the commercialization of these batteries in the early 1990s. These batteries were marketed by the electronics giant, the Sony Corporation, and they consisted of a lithium cobalt dioxide, $LiCoO_2$, cathode and a graphitic anode (Nazri & Pistoia, 2004). Lithium metal anodes were discarded earlier in favour of safer systems such as graphite due to the dangers associated with recharging.

In the years since the commercialization of lithium ion batteries, there have been many modifications to the three components of the cell: the anode, the cathode and the electrolyte. However, due to the low cost and relative efficiency of graphite as the anode system, very little work has been done in this area. Nonetheless, recently, silicon and germanium nanowires have demonstrated great potential as possible anode materials (Chan et al., 2008a; Chan et al., 2008b). It has long been known that silicon has a greater capacity for lithium ions than does graphite; however, previous attempts employing silicon particles and thin films have demonstrated significant degradation of the materials upon cycling. With these novel nanowires, this degradation is not observed. A small increase in the diameter of the nanowires occurs upon lithium intercalation, but this expansion is reversible upon the removal of the lithium ions. Research is still continuing into these materials due to their enhanced lithium storage capacities; however, these systems are plagued by their considerable expense when compared to graphite. Therefore,

much of the research into these battery systems has been directed towards the cathode and electrolyte systems.
The cathodes employed in modern lithium ion batteries are far more variable than the anodes. Although layered oxides are among the most common cathode systems, spinels (e.g. manganese spinel) and phospho-olivines (e.g. lithium iron phosphate) have been developed and are currently employed in commercial batteries (Nazri & Pistoia, 2004). These latter two systems are better suited to high current applications where safety is of high concern such as in power drills and electric vehicles.
Electrolytes in lithium ion batteries are traditionally composed of a lithium salt such as $LiClO_4$ or $LiPF_6$ suspended in an organic solvent which is usually a carbonate derivative (e.g. propylene carbonate). These electrolyte systems are by far the most common but there are significant disadvantages to using liquid electrolytes including safety concerns and inflexibility in terms of battery cell shapes. This has led to the development of lithium ion polymer batteries which consist of electrolytes where the lithium salt is suspended in a polymeric matrix. The use of these polymer electrolytes has led to the development of batteries which do not require a metal casing (Nazri & Pistoia, 2004). This in turn, enhances the energy density of the battery, allows for the construction of batteries of varying shape and flexibility, and improves the safety of the cells (Fergus, 2010). This criterion is are well-suited to the ever increasing portability of mobile devices that we are faced with today.

1.1 Solid polymer electrolytes

Traditionally, the polymers used as solid polymer electrolytes in lithium ion batteries were polyethyleneoxide (PEO) and polyacrylonitrile (PAN). However, in the past 20-25 years, numerous compounds have been examined in terms of their properties suitable to polymer electrolytes and it has been found that phosphazenes, such as poly[bis(methoxy-ethoxyethoxy)phosphazene] (MEEP), PEO derivatives, and polysiloxanes have all demonstrated significant ionic conductivities (Allcock et al., 1986; Bouridah et al., 1985; Xu et al., 2001). The properties of a material which typically relate well to potentially high conductivity values include fluidity, amorphousness, and the presence of electron rich centres throughout the material.
Based on molecular mechanics simulations, it is widely believed that lithium ions migrate through an electrolyte by diffusion enhanced by a process known as segmental motion (Meyer, 1998). In this process, lithium ions are thought to coordinate loosely to electron rich atoms or regions along a polymer chain and propagation along the chain occurs by coordinating to electron rich regions further down that chain or a neighbouring chain. Thus, the conductivity of the polymer is significantly affected by the mobility and fluidity of the polymer. Therefore, it is beneficial to explore the use of more fluid (or semi-solid) polymers in order to have a more highly conductive electrolyte. This flexibility correlates well with the glass transition temperatures (T_g) of the polymers. This transition defines the point where a polymer goes from a rigid, glass-like material to a more rubbery, flexible material. Therefore, polymers with low glass transition temperatures are highly desirable due to their more flexible, fluid nature. However, herein lies the problem. The benefits provided by solid polymer electrolytes, including improved safety and adaptable cell shapes, are present due to the solid nature of these materials (Fergus, 2010). Although these problems with mechanical stability are not nearly as severe in comparison to batteries containing liquid electrolytes, they are still significant for those with semi-solid polymer electrolytes. Ionic conductivity in solid polymer electrolytes is not as high as in liquid electrolytes and thus,

without advantages in terms of safety and processability, this form of electrolyte will never be used successfully in commercial batteries.

Improving the mechanical properties of the more highly conductive polymers has been a focal point of research in recent years. The goal is to improve the mechanical stability of the polymer without significantly hindering the conductive properties. Early approaches focussed on chemical cross-linking of the materials. Such methods have been applied to polymers such as MEEP; however, in addition to being far more rigid materials, they are also much less soluble which leads to processing problems (Tonge et al., 1989). In terms of the rigidity, the cross-linking significantly diminishes the mobility of the polymer chains and consequently, the conductivities have suffered greatly.

More recent approaches to improve mechanical stabilities are based on the addition of filler materials to the polymer matrices (Paul & Robeson, 2008). In some cases, especially in crystalline polymers, filler materials can actually increase the conductivity of the material due to the disruption of the crystalline regions resulting in a more amorphous material (Croce et al., 1998). It has been demonstrated that amorphous materials are far more conductive, which is likely due to the increased mobility of the polymer in a less rigid and ordered system. However, in most materials which are originally amorphous, the increased mechanical stability comes at the cost of reduced ionic conductivity (Tonge et al., 1989). The goal is to obtain adequate mechanical stability without sacrificing conductivity to an extent where the polymer is no longer viable as an electrolyte material. These requirements have led to a huge growth in research into nanocomposite materials for use as solid polymer electrolytes.

1.2 Exfoliated nanocomposite materials

With regards to lithium ion batteries, there are two common types of nanocomposite materials that have been explored: intercalated nanocomposites and exfoliated nano-composites (Figure 1). The latter class of materials are of particular interest due to the significant enhancement of the properties of the polymer materials upon inclusion of only a small quantity of filler material. Additionally, the property enhancement of the resulting material is often greater in these exfoliated compounds compared to their intercalated counterparts.

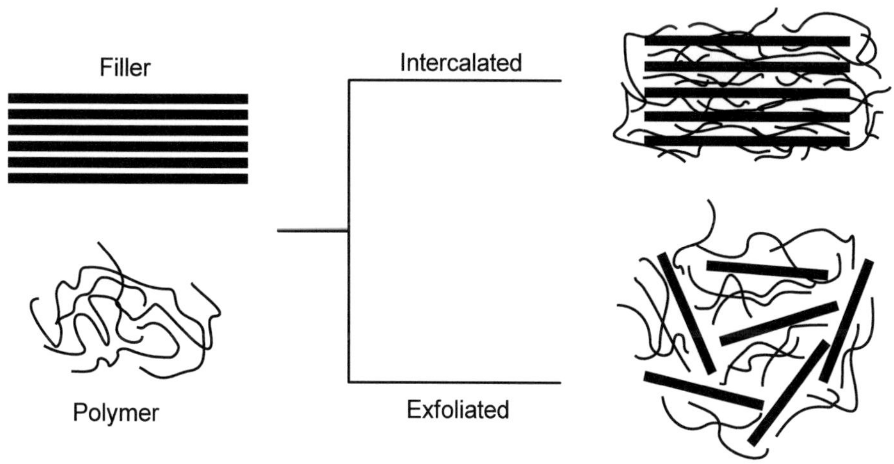

Fig. 1. Schematic representation of the two common forms of nanocomposite materials.

Exfoliated nanocomposites represent a relatively new class of materials (Paul & Robeson, 2008). One of the earliest reports of this class of compound was published by Toyota Central Research and Development Laboratories in 1993 (Kojima et al., 1993). The reported compound was a polymer-clay hybrid material consisting of an exfoliated aluminosilicate structure in the matrix of nylon-6. In comparison to nylon-6, these novel nanocomposites demonstrated improved heat distortion temperatures and elastic moduli.

Since this initial report, countless other exfoliated nanocomposite materials have been synthesized and characterized. These new materials often exhibit improved mechanical, thermal, or conductive properties dependent on the choice of material used as filler and the desired applications. With respect to polymeric materials, these filler materials have ranged from anything from other polymers, to ceramic powders, and more recently to the materials used in this research, carbon nanotubes. While one may not consider the dispersion of nanotubes within a polymer matrix to be a true exfoliated nanocomposite material, it is analogous in many ways which has led to the use of this terminology in this text.

Since their discovery in 1991, carbon nanotubes have been used extensively in countless areas of research (Iijima, 1991; Moniruzzaman & Winey, 2008). These materials (Figure 2) can be either single-walled (SWNTs) consisting of a lone nanotube or multi-walled (MWNTs) which are analogous to a Russian doll set with nanotubes of smaller diameters fitting neatly inside the larger ones (Iijima, 2002). The intertubular spacing is similar to the distance between the sheets of graphite in its 2-D layered structure with a separation of 3.3-3.6 Å. The key features responsible for the remarkable properties of these materials are the high aspect ratios and the highly conjugated network created by the series of fused benzene rings. CNTs have been noted to be stable in air up to temperatures of 700°C and in N_2, the networks are stable beyond 2000°C (Ahir, 2007).

Despite the noteworthy thermal properties, possibly the most attractive properties of CNTs are those pertaining to mechanical strength and stiffness. These properties have led to the belief that nanotubes could potentially be used in fibres for the hypothetical space elevator and are a major reason for the frequent use of these materials in nanocomposites. The Young's modulus, 1.4 TPa, and tensile strength, above 100 GPa, of CNTs are by far the greatest of any known material (Khare & Bose, 2005). For a good comparison, the Young's modulus and tensile strength of high-grade steel are only 200 GPa and 1-2 GPa, respectively. These characteristics could be highly beneficial for improving the mechanical properties of polymer nanocomposites. One such example was reported by Dalton et al. who

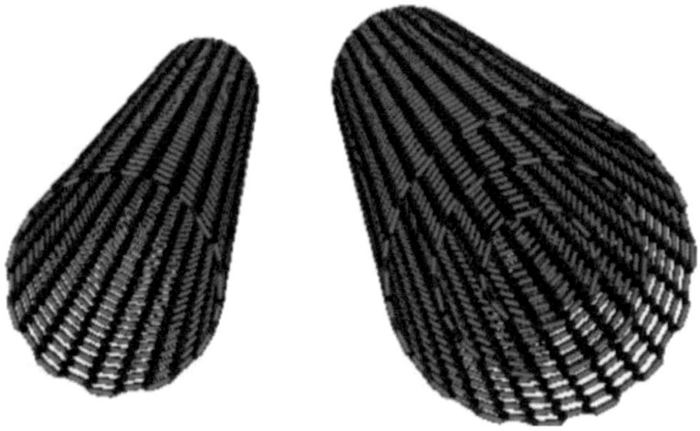

Fig. 2. Molecular model representation of SWNTs (left) and MWNTs (right).

synthesized nanocomposites with 60 wt% SWNTs in poly(vinyl alcohol) which displayed tensile strengths of 1.8 GPa (Dalton et al., 2003). However, ratios of nanotubes used in nanocomposites are often much lower than this in order to retain some desired properties of the polymer. Gao et al. recently synthesized SWNT/nylon-6 nanocomposites with only 5 wt% of the nanotubes (Gao et al., 2005). In this case, increases in Young's Modulus and tensile strength of 153% and 103%, respectively, were observed. The magnitude of the effect of the nanotubes is often related to dispersibility, which leads back to the need for evenly dispersed samples. This has led to varying levels of observed effects on mechanical stability in different studies, but a commonality among all studies has been the improvement of the mechanical properties of the polymer.

As noted, the remarkable thermal, mechanical, and conductive properties of carbon nanotubes make them ideal candidates to be used as filler materials in nanocomposite materials. However, due to the poor solubility of nanotubes, they are difficult to deal with and thus several different methods have been developed and utilized in the preparation of these types of nanocomposites (Ma et al., 2010). Depending on the desired application for the resulting material, the synthetic approach could differ significantly. For example, in a material where one would like to enhance the electrical properties of the material, mechanical processes such as ultrasonication would be better suited than functionalization of the nanotubes. By functionalizing the surface of the nanotubes, the conjugated network of π electrons would be disrupted and the electrical conductivity of the CNTs would decrease significantly. However, for the purposes of this research, this is a desired effect and thus, functionalization was a primary method used for improving the processibility of the nanotubes. By functionalizing the nanotubes, one can hinder the electrical properties of the material while maintaining some of the mechanical and thermal properties.

In addition to functionalization, ultrasonication was also used in the preparation of the materials. As previously mentioned, ultrasonication is ideal for systems where electrical properties are desired, as the conjugated system remains intact, and it has been demonstrated that this mechanical procedure provides sufficient energy to separate the bundles of CNTs to form homogeneous solutions. Care must be taken not to use excessive sonication, as this can significantly decrease the length of the tubes and have a detrimental effect on their properties (Ahir, 2007). Combining the effects from both functionalization and ultrasonication would improve the dispersion of the nanotubes, resulting in a more efficient synthesis since homogeneous dispersion of the filler material is essential to obtain optimal property enhancement in the resulting nanocomposite materials.

2. Synthetic methodology

2.1 Acid oxidation of MWNTs - method 1

The oxidation of carbon nanotubes was performed using commercially available short multi-walled nanotubes (Helix Material Solutions, diameter: 10-30 nm, length: 1-2 μm). The procedure was modified from the method developed by Rinzler et al., for single-walled carbon nanotubes (Rinzler et al., 1998). A sample of MWNTs (1.0 g) was weighed into a 100 mL round bottom flask. Nitric acid (60 mL, 60% by volume) was added to the flask with stirring to disperse the solid. The allotted reaction time was 3 hours. Following the refluxing process, the solid was isolated under reduced pressure and washed with deionized water to remove any traces of HNO_3. Trace moisture was then removed via lyophilisation. This whole process was then repeated once more to improve the solubility of the nanotubes in polar solvents.

2.2 Hummers' oxidation of MWNTs - method 2

The multi-walled nanotubes obtained from Helix Material Solutions were also oxidized via a procedure adapted from Hummers' Method for the synthesis of graphite oxide (Hummers & Offeman, 1958; Liu et al., 2000). A sample of MWNTs (1.0 g) was measured into a 600 mL Erlenmeyer flask to which sulphuric acid (23 mL, 98%) was subsequently added. This solution was stirred while being cooled to 0°C. The oxidizing agent, $KMnO_4$ (3.00 g, 19.0 mmol) was then added over a period of 10 minutes in small increments to prevent the temperature of the reaction from exceeding 20°C. Following the addition of the $KMnO_4$, the reaction flask was removed from the ice bath and allowed to warm to room temperature. Deionized water (23 mL) was then added gradually while ensuring the solution temperature remained below 98°C. The reaction mixture was then left to stir and cool to room temperature over a period of 15 minutes. Finally, deionized water (140 mL) was added in one quick addition followed by the rapid addition of H_2O_2 (10 mL). Upon the addition of H_2O_2, a black precipitate began to form.

The solution was left to stand overnight to allow the precipitate to settle. Once the precipitate had settled, excess solvent was decanted and the remaining solution was swirled to form a slurry. This slurry was then transferred to dialysis tubes to remove any impurities. The dialysis was deemed complete when no precipitate formed upon the addition of $BaCl_2$. Following dialysis, the oxidized nanotubes were isolated by lyophilisation.

2.3 Synthesis of poly[oligo(ethylene glycol)oxalate] (POEGO)

Benzene (99%) and oxalic acid dihydrate were purchased from Sigma-Aldrich while poly(ethylene glycol) (M_n: 380-420) was obtained from EMD Chemical. All chemicals were used without modification. Poly[oligo(ethylene glycol)oxalate] was prepared according to the procedure developed by Xu, et al. (Xu et al., 2001). PEG 400 (2.0 g, 5.0 mmol) was dissolved in benzene (50 mL) in a 250 mL round bottom flask. One molar equivalent of oxalic acid dihydrate (0.63 g, 5.0 mmol) was then added to the solution. The reaction was refluxed with magnetic stirring for a period of 3 days at which point the benzene was removed through rotary evaporation. The remaining product was then heated in a vacuum oven at 120°C for 2 days resulting in a highly viscous, pale-yellow polymer.

2.4 Synthesis of POEGO/LiOTf (LiPOEGO) complexes

POEGO, synthesized as previously described, was complexed with lithium triflate ($LiSO_3CF_3$, LiOTf) in various ratios. The polymer was dried extensively at which point a corresponding amount of LiOTf was added in the inert atmosphere of a nitrogen glove box. The mixture was then transferred to a vacuum oven where it was heated to 90°C overnight. The product was mixed lightly and heated for an additional two hours at 90°C in a vacuum oven.

2.5 Synthesis of MWNT/POEGO nanocomposites - method A

The exfoliated nanocomposites were prepared in both deionized water and acetone. POEGO (0.500 g) was dissolved in the desired solvent (15 mL) and stirred until dissolution was complete. Separately, a sample of oxidized nanotubes (0.0556 g) was dispersed in the desired solvent with the aid of ultrasonication for a period of 2 minutes. After sonication, the two solutions were mixed and the resulting solution was stirred for 24 hours before isolating the final product via lyophilisation (H_2O) or by removing the solvent under reduced pressure (acetone). This produced a polymer nanocomposite material with 10 wt%

MWNTs. This procedure was repeated to produce exfoliated nanocomposites containing 1 wt%, 5 wt%, 15 wt%, and 20 wt% MWNTs. This synthetic procedure was also employed in the preparation of MWNT/LiPOEGO nanocomposite materials.

2.6 Synthesis of MWNT/POEGO nanocomposites - method B

An *in situ*polymerization/nanocomposite preparation was also performed. A sample of oxidized nanotubes (0.111 g) was dispersed in 40 mL of benzene with 20 minutes of ultrasonication. This solution was then added to equimolar amounts of PEG 400 (1.00 g, 2.50 mmol) and oxalic acid dihydrate (0.315 g, 2.50 mmol). The polymerization procedure described in section 2.3 was then applied for the remainder of the synthesis. This yielded a polymer nanocomposite containing 10 wt% nanotubes relative to the quantity of PEG 400 that was used since mass ratios with respect to the polymer could not be determined accurately. This synthetic procedure was repeated to produce exfoliated nanocomposites containing 1 wt%, 5 wt%, 15 wt%, and 20 wt% MWNTs and was also used in the preparation of analogous lithiated samples.

3. Instrumentation

Powder X-ray diffraction (XRD) measurements were performed using a Bruker AXS D8 Advance diffractometer. The instrument was equipped with a graphite monochromator, variable divergence and antiscattering slits, and a scintillation detector. Cu(Kα) radiation (λ=1.542 Å) was used for the measurements. The samples were run in air under ambient conditions from 2-60° (2θ). Sample preparation involved the adhesion of the solid onto double-sided tape adhered to a glass substrate.

Thermogravimetric analyses (TGA) were performed using a TA Instruments TGA Q500 instrument. These analyses were performed under both dry air and dry nitrogen purges. For analyses performed in nitrogen, the furnace was allowed to purge with nitrogen for 20 minutes prior to the commencement of the runs at a rate of 60.00 mL/min. The analyses of these materials were performed at high resolution with a dynamic heating rate. This method of heating utilizes a rate of 10.0°C/min while the weight of the material is roughly constant and the heating rate decreases significantly as the material begins to decompose. The resolution number for these scans was 4.00 while the sensitivity value was 1.00.

Differential Scanning Calorimetry (DSC) analyses of the samples were carried out using a TA Instruments DSC Q100 instrument. Analyses were performed in aluminum pans under a dry nitrogen purge (50.00 mL/min). These analyses were performed using a DSC Heat/Cool/Heat Cycle which observes changes in heat flow versus increasing and decreasing temperatures. Heating rates of 10.00°C/min and cooling rates of 5.00°C/min were used for all samples.

A Bruker Equinox 55 FT-IR instrument with a resolution of 0.5 cm^{-1} was used in the measurement of FTIR spectra for the materials. All samples were run as pressed KBr pellets. A set of 64 background and sample scans were used in determining the spectra. The measurements took place in a chamber which was purged with nitrogen gas at a rate of 8 L/min to eliminate the presence of any CO_2 peaks in the spectra. The chamber was allowed to purge for 20 minutes prior to the commencement of each run.

Nuclear magnetic resonance (NMR) spectroscopy was employed to characterize the polymer samples. A Bruker 300 MHz NMR was used to perform the ^{13}C-NMR and ^{1}H-NMR scans. The measurement of the ^{13}C-NMR spectra involved 1024 scans while the ^{1}H-NMR experiments involved a set of 32 scans. $CDCl_3$ was used as the solvent for all samples.

A Jandel Multi-Height probe was used to perform electrical conductivity measurements. The spacing between the individual probes was 0.1 cm. All samples for measurements were pressed into pellets using a hydraulic press. A Keithley 2000 Multimeter set to 4-probe resistance measurements was then used to measure the resistance of these samples.
Ionic conductivities of the samples were carried out by AC impedance spectroscopy (IS). Due to the thick paste-like nature of these materials, a thin layer was spread evenly on a glass substrate with painted silver electrodes fitted on opposite ends. These samples were dried thoroughly either in a vacuum oven at 120°C for two days (Method 1B) or through lyophilisation for three days (Method 1A and Method 2A) to remove any traces of water which may lead to protonic conductivity. IS was carried out in a vacuum chamber and these samples were held under vacuum for a minimum of 20 hours prior to measurement to remove residual water adsorbed by the samples during their handling after the initial drying process. Throughout the IS measurements, the temperature was controlled using a Lakeshore 321 temperature controller and a Cryodyne 350CP refrigerator. The current flow was directed along the film (i.e. parallel to the substrate). The IS measurements were performed using a Solartron 1250 frequency response analyzer and a home-built accessory circuit for high impedance samples. The frequency range used for most samples was 5 kHz to 0.05 Hz.

4. Results and discussion

4.1 Synthesis and characterization of oxidized nanotubes

Prior to the preparation of the exfoliated nanocomposite materials, the synthesis and complete characterization of the filler material was required. Due to the electrically conductive nature and poor solubility of carbon nanotubes, the materials required processing before they could be incorporated into the polymer matrix for potential use as solid electrolyte materials. One such approach is that of oxidation which was carried out here. Two separate approaches were used in the oxidation of the CNTs. The first approach (Method 1) involved refluxing the nanotubes in concentrated acid solutions as previously described (Section 2.1), while the other approach (Method 2) involved an adaptation of Hummers' Method (Section 2.2) which was designed for the synthesis of graphite oxide.

Wavenumber (cm^{-1})	Designation
3431.6	O-H stretch
1721.4	C=O of ketone
1634.0	C=O stretch of carboxylic acid
1094.7	C-O stretch

Table 1. IR absorptions in Hummers' Oxidized MWNTs

The oxidized nanotubes prepared via Method 1 were characterized by FTIR spectroscopy; however, the spectrum did not provide sufficient evidence for oxidation. This may be expected due to the nature of the oxidation process. It is known that oxidation by refluxing in acidic solutions leads to the opening and shortening of the nanotubes with the oxidative functionalities occurring primarily at the tips of the nanotubes. Thus with minimal oxygen containing functional groups in a longer nanotube, IR spectroscopy was deemed an inefficient method of characterization. However, with regards to the nanotubes oxidized by the Hummers' Method adaptation, IR did prove to be more effective. The IR spectrum indicated several oxygen containing functionalities which are tabulated in Table 1.

Due to the inefficacy of IR measurements for the acid oxidized nanotubes, other methods of characterization were carried out. One such method was thermogravimetric analyses (TGAs) which were performed on the untreated nanotubes, the nanotubes that were oxidized twice by acid, and the Hummers' oxidized nanotubes. These TGAs were carried out in nitrogen as all nanotubes, regardless of the level of oxidation, would completely degrade in an oxygen atmosphere. As can be seen from the traces in the thermogram in Figure 3, the untreated nanotubes displayed very minimal decomposition with a total weight loss of less than 2% which can be attributed to impurities in the nanotubes from the method from which they were synthesized. The acid oxidized nanotubes demonstrated a greater weight loss with overall decomposition accounting for approximately 7% of the material, whereas the Hummers' oxidized nanotubes degraded further still with less than 60% of the material remaining after heating up to 800°C. This greater extent of oxidation is expected from the results previously described from the IR analyses. This greater level of oxidation is advantageous in terms of increased solubility and decreased electrical conductivity; however, the mechanical integrity of these nanotubes would likely be decreased.

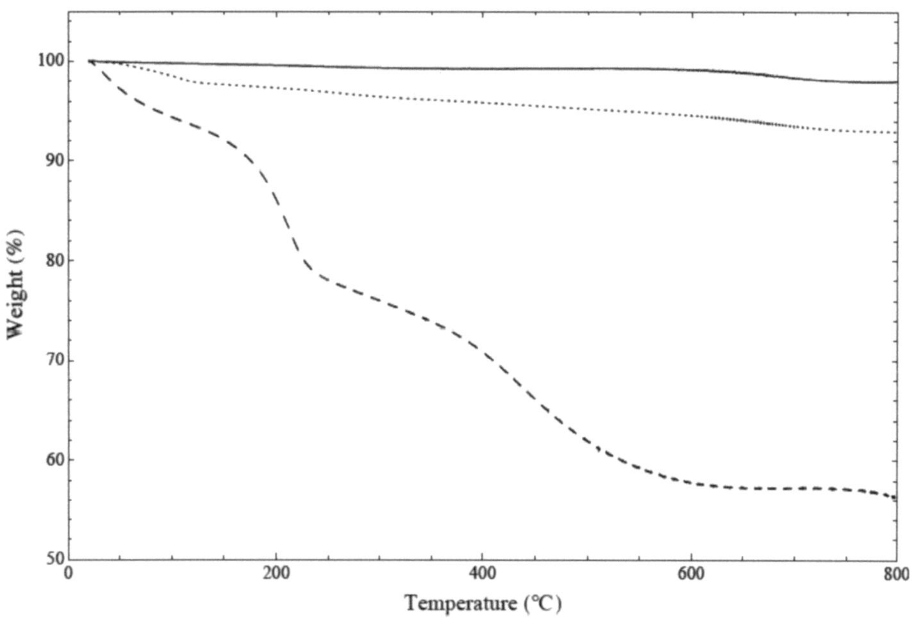

Fig. 3. TGA comparison of untreated MWNTs (solid), acid oxidized MWNTs (dotted), and Hummers' oxidized MWNTs (dashed).

Further structural information regarding these nanotubes was provided by XRD analyses. Again, the untreated, acid oxidized, and Hummers' oxidized nanotubes were characterized for comparative analysis (Figure 4). The untreated nanotubes display one main peak in the diffractogram with a d-spacing value of 3.4 Å. This value likely corresponds to the spacing between the individual nanotubes within a single MWNT. This peak is present in the acid oxidized nanotubes denoting that this oxidative method does not significantly alter the structure of the MWNTs. Considering acid oxidation results in oxidation primarily at the tips of the nanotubes, this is not unexpected. However, with the nanotubes oxidized using the harsher Hummers' method, this peak is still present, but not nearly as prominent as it was in the other two samples. Instead, the diffractogram is indicative of a primarily amorphous material. This likely indicates that the order of the MWNTs was disturbed in the oxidation process. Normally, the nanotubes are consistently separated by this intertubular distance of 3.4 Å; however, it is possible that these harsher oxidizing conditions could have introduced functionalities on the outer walls of the nanotubes which would disturb this regular spacing.

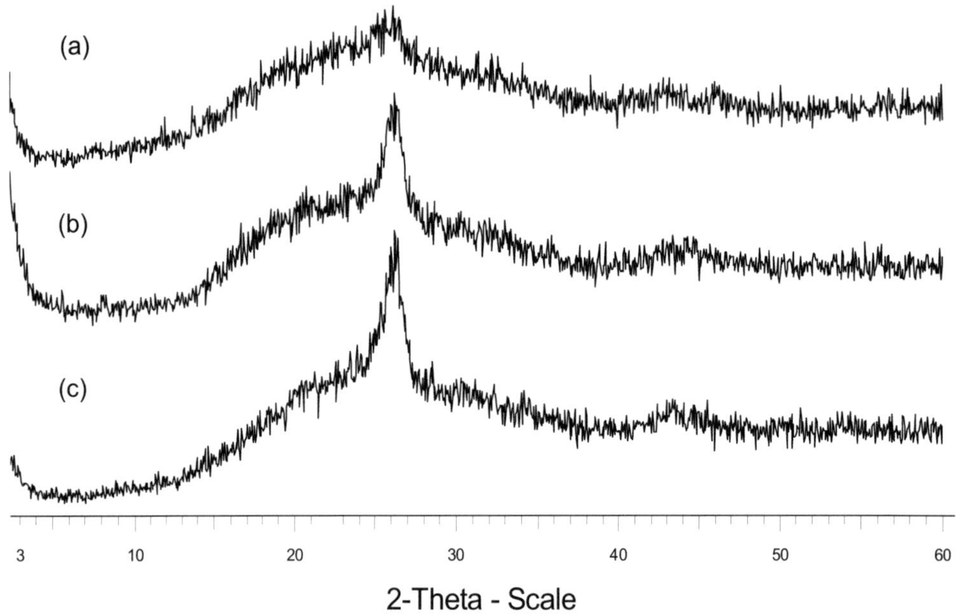

Fig. 4. Diffractogram of (a) Hummers' oxidized MWNTs, (b) Acid oxidized MWNTs, (c) Pure MWNTs.

The final, and potentially most important, method of characterization for these oxidized nanotubes was that of four-probe electrical conductivity measurements. As previously mentioned, due to the electrically conductive nature of the nanotubes, oxidation was required to disrupt the conjugated network of π electrons and inhibit conductivity in order for these nanocomposite materials to potentially be used as solid polymer electrolytes. In obtaining these measurements, pressed pellets of each of the materials were prepared using a hydraulic press and pressures of 6000 psi. Co-linear four-probe electrical conductivity measurements of each sample were then conducted, at room temperature, and the conductivity was determined by

$$\sigma = \ln(2)/\pi R w \tag{1}$$

where σ and R are the electrical conductivity and resistance of the sample, respectively, and w is the thickness of the pellet (Hall, 1967). Using this expression the conductivities of the untreated, acid oxidized, and Hummers' oxidized nanotubes were determined. These values are summarized in Table 2. It should be noted that as the measurements were made on pressed pellets, the presence of grain boundaries would lead to some error in the results. Nonetheless, it provides a good method for comparison to demonstrate the decreased conductivity in the oxidized samples.

Sample	Thickness (cm)	Resistance (Ω)	Conductivity (Scm^{-1})
MWNT pure	0.0500	4.40	1.0
Oxidized MWNT (Method 1)	0.0500	15.3	0.29
Oxidized MWNT (Method 2)	0.0087	1779	0.014

Table 2. Electrical Conductivity of MWNT samples

The nanotubes that were oxidized in acid (Method 1) displayed roughly a 3-fold decrease in conductivity, while those that were oxidized by Hummers' Method (Method 2), which demonstrated significant structural differences as evidenced by previous characterization, displayed a far greater 70-fold decrease. Despite the much greater decrease in conductivity for the oxidized nanotubes prepared by Method 2, many of the results for the nanocomposites are based on those prepared by Method 1, due to the current uncertainty regarding the mechanical and structural properties of those nanotubes that underwent the harsher oxidation method.

4.2 Characterization of poly[oligo(ethylene glycol)oxalate] (POEGO)

To date, the research conducted into the use of POEGO as a potential solid polymer electrolyte has been very limited, despite the fact that room temperature ionic conductivities of the polymer rival those of PEO. For this reason, we have decided to explore the conductive properties of this material upon complexation with lithium triflate (LiOTf), in addition to the effects on these properties, and others, due to the addition of the filler nanotubes.

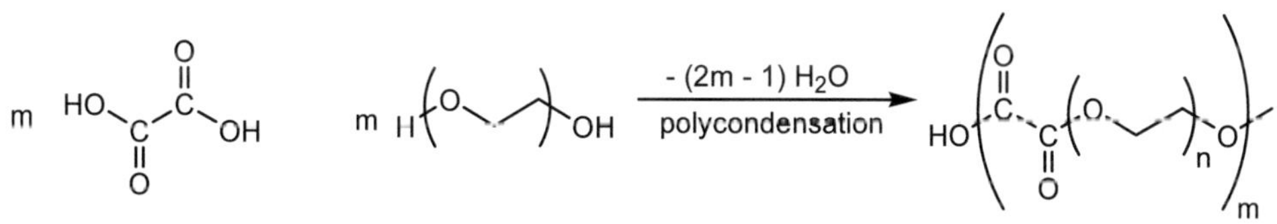

Fig. 5. Reaction scheme for the synthesis of poly[oligo(ethylene glycol)oxalate].

POEGO was synthesized according to the procedure outlined in the experimental section, after which it was characterized both structurally and thermally through a variety of techniques. These include FTIR, NMR, DSC, TGA, and XRD. Based on the reactants used in the preparation of the polymer, in addition to the expected structure (Figure 5), FTIR spectroscopy provides good evidence for the successful formation of the material through the presence of absorption peaks characteristic of hydroxyl groups, carboxyl groups, and esters among others. The characteristic peaks of the polymer are summarized in Table 3. It is known that unreacted PEG often remains following the synthesis of POEGO; however some key features which distinguish it from the starting polymer, PEG, include the carboxyl groups of the carboxylic acid as well as the esteric C=O groups.

^{1}H-NMR and ^{13}C-NMR were also employed in the characterization of the polymer providing further evidence of a successful synthesis. The results of these analyses are tabulated in Table 4. Some peaks that were present in the spectra were indicative of unreacted PEG 400 which is consistent with previous results in the literature (Xu et al. 2001).

Wavenumber (cm^{-1})	Designation
3489.9	OH stretch
2882.6	sp^3 C-H stretch
1765.2	C=O stretch of carboxylic acid
1760.3	C=O stretch of ester
1196.7/1099.6	C-O stretch of ester/ether/alcohol

Table 3. IR absorptions in POEGO

^{1}H-NMR	δ (ppm)	Designation
	4.42	-C(O)-C(O)-O-*CH_2*-
	3.78	-C(O)-C(O)-O-CH_2-*CH_2*-
	3.73	unreacted PEG 400
	3.65	-C(O)-C(O)-O-CH_2-CH_2-(O*CH_2CH_2*)$_{n-2}$-O-
^{13}C-NMR	**δ (ppm)**	**Designation**
	157.97	-*C*(O)-*C*(O) -
	72.90	unreacted PEG 400
	70.90	unreacted PEG 400
	70.62	-C(O)-C(O)-O-CH_2-CH_2-(OCH_2CH_2)$_{n-2}$-O-
	68.75	-C(O)-C(O)-O-CH_2-
	66.31	-C(O)-C(O)-O-CH_2-CH_2-
	62.01	unreacted PEG 400

Table 4. ^{1}H-NMR and ^{13}C-NMR Data for POEGO

The last form of structural analysis used in characterizing the polymer was a powder X-ray diffraction measurement. This was used to determine the crystallinity of the material. From this analysis, POEGO appeared to be an amorphous material which is consistent with its high ionic conductivity. This analysis was essential to verify the amorphous nature of the material considering its precursor, PEG/PEO, is known to be a semicrystalline material in some temperature ranges. As previously noted, amorphous materials are far more conductive than their crystalline counterparts due to the increased mobility of the polymeric chains in a less ordered system (Meyer, 1998).

With the IR, NMR, and XRD data confirming the successful synthesis of POEGO, the thermal analyses of the material provided data concerning the thermal properties of this polymer. Specifically, in terms of the flexibility or fluidity, and in turn, the conductivity of these materials, we are interested in the glass transition temperatures, T_g, of the polymeric materials, and how they are affected upon complexation with lithium salts or the addition of the nanotube filler material. Through DSC analysis, the glass transition temperature for POEGO was determined to be -56.7°C (Figure 6), which compares well to the literature value of -55°C. The minimal difference in these values can be accounted for by slight differences in the degree of polymerization as well as the method utilized to determine the onset of the glass transition temperature. It should be noted that the glass transition temperature was taken at the onset of the transition to be consistent with the literature. Later in the chapter, oftentimes only the midpoint of the T_g will be provided. This low glass transition temperature is ideal for an ionically conductive polymer since the polymeric chains would remain mobile and flexible even at low temperatures. Once the polymer is complexed with LiOTf or the nanocomposite materials with nanotubes are prepared, this glass transition temperature is expected to increase slightly; however, these results will be explored in depth later in this chapter.

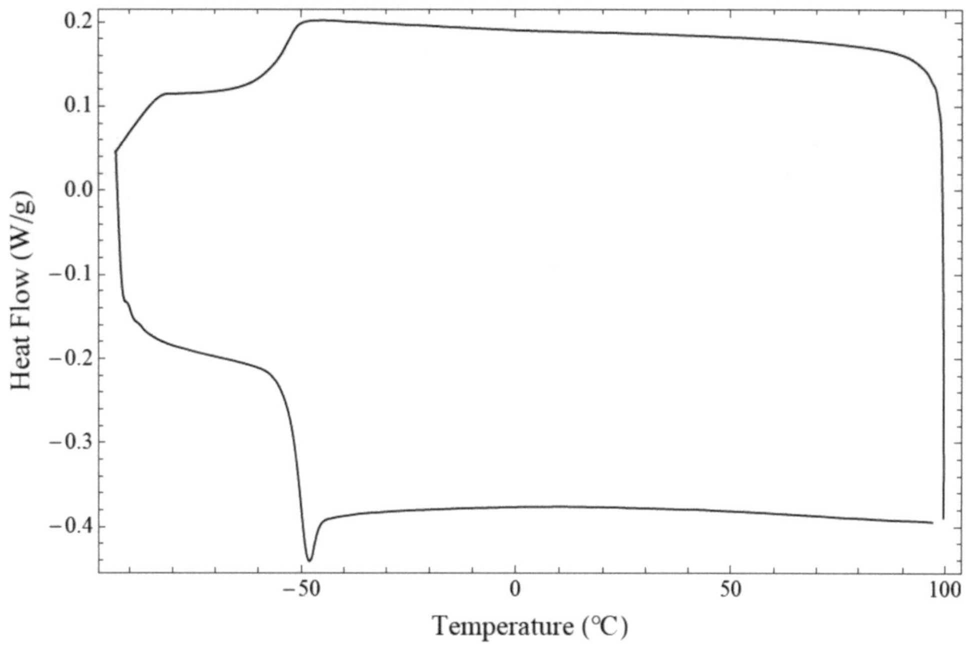

Fig. 6. DSC trace of poly[oligo(ethylene glycol)oxalate].

The final form of analysis used to characterize this polymer was thermogravimetric analysis to determine the thermal stability of the polymer for later comparison. These analyses were performed in both dry air and dry nitrogen atmospheres and it was found that the polymer was stable beyond 170°C in both environments. This level of thermal stability is more than adequate for a component of a battery system; however, despite the fact that this system was geared towards use as a solid polymer electrolyte, it could potentially find uses in other areas as well. Thus it is essential to determine the thermal stability of the material and what potential beneficial effects the filler nanotubes could provide to the resulting nanocomposite. The TGA traces conducted in air and N_2 indicated that the polymer is stable up to 221°C in a nitrogen atmosphere. This represents a 45°C increase over the initial temperature of decomposition, T_d, of POEGO in air. Beyond 400-500°C, the polymer had effectively decomposed entirely under both sets of conditions. These thermal stabilities are more than adequate for a material intended to be used as a solid polymer electrolyte.

4.3 Characterization of LiOTf/POEGO (LiPOEGO)

As previously mentioned, the data available in the literature on the thermal and conductive properties of POEGO is fairly limited. Much of the data concerning the ionic conductivity of the polymer is for higher molecular weights of POEGO which was complexed with lithium salts with larger anions such as lithium bis-(trifluoromethanesulfonyl)imide (LiTFSI); therefore, we chose to synthesize a range of lithiated polymers with different ratios of polymer to salt. From the published data and from knowledge of POEGO salt complexes from previous work done in our group, the ratios of [Li]/[EO] that were studied were 1:8, 1:12, 1:16, and 1:20 (Xu et al., 2001).[1]

[1] [EO] - the etheric unit of POEGO

These polymer salt complexes underwent characterization to determine the effects of the salt on the thermal and structural properties of the polymer. Of particular interest was the effect of the salt on the glass transition temperature of POEGO. One would expect that the T_g would increase upon the addition of salt which would have a detrimental effect on conductivity; however, adequate quantities of the salt would be required in order for conductivity to be sufficient. Thus these two competing effects were analyzed in the characterization of the polymer to determine the optimal ratio.

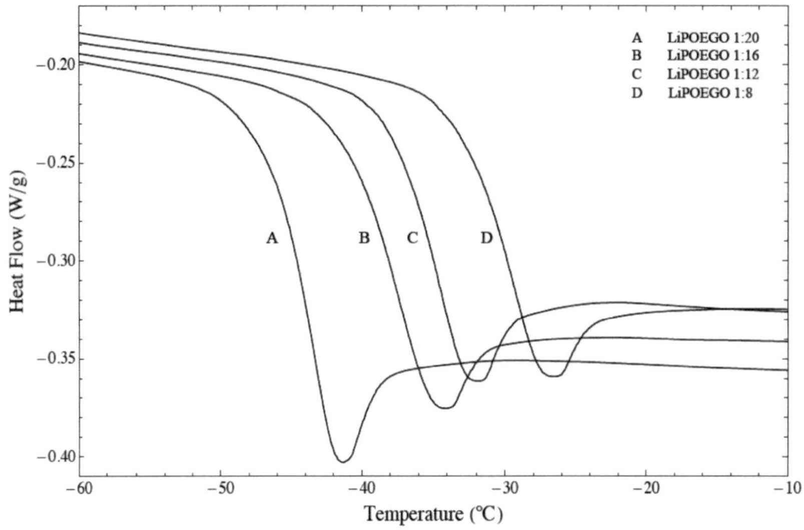

Fig. 7. DSC traces outlining the T_g of LiOTf/POEGO complexes of varying ratios.

In Figure 7, the DSC of the polymer/salt complexes for the four ratios of interest are displayed. As expected, we see an increase in the glass transition temperature upon the introduction of greater quantities of the salt. The midpoint of the T_g increases from -44.0°C for the 1:20 ratio up to -30.7°C for the sample with the highest salt concentration, 1:8. The results for the series of polymer complexes are listed in Table 5.

[Li]/[EO]	Onset of T_g (°C)	Midpoint of T_g (°C)
1:8	- 36.5	- 30.7
1:12	- 40.7	- 35.6
1:16	- 44.6	- 38.2
1:20	- 50.8	- 44.4
0	- 56.7	- 50.7

Table 5. Effect of Salt Concentration on T_g of POEGO

To ensure that complexation had been successful and the salt had efficiently dissociated within the polymer matrix, XRD analyses were carried out. Based on the nature of LiOTf, it is expected to be a highly crystalline material. This was confirmed by a series of sharp peaks present in the diffractogram (Figure 8). This was highly beneficial in determining the efficacy of the polymer salt complexes. With proper dissociation of the salt, combined with the amorphous nature of the polymeric material, it is expected that the diffractogram for the polymer/salt complexes should also be amorphous. The XRD for these materials confirmed this hypothesis with the samples of all ratios presenting a completely amorphous pattern.

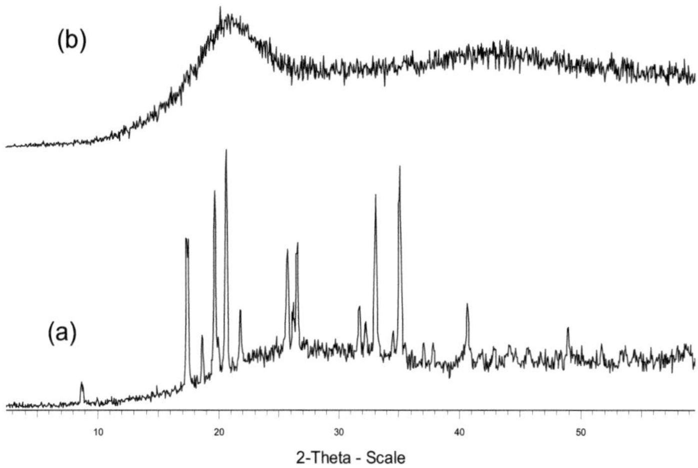

Fig. 8. XRD comparison of (a) crystalline LiOTf, and (b) amorphous LiPOEGO 1:8.

Thermal analyses of both the LiOTf and LiOTf/POEGO complexes were carried out by means of TGA to see what effect, if any, the salt had on the thermal stability of the polymer. It was found that lithium triflate begins to degrade at temperatures greater than 400°C; however, the effect of the salt on the onset temperature for the degradation of the polymer is negligible. In the polymer/salt complexes, we observe two main decomposition steps; the first occurring around 225°C denotes the degradation of the polymer while the second step begins around 418°C which corresponds largely to the decomposition of the triflate ion.
The final and most important method of characterization for the polymer/salt complexes was that of ionic conductivity measurements. Through these analyses, it was discovered that the optimal [Li]/[EO] ratio is 1:16 as expected from the literature results (Xu et al., 2001). With this ratio, the polymer had measured ionic conductivity values on the order of 10^{-6}-10^{-5} S/cm with temperatures ranging from 290-320 K which compares well to the previously reported values. These results are summarized below in Table 6. Using this information, the exfoliated nanocomposites synthesized for conductivity testing were prepared using the same 1:16 ratio.

[Li]/[EO]	Temperature (K)	Ionic Conductivity (S/cm)
1:16	290	2.9×10^{-6}
1:16	300	8.1×10^{-6}
1:16	310	1.9×10^{-5}
1:16	320	4.1×10^{-5}

Table 6. Ionic conductivity data for LiOTf/POEGO complex.

4.4 Characterization of the exfoliated nanocomposites

The most integral part of this research was developing the synthetic methodology and fully characterizing the polymer nanocomposites containing MWNTs as filler materials. Two main approaches were employed in the synthesis of these novel materials. The first method (Method A) is that which is commonly seen in the literature and involves the dispersion of the nanotubes in water via ultrasonication which is then mixed with a solution containing the desired polymer.

Mass Percent	ΔT_g (°C)			
	Method 1A[2]	Method 1A[3]	Method 2A[2]	Method 1B
1	- 0.71	+ 0.29	- 5.12	+ 1.12
5	- 1.46	+ 0.57	- 4.12	+ 0.55
10	- 0.57	- 0.42	- 1.70	+ 1.68
15	+ 1.62	+ 1.00	- 2.56	+ 0.69
20	+ 0.01	+ 0.77	- 0.57	+ 2.11

Table 7. ΔT_g (°C) of MWNT/POEGO complexes.

Samples prepared by this method used both acid oxidized nanotubes (Method 1A) and Hummers' oxidized nanotubes (Method 2A). Each of these samples was characterized by both DSC and TGA to determine the effect of the filler material on the thermal properties of the polymer. As previously demonstrated, when POEGO was mixed with LiOTf at any ratio, we saw an increase in the T_g of the polymer. Thus, it was expected that a similar increase would be observed upon the addition of the MWNTs to the polymer matrix. However, interestingly enough, this was not observed in some cases and instead decreases in T_g were recorded. Even at the higher loadings of MWNTs (i.e. 15-20 wt%), a slight decrease in T_g was observed in some cases as noted in Table 7.

As the glass transition temperature of POEGO can vary slightly with each synthesis, it was determined that ΔT_g would be a better method of comparison for the nanocomposites prepared by each of the different methods. With regards to the samples prepared by the *in situ* synthesis (Method 2A), the reference for comparison is the midpoint of the glass transition temperature previously recorded in this chapter (-50.7°C).

These decreases in the glass transition temperature led to further examination of the method of preparing the nanocomposites. First, the samples were prepared in a more volatile solvent (acetone) to determine whether or not trace water in the final products was causing this decrease in T_g. In addition to alternative solvents, an *in situ*preparation (Method 1B) was utilized not only to eliminate the use of water as a solvent, but also to potentially improve the dispersion of the nanotubes in the polymer matrix by adding them before the polymerization occurs. It should be noted that the *in situ*preparations were only performed with the nanotubes oxidized by acid (Method 1) due to the complete insolubility of the Hummers' oxidized nanotubes in benzene. The data appear to suggest that the samples prepared in water may be absorbing water which leads to the decrease in the glass transition temperature. However, it could also be a result of the disruption of the crystallinity of the unreacted PEG remaining in the polymer matrix. This disruption of the crystalline regions could lead to a decrease in the glass transition temperature counteracting the expected increase in the glass transition temperature due to the decreased mobility of the polymer chains in the more solid-like nanocomposite material (Zeleznak & Hoseney, 1987). This effect would be more dramatic in the materials prepared in water due to the much greater solubility of the oxidized nanotubes in water compared to acetone and benzene. This improved solubility would likely lead to better dispersion of the nanotubes in the polymer matrix and thus a more efficient disruption of the crystalline PEG.

A first glance at the table of data may raise some concerns as there does not appear to be any pattern in terms of the effect on the T_g of the nanocomposites. However, with several competing effects in terms of residual solvent content, disruption of crystallinity, and

[2] Samples prepared in H_2O
[3] Samples prepared in acetone

increased solid character, slight variations in any of these effects between samples would cause slight differences in the transition temperature. One might suggest that the only reason for the decreases in T_g is due to residual water; however, TGA analyses (see below) have demonstrated that all samples, regardless of the method of preparation, contain roughly equivalent amounts of solvent (~0.5-1.5 wt%). The key feature to note here is that none of the nanocomposite materials display major increases in the glass transition temperature which suggests that the ionic conductivity of the materials may not be significantly diminished due to the presence of the mechanically stabilizing nanotubes.

Another feature of these polymer nanocomposites that was examined was their thermal stability. In fact, this was the method used to determine the optimal conditions for the preparation of these materials. As previously mentioned, sonication can have a detrimental effect on the length and mechanical properties of the MWNTs. However, without adequate levels of sonication, the nanotubes do not disperse well in the solvents employed in this study, and thus thermal analyses of the resulting nanocomposites were performed to determine the optimal sonication time to provide adequate dispersion without significantly degrading the nanotubes. Sonication times of 0.5, 1, 2, 5, and 10 minutes were tested for nanocomposite materials containing 10 wt% MWNTs which were prepared by Method 1A. The results of these experiments are displayed in Figure 9. As demonstrated by the plot, the optimal sonication time was determined to be 2 minutes and thus, this length of sonication was utilized for preparation of all samples. The dashed line denotes the thermal stability of POEGO by itself (T_d = 220.92°C).

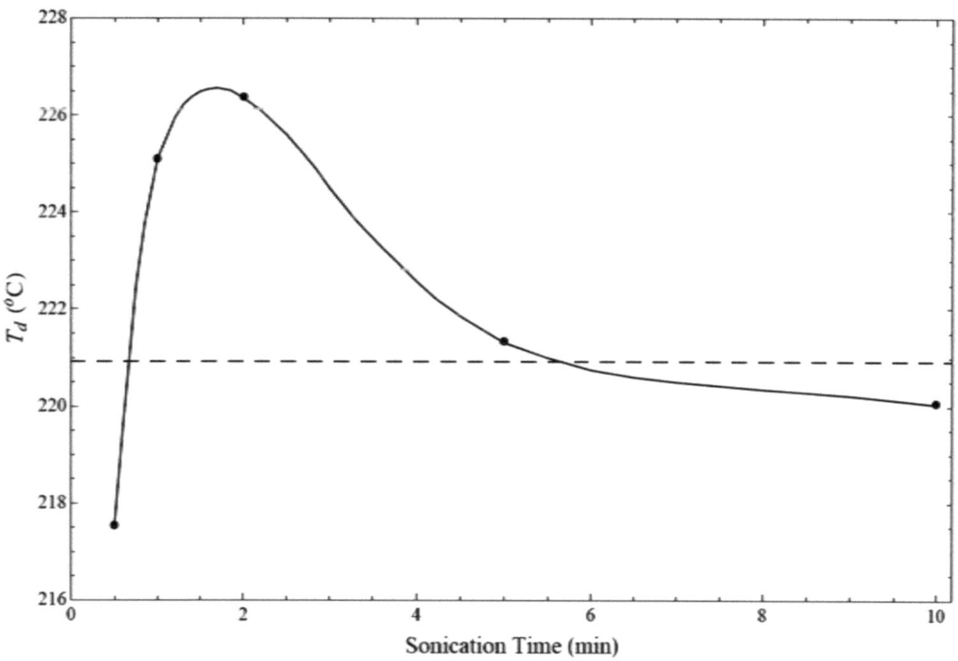

Fig. 9. Effects of sonication time on the thermal stability of polymer nanocomposites.

Using this optimized procedure, the samples of all methods were prepared and analyzed in terms of their thermal stabilities. These results are summarized in Table 8. As was the case with the DSC analyses, the parameter of interest here is ΔT_d due to slight variations in the decomposition temperature with each synthesis of POEGO. The reference for comparison for the nanocomposites prepared from Method 1B is that which was previously discussed in this text (T_d = 220.92°C).

Mass Percent	ΔT_d (°C)			
	Method 1A[4]	Method 1A[5]	Method 2A[4]	Method 1B
1	+ 1.04	+ 0.54	- 3.76	+ 4.19
5	- 1.51	- 0.53	+ 2.30	+ 2.38
10	+ 5.43	- 1.11	+ 0.39	+ 6.84
15	+ 0.65	- 1.86	+ 3.43	+ 4.95
20	+ 0.16	+ 0.03	+ 3.06	+ 7.22

Table 8. ΔT_d (°C) of MWNT/POEGO complexes.

Once again, there does not appear to be observable patterns in the thermal stabilities of these materials. One of the greatest increases in thermal stability (ΔT_d = 5.43°C) is that of the sample for which the process was optimized (Method 1A - 10 wt%). This sample was set as the benchmark for all of the others based on the fact that it was the sample with the lowest loading capacity which began to gain the dimensional stability of a solid. The samples containing 5 wt% of MWNTs appeared to be more solid than POEGO; however, they still had a slight tendency to flow like a glutinous material.

It is possible that if the process was optimized for every sample, the thermal stabilities of the materials would improve upon those which were observed here. The samples prepared by Method 1B appear to be the best in terms of thermal stability; however, these results must be interpreted with caution. There are two possible explanations for these results. The first is that of the improved dispersion of the nanotubes within the polymer matrix due to the presence of the nanotubes during the polymerization process. The other explanation is that with separate syntheses, the thermal stability of POEGO will vary to some extent and thus a portion of this improvement in thermal stability could be due to an increase in the stability of the POEGO itself.

While these thermal analyses are an important method of characterization for these materials, for a polymer nanocomposite which is thermally stable beyond 200°C, this is more than sufficient for a material with a desired application as a solid polymer electrolyte. If the desire was to significantly improve the thermal stability of a specific polymer, oxidation of the nanotubes would not be the ideal synthetic method unless subsequent reduction with hydrazine was conducted.

The final methods of characterization for these nanocomposite materials involved FTIR and XRD analyses to ensure that the nanotubes were not degraded during the synthesis of these materials and to observe whether interactions between the polymer and the filler occurred.

The IR analyses were conducted to observe what effect, if any, the filler nanotubes have on the vibrations of the functional groups within the polymer. There have been reports in the literature which have assessed the effects of a filler material by observing shifts in the vibrations in the IR spectrum (Zhang et al., 2007). Analogous analyses were conducted here as the vibrations of POEGO were compared to those in the exfoliated polymer nanocomposites. Untreated nanotubes would likely interact very weakly with the polymer; however, the oxidation process creates the possibility for hydrogen bonding between the MWNTs and the polymer matrix. Another property of these materials to consider is their

[4] Samples prepared in H_2O

[5] Samples prepared in acetone

increased rigidity due to the presence of this material. This rigidity could also shift the frequency of the vibrations which can be observed in their respective IR spectra. The summary of these results for a nanocomposite containing 10 wt% nanotubes prepared by Method 1A is listed in Table 9. Many of these shifts are between 5 and 20 wavenumbers as expected as the interactions between the filler and polymer would not be strong enough to cause major shifts. The vibration for the O-H stretch appears to be shifted significantly; however, this could be due to differences in residual water either in the materials themselves or in the KBr used to prepare the pellets for measurement.

POEGO ν (cm^{-1})	MWNT/POEGO ν (cm^{-1})	Δν (cm^{-1})	Designation
3489.9	3397.6	- 92.3	O-H stretch
2882.6	2877.7	- 4.9	sp^3 C-H stretch
1760.3	1745.7	- 14.6	C=O stretch of carboxylic acid
1196.7	1177.3	-19.4	C=O stretch of ester
1099.6	1104.4	+ 4.8	C-O stretch of ester/ether/alcohol

Table 9. FTIR wavenumber shifts in polymer nanocomposites.

As previously discussed, POEGO is an amorphous material while the oxidized nanotubes display a peak in the diffractogram corresponding to the intertubular spacing of 3.4 Å. The diffractograms for the nanocomposite materials also contain a peak with low intensity corresponding to this same spacing (Figure 10). TGAs of the nanocomposite materials were conducted in order to decompose the polymer and observe whether the nanotubes remain throughout the synthetic process. XRDs were conducted on the remaining material which showed significant increases in the intensity of the lone peak indicating that the structure of the nanotubes are still intact upon incorporation into the polymer nanocomposites. This change in intensity can be attributed to the decomposition of the amorphous polymer network and thus the relative strength of this reflection increases.

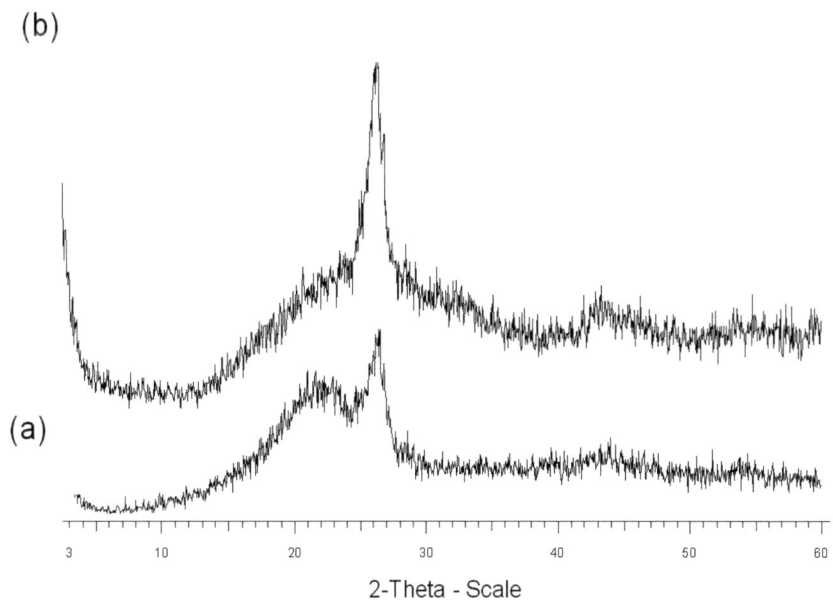

Fig. 10. Diffractogram of (a) 20 wt% nanocomposite before heating, (b) after heating to 800 °C.

4.5 Ionic conductivity results for polymer nanocomposites

The final method of characterization for these nanocomposite materials was that of AC impedance spectroscopy to determine their ionic conductivity properties. The samples were prepared and run as outlined in section 3. The thickness of each sample was estimated at 0.0055 cm (±0.001 cm). The sample width and length varied slightly with each sample and are summarized later in Table 10.

From these analyses, it was determined that the conductive properties of the materials were largely dependent on the method of preparation. The samples prepared by Method 1A appeared to be either electrical conductors or mixed electrical/ionic conductors. This was apparent from the Nyquist plot as the impedance Z went to the real axis and stayed there at lower frequencies, which is characteristic of an electrical conductor. (The resistance of the sample in parallel with cable capacitance produces a semi-circular impedance arc characteristic of a simple RC network). For an ionic conductor between blocking electrodes, the imaginary part of the impedance goes through a minimum known as the touchdown point and then begins increasing again as the frequency continues to decrease. The ionic resistance of the sample was taken to be the real part of the impedance at the touchdown point. Examples of both types of Nyquist plot are displayed in Figure 11.

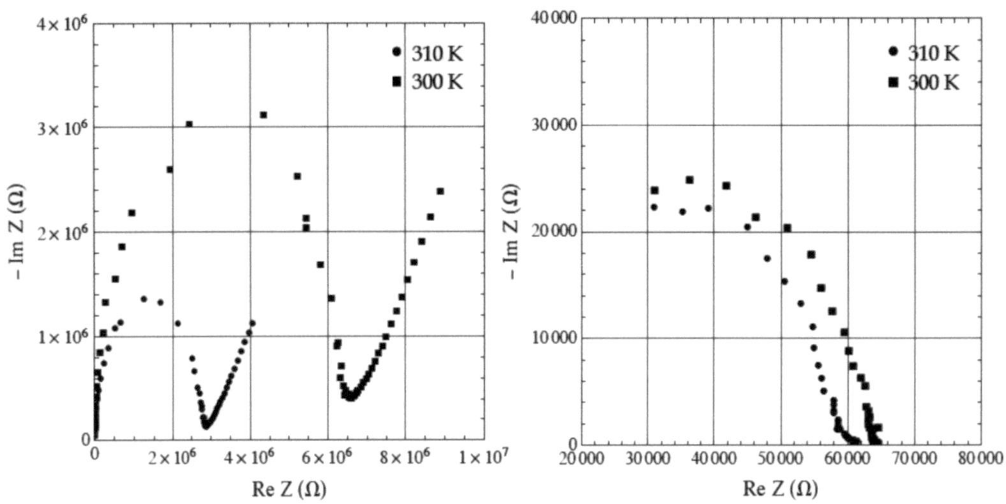

Fig. 11. Examples of Nyquist plots of ionic (left) and electrical/mixed conductors (right). High frequencies yield the points on the left of each plot; low frequencies are on the right.

After concluding that the samples prepared by Method 1A did have properties of electronic conductors through analyses of the 10 wt% and 20 wt% samples, no additional samples were measured by AC impedance spectroscopy. Although these materials could potentially be used in other fields, this electrical conductivity would prohibit their use as electrolyte materials, which are required to be electrical insulators. The samples prepared by Method 2A were more interesting. These samples, as previously mentioned, involved the mixing of the oxidized MWNTs during the polymerization process. At a low loading capacity of 5 wt%, this sample appeared to be an efficient ion conductor. This was evidenced in the previously shown Nyquist plot as well as the following plots of conductivity versus temperature. In this type of plot, the conductivity of the ionic conductors tends to increase almost exponentially with temperature, resulting in a linear trend when the data is displayed as a semi-log plot. Electrical conductors do not see as dramatic of an increase in conductivity with increasing temperatures. The conductivity results are summarized in Figure 12 as well as in Table 10. The box in the table outlines the conductivities of the samples which are solely ionic conductors.

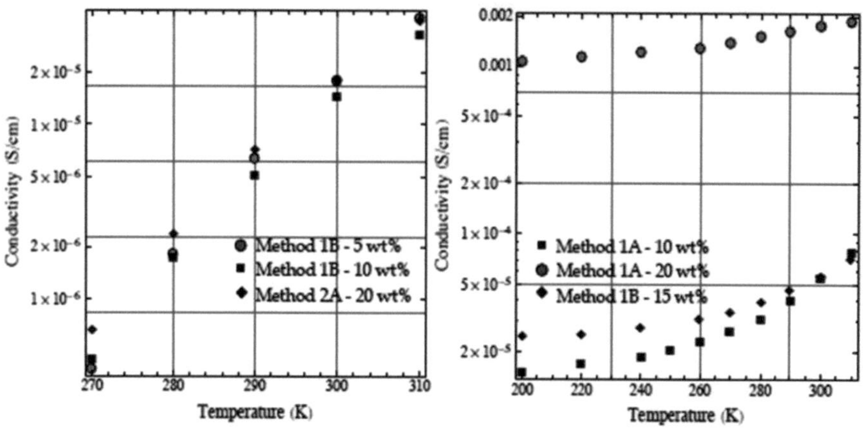

Fig. 12. Semi-log plots of conductivites of ionic conducting (left) and mixed conducting (right) polymer nanocomposite samples at various temperatures.

From the data presented in Figure 12, more evidence is provided that the samples prepared by Method 1A are electrical conductors as previously observed with the Nyquist plots. However, the plots of conductivity versus temperature are relatively flat at low temperatures, but as the temperature increases past 260 or 270K, the slopes of the lines increase observably. This suggests that these materials are mixed conductors. At higher temperatures, when the polymer is more fluid, ionic conductivity becomes a more important factor; however, at the lower temperature range when the mobility of the polymer chains decreases significantly, electrical conductivity becomes the major contributor to the conductivity of the samples.

Sample	Sample Dimensions (cm)		Conductivity (S/cm) (± 20%)		
	Width	Length	290 K	300 K	310 K
Method 1A - 10 wt%[6]	0.889	0.635	4.0×10^{-5}	5.4×10^{-5}	7.9×10^{-5}
Method 1A - 20 wt%[6]	1.003	0.610	1.6×10^{-3}	1.8×10^{-3}	1.9×10^{-3}
Method 2A - 20 wt%	0.876	0.584	**7.1×10^{-6}**	**1.8×10^{-5}**	**3.9×10^{-5}**
Method 1B - 5 wt%	0.980	0.610	**6.4×10^{-6}**	**1.8×10^{-5}**	**4.1×10^{-5}**
Method 1B - 10 wt%	0.927	0.610	**5.1×10^{-6}**	**1.5×10^{-5}**	**3.2×10^{-5}**
Method 1B - 15 wt%[6]	0.876	0.584	4.6×10^{-5}	5.6×10^{-5}	7.0×10^{-5}

Table 10. Conductivities of polymer nanocomposite materials.

When AC impedance spectroscopy was performed on the LiPOEGO samples (Table 6), the accuracy was estimated at ±50% due to the non-uniform thickness of the films and the estimation of the density of the materials. Here, the uncertainty was estimated at ±20% mainly

[6] Mixed electrical/ionic conductors.

due to the uncertainty in the thickness of the films as previously discussed. Three of the samples listed above (boxed, bolded area) are solely ionic conductors which makes them possible candidates to be used as solid polymer electrolytes. In comparison to LiPOEGO itself, the conductivities of these samples are very comparable. In fact, these values are actually higher than those previously recorded for LiPOEGO. However, when the uncertainties combined with the slightly greater possibility of proton conduction in these samples are taken into account, it is probable that the conductivity values of LiPOEGO and those of the polymer nanocomposites are very similar. This supports the hypothesis that the decreases in glass transition temperatures in the polymer nanocomposites were likely due to a disruption of any crystalline regions of the material. This would lead to an improvement in conductivity counteracting the expected decrease in conductivity due to the increased rigidity of the materials.

5. Conclusion

Exfoliated nanocomposite materials have generated plenty of interest due to the ability of a small quantity of filler material to significantly enhance the properties of the host polymer. We have demonstrated here the use of multi-walled carbon nanotubes oxidized by different methods as filler materials in the polymer matrix of poly[oligo(ethylene glycol)oxalate]. Both the host and filler material have been characterized thoroughly in terms of their thermal and structural properties. The nanotubes do not appear to have a significant impact on the thermal stability of the polymer. While some improvements were observed for certain filler loadings, no dramatic increases in thermal stability were observed. However, through analyses conducted using DSC, it was noted that the glass transition temperatures of these nanocomposite materials did not change significantly, contrary to expectations. In some cases, decreases in T_g were observed. This counterintuitive effect of the filler material on the polymer is attributed to several competing effects, with decreased crystallinity and residual solvent favouring a lowered T_g, while increased solid character and reduced flexibility of the polymer chains favour an increase. These competing effects result in insignificant changes in the conductivity of the polymer, suggesting that oxidized MWNTs may represent an ideal filler material to be used in polymer electrolytes.

Very little research has been performed using the polymer POEGO despite conductivities superior to many ionically conductive polymers. Using this methodology, several exfoliated nanocomposites could be prepared using different molecular weights of POEGO as well as with other highly conductive polymers such as poly[bis(methoxyethoxyethoxy)phosphazene] (MEEP), poly(oxymethylene-oxyethylene) (POMOE) and several polysiloxanes. Depending on the synthetic procedure for the desired polymer, the effectiveness of the *in situ* preparation of these exfoliated nanocomposites would vary. For the *in situ* preparation of the nanocomposites explored here, the solutions were heated in the refluxing process. This is an aspect of the synthesis for samples prepared by Method A that was not explored here. This heating could lead to improved solubility of the nanotubes and potentially improve the dispersion of the nanotubes in the polymer matrix.

The conductivity testing demonstrated a few important features of these polymer nanocomposites. It was noted that the method of preparation had a significant effect on the conductive properties of the materials. This suggests that heating could lead to improved dispersion as evidenced by the ionic conductivity of the samples prepared by the *in situ* preparation method. Furthermore, the level of oxidation is of particular interest. The samples prepared with 20 wt% of the Hummers' oxidized nanotubes (Method 1B) were ionically conducting whereas samples prepared with only 10 wt% of the acid oxidized nanotubes (Method 1A) were mixed conductors. This suggests that either using the Hummers' oxidation method or prolonging the reaction time for acid oxidation would lead to the best candidates for solid polymer electrolytes. Based on the data presented, the ideal materials for use as solid polymer electrolytes would be samples containing between 5-10 wt% nanotubes prepared by Method1B or samples prepared with 10-20 wt% nanotubes prepared by Method 2A. Each of these samples was ionically conducting and demonstrated a visible improvement in mechanical stability, although quantitative measurements of mechanical properties remain a topic for future work.

6. Acknowledgement

We would like to thank the Natural Sciences and Engineering Research Council of Canada (NSERC), Canada Foundation for Innovation (CFI), Atlantic Innovation Fund (AIF) and University of Prince Edward Island (UPEI) for funding.

7. References

Ahir, S.V. (2007). Polymers containing carbon nanotubes: active composite materials, In: *Polymeric Nanostructures and their Applications;* Nalwa, H.S. (Ed.), pp. 153-200, American Scientific Publishers, ISBN 978-158830692, California.

Allcock, H.R.; Austin, P.E.; Neenan, T.X.; Sisko, J.T.; Blonsky, P.M.; Shriver, D.F. (1986) Poly phosphazenes with etheric side groups: prospective biomedical and solid electrolyte polymers. *Macromolecules*, Vol. 19, No. 6, pp. 1508-1512, ISSN 0024-9297.

Bouridah, A.; Dalard, F.; Deroo, D.; Cheradame, H.; LeNest, J.F. (1985) Poly(dimethyl-siloxane)-PEO based networks used as electrolytes in lithium electrochemical solid state batteries. *Solid State Ionics*, Vol. 15, No. 3, pp. 233-240, ISSN 0167-2738.

Chan, C.K.; Peng, H.; Liu, G.; McIlwrath, K.; Zhang, X.F.; Huggins, R.A.; Cui, Y. (2008a). High performance lithium battery anodes using silicon nanowires. *Nature*, Vol. 3, No. 1, pp. 31-35, ISSN 1748-3387.

Chan, C.K.; Zhang, X.F.; Cui, Y. (2008b). High capacity Li ion battery anodes using Ge nano-wires. *Nano Lett.*, Vol. 8, No. 1, pp. 307-309, ISSN 1530-6984.

Croce, F.; Appetecchi, G.B.; Persi, L.; Scrosati, B. (1998) Nanocomposite polymer electrolytes for lithium batteries. *Nature*, Vol. 394, No. 6692, pp. 456-458, ISSN 0028-0836.

Dalton, A.B.; Collins, S.; Munoz, E.; Razal, J.M.; Ebron, V.H.; Ferraris, J.P.; Coleman, J.N.; Kim, B.G.; Baughman, R.H. (2003). Super-tough carbon-nanotube fibers. *Nature*, Vol. 423, No. 6941, ISSN 0028-0836.

Fergus, J.W. (2010). Ceramic and polymeric solid electrolytes for lithium-ion batteries. *J. Power Sources*, Vol. 195, No. 15, pp. 4554-4569, ISSN 0378-7753.

Gao, J.; Itkis, M. E.; Yu, A.; Bekyarova, E.; Zhao, B.; Haddon, R.C. (2005). Continuous spinning of a single-walled carbon nanotube-nylon composite fibre. *J. Am. Chem. Soc.*, Vol. 127, No. 11, pp. 3847-3854, ISSN 0002-7863.

Hall. R. (1967) Minimizing errors of four point probe measurements on circular wafers. *Journal of Scientific Instruments*, Vol. 44, No. 1, ISSN 0950-7671.

Hummers Jr., W.S.; Offeman, R.E. (1958). Preparation of graphitic oxide. *J. Am. Chem. Soc.*, Vol. 80, No. 6, pp. 1339, ISSN 0002-7863.

Iijima, S. (1991) Helical microtubules of graphitic carbon. *Nature,* Vol. 354, No. 6348, pp. 56-58, ISSN 0028-0836.

Iijima, S. (2001). Carbon nanotubes: past, present, and future. *Physica B,* Vol 323, No. 1, pp. 1-5, ISSN 0921-4526.

Khare, R.; Bose, S. (2005). Carbon nanotubes based composites - a review. *Journal of Minerals & Materials Characterization & Engineering,* Vol. 4, No. 1, pp. 31-46, ISSN 1559-2511.

Kojima, Y.; Usuki, A.; Kawasumi, M.; Okada, A.; Fukushima, Y.; Kurauchi, T.; Kamigaito, O. (1993). Mechanical properties of nylon 6-clay hybrid. *Mat. Res.,* Vol. 8, No. 5, pp. 1185-1189, ISSN 0884-2914.

Liu, P.; Gong, K.; Xiao, P.; Xiao, M. (2000). Preparation and characterization of poly(vinyl acetate)-intercalated graphite oxide nanocomposites. *J. Mater. Chem.,* Vol. 10, No. 4, pp. 933-935, ISSN 0959-9428.

Ma, P.-C.; Siddiqui, N.A.; Marom, G.; Kim, J.-K. (2010) Dispersion and functionalization of carbon nanotubes for polymer-based nanocomposites: a review. *Composites: Part A.,* Vol. 41A, No. 10, pp. 1345-1367, ISSN 1359-835X.

Meyer, W. (1998). Polymer electrolytes for lithium-ion batteries. *Adv. Mater.,* Vol. 10, No. 6, pp. 439-448, ISSN 0935-9648.

Mizushima, K.; Jones, P.C.; Wiseman, P.J.; Goodenough, J.B. (1980). Lithium cobalt oxide: a new cathode material for batteries of high energy density. *Mat. Res. Bull.,* Vol. 15, No. 6, pp. 783-789, ISSN 0025-5408.

Moniruzzaman, M.; Winey, K.I. (2006) Polymer nanocomposites containing carbon nanotubes. *Macromolecules,* Vol. 39, No. 16, pp. 5194-5205, ISSN 0024-9297.

Nazri, G.-A.; Pistoia, G. (Eds.). (2004) *Lithium Batteries: Science and Technology,* Kluwer Academic Publishers, ISBN 978-0387926742, Boston.

Paul, D.R.; Robeson, L.M. (2008). Polymer nanotechnology: Nanocomposites. *Polymer,* Vol. 49, No. 15, pp. 3187-3204, ISSN 0032-3861.

Rinzler, A.G.; Liu, J.; Dai, H.; Nikolaev, P.; Huffman, C.B.; Rodríguez-Macías, F.J.; Boul, P.J.; Lu, A.H.; Heymann, D.; Colbert, D.T.; Lee, R.S.; Fischer, J.E.; Rao, A.M.; Eklund, P.C.; Smalley, R.E. (1998). Large-scale purification of single-wall carbon nanotubes. Process, product, and characterization. *Appl. Phys. A,* Vol. 67, No. 1, pp. 29-37, ISSN 0947-8396.

Thackeray, M.M.; Davide, W.I.F.; Bruce, P.G.; Goodenough, J.B. (1983). Lithium insertion into manganese spinels. *Mat. Res. Bull.,* Vol. 18, No. 4, pp. 461-472, ISSN 0025-5408.

Tonge, J.S.; Shriver, D.F. (1989). Increased dimensional stability in ionic conducting polyphosphazenes systems. *J. Electrochem. Soc.,* Vol. 134, No. 1, pp. 269-270, ISSN 0013-4651.

Whittingham, M.S. (1976). Electrical energy storage and intercalation chemistry. *Science,* Vol. 192, No. 4244, pp. 1126-1127, ISSN 0036-8075.

Xu, W.; Belieres, J.-P.; Angell, C.A. (2001). Ionic conductivity and electrochemical stability of poly[oligo(ethylene glycol)oxalate]-lithium salt complexes. *Chem. Mater.,* Vol. 13., No. 2, pp. 575-580, ISSN 0897-4756.

Zeleznak, K.J.; Hoseney, R.C. (1987). The glass transition in starch. *Cereal Chemistry,* Vol. 64, No. 2, pp. 121-124, ISSN 0009-0352.

Zhang, R.; Hu, Y.; Li. B.; Chen. Z.; Fan. W. (2007). Studies on the preparation and structure of polyacrylamide/α-zirconium phosphate nanocomposites. *J. Mater. Sci.,* Vol. 42, No. 14, pp. 5641-5646, ISSN 0022-2461.

6

Mechanical Properties Evaluation of Bulk and Coated Material by Depth Sensing Indentation

J.V. Fernandes[1], N.A. Sakharova[1], J.M. Antunes[1,2] and M.C. Oliveira[1]
[1]*CEMUC, Department of Mechanical Engineering,*
University of Coimbra,
[2]*Escola Superior de Tecnologia de Abrantes,*
Instituto Politécnico de Tomar,
[1,2]*Portugal*

1. Introduction

The mechanical characterization of materials represents an important domain of research into development. Tensile and hardness tests are generally used for such effect. Samples for tensile test require specific geometries and sizes, which often limits its use, in particular when the available amount of material is not enough to carry out the test, as in case of micro-components. The hardness test, by its simplistic and not destructive character, becomes desirable to its utilization in the evaluation of the mechanical properties of materials.

The conventional hardness test consists of the application of a load on an indenter. This penetrates the sample along the direction of the normal of its surface. The load applied acts during a preset time, after which is removed, making possible the measurement of the size of the residual indentation, through optic resources. This allows the determination of the hardness, defined as the ratio between the applied load and area of the residual indentation.

The optical evaluation of the area of the residual indentation is limited, namely for very low applied loads and, consequently the resulting indentations have reduced dimension. However, during the last two decades, the development of the hardness equipments not only allowed to eliminate this limitation, but also to widen the application field of the hardness test. In fact, the advent of the Depth Sensing Indentation (DSI) equipments, making possible to plot the load - indentation depth curves, extended the application of the hardness test to scales close to the atomic one.

Moreover, the DSI equipments allow evaluating, not only the hardness, but also other mechanical properties, such as the Young's modulus (Doerner & Nix, 1986; Oliver & Pharr, 1992). Other mechanical properties, such as the yield stress and work-hardening coefficient (Dao et al., 2001; Antunes et al., 2007), of bulk and coated materials can be determined from the DSI results, by applying accurate models or inverse analysis procedures. Most of the achievements reached in this domain arise from the employment of numerical tools. For example, the possibility to carry out inverse analysis, i.e. to obtain the tensile curve of bulk and coatings materials, from the experimental load - indentation depth curve obtained by DSI, was possible due to the numerical simulation of the hardness test.

The aim of this chapter is to shed light on questions coupled with using of DSI tests for mechanical characterization of bulk and composite materials. In this context, this chapter reviews the reverse analysis procedures applied to experimental DSI results, with the aim to evaluate: (i) the stress - plastic strain curve of bulk materials; (ii) the Young's modulus of thin coatings; and (iii) the residual stress value of materials.

2. The depth – sensing indentation test

2.1 Load – indentation depth curves

The DSI equipments make possible to plot the load - indentation depth curve, also called indentation curve that represents the evolution of the applied load with the indentation depth, during the hardness test. This curve comprises a loading and an unloading part. In the meantime, a creep period can be carried on at the maximum load, which main purpose is to stabilize the deformation induced during loading. Also the accomplishment of a creep period is common during unloading, at the last or a relatively small load. This last creep period can be used for the purpose to correct the thermal drift of the equipment.

A schematic representation of the load - indentation depth curves of an elastic-plastic material is shown in Fig. 1. In this figure are indicated the maximum load applied, P_{max}, and the correspondent indentation depth, h_{max}; h_f represents the indentation depth after unloading and h_c corresponds to the contact indentation depth at the maximum load, which allows to calculate the correspondent contact area, A_c (this value is equal to the area of the residual indentation, represented by the dimension a in the Fig. 1 (b)), which is necessary for the evaluation of the hardness and the Young's modulus.

In case of purely elastic materials, the loading and unloading parts of the indentation curves are coincident, and the residual indentation depth, h_f, is equal to zero. For materials with a rigid-plastic behaviour, both indentation depths, at the maximum load and after unloading, are equal (h_f=h_{max}); this is because there is not elastic recovery, during unloading.

2.2 Contact area

For the Vickers and Berkovich indenters (both have pyramidal geometry, with square and triangular bases, respectively), the most used indenters in DSI tests, the contact area for the ideal geometry is given by:

$$A_c = 24.5h_c^2 \tag{1}$$

where the contact indentation depth, h_c, is determined from the unloading part of the indentation curve, using the following equation (Oliver & Pharr, 1992):

$$h_c = h_{max} - h_s = h_{max} - \varepsilon C P_{max} \tag{2}$$

where C represents the compliance, i.e. the inverse of the stiffness, S, at the maximum load $\left(C = 1/S = \left(dh/dP\right)_{P_{max}}\right)$ and ε is a parameter, depending on the indenter geometry, which defines the value of $h_s = \varepsilon C P_{max}$ (see Fig.1 (b)). The value of the geometrical parameter, ε, is between 0.72 and 1, depending on the indenter (see Fig. 1 (a)). For the Vickers and Berkovich indenters, ε is generally considered equal to 0.75 (Pharr & Bolshakov, 2002).

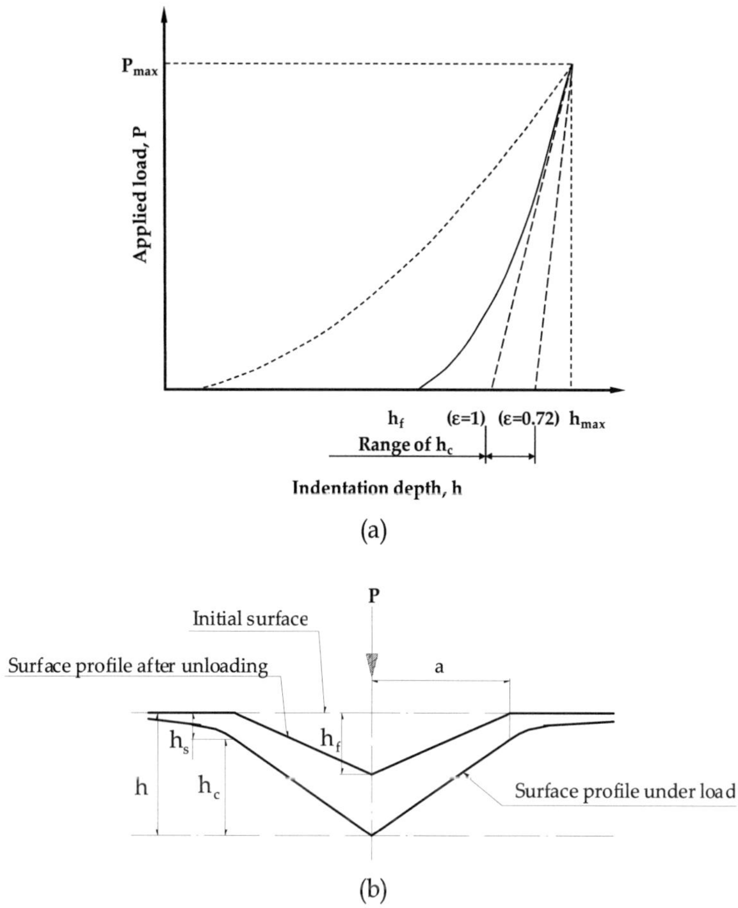

Fig. 1. Schematic representation of: (a) load – indentation depth curve (for simplicity, the creep periods are not shown); (b) correspondent geometrical parameters (Antunes et al., 2002).

2.3 Hardness and Young's modulus

The hardness, H_{IT}, is defined by the ratio between the maximum load and the contact area (ISO 14577, 2002):

$$H_{IT} = \frac{P_{max}}{A_C} \tag{3}$$

It is also possible to determine the Young's modulus of the material from the indentation curve. Its evaluation is based on the Sneddon equation (Sneddon, 1965), which establishes a linear relationship between the applied load, P, and the elastic deflexion of the sample surface, h_e, indented by a rigid circular flat punch, with radius a:

$$P = \frac{2E}{(1-\nu^2)} a h_e \tag{4}$$

where E and ν are, respectively, the Young's modulus and the Poisson ratio of the material. From the above equation, it is possible to obtain: $dP/dh_e = 2Ea/(1-\nu^2)$. By making equivalence between a circular flat punch, acting on the surface of a material submitted to elastic deflection, and a pyramidal indenter, at the beginning of unloading, such that $A_c=\pi a^2$ and $(dh/dP)_{P_{max}} = dh_e/dP$, it is possible to obtain an equation that relates the compliance at the maximum load, $C = (dh/dP)_{P_{max}}$, with the reduced Young's modulus, E_r, for the case of a pyramidal indenter (ISO 14577, 2002):

$$C = C_0 + \frac{\sqrt{\pi}}{2E_r}\frac{1}{\sqrt{A_c}}\frac{1}{\beta} \tag{5}$$

C_0 and β, which are absent in Eq. (4), are included in this equation for taking into account: C_0 - the compliance of the equipment; and β - the differences of geometry between the circular flat punch and the pyramidal indenter. A recent review concerning the β value (Oliver & Pharr, 2004), suggests that its value is between 1.023 and 1.085. Moreover, the β value is higher for the Berkovich than for the Vickers indenter (for example: 1.05 for Vickers e 1.08 for Berkovich (Antunes et al., 2006). The reduced Young's modulus, E_r, only depends on the elastic parameters of the tested material, if the indenter is considered rigid; however, in real cases, the reduced Young's modulus, E_r, depends also on the elastic parameters of the indenter:

$$\frac{1}{E_r} = \frac{(1-\nu^2)}{E} + \frac{(1-\nu_i^2)}{E_i} \tag{6}$$

where E and ν are, respectively, the Young's modulus and the Poisson ratio of the tested material (the reduced elastic modulus of the material is defined: $E^* = E/(1-\nu^2)$), E_i and ν_i are the correspondent elastic parameters of the material of the indenter. In case of diamond indenters (Vickers, Berkovich or conical): E_i=1140 GPa and ν_i=0.07.

3. Plastic properties of bulk materials

The objective to use the hardness test for characterization of the materials, in order to relate its results with the tension behaviour retraces to years 20 of the past century. L. Prandtl (Prandtl, 1920) was the one of the pioneers in this area, when relating the hardness of the material with its resistance to the entrance in the plastic regimen, using the slip-line method, for the case of a flat punch. Afterward, a study concerning the indentation of ductile materials with resource to the spherical cavity model, allowed estimating a relationship between the hardness and the yield stress (Bishop & Mott, 1945). Based on hardness results performed in mild steel and copper, Tabor (Tabor, 1951) found a linear relationship between the hardness and the representative stress, σ_r (coupled with a value of plastic strain in tension, ε_r, i.e. the so-called representative plastic strain). This means that different materials with tensile curves crossing each other at the same point (σ_r, ε_r) show the same hardness value. The equation between the hardness and the representative stress was deduced based on former studies (Hill et al., 1947), and can be expressed (Tabor, 1951):

$$H_{IT} = \eta\,\sigma_r \tag{7}$$

where H_{IT} is the hardness, η is a parameter (η=3.3, for the Vickers indenter) and σ_r is the representative stress, corresponding to a representative plastic strain, ε_r, equal to 0.08. Also, based on previous studies (Bishop& Mott, 1945; Hill, 1950; Marsh, 1964; Hirst & Howse, 1969), K.L. Johnson (Johnson, 1970) proposed the following relationship between the hardness, H_{IT}, and the representative stress, σ_r, for elastic-plastic materials:

$$\frac{H_{IT}}{\sigma_r} = \frac{2}{3}\left[1 + \ln\left(\frac{E}{6\sigma_r(1-\nu)}\mathrm{tg}\alpha + \frac{2(1-2\nu)}{3(1-\nu)}\right)\right] \tag{8}$$

where E and ν are the Young's modulus and the Poisson ratio of the material, respectively, and α is the angle between the sample surface, normal to the loading axis, and the surface of the indenter. The representative stress, σ_r, is related to the representative plastic strain, ε_r, which depends on the apical indenter angle, $\phi=(\pi/2-\alpha)$, through the equation (Johnson, 1970):

$$\varepsilon_r = 0.2\mathrm{cotg}\varphi \tag{9}$$

Subsequently, bi- and tri-dimensional numerical simulation results of the hardness test of several materials, using a conic indenter with an apical angle equal to 70.3° (for which the contact area is consistent with Eq. (1)), allow establishing dimensionless Π functions for characterizing the plastic behaviour (Dao et al., 2001). The Π functions allow establishing an inverse analysis procedure for determining the stress - plastic strain curve of materials, from the indentation curves.

The stress - plastic strain curve can be defined from the representative stress and strain and the work-hardening coefficient, n, when described by the Swift law (Swift, 1952):

$$\sigma = k(\varepsilon_0 + \varepsilon)^n \tag{10}$$

where σ and ε are the stress and the plastic deformation, respectively; n (work-hardening coefficient), k and ε_0 (which value is very low: close to $\varepsilon_0 = \sigma_0/E$, where σ_0 is the yield stress) are material constants. The knowledge of the pair (σ_r, ε_r), ε_0 and n allows determining the constant k of the Swift law: $k = \sigma_r/(\varepsilon_0 + \varepsilon_r)^n$. The yield stress can be determined as: $\sigma_0 = k\varepsilon_0^n$.

For determining the representative stress, the following dimensionless function, Π_1, was proposed (Dao et al., 2001):

$$\frac{K}{\sigma_r} = \Pi_1\left(\frac{E^*}{\sigma_r}\right) = -1.311\left[\ln\left(\frac{E^*}{\sigma_r}\right)\right]^3 + 13.635\left[\ln\left(\frac{E^*}{\sigma_r}\right)\right]^2 + \\ +30.549\left[\ln\left(\frac{E^*}{\sigma_r}\right)\right] + 29.267 \tag{11}$$

where K is the constant of the Kick law (Kick, 1885), which describes the loading part of the indentation curve (Kick law: $P = Kh^q$; in absence of indentation size effect, as in assumed by (Dao et al., 2001): $q = 2$), and E_r is the reduced elastic modulus of the material ($E^* = E/(1-\nu^2)$).

In Eq. (11), the representative stress, σ_r, and the correspondent representative plastic deformation, $\varepsilon_r = 0.033$, concern materials with equal reduced elastic modulus and with tensile stress - plastic strain curves that cross each other at the same point ($(\sigma_r, \varepsilon_r)$), even if with different values of the work-hardening coefficient. That is, the evolution of the ratio K/σ_r *versus* E^*/σ_r is independent of the work-hardening coefficient, for materials with the same representative stress, σ_r, corresponding to a representative plastic strain ε_r=0.033, as shown in Fig.2. In case of $\varepsilon_r \neq 0.033$, Eq. (11) is not unique, i.e. the evolution of K/σ_r *versus* E^*/σ_r depends on the work-hardening coefficient of the material.
The full description of the tensile curve needs also the estimation of the work-hardening coefficient. From the knowledge of the values of the reduced elastic modulus ($E^* = E/(1-\nu^2)$) and the representative stress ($\sigma_r = \sigma_{0.033}$, corresponding to ε_r=0.033) of the material, the work-hardening coefficient can be determined from one of the following functions, Π_2 or Π_3 (Dao et al., 2001) as it is given by Eqs. (12) and (13).

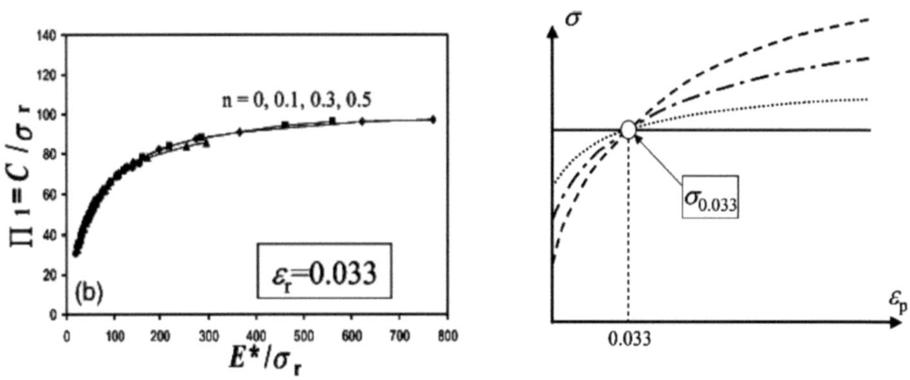

(a) (b)

Fig. 2. Π_1 (Eq. (11)) *versus* E^*/σ_r (numerical results), for the case: $\varepsilon_r = 0.033$ (on the left). Tensile stress – plastic strain curves (on the right) (Dao et al., 2001).

$$\frac{1}{E^* h_{max} C} = \Pi_2\left(\frac{E^*}{\sigma_{0.033}}, n\right) = (-1.40557n^3 + 0.77526n^2$$
$$+0.15830n - 0.06831\left[\ln\left(\frac{E^*}{\sigma_{0.033}}\right)\right]^3 + (17.93006n^3$$
$$-9.220911n^2 - 2.37733n + 0.86295)\left[\ln\left(\frac{E^*}{\sigma_{0.033}}\right)\right]^2 \tag{12}$$
$$+(-79.99715n^3 + 40.556201n^2 + 9.00157n - 2.54543)$$
$$\left[\ln\left(\frac{E^*}{\sigma_{0.033}}\right)\right] + (122.65069n^3 - 63.884181n^2 - 9.58936n$$
$$+6.20045)$$

$$\frac{h_f}{h_{max}} = \Pi_3\left(\frac{E^*}{\sigma_{0.033}}, n\right) = (0.010100n^2 + 0.0017639n$$
$$-0.0040837)\left[\ln\left(\frac{E^*}{\sigma_{0.033}}\right)\right]^3 + (0.14386n^2 + 0.018153n$$
$$-0.088198)\left[\ln\left(\frac{E^*}{\sigma_{0.033}}\right)\right]^2 + (0.59505n^2 + 0.034074n \tag{13}$$
$$-0.65417)\left[\ln\left(\frac{E^*}{\sigma_{0.033}}\right)\right] + (0.58180n^2 - 0.088460n$$
$$-0.67290)$$

where h_{max} and h_f (see Fig. 1) are the indentation depths at the maximum load and after unloading, respectively, and C is the material compliance (in (Dao et al., 2001) it is used a rigid indenter in their simulations; so, in real cases, the compliance value, C, to be considered in this equation is (C-C_0), i.e. the compliance simply due to the material).

However, this inverse analysis procedure has restrictions concerning the estimation of a unique solution for the work-hardening coefficient, namely in case of materials for which $\sigma_0 / E_r \geq 0.033$ and $n \geq 0.3$ (σ_0 and n are the yield stress and the work-hardening coefficient of the material, respectively). In this case, for different n values the curves are quite close or even cross each other for low values of the ratio E^* / σ_r (Dao et al., 2001).

Other inverse analysis procedures were proposed, similar to the above mentioned (Dao et al., 2001), but making use of experimental indentation curves obtained using multiples indenters, with different values of the apical angle. The use of two or more indenter with different equivalent apical angles allows obtaining a unique solution for the work-hardening coefficient as in (Bucaille et al., 2003; Cao et al., 2005; Casals & Alcalá, 2005; Swaddiwudhipong et al., 2005), for example. The main difference between these procedures is the type of proposed functions. The inconvenience, common to all these procedures, is the experimental use of multiple indenters.

Recently, a new reverse analysis procedure, making use of results obtained with single indenter geometry, was proposed by (Antunes et al., 2007). The methodology, based on three-dimensional numerical simulations of the hardness test, is constructed in order to ensure unique results for the representative stress and the work-hardening coefficient. In a first step, it extracts the representative stress and plastic strain, which have a slight dependence on the elastic modulus, as shown in Fig.3.

This figure shows that for each value of the representative stress, the evolution of the associated plastic strain with the reduced Young's modulus is approximately linear, for values of this modulus below 450 GPa. In this case, the representative plastic strain slightly increases from a value close to 0.034, to values that depend on the representative stress, reaching a maximum of 0.042 when the representative stress is equal or higher than 3 GPa. For values of the reduced Young's modulus higher than 450 GPa, the representative plastic strain only depends on the representative stress of the material. In the second step, the hardening exponent is deduced from the unloading stiffness.

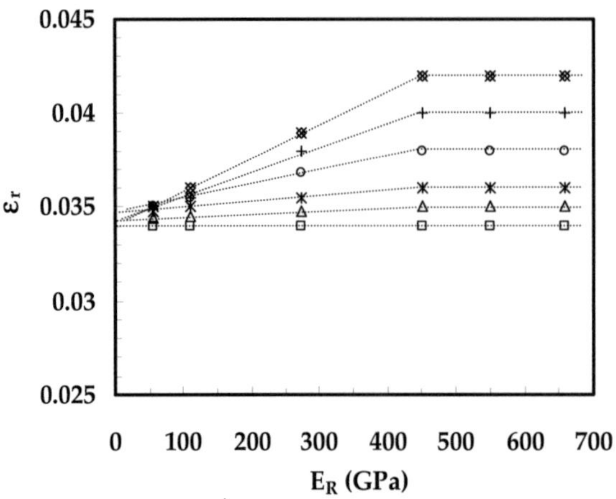

Fig. 3. Representative plastic strain, ε_r, *versus* the material reduced elastic modulus, E^* (in the figure E_R has the same meaning that E^*), for different values of the representative stress, σ_r (Antunes et al., 2007).

Concerning the first step, it was observed a quasi-linear relationship between E^*/H_{IT} and E^*/σ_r, independently on the work-hardening coefficient:

$$\frac{E^*}{H_{IT}} = 0.231\left(\frac{E^*}{\sigma_r}\right) + 4.910 \tag{14}$$

The use of this equation, for estimation of the representative stress, σ_r, presume the previous experimental determination of the hardness, H_{IT}, and the reduced elastic modulus of the material, E^*, from the hardness test. The representative stress obtained by Eq. (14) can still be optimized by comparing the experimental and numerical indentation curves, by comparing the correspondent values of the maximum load P_{max}^{exp} (experimental maximum load) and P_{max}^{num} (numerical maximum load), obtained at the maximum indentation depth, h_{max}. In each iteration, the value of the representative stress used as input in the numerical tests, is altered until coincidence of the numerical curve with the experimental one is obtained, i.e. $P_{max}^{num} = P_{max}^{exp}$. A possible optimization strategy uses the ratio of the maximum loads, $P_{max}^{num}/P_{max}^{exp}$ to update the value of the representative stress:

$$\sigma_r(i+1) = \sigma_r(i)\left[\frac{P_{max}^{exp}}{P_{max}^{num}}\right] \tag{15}$$

where $\sigma_r(i+1)$ and $\sigma_r(i)$ are the (i+1) and (i) order values of the representative stress to be tested. The iterative process ends when the ratio between maximum loads approximates one, with the desired error.

Fig. 4 shows a numerical example of reverse analysis for determination of representative stress. The load-unloading curves, experimental and numerical, obtained during the optimization process of the representative stress, are shown. Apparently, all curves are identical. The initial approximation of the representative stress (1st iteration), obtained with Eq. (14) was 0.227 GPa, which corresponds to an error in the maximum load of about 1%. Fig. 4b shows that, in the 3rd iteration of the stress optimization, identical values of maximum load (error less than 0.1%) for curves, experimental and numerical, were obtained (material with representative stress, σ_r=0.229. GPa)

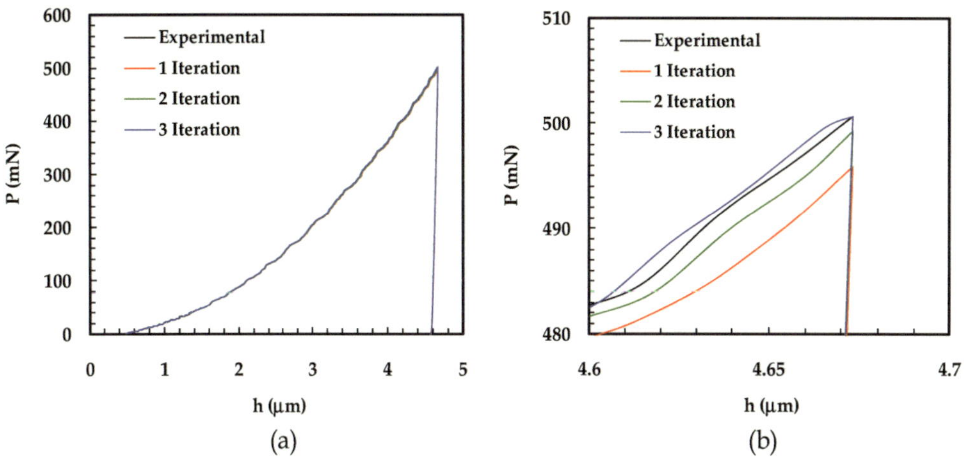

Fig. 4. Example of iterations, in order to make coincident the loading part of the indentation curves (Antunes et al., 2007): (a) full load-unloading curves; (b) detail of the region close to the maximum load.

Concerning the second step, for evaluating the work-hardening coefficient, this inverse analysis procedure requires the previous experimental determination of the stiffness, $(S = 1/C = (dP/dh)_{P_{max}})$, determined from the unloading part of the indentation curve (in addition to the representative stress, determined in the first step). This stage of the inverse analysis consists on the comparison of the experimental S value, with the ones obtained by numerical simulation tests on materials, having the reduced elastic modulus, E^*, experimentally determined, as for the previous step, and the representative stress, σ_r, determined in the first step, but with different values of the work-hardening coefficient, n (Swift law (Eq. (10)), varying in a previously chosen range. Examples of numerical simulation results are shown in Fig. 5.

Fig. 6 shows that the unloading part of the indention curve depends on the work-hardening coefficient, when the tensile curves cross each other at the same point $(\sigma_r, \varepsilon_r)$ and the material has the same reduced elastic modulus. The most efficient way to compare the unloading part of the experimental and numerical indentation curves is using the value of the stiffness in the maximum load. So, the plotting of the stiffness, S^{num}, numerically obtained as a function of the work-hardening coefficient, n, allows the comparison with the experimental stiffness, S^{exp}, as shown in the example of Fig. 6. The evolution of S *versus* n follows, for the studied cases, a straight line, which makes easier the comparison between experimental and numerical results (two or three numerical simulations are enough for describing such linear behaviour). For the case of Fig. 6, the work-hardening coefficient of the material is close to 0.27.

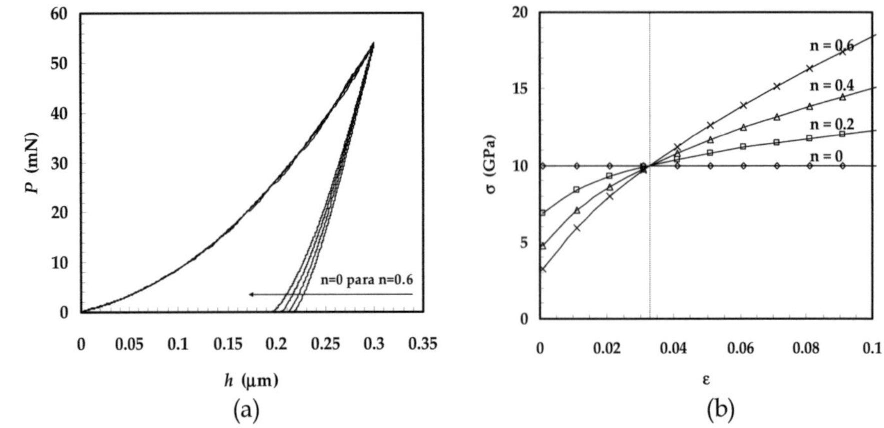

Fig. 5. (a) Examples of indentation curves obtained by numerical simulation of materials with the same reduced elastic modulus and stress - plastic strain curves crossing each other in the same point $(\sigma_r, \varepsilon_r)$; (b) Schematic representation of correspondent stress - plastic strain curves, which only depend on the work-hardening coefficient (Antunes et al., 2007).

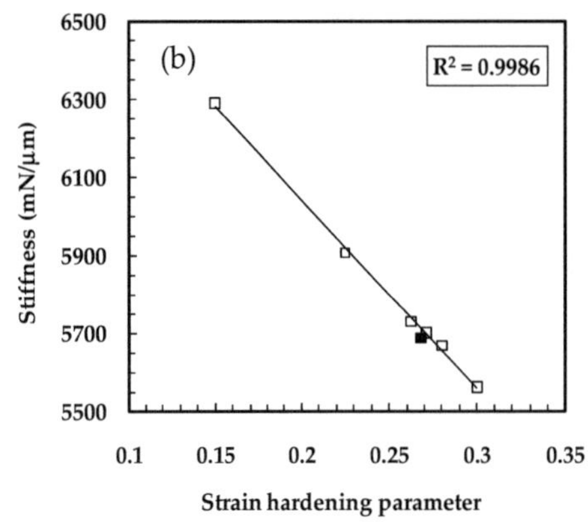

Fig. 6. Example of the evolution of the stiffness of the unloading indentation curve at the maximum load as a function of the work-hardening coefficient (open symbols: numerical results; solid symbol: experimental result) (Antunes et al., 2007).

When rigid indenters are used in the numerical simulations, the experimental and numerical stiffness values cannot be directly compared. In fact, in case of experimental unloading, the measured stiffness depends on the material and on the experimental equipment; in case of the numerical unloading, the stiffness depends only on the material. So, when using rigid indenters, previous to plot the experimental stiffness value on Fig. 6, an adjustment is required, in order to determined the equivalent value of the stiffness ($S_{equi} = 1/C_{equi}$). The equivalence between the compliance values, C_{equi} and C_{exp}, can be established by means of following equation (Antunes et al., 2007):

$$C_{equi} = \left(\sqrt{\frac{A_{exp}}{A_{num}}}\right)\left(C_{exp} - C_0 - \frac{1}{\beta}\frac{\sqrt{\pi}}{2}\frac{1}{\sqrt{A_{exp}}}\frac{(1-\nu_i)}{E_i}\right) \quad (16)$$

where C_0 is the compliance of the equipment, β is the correction factor of the indenter geometry, A_{exp} and A_{num} are the contact areas experimentally and numerically determined, respectively; ν_i and E_i are the Poisson's ratio and the Young's modulus of the experimental indenter, respectively.

4. Elastic modulus of thin coatings

The determination of the Young's modulus of thin coatings is significantly affected by the presence of the substrate. Indeed, the indentation region in elastic domain is not confined to the coating, but is extended to the substrate in particular for small thicknesses, almost from the beginning of the indentation. In this case, the determination of the Young's modulus is problematic, especially when there is a significant difference between the values of the modulus of the coating and of the substrate. The usual solution for this problem consists of separating the contribution of the thin coating from the composite Young´s modulus results. In general, Young's modulus of thin coatings is extracted using analytical models (Doerner & Nix, 1986; King, 1987; Gao et al., 1992; Menčík et al., 1997; Saha & Nix, 2002; Antunes et al., 2007), which validity and restrictions seem to depend on the materials of the composite and on the test conditions.

Recently, a reverse analysis methodology, which allows avoiding the use of such analytical functions, for determination of the films' Young's modulus was proposed (Antunes et al., 2008). This methodology is based in a comparison of the composite Young's modulus, numerically evaluated by hardness tests, using a pyramidal indenter, with that obtained in the numerical simulation with flat punch indenters (with different areas), for the same values of equivalent relative contact indentation depth h/t (t is the thickness of the coating; in case of pyramidal indenter, h has the same meaning than h_c in Eqs. (1) and (2), in case of the flat punch indenter, the equivalent contact indentation depth $h = \sqrt{A/24.5}$, where A is the flat punch area). This comparison showed that the composite' Young's modulus results do not depend on the geometry of the indenter, but only on the contact area, as shown in Fig. 7.

The reverse analyses methodology consists on the experimental determination, by using a pyramidal indenter, of the contact area and of the reduced Young's modulus, E^*, for a given relative experimental contact indentation depth h/t. Subsequently, numerical simulations of the indentation test with a flat punch indenter are performed. For these simulations, the flat punch indenter must have the same equivalent relative contact area (or depth) that in case of the experimental test on the composite, with the pyramidal indenter; and the value of the substrate reduced Young's modulus is equal to the experimentally determined for the substrate, E^* which must be done previously. For the first simulation, a value for the films Young's modulus, $E_f^*(\text{punch})$, is chosen: $E_f^*(\text{punch})$ must be higher than E^* in case of $E^* > E_s^*$ or lower than E^* in case of $E^* < E_s^*$. The value of the composite Young's modulus, $E^*(\text{punch})$, numerically determined with the flat punch indenter is now compared with the one, E^*, experimentally determined with the pyramidal indenter. If the two values of the reduced modulus, $E^*(\text{punch})$ and E^*, are enough similar (within a predefined range of accuracy), then the reduced elastic modulus, E_f^*, can be estimated as being equal to $E^*(\text{punch})$. If there is a too great difference between the experimental (pyramidal) and numerical (flat punch) composite reduced elastic modulus (this is generally the case), then an iterative method will be used for optimization. This is, new values for $E_f^*(\text{punch})$ are chosen for numerical simulations with the flat punch indenter until the numerical and experimental values of the composite reduced modulus, $E^*(\text{punch})$ and E^*, are within the delimited range of accuracy.

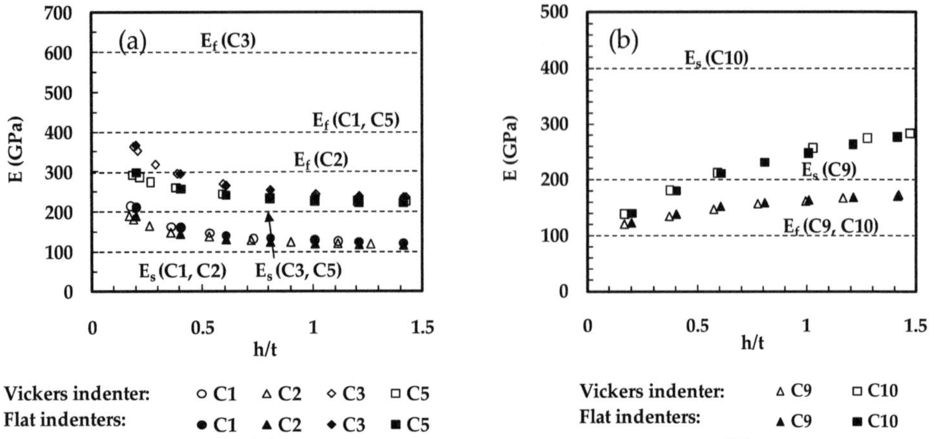

(a) (b)

Fig. 7. (a) Composite Young's modulus, E, *versus* the normalized contact indentation depth, h/t (C1, C2, C3, C4, C5, C6, C7, C8, C9 and C10 concern composites with values of E_f/E_s of 4.00, 1.50, 3.00, 2.00, 2.00, 2.00, 1.00, 1.00, 0.50, 0.50, 0.25, respectively). Results obtained with Vickers and squared flat punch indenters (in this case the contact indentation depth is equal to the one of the Vickers indenter with the same contact area). (Antunes et al., 2007, 2008).

This inverse analysis methodology is accurate and easily performed. Moreover, the use of flat punch indenter ensures that no plastic deformation occurs in the composite, for small indentation depths, which excludes the requirement of knowledge the plastic properties of the film and the substrate, for performing the numerical simulations. In fact, the numerical simulation results of indentation tests with flat punch indenters show a linear relationship between the load, P, and the elastic deflexion of the sample surface, h_e, as described by Eq. 4 and shown in Fig. 8. This linear relationship allows us to calculate the elastic modulus of the composite, by using Eq. (4).

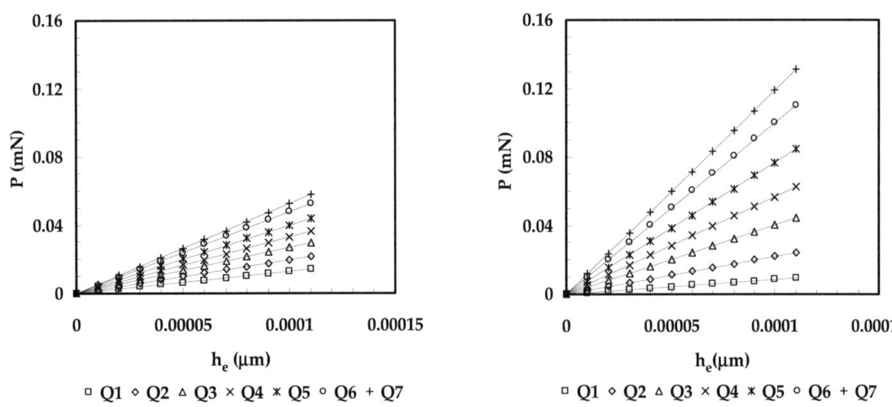

(a) (b)

Fig. 8. (a) Load, P, *versus* the elastic deflexion of the sample surface, h_e, results obtained by numerical simulation, using squared flat punch indenters, with different areas (Q1: 0.25 μm²; Q2: 1.00 μm²; Q3: 2.25 μm²; Q4: 4.00 μm²; Q5: 6.25 μm²; Q6: 9.00 μm²; Q7: 12.25 μm²): (a) E_f = 400 GPa e E_s = 100 GPa; (b) E_f = 100 GPa e E_s = 400 GPa (Antunes et al., 2008).

5. Residual stresses

The evaluation of residual stresses is a topic of major interest for many engineering applications. In fact, the presence of surface residual stresses usually modifies the mechanical properties and performance of mechanical devices, such as the resistance to fatigue, fracture and corrosion (Golovin, 2008). For that reason, whereas compressive residual stresses are usually favorable and can even be intentionally induced to improve the mechanical properties, tensile residual stresses generally reduce the quality of performance, causing early failure of the products manufactured (Bocciarelli & Maier, 2007). Moreover, when the existence of residual stress is ignored, the measured mechanical properties do not correspond to the materials tested by the indentation (Golovin., 2008).

Several works have been performed in order to understand the effects of residual stresses on depth sensing indentation data, and procedures for extracting residual stresses have been proposed. It has been demonstrated that the indentation curve and thus the measured hardness depends on the residual stress level (Tsui et al., 1996; Bolshakov et al., 1996). Later, some other authors found similar results and proposed methodologies for extracting the residual stress (Suresh & Giannakopoulos, 1998; Chen et al., 2006) from depth sensing indentation data. Most of the times, these methodologies concern the case of in-plane surface equibiaxial residual stresses (although the case of uniaxial residual stress has also been studied; see for example (Zhao et al., 2006)) and consider the residual stresses uniform over the depth of influence of the indenter. A recent paper (Jang, 2009) critically reviewed and discussed issues and methodologies involved with residual stress estimation and its effects on depth sensing indentation results.

Important results characterize the influence of residual stress on the materials' mechanical properties determined by nanoindentation tests. The load-indentation depth curves obtained for samples subjected to compressive residual stresses are situated above the curve for material without residual stresses, and for samples subjected to tensile residual stresses the curves are positioned below the curve obtained for material without residual stresses. When the absolute value of residual stresses decreases, the load indentation curves tend towards the one obtained for material without residual stresses. Examples of load-indentation depth curves, obtained by numerical simulation of the nanohardness tests for materials subjected to different levels of compression and tension residual stresses are shown on Fig. 9 (Sakharova et al., 2011). Such behaviour is qualitatively representative for materials subjected to residual stresses and have been reported by various authors (Suresh & Giannakopoulos, 1998; Bocciarelli & Maier, 2007; Sakharova et al., 2011). The presence of residual stresses also affects the measured hardness: this hardness decreases for the samples with tensile residual stresses and increases for the samples with compressive residual stresses, although the influence of the residual stress is less important in compressive than in tension residual stress. Additionally, the maximum indentation load is sensitive to the residual stress value. A linear relationship was found between the relative maximum load ($(P_r-P_{wr})/P_{wr}$), where P_r and P_{wr} are values of maximum load with and without residual stress, and the corresponding value of residual stress normalized by yield stress (σ_r/σ_0), although the trend-lines have different slopes for each case in a given material (Sakharova et al., 2011).

Basing on these results, a reverse analysis methodology for the determination of sign and value of equibiaxial residual stresses in the surface of materials, from depth sensing indentation results obtained with pyramidal indenters, was very recently proposed (Sakharova et al., 2011). This reverse analysis methodology consists of obtaining coincidence between the loading part of numerical and experimental indentation curves for surface of materials with residual stresses. The particularity of this approach is to rely exclusively on loading part of the indentation curve. In fact, the maximum load value is enough sensitive to detect the presence of residual stress, particularly in the cases of materials with low values of the E/σ_0 ratio.

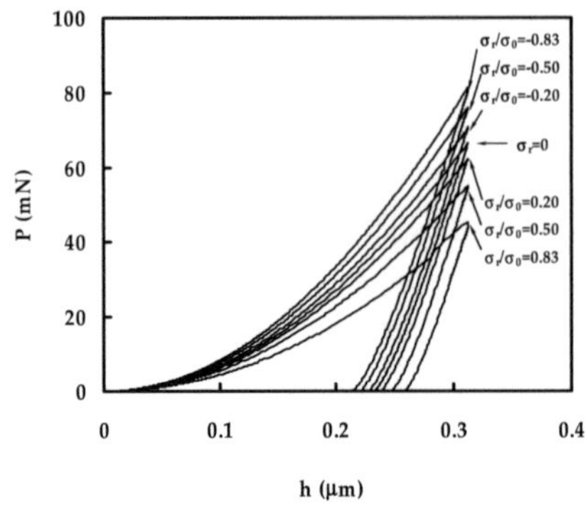

Fig. 9. Example of the load-indentation depth curves obtained with and without residual stresses of tension and compression for material with $E/\sigma_0 = 40$ (Sakharova et al., 2011).

The first step of this reverse analyses consists in carrying out experimental indentation tests up to a given value of maximum indentation depth, in two different regions of the sample: (i) a region with residual stress and (ii) a region far from the area subjected to residual stresses (on the edge of the sample, for example). The comparison of these curves permits the sign of residual stresses to be identified: tension or compression. For residual tension, the level of load-indentation depth curve obtained in region with residual stress is inferior to the one of the curve obtained in the region without residual stress; the opposite occurs for residual compression. In order to proceed with the reverse analyses, elastic (E and ν) and plastic (σ_0 and n) properties of the material are required, for input in numerical simulations. These properties can be determined independently (for example, by a tension test) or with an indentation test on a surface region without residual stresses, which permits both the elastic and plastic properties of the material to be determined, using the reverse analysis procedure as described in Section 3.

The next step, numerical simulations of indentation tests introducing different values of residual stresses in the programme must be carried out to estimate the residual stress value. Therefore, this reverse analysis methodology consists on the comparison of indentation curves: (i) numerically generated ones with residual stresses and (ii) experimental ones, both types obtained in regions of the sample with residual stresses. The best way to do this comparison is to build graphs, which show the evolution of residual stresses used as input in the numerical simulations, σ_{Nr}, as a function of the relative difference $(P_{Nr} - P_{expr})/P_{expr}$ between maximum loads obtained in the corresponding numerical test with residual stresses, P_{Nr}, and the one experimentally obtained in the region with residual stresses, P_{expr}. Three or four numerical tests can be sufficient, depending on the linear correlation obtained between σ_{Nr}, and $(P_{Nr} - P_{expr})/P_{expr}$.

Fig. 10. shows a schematic example of the determination of residual stress value. In this figure, the experimental load-indentation curves (with residual stresses) for the material chosen as example and the load-indentation depth curves obtained by numerical simulation for the same material with different level of residual stresses are shown. Basing on the results from this figure, the evolution of σ_{Nr} (residual stress introduced in the numerical

simulations) as a function of relative difference, $(P_{Nr} - P_{expr}) / P_{expr}$, was built as shown in Fig. 10b. The interception of the linear trend-line with the vertical axis allows the value of residual stresses, σ_{E_r}, to be estimated, since it occurs at the condition $P_{Nr} = P_{expr}$.
A real example of the determination of residual stress value is shown in Fig. 11a, for the case of equibiaxial compressive residual stress in commercial stainless steel SUS304-CSP specimens (Wang et al., 2006). The results obtained for the residual stress is σ_{E_r} = 618 MPa (see Fig. 11a) very close to the mentioned by authors of (Wang et al., 2006), which was calculated by using their own method: σ_{E_r} = 631 MPa.

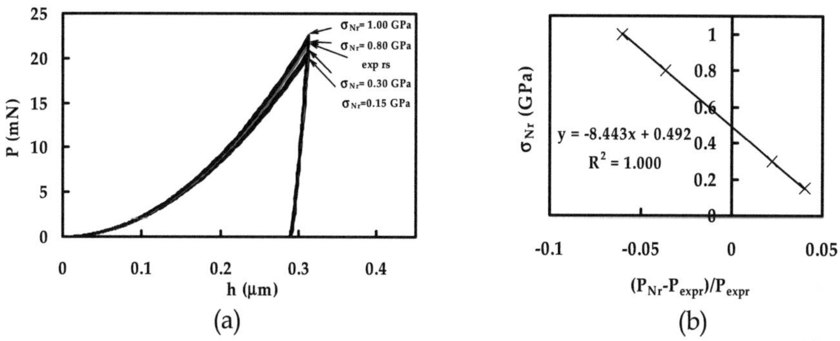

Fig. 10. (a) Load-indentation depth curves, obtained with residual stresses, experimental (exp rs) and numerically (σ_{Nr}) for material with σ_0 = 1.50 GPa, n = 0.30 and E = 600 GPa; (b) evolution of σ_{Nr} with the value of $((P_{Nr}-P_{expr})/P_{expr})$, based on graph on (a), for the determination of σ_{Er} for this material (Sakharova et al., 2011); σ_r is the value of residual stresses considered; σ_{Er} - estimated residual stresses values obtained by inverse analysis.

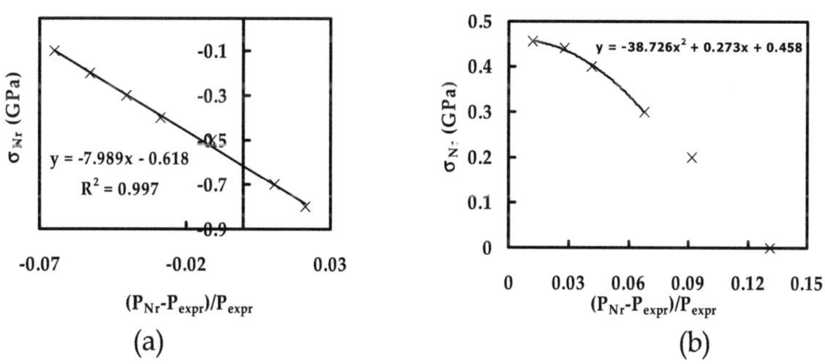

Fig. 11. Evolution of σ_{Nr} with the value of $((P_{Nr}-P_{expr})/P_{expr})$ for the determination of σ_{Er} for (a) stainless steel SUS304-CSP specimens (Wang et al., 2006); (b) for API X65 steel plate specimens (Lee et al., 2004).

Furthermore, this reverse analysis procedure can be extended for indentation tests with any indenter geometry (Sakharova et al., 2011). In fact, this methodology was also successfully used for a non-pyramidal indenter, the Brale indenter, for the case of equibiaxial tensile residual stress in small-grained API X65 steel plate specimens (Lee et al., 2004; Bocciarelli & Maier, 2007). In this case, the estimated mean value of residual stress is σ_{E_r} = 458 MPa (considering the error of the experimental maximum load from the (Lee et al., 2004) results, the estimated residual stress value is in the range: 445 to 475 MPa), which is close to the experimentally determined by (Lee et al., 2004): σ_{E_r} = 440 MPa.

It must be mentioned, that in Fig. 11b, where the results of the reverse analysis procedure are shown, a lack of linearity between σ_{Nr} and $(P_{Nr} - P_{expr})/P_{expr}$ occurred, probably due to the type of indenter geometry (Brale indenter instead of pyramidal indenter). For this reason, the extrapolation of σ_{Nr} for $(P_{Nr} - P_{expr})/P_{expr} = 0$, was carried out using a second degree polynomial equation.

6. Conclusion

The mechanical characterization of materials represents an important domain of research into development. The hardness test, by its simplistic and not destructive character, becomes advantageous for utilization in the evaluation of the mechanical properties of materials. Depth Sensing Indentation (DSI) equipments, intensively developed during the last two decades, making possible to plot the load - indentation depth curves, extended the application of the hardness test to scales close to the atomic one. Moreover, the DSI equipments allow evaluating, not only the hardness, but also other mechanical properties, such as the Young's modulus. Other mechanical properties, such as the yield stress and work-hardening coefficient, of bulk and coated materials can be determined from the DSI results, by applying reverse analysis procedures. Most of the achievements reached in this domain arise from the employment of numerical tools. The examples mentioned in this review concern the option of carrying out reverse analysis, i.e. to obtain the tensile curve of bulk materials and the Young's modulus of thin coatings, from the experimental load - indentation depth curve obtained by DSI, with resource of numerical simulation of the hardness test. Also, numerical simulations of indentation tests are an important tool to determine the surface residual stresses, using only the loading part of the indentation curve to achieve the results. In generally, the above mentioned reverse analysis methodologies are simple, fast and accurate procedures for mechanical properties determination. Finally, it is important to note that no special software is needed to perform the numerical simulations of the hardness tests. A large number of commercial codes can be used with this objective.

7. References

Antunes, J.M.; Cavaleiro, A.; Menezes; L.F., Simões, M.I. & Fernandes, J.V. (2002). Ultra-microhardness testing procedure with Vickers indenter, *Surface & Coatings Technology*, Vol. 149, No. 1, (1 January 2002), pp. 27-35, ISSN 0257-8972

Antunes, J.M.; Menezes, L.F. & Fernandes, J.V. (2006). Three-dimensional numerical simulation of Vickers indentation tests, *International Journal of Solids and Structures*, Vol. 43, No. 3-4, (February 2006), pp. 784-806, ISSN 0020-7683

Antunes, J.M.; Menezes, L.F. & Fernandes, J.V. (2007). On the determination of the Young's modulus of thin films using indentation tests, *International Journal of Solids and Structures*, Vol. 44, No. 25-26, (15 December 2007), pp. 8313-8334, ISSN 0020-7683

Antunes, J.M.; Fernandes, J.V.; Sakharova, N.A. & Menezes L.F. (2008). Reverse analysis in depth-sensing indentation for evaluation of the Young's modulus of thin films, *Philosophical Magazine*, Vol. 88, No. 3, (21 January 2008), pp. 313–325, ISSN 1478-6435

Bishop, R.F. & Mott, N.F. (1945). The theory of indentation and hardness tests, *Proceedings of the Physical Society of London*, Vol. 57, No. 321, pp. 147-159.

Bocciarelli, M. & Maier, G. (2007). Indentation and imprint mapping method for identification of residual stresses, *Computational Materials Science*, Vol. 39, No.2, (April 2007), pp. 381–392, ISSN 0927-0256

Bolshakov, A.; Oliver, W. C. & Pharr, G. M. (1996). Influences of stress on the measurement of mechanical properties using nanoindentation. 2. Finite element simulations, *Journal of Materials Research*, Vol. 11, No. 3, (March 1996), pp. 760-768, ISSN 0884-2914

Bucaille, J.L.; Stauss, S.; Felder, E. & Michler, J. (2003). Determination of plastic properties of metals by instrumented indentation using different sharp indenters, *Acta Materialia*, Vol. 51, No. 6, (2 Abril 2003), pp. 1663-1678, ISSN 1359-6454

Cao, Y.P.; Qian, X.Q.; Lu, J. & Yao, Z.H. (2005). An energy-based method to extract plastic properties of metal materials from conical indentation tests, *Journal of Materials Research*, Vol. 20, No. 5, (May 2005), pp. 1194-1206, ISSN 0884-2914

Casals, O. & Alcalá, J. (2005). The duality in mechanical property extractions from Vickers and Berkovich instrumented indentation experiments, *Acta Materialia*, Vol. 53, No. 13, (August 2005), pp. 3545-3561, ISSN 1359-6454

Chen X.; Yan J. & Karlsson, A. M. (2006). On determination of residual stress and mechanical properties by indentation, *Materials Science and Engineering: A*, Vol. 416, No. 1-2, (25 January 2006), pp. 139-149, ISSN 0921-5093

Dao, M.; Chollacoop, N.; Van Vliet; K.J., Venkatesh, T.A. & Suresh, S. (2001). Computational modelling of the forward and reverse problems in instrumented sharp indentation, *Acta Materialia*, Vol. 49, No. 19, (November 2001), pp. 3899-3918, ISSN 1359-6454

Doerner, M.F. & Nix, W.D. (1986). A method for interpreting the data from depth-sensing indentation instruments, *Journal of Materials Research*, Vol. 1, No.4, (April 1986), pp. 601-609, ISSN 0884-2914

Gao, H.; Cheng-Hsin, C. & Jin, L. (1992). Elastic contact versus indentation modelling of multi-layered materials, *International Journal of Solids and Structures*, Vol. 29, No. 20, pp. 2471-2492, ISSN 0020-7683

Golovin, Y.I. (2008). Nanoindentation and mechanical properties of solids in submicrovolumes, thin near-surface layers, and films: a review, *Physics of the Solid State*, Vol. 50, No. 12, (December 2008), pp. 2205-2236, ISSN 1063-7834

Hill, R. (1950). *The mathematical theory of plasticity*, Clarendon Press, Oxford, United Kingdom

Hirst, W. & Howse, M.G.J.W. (1969). Indentation of materials by wedges, *Proceedings of the Royal Society of London. Series A-Mathematical and Physical Sciences*, Vol. 311, No. 1506, pp. 429-444

Hill, R.; Lee, E.H. & Tupper, S.J. (1947) The theory of wedge indentation of ductile materials, *Proceedings of the Royal Society of London. Series A-Mathematical and Physical Sciences*, Vol. 188, No. 1013, pp. 273-289

ISO 14577 (2002). *Metallic materials – Instrumented indentation tests for hardness and materials parameters*, ISO Central Secretariat, Genebra, Suiça

Jang, J. I. (2009). Estimation of residual stress by instrumented indentation: a review, *Journal of Ceramic Processing Research*, Vol. 10, No. 3, pp. 391-400, ISSN 1229-9162

Johnson, K.L. (1970). The correlation of indentation experiments, *Journal of the Mechanics and Physics of Solids*, Vol. 18, No. 2, (April 1970), pp. 115-126, ISSN 0022-5096

Kick, F. (1885). *Das Gesetz der proportionalen Widerstände und seine Anwendungen*, Felix-Verlag, Leipzig, German

King, R.B., (1987). Elastic analysis of some punch problems for a layered medium, *International Journal of Solids and Structures*, Vol. 23, No. 12, (December 1987), pp. 1657-1664, ISSN 0020-7683

Lee, Y. H.; Takashima, K.; Higo, Y. & Know, D. (2004). Prediction of stress directionality from pile-up morphology around remnant indentation, *Scripta Materialia*, Vol. 51, No. 9, (November 2004), pp. 887-891, ISSN 1359-6462

Marsh, D.M. (1964). Plastic flow in glass, *Proceedings of the Royal Society of London. Series A-Mathematical and Physical Sciences*, Vol. 279, No. 137, pp. 420-435

Menčík, J.; Munz, D.; Quandt, E.; Weppelmann, E.R. & Swain, M.V. (1997). Determination of elastic modulus of thin layers using nanoindentation, *Journal of Materials Research*, Vol. 12, No. 9, (September 1997), pp. 2475-2484, ISSN 0884-2914

Oliver, W.C. & Pharr, G.M. (1992). An improved technique for determining hardness and elastic-modulus using load and displacement sensing indentation experiments, *Journal of Materials Research*, Vol. 7, No. 6, (June 1982), pp. 1564-1583, ISSN 0884-2914

Oliver, W.C. & Pharr, G.M. (2004). Measurement of hardness and elastic modulus by instrumented indentation: Advances in understanding and refinements to methodology, *Journal of Materials Research*, Vol. 19, No. 1, (January 2004), pp. 3-20, ISSN 0884-2914

Pharr, G.M. & Bolshakov, A. (2002). Understanding nanoindentation unloading curves, *Journal of Materials Research*, Vol.17, No. 10, (October 2002), pp. 2660-2671, ISSN 0884-2914

Prandtl, L. (1920). Uber die Harte plastischer Korper, *Nachrichten von der Koniglichen Gesellschaft der Wissenschaften zu Gottingen: Mathematish-physikalishce Klasse*, Vol. 1920, (February 1920), pp. 74-85

Sakharova, N. A.; Prates, P. A.; Oliveira, M. C.; Fernandes, J. V. & Antunes, J. M. (2011). A Simple Method for Estimation of Residual Stresses by Depth-Sensing Indentation, *Strain* (in press), online ISSN 1475-1305

Saha, R. & Nix, W.D. (2002). Effects of the substrate on the determination of thin film mechanical properties by nanoindentation, *Acta Materialia*, Vol. 50 No. 1, (8 january 2002), pp. 23-38, ISSN 1359-6454

Sneddon, I.N. (1965). The relation between load and penetration in the axisymmetric Boussinesq problem for a punch of arbitrary profile, *International Journal of Engineering and Science*, Vol 3, No. 1, (May 1965), pp. 47-57, 0020-7225

Suresh, S. & Giannakopoulos, A. E. (1998). A new method for estimating residual stresses by instrumented sharp indentation, *Acta Materialia*, Vol. 46, No. 16, (9 October 1998), pp. 5755-5767, ISSN 1359-6454

Swaddiwudhipong, S.; Tho, K.K.; Lui, Z.S. & Zeng, K. (2005). Material characterization based on dual indenters, *International Journal of Solids and Structures*, Vol. 42, No. 1, (January 2005), pp. 69-83, ISSN 0020-7683

Swift, H.W. (1952). Plastic instability under plane stress, *Journal of the Mechanics and Physics of Solids*, Vol. 1, No. 1, (October 1952), pp. 1-18, ISSN 0022-5096

Tabor, D. (1951). *The Hardness of Metals*, Clarendon Press, Oxford, United Kingdom

Tsui, T. Y.; Oliver, W. C. & Pharr, G. M. (1996). Influences of stress on the measurement of mechanical properties using nanoindentation. 1. Experimental studies in an aluminum alloy, *Journal of Materials Research*, Vol. 11, No. 3, (March 1996), pp. 752-759, ISSN 0884-2914

Wang, Q., Ozaki, K.; Ishikawa H.; Nakano, S. & Ogiso, H. (2006). Indentation method to measure the residual stress introduced by ion implantation, *Nuclear Instruments and Methods in Physics Research Section B: Beam Interactions with Materials and Atoms*, Vol. 242, No. 1-2, (January 2006), pp. 88-92, ISSN 0168-583X

Zhao, M.; Chen, X.; Yan, J. & Karlsson, A. M. (2006). Determination of uniaxial residual stress and mechanical properties by instrumented indentation, *Acta Materialia*, Vol. 54, No. 10, (June 2006), pp. 2823-2832, ISSN 1359-6454

7

Composites of Engineering Plastics with Layered Silicate Nanofillers: Preparation and Study of Microstructure and Thermomechanical Properties

Petroula A. Tarantili
National Technical Univ. of Athens,
School of Chemical Engineering, Polymer Technology Lab.,
Greece

1. Introduction

1.1 Polymer nanocomposites

Polymer/layered silicate nanocomposites have recently attracted a great deal of interest because of their unique properties, such as enhanced mechanical property, increased thermal stability, improved gas barrier properties and reduced flammability. According to the arrangement of silicate layers within the polymer matrix, two types of morphology can be achieved in nanocomposites: namely intercalated or exfoliated structures. Exfoliated structures have been well recognized as superior morphology for high performance at lower clay loadings, but are difficult to achieve. Attempts to improve processability and ensure efficient dispersion of the above fillers have led to modifications, e.g. with quaternary ammonium surfactants, which is expected to increase the inter-gallery spacing and provide enough hydrophobicity to clay particles in order to make them miscible with the polymer matrix.

A typical example is sodium montmorillonite (MMT), one of the most commonly used clays, which is hydrophilic and therefore shows restricted compatibility with many polymers. To obtain good interfacial adhesion for improved mechanical properties, the clay needs to be modified prior to incorporating into the usual organophilic polymer matrices. Clay modification can be achieved by an ion exchange reaction with organophilic cations (Utracki, 2004). There are two reasons for this modification: (i) the addition of an intercalating agent to increase the space between the layered silicates and make it more uniform and (ii) the addition of small organic molecules bonded to silicates to make MMT more miscible with the polymer matrix. Therefore, polymer molecules are allowed to enter the enlarged intergallery of silicates for further intercalation or exfoliation. In general the intercalated agents are small molecules of cationic surfactant, such dodecyl ammonium chloride and 1-hexadecyl ammonium blomide.

Clay possesses net negative charge on its lamellar surface and, therefore, it can absorb cations, such as Na^+ or Ca^{2+}. Alkyl ammonium ions can replace metal cations through a cation exchange process and occupy the gallery space between nanoscaled layers of the clay to alter the original silicate surface from hydrophilic to organophilic (Burnside and

Giannelis, 1995). Because of the negative charge of the silicate layer, the cations head group of an alkyl ammonium molecule preferentially resides at the layer surface with the aliphatic tails being removed from the surface (Chen and Yoon, 2005).
Nanocomposites can be prepared by solvent casting, by in situ polymerization or by melt compounding (Pinnavaia & Beall, 2002). So far, melt intercalation method has been the most commonly used procedure because of advantages especially in terms of commercial versatility and mass production ability.
Three types of polymer/clay composites are usually recognized: a)"conventional composites", in which the clay is added as a common filler, b)"intercalated nanocomposites", where a small portion of the polymer is inserted into the interlayer spacing between the layered silicates, and c)"exfoliated or delaminated nanocomposites", in which the silicate layers are almost or completely separated and dispersed in a continuous matrix.
While preparing nanocomposites by melt mixing, the exfoliation and dispersion of nanoclays in a polymer matrix depend on the type of organic modifier in the nanoclay, the initial interlayer spacing, the compatibility of the polymer matrix towards the nanoclay, the type and concentration of an added compatibilizer, the viscosity of the resin and the operational conditions. Although shear is, in general, a very important factor for efficient filler dispersion during mixing processes, it does not enable by itself to provide nanometric dispersion of the clay. Interfacial adhesion needs to be high enough in order to achieve better compatibility and, therefore, improved performance of the composite (Patiño-Soto et al., 2008).
Regarding melt processing, for polymers that require high melt temperatures, the thermal stability of the organic component of modified clay becomes a significant factor because it sets the top limit temperature for polymer processing.

1.2 Nanocomposites of poly(acrylonitrile-butadiene-styrene)

Poly(acrylonitrile-butadiene-styrene) (ABS) is a popular engineering thermoplastic because of its unique properties, including excellent mechanical response, good processing characteristics, chemical resistance and fine surface appearance. It consists of styrene-acrylonitrile (SAN) continuous phase partially grafted to a dispersed polybutadiene (PB) phase of elastomeric nature. The commercial ABS is produced in the main by emulsion copolymerization of styrene-acrylonitrile in the presence of rubber, and the final product is usually constructed with the physical mixture of styrene-acrylonitrile copolymer as well as the graft copolymer of styrene-acrylonitrile into the rubber. SAN exhibits the combined properties of the ease of processing of polystyrene and the rigidity and chemical resistance of acrylonitrile, whereas the incorporation of butadiene rubber into SAN imparts the high impact strength characteristic of ABS polymer. ABS is widely used in the automotive industry, telecommunications, business machines and consumer market, mainly because its property-price profile is intermediate between the lower priced commodity thermoplastics and the more expensive high performance engineering plastics. It is also used with the appropriate reinforcements that may be incorporated to the plastic matrix in order to enhance functional properties. One of the main drawbacks of ABS is its lower thermal stability and inherent flammability (Karahaliou & Tarantili, 2009). Therefore there is a need to increase its thermal stability and flame retardant properties (Owen & Harper, 1999).

The use of halogenated fire retardants (FR) are limited by the European Union legislation. The alternative inorganic FR's are often incorporated at such high loadings, i.e. up to 50 wt%, that there is a significant compromise in the final balance of properties for the composite, particularly toughness. Formation of ABS nanocomposite by addition of montmorillonite clay platelets, are believed to be a new approach in fire retardancy, which has been shown to be effective at low loadings and may lead to better balance of physical properties (Stretz et al., 2005).

1.2.1 Preparation

As already mentioned, melt blending is by far the most common method for preparation of polymer nanocomposites, because it involves the processing operations commonly adopted for the parent polymer.

A quick review of the related literature shows that ABS/MMT composites can be prepared by melt processing in a twin screw extruder (Wang et al., 2004b; Stretz et al., 2005; Kim et al., 2008; Patiño-Soto et al., 2007) or using a two-roll mill (Xiao et al., 2007; Yeh et al., 2005) or an internal mixer (Aalaie et al., 2007).

However, the high temperature, shear stress and local overheating induced by the shear itself can affect the clay through thermal degradation of the organic modifier, phase separation of the clay and possibly reduction of aspect ratio of the layered silicate (Dennis et al., 2001). In order to overcome these problems, a low temperature solvent/non-solvent method for preparation of ABS/clay nanocomposites has been developed by Pourabas and Raeesi, 2005, in which the organoclays were just dispersed in ethanol, a non solvent for ABS. Then the clay dispersion in ethanol was introduced to the ABS/THF solution and the ABS/clay nanocomposite was obtained after precipitation. The method utilized a special mechanical tool known as homogenizer, which consists of two co-axial cylindrical stator and rotating parts with a small gap between them. The homogenizer operates as a wet twin-roll mill, and the precipitating polymer together with OMMT undergo high shear stresses when processed between the rotor and stator cylinders. In that way, it is said that intercalated structure nanocomposites with a uniform interlayer spacing of the silicate layers were obtained. The above authors also mentioned that this method is appropriate for polymers with moderate solubility in common solvents. Higher solubility will cause difficulties in the precipitation step as enormous volume of non-solvent will be needed.

A new two-step technique, namely SOAM method, was established by Lim et al., 2010, where the polymer matrix and organoclay were blended in solution and then compounded in melt in a torque rheometer. Comparative studies with nanocomposites prepared via solution blending showed that SOAM method is more effective in providing high degree of dispersion of organoclay in the polymer matrix.

Masterbatch approaches, as a means of improving exfoliation, have also been investigated by Wilkie & Zheng, 2003 as well as by Guo et al., 2004. Poly(ε-caprolactone) (PCL) nanocomposites were prepared either by melt blending of PCL with OMMT or by in situ ring opening polymerization of ε-caprolactone (CL) with organically modified clay. The new organically-modified clay was then blended with ABS melt and exfoliated structures were produced (Zheng & Wilkie, 2003). Guo et al., 2004 used amphiphilic poly(styrene-*b*-ethylene oxide) block copolymers, prepared by anionic copolymerization, to modify layered silicates via melt intercalation. XRD and microscopy studies on ABS/intercalated clay nanocomposites prepared by melt blending demonstrated a good dispersion of silicate layers in the polymer matrix. The above researchers attributed this behaviour to strong interactions from hydrogen bonds between the layered silicate and PEO block chain segments, making the block copolymer to adhere to the interlayer and act as intercalating agent. They also observed that ABS nanocomposites showed considerable improvement in its tensile strength, modulus and thermal stability.

Emulsion and in situ polymerization processes for the preparation of ABS/clay nanocomposites have been reported by Jang et al., 2001 and Kim et al., 2011. Jang et al., 2001 prepared ABS/MMT-Na^+ nanocomposites by direct intercalation through one-step emulsion polymerization. The mixture of acrylonitrile and styrene monomers and 1,3-polybutadiene (PBD) rubber latex was copolymerized in the presence of MMT-Na^+. By SEM and TEM images it was observed homogeneous dispersion of MMT particles in the polymer matrix. TGA measurements indicated enhanced thermal stability, whereas DSC thermogram does not exhibited any thermal transition.

1.2.2 Effect of type of clay on the nanocomposites morphology

The effect of clay size on the dispersion morphology and emulsion stability during preparation of ABS/layered silicate nanocomposites through in situ emulsion polymerization was studied by Kim et al., 2011. Three types of silicates: Laponite, Cloisite-Na and Kunipia-F with particle sizes about 20-30, 70-150 and 300-500 nm, respectively were used. The obtained results indicated that, ABS layered silicate nanocomposites produced by using Cloisite-Na and Laponite showed exfoliated structures and stable emulsions. On the other hand, nanocomposites with Kunipia-F contain aggregated particles and separated phases. The nanocomposite reinforced with a mixture of Kunipia-F/Cloisite-Na or Kunipia-F/Laponite showed stable emulsion. This behaviour was attributed to the small sizes of Cloisite-Na and Laponite which also showed good efficiency to stabilize the emulsion particle during polymerization.

As stated above, the compatibility of polymer matrix towards the organic modification of clay reinforcement is a very critical parameter affecting the nanocomposite structure and the enhancement of final properties. Four different commercial clays (Cloisite-Na, 10A, 20A, 30B) were used by Patiño-Soto et al., 2008 for preparation of ABS matrix nanocomposites by processing in a twin-screw extruder. The above authors observed that Cloisite 20A and Cloisite 30B (the first owing greater initial intergallery spacing but lower polarity and the second with smaller intergallery spacing but greater polarity), produce ABS nanocomposites with maximum intergallery spacing, better thermal stability and fire retardancy.

On the other hand, Lim et al., 2010 have studied the solution technique followed by melt blending process for preparation of ABS nanocomposites and reported that the system ABS/Cloisite 30B gave more exfoliated structures and better thermal and mechanical properties in comparison with the results obtained for Cloisite 10A and Cloisite 25A reinforcements. The Flory-Huggins interaction parameter of the system ABS and Cloisite 30B was calculated and found smaller than that for ABS/Cloisite 10A as well as for ABS/Cloisite 25A. Therefore, thermodynamic interaction between ABS and Cloisite 30B was predicted more favourable than that of the other two systems.

1.2.3 Clay modification

Su et al., 2004 prepared new organically modified clays using copolymers of styrene (COPS) and methacrylate (MAPS) with vinylbenzyl chloride. Both these polymeric clays offer advantages in thermal stability when compared to those treated with conventional ammonium salts and this may enable melt processing of polymers requiring higher temperatures. From TEM images it is clear that in styrene copolymer-containing clay/ABS nanocomposites individual clay layers are present, while the methacrylate copolymer gives some mixtures of immiscible and intercalated structures. Thermal stability is higher for the COPS systems, since styrene is inherently more thermally stable than methyl methacrylate.

In another work an oligomerically modified clay (with styrene, vinylbenzyl chloride and lauryl acrylate), has been used to fabricate ABS nanocomposites by melt blending and mixed delaminated/intercalated structures were obtained (Zhang et al., 2006a). The plasticizing effect from the oligomer was successfully suppressed by increasing the inorganic clay content to 50%. The decreased organic content of clays reduces plasticization, giving better mechanical properties, but also it creates an adverse effect on the dispersability of clay in the polymer, leading to morphological changes from nanocomposite to microcomposites (Zhang et al., 2006b).

A modified clay with enhanced thermal stability and larger d-spacing compared to commercially available clays, was developed by Chigwada et al., 2006, who used an ammonium salt containing 4-acetylbiphenyl (BPNC16). BPNC16 modified clay has enhanced thermal stability and large d-spacing compared to some commercially available clays. The nanocomposites prepared with this clay showed improved thermal stability and a significant reduction in the peak heat release rate from cone calorimetric measurements.

ABS/clay nanocomposites containing imidazolium-modified montmorillonite were prepared by Modesti et al., 2008, via the classic melt intercalation and a low-temperature solution process. WAXD and TEM show that with both processes, a mixed intercalated/exfoliated structure was obtained, however confocal microscopy showed that the clay is not homogeneous dispersed and that micrometer aggregates of clay tactoids are present. The sonication process reduces the size of these aggregates as compared to melt blending and improves the degree of dispersion. This effect was confirmed by the values of elongation at bleak and flexural elastic modulus obtained from composite specimens. Fluorescence spectroscopy has also been used to investigate the distribution of clay within the composite and results indicate that clay layers in ABS were preferentially located in the SAN phase, independently of the dispersion process.

Based on the structure characteristic of ABS resin and the principle "like dissolves like", hexadecyl triphenyl phosphonium bromide (P16) and cetyl pyridium chloride (CPC) were used by Cai et al., 2010 to prepare a high thermal stability OMMT. Intercalated structure for ABS/OMMT-P16 hybrids and intercalated-exfoliated structure for ABS/OMMT-CPC nanocomposites were respectively formed by melt mixing method. TGA results revealed that the onset temperature of thermal degradation and charred residue at 700°C of the ABS nanocomposites were remarkably increased compared to pure ABS. It was also found from Cone calorimetry tests, that the peak of heat release rate (PHRR) decreased significantly, contributing to reduced flammability. In addition, DMA measurements indicated that the loading of silicate clays improved the storage modulus of ABS resin. Partial exfoliation of OMMT-CPC nanocomposites was, therefore, advantageous for increasing thermal stability, decreasing flammability and improving mechanical properties.

Other researchers, treated commercial OMMT (Cloisite25A) with bis(3-triethoxysilylpropyl) tetrasulfane (TSS), in order to enhance interactions between clay and ABS through the chemical reaction of tetra sulfide groups with vinyl groups present in ABS (Kim et al., 2008).

Intercalated/exfoliated coexisting structures were observed by TEM. The ABS/clay composites exhibited much higher tensile strength and elongation at break than that of neat ABS and especially the elongation at break of ABS/TSS-C25A composite was 5 times higher than that of neat ABS.

1.2.4 ABS modification to improve compatibility

One approach to induce attractive interactions between a polymer and layered silicates is to employ a third component, commonly referred as compatibilizing agent, which can help improve compatibility between the polymer and the organoclay. Specifically Lim & Park, 2001 employed polystyrene grafted with maleic anhydride, PS-g-MA, as compatibilizing agent to prepare nanocomposites consisting of PS and organoclays. Polypropylene grafted with MA, PP-g-MA, was employed as compatibilizing agent to prepare nanocomposites of PP and organoclays (Solomon et al., 2001; Galgali et al., 2001). A similar attempt to improve compatibility between clay reinforcement and ABS matrix has also been made by Ma et al. 2006, who tried to increase ABS polarity by appropriate modification. They performed grafting of maleic acid anhydride (MAH) to ABS and investigated the effect of this modification on the morphology, thermal stability, flammability and dynamic mechanical properties of the obtained ABS/OMMT nanocomposites. Maleic anhydride was successfully grafted into butadiene chains of the ABS backbone in the molten state using dicumyl peroxide as initiator and styrene as the comonomer. TEM images showed that the size of the dispersed rubber domains of ABS-*g*-MAH increase and the dispersion is more uniform than that taking place in pure ABS resin. Intercalated/exfoliated structures were formed in ABS-*g*-MAH/OMMT nanocomposites and the rubber phase intercalated into clay layers that were distributed in both SAN and rubber phase. The intercalation of clay in rubber phase is probably due to grafting of MAH to butadiene chain, thus improving the polarity of this phase, which results in stronger affinities between clay and PB and, therefore, rubber phase can intercalate into clay layers. These nanocomposites exhibit better barrier properties, thermal stability and significantly reduced flammability as compared with ABS/OMMT systems. The improvement of flame retardancy of ABS-*g*-MAH/OMMT nanocomposites can be ascribed to the better dispersion of clay layers and intercalation of clay in both PAN and PB phases. Moreover, T_g of ABS-g-MAH/OMMT was 1.5°C higher than that of ABS/OMMT nanocomposites, which suggests that intercalated/exfoliated structure is more effective in improving rigidity and barrier ability than intercalated structures of ABS/clay hybrids.

Other types of treatment were studied by Kim et al. 2003, who reported the synergistic effect of triphenyl phosphate (TPP) and nanosilicates on the enhancement of thermal stabilization of ABS/clay nanocomposites.

1.2.5 Effect of ABS monomers composition in the nanocomposites morphology

It is well known that among the three initial monomers of ABS, acrylonitrile (AN) is the most polar one and its content would affect polar interactions with modified clays in nanocomposite. The effect of AN content on the structure of ABS nanocomposites was studied by Patiño-Soto et al., 2007 and the obtained results confirmed that better intercalation and/or exfoliation can be obtained when using ABS with higher AN content, due to the higher content of polar groups, which allows more homogeneous dispersion and intercalation of polymer chains within OMMT layers and, furthermore, some exfoliation of the nanoclay. The study of properties using TGA and flammability tests showed that the nanodispersed silicate layers enhanced the thermal stability of the ABS matrix and that a grade of ABS with higher AN content was more effective in providing fire retardancy. DMA showed that the storage modulus of both, low and higher AN content ABS polymers, was quite similar. The incorporation of organoclay increases the modulus of both examined ABS and this effect is slightly stronger when ABS with higher AN content was used. Similar behaviour was observed for T_g calculated from the maximum of tan δ (Patiño-Soto et al., 2008).

As mentioned earlier, ABS is a complex system, with the rubber particles (polybutadiene) dispersed in the matrix of poly(styrene-co-acrylonitrile) (SAN). Therefore the addition of a third component, i.e. the nanoclay filler, makes the system more complicated. SAN is the matrix phase of the multiphase ABS, thus SAN/MMT could be considered a model system useful for understanding ABS/MMT compatibility. Stretz at al., 2005 concluded that clay particles reside predominantly in the SAN phase with some accumulation at the rubber particle surfaces. The effect of reinforcing was somewhat lower in ABS/clay nanocomposites when compared to SAN/clay hybrids. The reason has been attributed to the different moduli of ABS and SAN matrixes and to the random orientation of clay particles in ABS/MMT composites at the rubber particle surface. They concluded that the SAN/MMT composites represent a good model system for studying the more complex ABS/MMT composites.

In order to explore the effect of different phases of ABS on the preparation of ABS nanocomposites, an indirect method of preparation of SAN/clay nanocomposites followed by the addition of ABS, was studied in comparison with the direct method by which nanocomposites were prepared directly using commercial ABS. It was observed that with the indirect method, the compatibility between the components was reduced because clay layers were already dispersed inside the SAN matrix. In nanocomposites prepared by the direct method, less dispersion of the layers and accumulation of the clay particles in the interphase regions was observed (Pourabbas and Azimi, 2008). Kinetic analysis carried out with the data from dynamic TGA showed that the addition of *g*ABS to SAN nanocomposites reduced the activation energy (E_a) but decreased the reaction order (n) on the global degradation reaction. Relatively higher E_a values were obtained for indirectly prepared ABS nanocomposites, however the steady values of n indicated that the thermal degradation mechanism remains unchanged independent of the preparation methods employed. It was also found that reduction of E_a and n, by increasing the clay content in the nanocomposites. Even though the effect of enhancement of thermal stability due to clay nanofillers, (observed in most of the polymeric nanocomposites and attributed to the barrier property and char promoting effect of the clay, which protects the inner materials from further decomposition or combustion), the reduced E_a values by increasing clay content showed their accelerating effect on thermal degradation reactions. Pourabbas & Azimi, 2008 concluded that clay particles or layers take part in thermal degradation mechanisms and they must not be regarded as inert material.

Many studies on ABS nanocomposites are concentrated on the scope to balance their mechanical and improve thermal and flammability properties. In a recent work (Stretz at al., 2005), detailed results on mechanical properties of ABS/clay nanocomposites were reported, showing a remarkable decrease in impact strength. Since the impact strength of ABS/clay nanocomposites is very important mechanical property for technical applications, the optimization of formulation seems to be necessary. One approach to improve the impact strength is to increase rubber content in ABS/clay nanocomposites. Kim et al., 2009 investigated the effect of rubber content in ABS on the mechanical properties of ABS/clay nanocomposites, in an attempt to find the optimum content with good balance of mechanical properties. Three different types of ABS with different rubber content (25, 30 and 35 wt%) were tested and it was concluded that for addition of 3 % wt clay, ABS nanocomposite with 35% content of rubber displayed the highest reinforcing effect, in terms of tensile properties and impact strength. Furthermore, the addition of UV stabilizer further improved the impact strength of the optimum nanocomposite composition. The above researchers mentioned that increasing rubber content may lead to a lower degree of dispersion of clay and/or to a lower degree of orientation and this effect became stronger with increasing clay content.

1.2.6 Characterization of ABS/organoclay nanocomposites

Enhancement of improvement of the thermal stability and mechanical strength was observed by Yeh et al. 2006, by testing samples of a series of injection molded ABS/organoclay nanocomposites at low clay loading. In addition, they observed that incorporation of organoclay platelets into ABS matrix resulted in an increase of T_g, associated with the confinement of the intercalated polymer chains within the organoclay galleries that prevent the segmental motions of polymer chains. They also measured an increase of hydrophobicity of nanocomposites, based on the contact angle measurements.

Studies of thermal properties in intercalated/exfoliated ABS/OMMT nanocomposites by Wang et al., 2002 using thermogravimetric analysis showed that the pyrolysis of ABS was divided into two steps. They explained the improved thermal stability of ABS matrix in the formation of carbonaceous-silicate char on the surface during burning, which insulated the underlying material and slowed the escape of the volatile product generated during decomposition. The nanodispersed lamellae of clay in polymer matrix will result in a spatially more uniform and thicker char on decomposition.

The increase in decomposition temperature during TGA in ABS nanocomposites might be taken as measure of the occurrence of clay intercalation-exfoliation, since the intercalated or exfoliated clay will act as a barrier for the diffusion of atmospheric oxygen into the material, inhibiting the polymer decomposition (Morawiec et al., 2005). Since TGA is a dynamic experiment, another possible reason for the increase of decomposition temperature of the ABS nanocomposites, is the increase of viscosity due to the addition of clay, which will increase the difficulty for the inward diffusion of oxygen and the outward diffusion of the degradation by-products (Valdes et al., 2006).

Patiño-Soto et al., 2007, observed that in all cases ABS nanocomposites produce more char than pure ABS. Several studies have shown that the lower flammability of polymer-clay nanocomposites is related to the formation of char with a multilayer carbonaceous-silicate structure. This char builds up on the surface during burning, insulating the underlying material and slowing the escape of the volatile products generated during thermal decomposition. The observed increase in char formation in the ABS nanocomposites indicates that the clay enhances the formation of char on the surface of ABS and as a consequence, reduces the rate of decomposition and improves thermal stability.

As mentioned above, flammability of ABS restricts its application and, therefore, many halogenated organics have been used as flame retardant additives. Nowadays, a lot of emphasis has been placed on the replacement of halogen-type flame retardants by non-halogen types and the flame retardants based on phosphorus-containing compounds are known to be the most acceptable candidates for this purpose, still having some problems with toxicity. Triphenyl phosphate (TPP) and its analogues are known to be the most effective candidate for many polymers including ABS. On the other hand, polymer/clay nanocomposites are believed to be a new promising approach to fire retardancy.

Kim et al., 2003 reported the synergistic effect of nano triphenyl phosphate and nanosilicates on the enhancement thermal stabilization of ABS/clay nanocomposites. Epoxy resin and silane coupling agents were incorporated as flame co-retardants. A very large increase in the limiting oxygen index (LOI) value was observed with epoxy addition and further enhancement in thermal stability was obtained for ABS compounds containing small amounts of coupling agent. It was also found that the enhancement was closely related to the morphologies of the chars formed after combustion.

TTP was also used by He et al., 2010, who tried to improve the thermal properties and flame retardancy of ABS/OMMT nanocomposites. The results by cone calorimetry analysis indicated that the reduction of peak heat release rate for ABS/OMMT and ABS/OMMT/TTP nanocomposites was 47.3 and 57.3% respectively, while the total heat released became lower to about 10 and 17% and the mass loss rate decreased by about 41 and 53% respectively, as compared to pristine ABS. The Limiting Oxygen Index (LOI) value of ABS/OMMT and ABS/OMMT/TTP nanocomposites increased to 22.8 and 28 from 18% respectively with respect to pristine ABS. The vertical burning test (UL-94V) results revealed that the nanocomposites could achieve V-0 grade.

Rheological investigation by Aalaie & Rahmatpour, 2007, on ABS/OMMT hybrids has shown that the prepared nanocomposites and their pristine counterpart have shear-thinning behaviour, obeying the power law equation. At low shear rates, the steady shear viscosity and shear stress of the nanocomposites increase with increasing the nanoclay content. However, at high shear rates they behave similar to pure ABS. It was shown that the flow activation energy (E_a) values increase with increasing nanoclay content. Mechanical tests showed that the flexural moduli of nanocomposites increase with increasing nanoclay loading, but flexural strength as well as tensile and impact properties decrease with the increase of nanoclay content.

Viscoelastic properties of ABS/organoclay nanocomposites prepared by the SOAM method and the solution blending method were studied by Lim et al., 2010, who measured the rheological properties under oscillatory shear condition. They observed that the storage modulus G′ became less dependent on the frequency with increasing clay loading and this was due to the formation of network structure of the clay plates, indicating solid-like characteristics. Furthermore, nanocomposites prepared by the SOAM method showed more solid-like behaviour and exhibited higher increase of storage modulus throughout the frequency range, than those prepared by solution blending.

From the results of linear and nonlinear viscoelastic measurements in ABS/clay melts it was found that, in addition to the physical network originated from interconnection of rubber particles, a 3D network is formed between organoclay platelets and/or tactoids that causes reduced temperature dependence of linear viscoelastic properties of nanocomposite samples as compared with ABS matrix (Saadat et al., 2010).

Extrudate swell, also called die swell, is one of the most important viscoelastic responses of polymer melts and play a key role in processability of polymers and die design in extrusion process. According to Saadat et al., 2010 the presence of organoclay reduced die swell of

ABS/clay nanocomposites and this was explained in terms of great surface area and anisometric nature of organoclay tactoids and/or platelets, which promote energy consumption and less energy to be stored in chains.
The OMMT/ABS nanocomposite materials were subjected to a series of five succesive extrusion cycles, in a co-rotating twin screw extruder and the effect of reprocessing on the properties of this compound was investigated by Karahaliou & Tarantili, 2010. By ATR/FTIR analysis it was observed that the above reprocessing of OMMT/ABS composites resulted in a progressive increase of oxidation products. More specifically, the concentration of carbonyl groups increased gradually during the extrusion processing, but at a lower extent compared with the increase of ester-groups. On the other hand, the ester-groups showed a dramatic increase after the second extrusion process, followed by lower changes up to the fifth pass. The band corresponding to -C≡N group, present in AN, remained almost stable and, therefore, it can be concluded that this group is relatively resistant against these processing conditions. The color of samples was rather constant through the successive extrusion cycles. On the other hand, the melting temperature was slightly decreased, whereas glass transition temperature (T_g) was almost stable. Extrusion reprocessing did not have any significant effect on the temperature of the maximum thermal degradation rate (T_{max}), whereas a slight decrease in the onset temperature of thermal degradation (T_{onset}) as it was observed by TGA. Finally, the rheological and tensile properties of the investigated composites did not display any significant change by increasing the number of reprocessing cycles. The above researchers concluded that OMMT nanoparticles did not only act as reinforcement for ABS but also seemed to play a role on the ageing mechanism taking place during repeated melt processing. Therefore, the incorporation of the above nanofiller into ABS/PC blends could be considered as an alternative option for the mechanical recycling of ABS/PC wastes, providing low cost and easy processing.

1.3 Nanocomposites of polycarbonates

1.3.1 Basic aspects and preparation

Polycarbonate (PC) is commonly used as high-performance amorphous engineering thermoplastic because of its distinct properties, such as high-impact strength, transparency, heat resistance and dimensional stability. In addition, this family of polymers show excellent electrical properties, colorability, high-gloss, flame retardancy and high-heat distortion temperature (HDT). However, it is notch sensitive and, moreover, difficult to process, because its high-melt viscosity hinders fluidity and the residual stresses resulting from the process could be a potential hazard for fracture formation. Approaches were developed to decrease the notch sensitivity and enhance processability without reducing the superior toughness of PC. Polycarbonate has been modified and tailored in many different ways, particularly with other polymers for use in demanding applications particularly when its outstanding impact strength is a critical factor. Recently, nanocomposite technology has been applied to modifying PC. Polycarbonate nanocomposites, because of the nanoscopic length scale, offer the promise of improved physical properties, such as strength, modulus and scratch resistance without sacrificing optical clarity and toughness.
Recently, Huang et al. reported on polycarbonate-layered silicate nanocomposites prepared by two different methods. A partially exfoliated cyclic oligomer (OLS) nanocomposite was obtained by mixing cyclic carbonate and ditallow dimethyl exchanged MMT in a brabender

mixer (Huang et al., 2000). Subsequent ring-opening polymerization of the cyclic oligomers preserved the exfoliated structure affording a route to polycarbonate-OLS nanocomposites. An analogous mixing experiment with linear polycarbonate and OLS only produced an intercalated structure. Conventional melt or solution processing of the above mentioned OMMT with either linear polycarbonate or cycling oligomers yielded intercalated nanocomposites. These results demonstrate that consideration of molecular architecture (cyclic versus linear) and kinetics (medium viscosity and shear) are critical for nanocomposite formation.

1.3.2 Types of fillers and mixing techniques

In another work, the influence of various processing variables and that of organoclay structure on the degradation process and color formation of PC nanocomposites was studied by Yoon et al, 2003b, who used two different twin screw extruders and a series of organoclays based on sodium MMT, with relatively high iron content, exchanged with various amine surfactants. It was found that the extruder with longer residence time and broader residence time distribution was the most effective equipment for dispersing the clay but gave products with more color. Molecular weight degradation of the PC matrix during melt processing produces phenolic end groups and more extended dispersion of the clay generally lead to higher reduction in molecular weight due to the increased surface area of clay exposed. Double bonds in the hydrocarbon tail of the surfactants lead to more darkly colored materials than saturated surfactants. The most severe color change was observed when using a surfactant containing hydroxyethyl groups and a hydrocarbon tail derived from tallow.

A selected series of organoclays based on the synthetic clay Laponite® and calcium MMT from Texas (TX-MMT) were also prepared in an attempt to explore the effect of the clay structure on the overall behavior of final nanocomposite. Laponite® and TX-MMT produce less color polluted products of MC nanocomposites as compared with MMT, probably due to the lower content of iron (Yoon et al., 2003b).

Lee & Han, 2003 studied the linear dynamic viscoelastic properties and non-linear transient rheology of PC/organoclay (Cloisite 30B) and PC/natural clay nanocomposites prepared by melt blending in a twin-screw extruder. The results obtained by FTIR spectroscopy on extrudates, at temperatures ranging from 30 to 280°C, showed that carbonyl groups in PC and hydroxyl groups of Cloisite 30B in PC/Cloisite 30B nanocomposites can form hydrogen bonds, while no evidence of hydrogen bonding was observed in PC/Cloisite Na^+ nanocomposites. TEM images showed that organoclay platelets are well dispersed in PC/Cloisite 30B nanocomposites, whereas the untreated clay platelets are poorly dispersed in PC/Cloisite Na^+ nanocomposites. The rheological measurements supported the conclusion based on the results of XRD, TEM and FTIR spectroscopy. An increase in dynamic moduli of nanocomposites was observed with increasing concentrations of Cloisite 30B. This was attributed to an increase of the surface area of dispersed layered silicates of Cloisite 30B platelets as the concentration of clay increases and, consequently, an increase of the number of sites available for hydrogen bonding between carbonyl groups in PC and the hydroxyl groups in MT2EtOH of Cloisite 30B takes place.

An alternative technique for making PC nanocomposites was based on the melt polymerization of bisphenol-A with diphenyl carbonate. The prepolymers were intercalated with OMMT by melt mixing and solution mixing. Subsequently, micro-wave solid-state polymerization converted the pre-intercalated into exfoliated nanocomposite, whereas conventional solid-state polymerization using oil heating only increased the gallery size of MMT (Yoo et al., 2004).

Carrión et al. 2008, prepared a new PC nanocomposite containing 3 wt% of OMMT bentone 2010 and shaped some samples from this system by extrusion and injection moulding. Then, its tribological properties were determined under a pin-on-disc configuration against stainless steel. The nanocomposite presented a 88% reduction of friction and up to two orders of magnitude reduction of wear rate with respect to the base polymer. The good tribological performance of the new composite was attributed to the uniform microstructure and to the increase of nanoclay stacking distance.

Wu et al., 2007, prepared PC/clay nanocomposites by melt intercalation using epoxy resin as a compatibilizer. The above researchers reported that the addition of epoxy can improve clay dispersion, enhancing the low-frequency viscoelastic responses, while high loadings of epoxy lead to severe degradation of PC matrix, decreasing the high frequency responses and, furthermore, creating a plasticizing effect due to excessive epoxy. Moreover, all samples showed high sensitivity to both, quiescent and large amplitude oscillatory shear deformation, despite enhanced percolation of tactoids due to the compatibilization of epoxy.

1.3.3 Thermal and mechanical properties

As already mentioned, OMMT is unstable at high processing temperature, tends to degrade and forms agglomerates on the high processing temperatures of PC, i.e. at about 220-260°C. Thus, it could suppress the dispersion of clays in the PC matrix and subsequently weaken the properties of composites. For achieving a good PC/MMT nanocomposite, two criteria should be met, that are good thermal stability of MMT and good compatibility between PC and MMT.

Severe et al., 2002 studied the thermal stability of polycarbonate nanocomposites formed in a twin screw extruder, using phosphonium exchanged montmorillonite and synthetic clays. They found that the former provided better thermal stability in comparison with synthetic clays. Also, a series of flame retardant formulations of solid bisphenol A bis(diphenyl phosphate) (S-BDP) and OMMT were prepared by melt compounding procedures. OMMT was well dispersed into the matrix showing an intercalated-exfoliated morphology. S-BDP and OMMT exhibit a synergistic effect in the vertical burning test (UL-94) but an antagonistic effect in the limiting oxygen index (LOI) evaluation. The introduction of OMMT could especially enhance the thermal-oxidative stability of the material (Feng et al., 2010).

Organoclay nanocomposites based on medium molecular weight PC (MMW-PC) and high molecular weight PC (HMW-PC) were prepared using a twin screw extruder (Yoon et al, 2003a). Tensile tests showed that HMW-PC gives better stiffness and ductility to the nanocomposite than MMW-PC. This effect was attributed to the higher shear stress generated during melt processing. The effect of chemical composition of the surfactant on the morphology and physical properties of nanocomposites were explored for systems formed from HMW-PC. The surfactant having polyoxyethylene and octadecyl alkyl tails produces the most significant improvement of modulus and lead to partially exfoliated platelets and this was attributed to the miscibility of poly(ethylene oxide) tail with PC. However, nanocomposites formed from a range of other organoclays contained both intercalated tactoids and collapsed clay particles with few, if any, exfoliated platelets.

Chow & Neoh, 2009 used three different loadings of 3-aminopropyl-triethoxysilane (APS) to modify the $MMTNa^{+}$ via cation exchange technique. Nanocomposites were prepared by melt compounding and their morphology was characterized by using XRD and TEM, which revealed partial intercalation and exfoliation of silane-treated montmorillonite (STMMT) in the PC matrix. Increase of APS concentration significantly enhanced the storage modulus (E') and improved the thermal stability of PC nanocomposites. Their results indicated that APS was effectively interacting with PC resins and clay, and subsequently enhancing the thermal stability of PC nanocomposites. This was due to the crosslinked structure between APS and clay which made the degradation more difficult. The improvement in storage modulus could be associated to the stiffness of these layers on molecular motion of polymer chains, the high aspect ratio and degree of dispersion of STMMT. In highly intercalated and exfoliated nanocomposites, individual clay layers with high aspect ratio are dispersed homogeneously in the polymer matrix. Their results indicated that the stiffness of the nanocomposite was contributed by the clay loading instead of the concentration of the modifier.

A series of PC/organoclay nanocomposites from two commercial surface-modified MMT were prepared by melt compounding using a twin screw extruder (Nevalainen et al., 2009). The PC nanocomposites showed rather good dispersion of nanoclay, with a mixture of exfoliated, intercalated and confined morphology. The type of surfactant did not have a significant effect on the overall dispersion of nanoclay. The effect of clay reinforcement on the mechanical response of specimens subjected to tensile and impact test demonstrated an increase of Young's modulus and yield strength. A transition from ductile to brittle deformation Polycarbonate upon tensile loading occurred at the investigated concentrations of clay loadings. Notched impact strength experiments supported this behaviour, showing that impact strength decreases significantly as nanoclay content increases from 1 to 5%, regardless of the performed nanoclay surface modification. According to TGA results, PC nanocomposites showed a complex behaviour in comparison with pure PC. Incorporation of nanoclay lowers the initial decomposition temperature but at the final stages of decomposition the weight loss of PC nanocomposites shifted to some higher temperatures than those for with pure PC. Furthermore, T_g dropped as a function of clay loading probably because of the enhanced chain mobility in the presence of filler. The tribological properties of selected specimens were evaluated using a pin-on-disc and it was observed that nanoclay did not significantly change the tribological behavior of PC, possibly due to the tendency of the studied nanoclays to form micro-sized agglomerates in the PC.

1.4 Nanocomposites of ABS/PC blends

1.4.1 General characteristics and morphology

It also should be noted that ABS can be easily mixed with polycarbonate (PC) polymer and the obtained alloy is widely used in the production of electric and electronic equipment, due to a balanced combination of properties of the two components. ABS/PC blends are used in the computer and business equipment industry to mould housings, such computer monitors, printer peripherals, laptop cases and, more recently, flat panel LCD television bezels and housings. PC has high thermal stability and good impact behaviour whereas ABS shows easy processability and economical benefits. When PC is blended with ABS, the polycarbonate matrix contributes to mechanical and thermal properties, whereas the ABS contributes to processability, economy and more reliable impact strength. The desired mechanical properties and thermal stability properties can be achieved by changing

PC/ABS ratio. Thermoplastics including PC/ABS blends are easily combustible and, however, have several applications in electronics, electrical and car industries, where plastic parts must display low flammability. To achieve an optimum level or fire retardancy, a large amount of non-halogenated type flame retardants (FRs) is required in the formulation. However, the addition of large amounts of FRs could decrease mechanical properties of resin and also affects its rheological behaviour.

Extensive investigation of the morphology of PC/ABS blends throughout the entire composition range was carried out by Lee et al., 1992. The PC-rich blends showed that dispersed ABS obtained a "bead-and-string" structure. With increasing ABS content the bead-and-string structures became more densely arrayed and more interconnected. A transition was observed between the 70/30 and 60/40 compositions, from the bead-and-string structure to a coalesced configuration. For the 40/60 compositions, PC formed a dispersed phase with SAN inclusions (Lee et al., 1992).

From studies on different compositions of ABS/PC blends it was observed that ABS seems to act as plasticizer when incorporated to PC at concentrations above 20%, probably due to the dispersion of polybutadiene phase into the thermoplastic matrix of the blend (Tarantili et al., 2010). Differential Scanning Calorimetry tests showed dependence of the glass transition temperature (T_g) on the composition of the examined blend, which suggests certain interactions among the components of the mixture. Therefore, it was concluded that the ABS/PC system shows a wide range of miscibility of its components up to a minimum of 20%ABS concentration. The change of T_g was much more intense for PC phase, which was interpreted assuming the plasticizing effect of ABS on PC. Furthermore, this behaviour was confirmed by the determination of rheological characteristics of the blend via Melt Flow Index (MFI) measurements. A decrease in mechanical properties, in terms of tensile strength and elongation, was also recorded, whereas the modulus of elasticity showed some improvement.

As mentioned above PC/ABS mixtures display partial miscibility and, therefore, modification for the improvement of their performance is necessary. Elmaghor et al., 2004 studied the effect of maleic anhydride grafted ABS (ABS-g-MA)/PC blends. The above authors observed improved dispersion of ABS-*g*-MA in PC, resulting in excellent toughening. However, it was shown that the effect of grafting of MA to ABS was not the promotion of compatibility, but the creation of special morphology of ABS domains dispersed in PC matrix. At a certain PC/ABS-g-MA weight ratio, the ABS domains connected together to form a network and gave rise to a maximum of the notched impact strength.

Organoclay nanocomposites have attached considerably interest as an alternative method to improve FR performance and mechanical properties of polymers. Wang et al., 2003 prepared and studied nanocomposites based on alloys of ABS/PC blends reinforced with MMT nanoparticles. Intercalated structure was obtained and improvement in the thermal stability of the mixture ABS/PC was reported by the incorporation of MMT. The morphology of these alloys nanocomposites indicate that it is mainly the ABS molecules which are intercalated into the clay layers rather those of PC. Further study by Wang et al., 2004b on the dynamic self-assembly of clay layers in ABS/PC phases, made in samples of nanocomposites prepared by melt mixing in a twin screw extruder, revealed that in the interphase region, a higher density of the dispersed clay particles is observed as compared to the density inside the ABS phase. The excess silicate layers in the area of interphase is presumably the result of segregation of the clay particles from the surface, attributed to the surface tension or from the crystal growth front attributed to the exclusion. This segregation phenomenon is greatly dependent on the choice of polymers and, more specifically, the more compatible or the greater the intercalation of polymer with OMMT, the more the density of clay layers is dispersed. They also reported that with the increase of melt-mixing time, the clay layers were self-organized from the PC phase to the ABS phase. It is well known that the presence of acidic or basic impurities will enhance depolymerization of carbonic ester in PC. However, OMMT can produce Lewis or Brönsted acid sites in the aluminosilicate when heated over 200°C, which have a positive contribution to the thermal degradation of PC. Thus, the self-organization of clay layers in ABS phase may be helpful for avoiding the enhancing effect of clay on PC during decomposition and enhance the thermal stability of ABS/PC.

Studies of the morphology on ABS/PC (70/30 wt%) clay nanocomposites by Hong et al., 2007, by TEM confirmed the fact that clay existed in the ABS phase and the interphase of ABS and PC. With further study using SEM, they mentioned that droplet size of the PC in the ABS/PC/clay nanocomposites did not change significantly with the presence of clay and has been found to be from 1.0 to 1.5 μm when the clay was added up to 5 phr. The small difference of the droplet size of the PC was attributed to the similar values of viscosities of the dispersed phase (PC) and continuous phase (ABS). From the results obtained by Dynamic Mechanical Analysis (DMA) of samples taken from ABS/PC clay nanocomposites, it was observed that the storage modulus was increased by the addition of clay at the rubbery state, i.e. at temperatures between 120 and 150°C. From the results obtained for tanδ of the ABS/PC/clay nanocomposites, it was shown that double tanδ peaks were observed. The height of the lower temperature tanδ peak was decreased from 4.5 to 3.3 when clay concentration was increased up to 5 phr, which suggests that the motion of ABS chain was restricted by the incorporation of clay in the ABS phase. The restriction of the ABS chain mobility by the clay was attributed to same conclusion drawn about the system morphology, i.e. to the fact that the clay accumulates within ABS and at the interface of ABS and PC.

1.4.2 Flame retardancy and thermal stability

In another work by Zong et al., 2204, PC/ABS/clay nanocomposites were prepared through direct melt intercalation in a twin screw extruder and detailed kinetic analysis of their thermal stability has been performed using thermogravimetric analysis. The activation energies were determined using the Kissinger method and the Flynn-Wall-Ozawa method. It was clearly observed that activation energies of the thermal degradation for nanocomposites are greater than those of the pure polymer. These data indicate an important role of clay in improving the flame retardancy of the alloy. This increasing tendency coincides with the results obtained by thermal analysis, indicating that the polymer/clay nanocomposite has higher thermal stability and lower flammability. The above authors confirmed previous studies which have shown that the lower flammability of polymer/clay nanocomposites is not due to retention of a large fraction of fuel, but rather to the form of carbonaceous char, in the condensed phase. The nano-dispersed lamellae of clay in polymer matrix may change the decomposition process of polymer since the nano-dispersed silicate layers acted as thermal hinder in polymer matrix. The nano-dispersed silicate layers slow the decomposition rate and increase the temperature of degradation by acting as an excellent thermal insulator and mass transport barrier (Zong et al., 2004).

Zong et al., 2005, studied the pyrolytic and thermo-oxidative degradation of ABS/PC alloy and ABS/PC/MMT nanocomposites by thermogravimetric analysis. The kinetic evaluations were performed by the model-free kinetic analysis and the multivariate non-linear regression. The investigation shows that thermal degradation behavior and kinetic parameters were greatly dependent on the nature of the purge gas as well as on the montmorillonite loadings. The activation energy values of samples in air were smaller than that in nitrogen, indicating that the thermal degradation mechanism was different in the above two gases. Oxygen seemed to play a key role in initiating depolymerization during the thermo-oxidative degradation of the material. This promotional effect is in agreement with the lower activation energy. In the case of nanocomposites, clay acts as a char promoter slowing down degradation and providing a transient protective barrier to both mass and energy transport in the nanocomposites. The experimental results of the above authors showed that there is an obvious tendency for the apparent activation energy of nanocomposites to increase. This increasing tendency is consistent with the thermal analysis results, suggesting that the polymer/clay nanocomposites have higher thermal stability and lower flammability. The kinetic analysis also indicated that pyrolytic degradation and thermo-oxidative degradation of PC/ABS alloy and ABS/PC/OMMT nanocomposites are two kinds of different reaction models. Pyrolytic degradation reaction of the polymer follows a two-step parallel reaction model: *n*th-order reaction model, and the *a*th-degree autocatalytic reaction with an *n*th-order reaction autocatalytic reaction, whereas the thermo-oxidative degradation reaction of the polymer is a two-step following reaction model: A→B→C of *n*th-order reaction model, and autocatalytic reaction model.

In the literature, it is reported that when polymer-layered silicate nanocomposites are evaluated for their FR properties by some test methods, such as limited oxygen index and vertical burning test (UL-94), they do not perform better than polymers without nanocaly reinforcement. Because of the aforementioned drawbacks, researchers have been focused on the combined use of nanoclay and conventional FRs systems (Wang et al., 2004a).

1.4.3 The effect of various additives

The effect of triphenyl phosphate (TPP) and MMT nanoclay on the morphology and rheological properties of ABS/PC blends was investigated by Feyz et al., 2010a. Improvement of the polymer intercalation was achieved by the presence of TPP. The results of complex viscosity (η^*) with angular frequency (ω) showed good agreement with those obtained with Carreau model. A plateau modulus (G_N^0) was observed for blends containing nanoclay, indicating the formation of network structure that increases the modulus as the result of intercalation of nanoclay. The complex viscosity is increased with the inclusion of nanoclay and TPP, but this effect was more evident with nanoclay. The above authors also observed that the addition of nanoclay enhances the non Newtonian behavior of ABS/PC blends, particularly at low-frequencies. They concluded that for composites containing both, nanoclay and TPP, the rheological behaviour of the blend was mainly controlled by the presence of nanoclay than TPP.

The thermal stability of ABS/PC/TPP/organoclay was studied using TGA and the degradation kinetic parameters were determined using the Kissinger, Flynn-Wall-Ozawa and Coats-Redfern methods (Feyz et al., 2010b). It was found that samples containing both TPP and nanoclay have the highest activation energy. Cone calorimetry, limited oxygen index (LOI) and UL 94 methods were used to investigate the fire behaviour and flammability of materials. The reduced mass loss rate (MLR), peak heat release rate (PHRR) and enhanced LOI of the composite containing a mixture system, confirmed the synergistic effect of TPP and nanoclay. This formulation showed immediate self-extinguishing properties with V-0 classification.

2. Results and discussion

2.1 Introduction

From the above presented literature survey as well as from the discussion of experimental results, it was clearly shown that clay nanoparticles do no naturally delaminate into various polymeric matrices and this may be attributed to their chemical structure or to processing conditions. On the other hand, it is obvious that in order to achieve the expected property improvement in nanocomposites, stacked clay platelets must be dispersed into the polymeric mass and strong interactions between the polymer and mineral phase must be promoted. Therefore, the existing commercial grades of nanoclays are usually modified by several organic reagents (e.g. with quaternary ammonium surfactants) as a means to increase the inter-gallery spacing and to achieve enough hydrophobicity that further promotes miscibility with the polymer matrix. In addition to the above approach, the processing parameters during the incorporation of nanofillers into polymers are often adjusted, so that the appropriate shear is developed in order to create maximum exfoliation and adequate dispersion of the clay nanoparticles. However, as mentioned above shear alone does not enable to provide an efficient and nanometric dispersion of the clay. Interfacial adhesion needs to be high enough in order to achieve better compatibility and, therefore, improved performance of the composite. Some other factors influencing the obtained microstructure of the system are the initial spacing between layers, the affinity between polymer matrix and filler, the type and concentration of modifiers etc.

As a result, the experimental design of the study of the related nanocomposites was based on the investigation of the combined effect of processing characteristics as well as of the type and amount of nanofiller on the structure and final properties of the obtained nanocomposite. More specifically, systems processed by melt blending techniques were investigated for the preparation of organically modified montmorillonite (OMMT)/ABS nanocomposites. A commercial OMMT named Cloisite 30B was used, being a natural montmorillonite modified with a surfactant, namely (MT2EtOH): methyl, tallow, bis-2-hydroxyethyl, quaternary ammonium chloride. Tallow (T) consists of ~65% C18, ~30% C16 and ~5% C14.

Mixing took place in a powerful and efficient machine, i.e. a twin screw extruder suitable for processing of thermoplastics. The following factors that affect nanocomposite structure were examined: (i) the extruder rotation speed and (ii) the clay content, in an attempt to elucidate the interrelations of composition and processing conditions. Characterization of nanocomposites was performed using X-ray diffraction (XRD), Scanning Electron Spectroscopy (SEM), Attenuated Total Reflectance Fourier Transform Infrared Spectroscopy (ATR-FTIR) analysis, Thermogravimetric Analysis (TGA) and Differential Scanning Calorimetry (DSC). Rheological properties of the above blends were investigated via Melt Flow Index (MFI) tests, as a measure of their processability during melt mixing and moulding processing. Furthermore, the mechanical strength of the obtained hybrids was explored. Intercalated or/and partial exfoliated composite structures of OMMT/ABS can be prepared by melt mixing in a twin screw extruder. Screw rotation speed does not have

significant effect on the composite structure; however the second extrusion cycle improves the dispersion efficiency of the system.
In addition to the above, ABS/PC/OMMT nanocomposites were prepared and the effect of PC incorporation was assessed in comparison with unfilled ABS/PC blends. The observed interactions were interpreted in terms of the influence of each component on the functional properties of the final mixture. This attempt might also be a contribution to the effective recycling of ABS/PC blends via upgrading of the relevant waste stream using layered silicate reinforcement.

2.2 ABS nanocomposites

2.2.1 X-ray diffraction (XRD)

XRD analysis was performed to assess the status of clay dispersion within the polymer matrix. Further analysis of the obtained diffractographs, by the use of Bragg's rule, i.e. $2d\sin\theta=n\lambda$, can give the basal spacing for the clay layers.
Cloisite 30B pattern in Figure 1, shows an intense peak around $2\theta=4.89°$, which denotes that basal spacing in silicate layers is 1.8 nm. It can also be seen that, after melt mixing, the three studied screw rotation speeds (35, 100 and 200 RPM) does not have any significant effect on the dispersion of clay into the ABS matrix. Similarly, no significant effect of screw rotation speed on the obtained morphology of nanocomposites was observed.

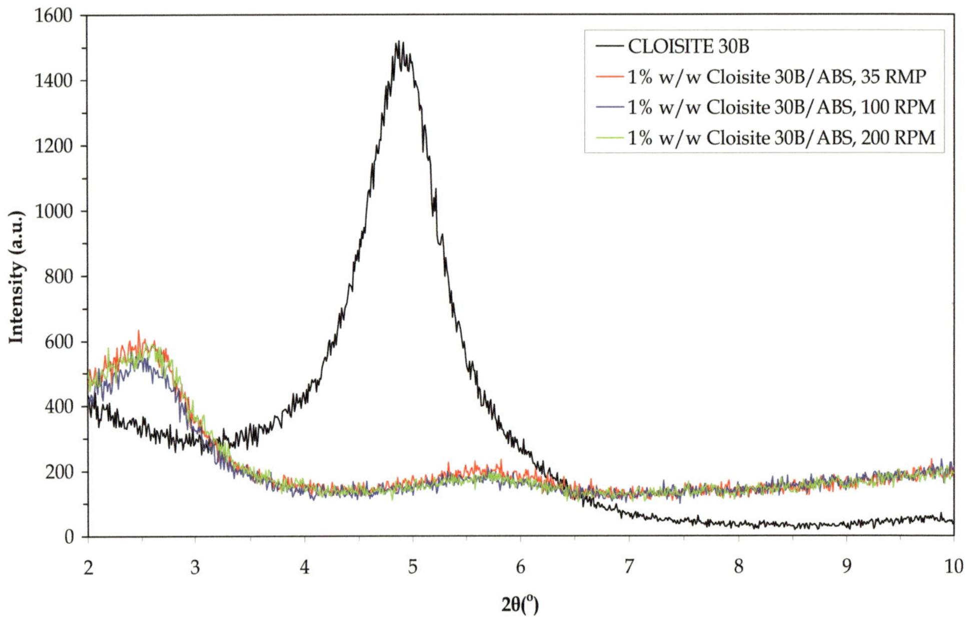

Fig. 1. XRD patterns of Cloisite 30B and its composites with ABS using different screw rotation speeds.

From Figure 2 it can be observed that the sample with 2% w/w clay shows a well defined peak at 2.61 °, with decreased intensity and intergallery spacing of 3.38 nm. The sample with 1% w/w Cloisite 30B shows a significant decrease in intensity of the diffraction peak, with an intergallery spacing of 3.55 nm. This probably suggests the formation of mixed exfoliated/intercalated silicate nanolayers of the clay particles dispersed within the ABS matrix.

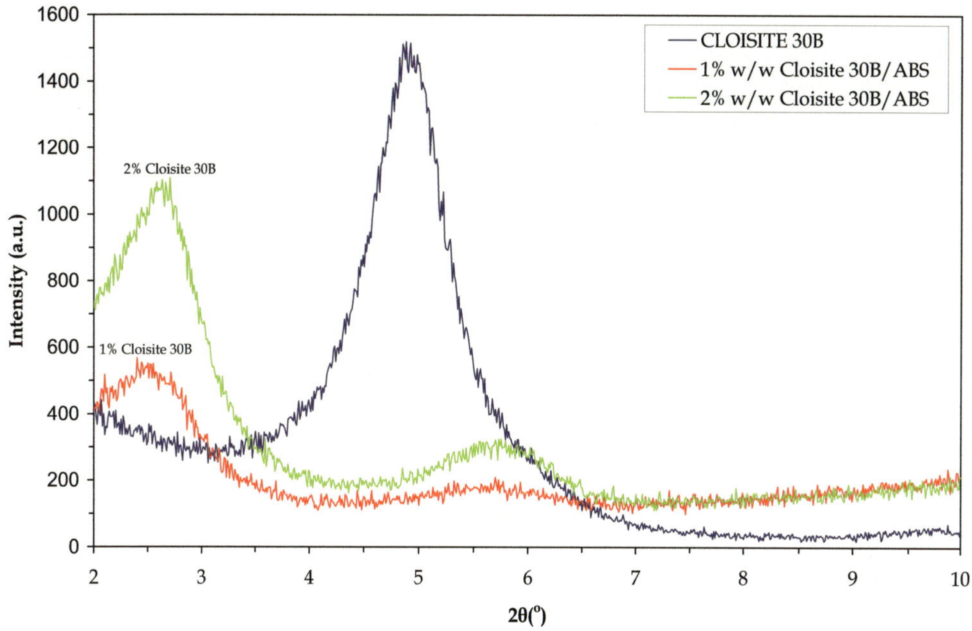

Fig. 2. XRD patterns of Cloisite 30B/ABS nanocomposites as function of clay content, at 200 RPM.

However, reprocessing in the twin screw extruder improves dispersion and intercalation of clay in the polymer matrix, after this second cycle, as can be clearly seen in Figure 3.

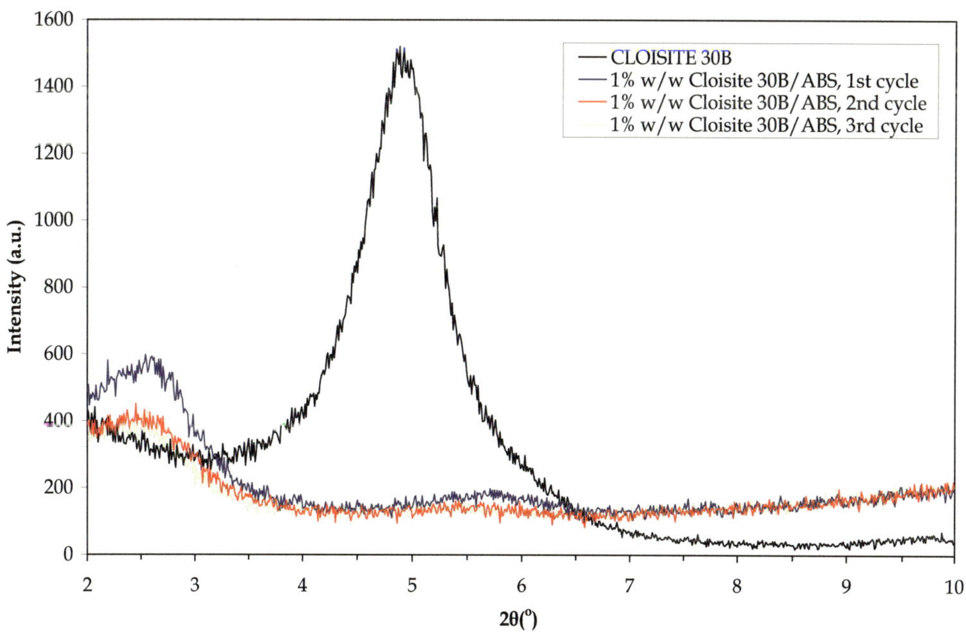

Fig. 3. XRD patterns of 1% w/w Cloisite 30B/ABS nanocomposites after extrusion reprocessing at 100 RPM.

Vaia et al., 1993 proposed that an increase in the intergallery spacing, which is related to the intercalation of the polymer through the clay galleries, results in a new diffraction pattern, corresponding to the increased spacing of the clay galleries. According to these authors, the degree of intercalation in the hybrid composite could be determined by changes in sharpness and intensity of the corresponding reflection peaks.

2.2.2 MFI

The incorporation of 1% w/w nanoclays in ABS results in a decrease of the flow rate of the mixture compared with pure ABS, which is due to the increase of viscosity (Table 1).

	MFI (g/10 min)
ABS	1.507 ±0.181
1% w/w Cl 30B/ABS	1.161 ±0.165

Table 1. MFI measurements of OMMT/ABS composites.

Similar results were reported by Aalaie & Rahmatpour, 2007 who studied the rheological behaviour of ABS nanocomposites at low shear rates. Confinement of polymer chains motion caused by organoclay platelets and tactoids in the ABS matrix, together with interactions between polar groups of ABS and oxygen groups of Cloisite 30B, may cause this effect. However, at high shear rates, flow properties of the composite are comparable to those of pure ABS. This behaviour can be attributed to preferential orientation of clay layers, or even anisotropic tactoids, parallel to flow direction.

2.2.3 ATR-FTIR

Representative ATR spectra of Cloisite 30B, ABS and 1% w/w Cloisite 30B/ABS composites, are presented in Figure 4a. Cloisite 30B reveals absorption bands at 3630 and 3367 cm^{-1} deriving from O-H stretching for the silicate group and water respectively, 1643 cm^{-1} (related to O-H bending), 1018 cm^{-1} (stretching vibration of Si-O-Si from silicate) and 920 cm^{-1} (Al-OH-Al deformation of aluminates) (Bora et al., 2000). The organic modification is responsible for the bands located at 2927, 2853 and 1470 cm^{-1}, which were assigned to C-H vibrations of methylene groups (asymmetric stretching, symmetric stretching and bending respectively).
In Figure 4b&c are also presented the spectra of 1% Cloisite 30B/ABS composites and pure ABS. The aromatic and aliphatic C-H stretching can clearly be seen in the range 3200-3000 cm^{-1} and 3000-2800 cm^{-1}, respectively. The intense and well-defined C≡N stretching from acrylonitrile (AN) is located at 2237 cm^{-1}. The scissoring mode of CH_2 groups is at 1457 cm^{-1} and the intense ring bends are visible at 798 and 698 cm^{-1}. Typical dienic rubber bands are easy to identify in the range 1000-900 cm^{-1} as follows: absorption peak at 998 cm^{-1} corresponds to off-frequency vibrations of CH in CH=CH groups, band at 967 cm^{-1} is due to off-frequency vibrations of CH in *trans*-CH=CH- moieties, band at 913 cm^{-1} is determined by off-frequency vibrations for $=CH_2$. The deformation vibration for $=CH_2$, which usually appears at wavelengths under 1465 cm^{-1}, shifts to 1457 cm^{-1} due to the influence of the double bond. The bands in the interval 2950-2800 cm^{-1} can be attributed to asymmetric and symmetric $-CH_2$ groups. The hydroxyl peak appears as a broad signal with a maximum at ≈3650 cm^{-1}, typical of an OH stretch. The hydroxyl peak appears as a broad signal with a maximum at ≈3650 cm^{-1}, typical of an OH stretch.

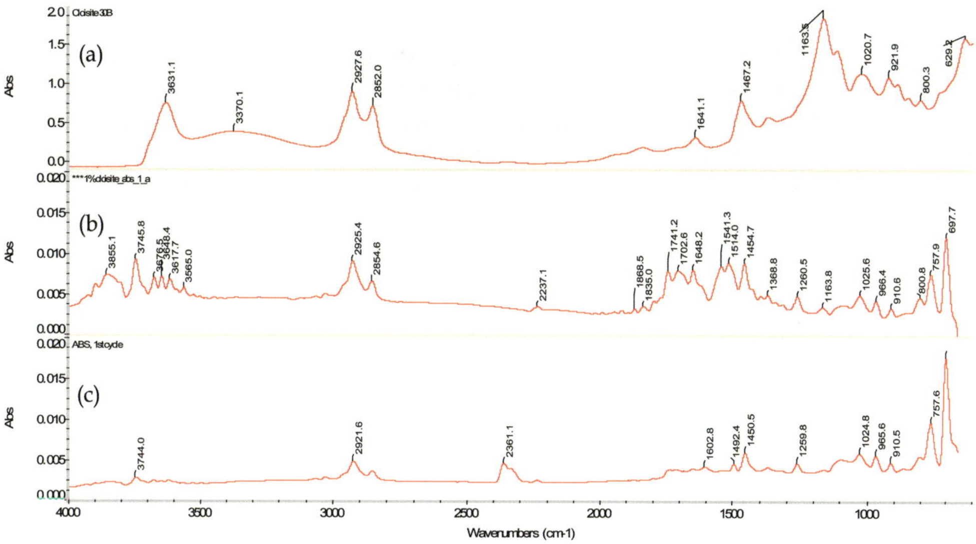

Fig. 4. ATR-FTIR spectra of (a) Cloisite 30B, (b) ABS and 1% w/w Cloisite 30B/ABS composite and (c) pure ABS.

2.2.4 Color test

The color changes in terms of Lab scale were presented in Table 2. The incorporation of montmorillonite has a clear effect by changing the color texture of pure ABS. The observed changes in red saturation index a, yellow saturation index b and total color change ΔE, may also be the result of degradation phenomena of organic modification of nanoclay occurred during melt mixing.

	ΔL	Δα	Δb	ΔE
ABS	-2.73 ±1.06	0.60 ±0.15	5.25±0.71	6.05 ±0.61
1% w/w Cl 30B/ABS	-5.91 ±1.94	3.62 ±0.66	13.176 ±2.20	14.92 ±2.79

Table 2. The change of total color (ΔE) of 1% w/w OMMT/ABS composites.

2.2.5 DSC

The glass transition temperatures of ABS as derived from the DSC runs, are listed in Table 3. The glass transition of SAN phase was found to be at ~103°C. The DSC thermographs revealed a fairly constant glass transition temperature (T_g) of 103 (±1) °C for ABS nanocomposites, which is expected for the SAN-phase of ABS.

	T_g (°C)	T_{onset}(°C)	T_{max}(°C)	Residue %
ABS	102.38	407.29	428.33	2.893
1% w/w Cl 30B/ABS	103.50	403.03	427.33	3.574

Table 3. DSC & TGA results of ABS and 1% OMMT/ABS composites.

2.2.6 TGA

The decomposition temperature of Cloisite 30B is above 300°C and seems to be initiated by decomposition of the organic modifier. This suggests that during melt processing, which

was carried out around 200ºC, the organic modifier in the clay remains active, facilitating dispersion and further assisting the formation of exfoliated or intercalated structures, as it was somehow confirmed by the XRD patterns. Finally, the char residue increases to about 0.7%, which is obviously a result of the incorporation of the inorganic clay that remains as residue and contributes to the final weight (Table 3). It should be noted that, TGA studies of Yeh et al., 2006 made with ABS nanocomposites in air, have also shown an enhancement of thermal stability as well as an increase of the remaining residue.
Similarly, Patiño-Soto et al., 2007 observed that dispersed clay enhances the thermal stability of ABS, since the intercalated or exfoliated clays display barrier properties against the diffusion of atmospheric oxygen into the material, inhibiting oxidative attack of the system. Similar results were reported by Kim et al., 2008 and the observed effect was attributed to the shielding effect of clay layers. According to Zanetti et al., 2001 the layers of organo-modified silicate are impermeable from small molecules, generated during the thermal degradation process and, therefore, degradation products have to take a long way around the clay layers. According to these authors, the addition of organoclay to polymer matrix is expected to slow down the release rate of decomposed products to endow the degradation products more chance to bind back to undecomposed original polymer matrix. Moreover, given that TGA is a dynamic experiment, another possible reason for the increase in the decomposition temperature of ABS nanocomposites is the increase in viscosity due to the addition of clay, which will inhibit transmission, i.e. it will increase the difficulty for the inward diffusion of oxygen and the outward diffusion of the degradation by-products.
The ABS samples with higher AN content produce a greater increase of decomposition temperature, which was attributed to better clay intercalation-exfoliation, due to the higher polarity brought about by the higher AN content.
Based on the above review of the related literature, we can draw the conclusion that our nanocomposites did not display a clear improvement of their thermal stability because of the insufficient dispersion of nanofiller taking place under the experimental conditions followed in our experiments. In fact, the data obtained from XRD analysis revealed that melt mixing with the operational parameters of this work, did not succeed to produce fully exfoliated structures, which would probably have led to essential improvement of thermal resistance and prevention of thermal degradation during TGA experiments.

2.2.7 Tensile test

From Table 4 it can be seen that the incorporation of Cloisite 30B in ABS does not have any significant effect in tensile strength, but increases the tensile modulus and decreases the elongation at break of the reinforced samples in comparison with pure ABS.

	Tensile strength (MPa)	Modulus of elasticity (MPa)	Strain at break (%)
ABS	50.33 ±1.794	1939.75 ±103.84	12.15 ±5.805
1% w/w Cl 30B/ABS	50.78 ±0.619	2117.91 ± 159.05	6.69 ±2.581

Table 4. Tensile test results of 1% w/w OMMT/ABS composites.

Aalaie & Rahmatroup, 2007 reported a decrease of tensile strength and elongation of ABS/Cloisite 30B nanocomposites , in comparison with specimens of pure ABS. Su et al., 2004 reported a decrease of tensile properties of ABS/clay hybrids despite the intercalated/exfoliated structure formation. On the other hand, Zhang et al., 2006a claim

that nanocomposite formation did not improve the mechanical properties of ABS reinforced with oligomerically-modified clay with high organic content. They observed in most cases decrease in tensile strength, Young's modulus and elongation at break. In addition, they reported that a plasticization effect of organically modified clay may reduce the mechanical properties of nanocomposites (Zhang et al., 2006b). In case the plasticization from the oligomeric surfactant was suppressed the tensile properties were improved, compared to similar oligomerically-modified clays with higher organic content.
Using different types of surfactant-modified clays for ABS reinforcement, Stretz et al., 2005 observed that the elongation at break decreases and the yield strength remains about the same. However, for some clays, increased MMT content resulted in a slight increase of yield strength. Yeh et al., 2006 found that the incorporation of organoclay platelets into polymer matrix increased the tensile strength of ABS for concentrations up to 3 wt %. Further amounts of organoclay loading (e.g. 5 wt %) were found to decrease tensile strength of the nanocomposite material. On the other hand, elongation at break was decreasing as the organoclay loading increased in the composite.
Interestingly, Kim et al., 2008 reported increase in tensile strength and elongation at break of ABS composites reinforced with unmodified Cloisite 25A and modified with tetrasulfane.

2.3 ABS/PC blends nanocomposites

2.3.1 XRD

The XRD patterns of Cloisite 30B and the 1phr Cloisite 30B/ABS/PC nanocomposite are shown in Figure 5.

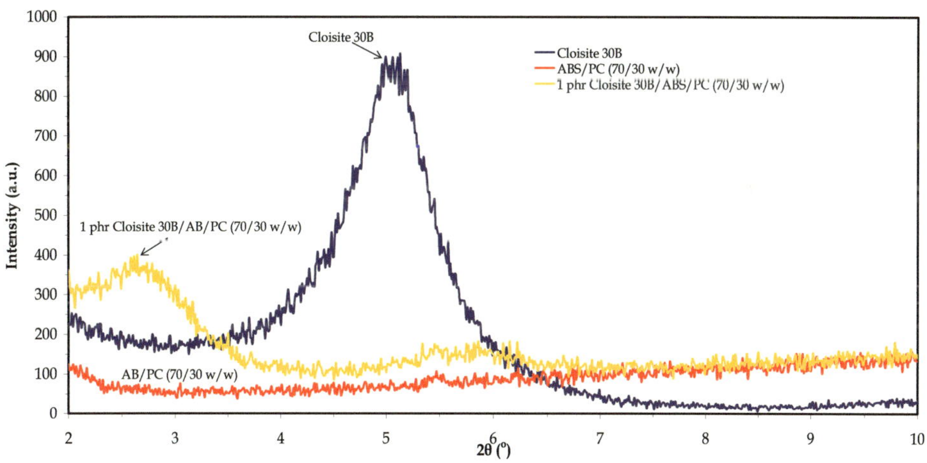

Fig. 5. XRD patterns of Cloisite 30B, ABS/PC and 1% w/w Cloisite 30B/ABS nanocomposite.

It can be observed that the diffraction peaks of silicate clay in the ABS/PC blend nanocomposite shift to lower angle values. The corresponding average basal spacing of MMT increased from 1.8 to 3.36 nm. The XRD results indicated that ABS/PC blend intercalates into the silicate clay layers and enlarges the interlayer spacing of OMMT.

2.3.2 SEM

Microscopic investigation of gold coated surfaces deriving from cryogenic fracture of ABS/PC specimens, as well as of their nanocomposite system containing 1 phr Cloisite 30B, revealed a slight increase of surface roughness of the later. This probably suggests inhomogenity of filler dispersion, which is consistent with the fact of preferential accumulation of clay nanoparticles in the ABS phase (Hong et al., 2007).

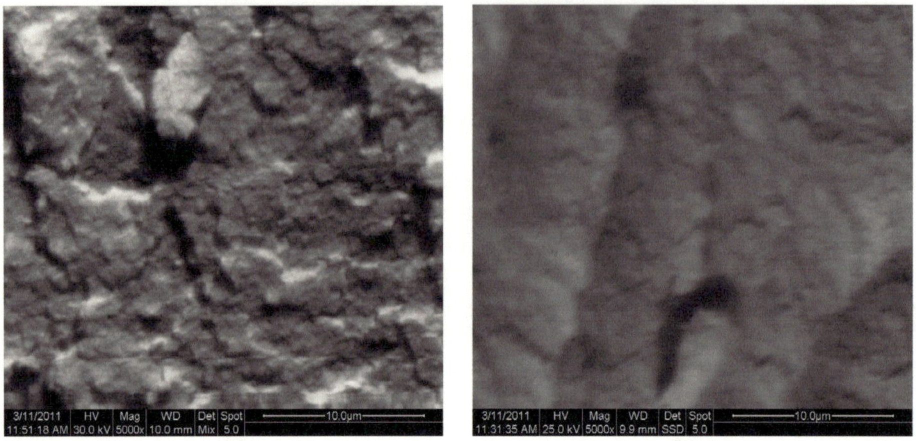

Fig. 6. SEM images of (a) ABS/PC (70/30, w/w) blend and (b) its 1phr Cloisite 30B nanocomposite.

2.3.3 DSC

DSC studies on ABS/PC blends and their nanocomposite containing 1phr Cloisite 30B show a slight increase of the T_g of PC phase, whereas the T_g of ABS phase remains stable (Figure 7).

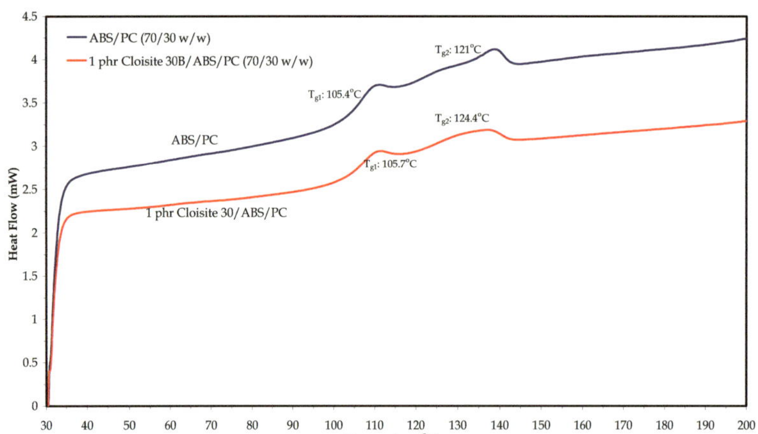

Fig. 7. DSC curves of ABS/PC (70/30 w/w) and 1 phr Cloisite 30B/ABS/PC (70/30 w/w) composite.

Studies on ABS/PC nanocomposites have shown that most of the nanoclay accumulates in the ABS phase and the T_g of this phase is influenced, due to restriction of ABS chain mobility by the presence of reinforcement. Investigation of the T_g changes, performed with Dynamic Mechanical Analysis, revealed a decrease on the height of peak corresponded to ABS phase (Hong at al., 2007).

2.3.4 TGA

The thermal stability is an important characteristic strongly influenced by the nanocomposite morphology. The decomposition temperature of ABS is about 300-450°C, while PC shows higher thermal stability and decomposes at 500-600°C. It is believed that the thermal degradation of PC/ABS blends consists of several complex processes, such as hydrolysis and thermal degradation, each becoming predominant during different stages of the overall process (Balart et al., 2005).

TGA results for ABS/PC blend, at 70/30 w/w composition, showed that the thermal degradation behaviour of mixture is mainly controlled by the content of ABS, which is the main constituent of the mixture. It can also be observed from Figure 9 and Table 5 that the incorporation of 1 phr montmorillonite in ABS/PC blends improves the thermal resistance especially in the high temperatures zone of thermal degradation. In the case of nanocomposite, an increase in the temperature of maximum thermal degradation rate to about 2°C and a shifting of thermal degradation to higher temperatures were observed. The increase of residue in the case of nanocomposite is related with the presence of clay in polymer matrix that may change the decomposition process of polymer, since the nanodispersed silicate layers contribute to hinder heat transfer within the polymer matrix.

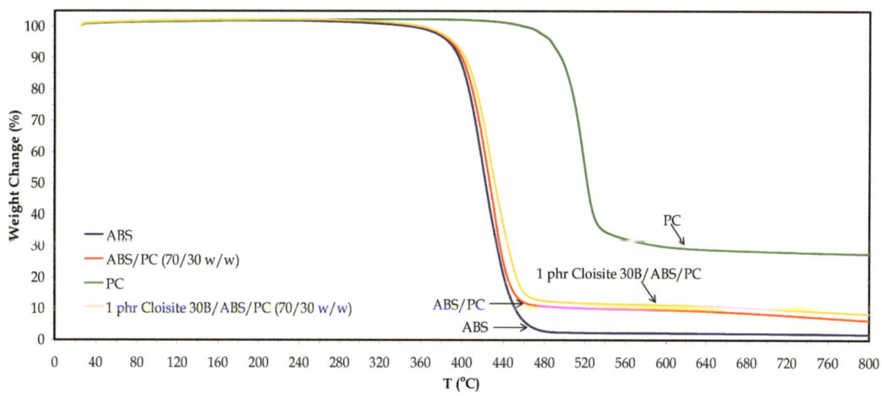

Fig. 8. TGA curves of ABS, PC, of ABS/PC (70/30 w/w) blend and 1 phr Cloisite 30B/ABS/PC (70/30 w/w) nanocomposite.

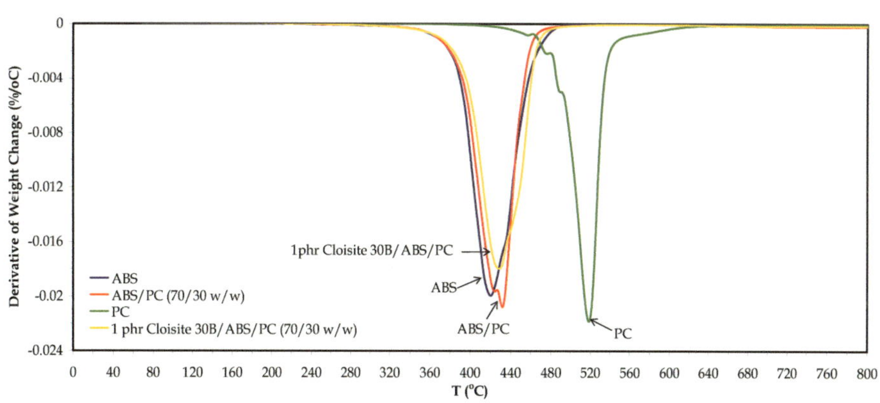

Fig. 9. Derivative of weight change versus temperature of ABS, PC, ABS/PC (70/30 w/w) blend and 1 phr Cloisite 30B/ABS/PC (70/30 w/w) nanocomposite.

Two distinct peaks were observed in TGA graphs obtained from ABS/PC nanocomposites at 434 and 500°C, whereas ABS/PC showed a main peak at 433°C (Wang et al., 2003). The higher peaks were attributed to the nanodispersed silicate layers, which could slow the decomposition and increase the temperature of degradation by acting as an excellent thermal insulator and mass transport barrier, as stated above.

One step of thermal degradation of PC/ABS/OMMT hybrids under pyrolytic conditions was reported by Zong et al., 2005. The onset temperature of thermal degradation was slightly higher for the nanocomposites than for virgin PC/ABS alloy. Under thermo-oxidative conditions PC/ABS alloy and PC/ABS/OMMT hybrids degrade in two steps. In that case, the onset temperature of thermal degradation was not enhanced but charring was increased. The above researchers pointed out the important role of oxygen in the degradation of ABS/PC alloy and ABS/PC/OMMT hybrids. They observed that the onset temperature corresponding to the main degradation was higher under nitrogen than in air for the alloy and nanocomposites. The calculated activation energy for samples in air was smaller than that in nitrogen, indicating that the thermal degradation mechanism was different in these two gases. Enhanced residues in air were also observed in the temperature range 470-540°C, suggesting that oxygen reacted with PC/ABS alloy and/or its degradation products to yield char. Oxygen initiates depolymerization leading to the formation of hydroperoxides, which can yield char. This char undergoes subsequent degradation at high temperatures and as a result, the reaction of oxygen with the polymeric matrix initiates earlier degradation of the material and then, yielding more char, stabilization takes place in a particular temperature range. In the case of nanocomposites, the obtained results suggest that the role of clay is to promote char formation, during the thermo-oxidative degradation of the material, slowing down degradation and providing a transient protective barrier to both mass and energy transport in nanocomposites.

	T_{onset} (°C)	T_{peak} (°C)	Residue (%)
PC	503.8±0.785	524.42±0.099	27.59±0.118
ABS	402.23±0.106	422.64±0.502	0.91±1.280
ABS/PC (70/30 w/w)	408.32±0.806	429.8±0.028	6.47±0.017
1 phr Cloisite 30B/ABS/PC (70/30 w/w)	408.01±0.884	431.3±0.856	8.41±0.035

Table 5. TGA results in PC, ABS, ABS/PC (70/30 w/w) and 1 phr Cloisite 30B/ABS/PC (70/30 w/w) nanocomposite.

The synergistic effect of triphenyl phosphate (TPP) flame retardant and nanoclay reinforcement in improving the thermal resistance of ABS/PC blends was observed by Feyz et al., 2010 using TGA. They also observed that decomposition of PC/ABS is a single step process, whereas decomposition of the flame-retarded PC/ABS having nanoclay is a two step process: the first step takes place in the range of 400-500°C and the second in the range of 500-600°C. The two step decomposition determined with TGA experiments, was attributed to the interaction between nanoclay and the polymeric matrix, leading to changes of the thermal stability of ABS resin. The initial weight loss of ABS/PC/nanoclay occurring at 200°C is related to the thermal decomposition of alkyl ammonium salt present in the nanoclay as a modifier. The decomposition of alkyl ammonium salt causes a delay in the thermal decomposition of the nanocomposite at higher temperatures and by doing this, an increase of the activation energy of the matrix decomposition can be observed in the subsequent steps.

2.3.5 Tensile test

The incorporation of montmorillonite in ABS/PC blend has a positive effect in tensile strength, whereas the modulus of elasticity and strain at break remain almost stable.

	Tensile strength (MPa)	Modulus of elasticity (MPa)	Strain at break (%)
ABS/PC (70/30 w/w)	51.96±0.62	2181.15±103.40	13.04±1.22
1 phr Cloisite 30B/ABS/PC (70/30 w/w)	53.45±0.35	2108.78±129.46	12.44±0.80

Table 6. Tensile test results of ABS/PC (70/30 w/w) and 1 phr OMMT/ABS/PC (70/30 w/w) nanocomposite.

3. Conclusions

The above literature review on polymeric nanocomposites and the presented results deriving from the experimental work on the related systems, can lead to the following conclusions:

The use of nanostructured montmorillonite as a filler for making composites with high performance polymeric systems, such as ABS, PC and their mixtures, is a promising procedure, leading to products with increased scientific and technical significance.

Mixed intercalated or/and partial exfoliated composite structures of organic modified montmorillonite/ABS can be prepared by melt mixing in a twin screw extruder, especially designed for processing thermoplastic melts. Such machines, are today the most powerful melt mixing system for continuous work, whereas internal mixers could be tested alternatively, as batch systems.

Screw rotation speed does not have significant effect in the composite structure, however the second extrusion cycle improves the dispersion efficiency of the system. This suggests, that longer residence time, i.e. a higher L/D ratio in the extruder design would probably be beneficial. It should be noted that the screw geometry must be considered stable for those experiments and the same applies for screw rotation speed, since it controls shear stresses.

No obvious effect of the nanoclay loadings on the T_g of SAN phase in ABS terpolymer was observed. A decrease of the temperature of thermal degradation in the case of nanocomposites was observed and this effect was attributed to the degradation of the organic modification of clay. However, the increase of the amount of residue in the case of nanocomposite indicates a change in the thermal degradation mechanism of ABS, due to the contribution of nanoclay, that seems to have a protective action to the polymer. Improvement of the modulus of elasticity was also observed as a result of organoclay reinforcement.

The incorporation of PC in ABS shifted the thermal degradation of the latter to higher temperature range and this behaviour was further enhanced with the addition of 1 phr nanoclay. DSC thermographs in ABS/PC blend reinforced with 1 phr OMMT showed an increase in the T_g of PC phase, whereas the lower temperature T_g, corresponding to the ABS phase, does not show any change.

The mechanical properties of ABS/PC nanocomposite present an increase, in terms of tensile strength but without any obvious effect in the modulus of elasticity.

All the above results are considered as a very promising option for improving applications of ABS and ABS/PC blends, towards the design of thermally stable materials with improved mechanical performance through the preparation of nanocomposites with hydrophilic clay by continuous melt blending in a twin screw extruder. Based on the above, the OMMT nanoparticles have shown to act not only as reinforcement for ABS, but also they seemed to play a significant role on the interactions taking place during the thermo-mechanical ageing of ABS that naturally occurs when it takes repeated cycles of melt processing. Therefore, this technology could also be applied for upgrading plastic waste by mechanical recycling.

In future research work, improvement of the nanoclay intercalation in pure ABS as well as in ABS/PC blends should be examined, with special focus on parameters such as the content of ABS terpolymer and molecular weight of PC. In addition the incorporation of a polar compatibilizer like maleic anhydride grafted ABS (MA-*g*-ABS), might also have an interesting effect and, therefore, it also should be examined.

4. Acknowledgment

Special thanks are given to the researchers Ms. E. Karahaliou and Ms. M. Triantou, for their assistance in some of the experimental part of this work.

5. References

Aalaie, J.; Rahmatroup, A. (2007). Study on Preparation and Properties of Acrylonitrile-Butadiene-Styrene/Montmorillonite Nanocomposites. *Journal of Macromolecular Science Part B: Physics,* Vol. 46, No. 6, (November 2007), pp. 1255-1265, ISSN 0022-2348

Bora, M.; Ganguli, J.N.; Dutta, D.K. (2000). Thermal and spectroscopic studies on the decomposition of [Ni{di(2-aminoethyl)amine}$_2$]- and [Ni(2,2':6',2''-terpyridine)$_2$]-Montmorillonite intercalated composites *Thermochimica Acta 2000,* Vol. 346, No. 1-2, (March 2000), pp. 169-175, ISSN: 0040-6031

Burnside, S.D.; Giannelis, E.P. (1995). Synthesis and properties of new poly(dimethylsiloxane) nanocomposites. *Chemistry of Materials,* Vol. 7, No. 9, (September 1995), pp. 1597-1600, ISSN 0897-4756

Cai, Y.; Huang, F.; Xia, X.; Wei, Q.; Tong, T.; Wei. A.; Gao, W. (2010). Comparison Between Structures and Properties of ABS Nanocomposites Derived from Two Different Kinds of OMT. *Journal of Materials Engineering Performance,* Vol. 19, No. 2, (March 2010), pp.171-176 ISSN 1059-9495

Carrión, F.J.; Arribas, A.; Bermúdez, M.-D.; Guillamon, A. (2008). Physical and tribological properties of a new polycarbonate-organoclay nanocomposite. *European Polymer Journal,* Vol. 44, No. 4, (April 2008), pp.968-977, ISSN 0014-3057

Chen, G.X.; Yoon, J.S. (2005). Nanocomposites of poly[(butylene succinate)-*co*-(butylene adipate)] (PBSA) and twice-functionalized organoclay. *Polymer International,* Vol. 54, No. 6, (June 2005), pp. 939-945, ISSN 1097-0126

Chigwada G.; Wang, D.; Jiang, D.D.; Wilkie, C.A. (2006). Styrenic nanocomposites prepared using a novel biphenyl-containing modified clay. *Polymer Degradation and Stability,* Vol. 91, No. 4, (April 2006), pp. 755-762, ISSN 0141-3910

Chow, W.S.; Neoh, S.S. (2009). Dynamic Mechanical, Thermal, and Morphological Properties of Silane-Treated Montmorillonite Reinforced Polycarbonate Nanocomposites. *Journal of Applied Polymer Science,* Vol. 114, No. 6, pp. 3967-3975, ISSN 1097-4628

Dennis, H.; Hunter, D.L.; Chang, D.; Kim, S.; White, S.; White J.L.; Cho, J.W.; Paul, D.R. (2001). Effect of melt processing conditions on the extent of exfoliation in organoclay-based nanocomposites. *Polymer,* Vol. 42, No. 23, (November 2001), pp.9513-9522, ISSN 0032-3861

Elmaghor, F.; Zhang, L.; Fan, R.; Li, H. (2004). Recycling of polycarbonate by blending with maleic anhydride grafted ABS. *Polymer,* Vol. 45, No. 19, (September 2004), pp. 6719-6724, ISSN 0032-3861

Feng, J.; Hao, J.; Du, J.; Yang, R. (2010). Flame retardancy and thermal properties of solif bisphenol A bis(diphenyl phosphate) combined with montmorillonite in polycarbonate. *Polymer Degradation and Stability,* Vol. 95, No. 10, (October 2010), pp. 2041-2048, ISSN 0141-3910

Feyz, E.; Jahani, Y.; Esfandeh, M.; Ghafelehbashi, M.; Jafari, S.H. (2010a). Study of the viscoelastic properties of PC/ABS blend containing triphenyl phosphate and nanoclay and its correlation with morphology. *Journal of Applied Polymer Science,* Vol. 19, No. 3, (November 2010), pp. 1796–1804, ISSN 1097-4628

Feyz, E.; Jahani, Y.; Esfandeh, M. (2010b). Comparison of the Effect of an Organoclay, Triphenylphosphate , anf a Mixture of Both on the Degrafation and Combustion Behavior of PC/ABS Blends. *Macromolecular Symposia,* Vol. 298, No. 1, pp. 130-139, ISSN 1521-3900

Galgali, G.; Ramesh, C.; Lele, A. (2001). A Rheological Study on the Kinetics of Hybrid Formation in Polypropylene Nanocomposites. *Macromolecules,* Vol. 34, No. 4, (February 2001), pp. 852-858, ISSN 0032-3861

Guo, T.; Hao, G.; Song, M.; Zhang, B. (2004). Amphiphilic poly(styrene-b-ethylene oxide)-block-copolymer-intercalated layered silicate and its nanocomposites with acrylonitrile–butadiene– styrene resin. *Journal of Applied Polymer Science,* Vol. 94, No. 1, (September 2004), pp. 238-242, ISSN 1097-4628

He, X.J.; Wang, L.J.; Xie, X.L.; Zhang, K. (2010). Investigation of thermal property and flame retardancy of ABS/montmorillonite nanocomposites. *Plastics Rubber and Composites,* Vol. 39, No. 2, (March 2010), pp. 54-60, ISSN 1465-8011

Hong, J.H.; Sung, Y.-T.; Song, K.H.; Kim, W.N.; Kang, B.I.; Kim, S.L.; Lee, C.H. (2007). Morphology and dynamic mechanical properties of poly(acrylonitrile-butadiene-styrene)/polycarbonate/clay nanocomposites prepared by melt mixing. *Composite Interfaces,* Vol. 14, No. 5-6, pp. 519-532, ISNN 0927-6440

Huang, X.; Lewis, S.; Brittain, J.; Vaia, R.A. (2000). Synthesis of Polycarbonate-Layered Silicate Nanocomposites via Cyclic Oligomers. *Macromolecules,* Vol. 33, No. 9, (March 2000), pp. 2000-2004, ISSN 0032-3861

Jang, L.W.; Kang, C.M.; Lee, D.C. (2001). A New Hybrid Nanocomposite Prepared by Emulsion Copolymerization of ABS in the Presence of Clay. *Journal of Polymer Science: Part B: Polymer Physics,* Vol. 39, No. 6, pp. 719-727, ISSN 1099-0488

Karahaliou, E.-K.; Tarantili, P.A. (2009). Stability of ABS compounds subjected to repeated cycles of extrusion processing. *Polymer Engineering and Science,* Vol. 49, No. 11, (November 2009), pp. 2269-2275, ISSN 1548-2634

Karahaliou, E.-K.; and Tarantili, P.A. (2009). Preparation of Poly(Acrylonitrile-Butadiene-Styrene)/Montmorillonite Nanocomposites and Degradation Studies During Extrusion Reprocessing, *Journal of Applied Polymer Science,* Vol. 113, No. 4, (August 2009), pp. 2271-2281, ISSN 1097-4628

Kim. J.; Lee, K.; Lee, K.; Bae, J.; Yang, J.; Hong, S. (2003). Studies on the thermal stabilization of ABS; synergistic effect of thiphenyl phosphate nanocomposite, epoxy resin, and silane coupling agent mixtures. *Polymer Degradation and Stability*, Vol. 79, No. 2, pp. 201-207, ISSN 0141-3910

Kim, H.-S.; Park B.H.; Choi, J.H.; Yoon, J.-S. (2008). Preparation and Mechanical Properties of Acrylonitrile-Butadiene-Styrene Copolymer/Clay Nanocomposites. *Journal of Applied Polymer Science*, Vol. 107, No. 4, (November 2008), pp. 2539-2544, ISSN 1097-4628

Kim, H.-K.; Kim, G.-H.; Cho, B.-M.; Ha, C.-S. (2009). Effect of Rubber Content of ABS on the Mechanical Properties of ABS/Clay Nanocomposites. *Composite Interfaces*, Vol. 16, No. 4, pp. 337-346, ISNN 0927-6440

Kim, K.; Kim, J.-K.; Chung, I.J. (2011). Effect of the clay size on the dispersion morphology and emulsion stability of ABS/layered silicate nanocomposites. *Journal of Applied Polymer Science*, Vol. 19, No. 3, (February 2011), pp. 1287–1296, ISSN 1097-4628

Lee, M.-P.; Hiltner, A.; Baer, E. (1992). Phase morphology of injection-moulded polycarbonate/acrylonitrile-butadiene-styrene blends. *Polymer*, Vol. 33, No. 4, pp. 685-697, ISSN 0032-3861

Lee, K.M.; Han, C.D. (2003). Effect of hydrogen bonding on the rheology of polycarbonate/organoclay nanocomposites. *Polymer*, Vol. 44, No. 16, (July 2003), pp. 4573-4588, ISSN 0032-3861

Lim, Y.T.; Park, O.O. (2001). Phase morphology and rheological behavior of polymer/layered silicate nanocomposites. *Rheologica Acta*, Vol. 40, No. 3, (May 2001), pp. 220-229, ISSN 1435-1528

Lim, S.-K.; Hong, E.-P.; Song, Y.-H.; Park, B.J.; Choi H.J.; Chin, I.-J. (2010). Preparation and interaction characteristics of exfoliated ABS/organoclay nanocomposite. *Polymer Engineering and Science*, Vol. 50, No. 3, (March 2010), pp.504-512, ISSN 1548-2634

Ma, H.; Xu Z.; Tong, L.; Gu A.; Fang, Z. (2006). Studies of ABS-graft-maleic anhydride/clay nanocomposites: Morphologies, thermal stability and flammability properties. *Polymer Degradation and Stability*, Vol. 91, No. 12, (December 2006), pp. 2951-2959, ISSN 0141-3910

Modesti, M.; Besco, S.; Lorenzetti, A.; Zammarano, M.; Causin, V.; Marega, C.; Gilman, J. W.; D. Fox, M.; Trulove, P.C.; De Long H.C.; Maupin, P.H. (2008). Imidazolium-modified clay-based ABS nanocomposites: a comparison between melt-blending and solution-sonication processes. *Polymer for Advanced Technologies*, Vol. 19, No. 11, (November 2008), pp.1576-1583, ISSN 1099-1581

Morawiec, J.; Pawlak A.; Slouf, M.; Galeski A.; Piorkowska, E.; Krasnikowa, N. (2005). Preparation and properties of compatibilized LDPE/organo-modified montmorillonite nanocomposites. *European Polymer Journal*, Vol. 41, No. 5, pp.1115-1122, (May 2005), ISSN 0014-3057

Nevalainen, K.; Vuorinen, J.; Villman, V.; Suihkonen, R.; Järvelä, P.; Sundelin, J.; Lepistö, T. (2009). Characterization of Twin-Screw-Extruder-Compounded Polycarbonate Nanoclay Composites. *Polymer Engineering and Science*, Vol. 49, No. 4, (April 2009), pp.631-640, ISSN 1548-2634

Owen, S.R.; Harper, J.F. (1999). Mechanical, microscopical and fire retardant studies of ABS polymers. *Polymer Degradation and Stability*, Vol. 64, No. 3, (June 1999), pp. 449-455, ISSN 0141-3910

Patiño-Soto A.P.; Sánchez-Valdes S.; Ramos-Devalle L.F. (2007). Morphological and Thermal Properties of ABS/Montmorillonite Nanocomposites Using ABS with different AN Contents. *Macromolecular Materials Engineering*, Vol. 292, No. 3, pp. 302-309, ISSN 1439-2054

Patiño-Soto A.P.; Sánchez-Valdes S.; Ramos-Devalle L.F. (2008). Morphological and Thermal Properties of ABS/Montmorillonite Nanocomposites Using Two Different ABS Polymers and Four Different Montmorillonite Clays. *Journal of Polymer Science: Part B: Polymer Physics,* Vol. 46, No.2, pp. 190-200, ISSN 1099-0488

Pinnavaia, T.J.; Beall, G.W. (2002). *Polymer-clay nanocomposites,* J. Wiley & Sons Ltd, ISBN: 0-471-63700-9 UK

Pourabas, B.; Raeesi, V.; (2005). Preparation of ABS/montmorillonite nanocomposite using a solvent-non-solvent method. *Polymer,* Vol. 46, No. 15, pp. 5533-5540, ISSN 0032-3861

Pourabas, B.; Azimi, H. (2008). Indirect Synthesis of ABS/Clay Nanocomposites, Comparison and Thermal Properties. *Journal of Composite Materials,* Vol. 42, No. 23, (December 2008), pp. 2499-2522, ISSN 0021-9983

Saadat, A.; Nazockdast, H.; Sepehr F.; Mehranpour M. (2010). Linear and nonlinear melt rheology and extrudate swell of acrylonitrile-butadiene-styrene and organoclay-filled acrylonitrile-butadiene-styrene nanocomposite. *Polymer Engineering and Science,* Vol. 50, No. 12, (December 2010), pp. 2340-2349, ISSN 1548-2634

Severe, G.; Hsieh A,J.; Koene, B.E. Effect of Layered Silicates on Thermal Characteristics of Polycarbonate Nanocomposites. 58th ANTEC 2000, 2, pp. 1523-1526, ISBN 9781566769570, Orlando, Florida, USA, May 8-9, 2000

Solomon, M.J.; Almusallam, A.S.; Seefeldt, K.F.; Somwangthanaroj, A.; Varadan, P. (2001). *Macromolecules,* Vol. 34, No. 6, (March 2001), pp. 1864-1872, ISSN 0032-3861

Stretz, H.A.; Paul, D.R.; Cassidy, P.E. (2005). Poly(styrene-co-acrylonitrile)/montmorillonite organoclay mixtures: a model system for ABS nanocomposite. *Polymer,* Vol. 46, No. 11, pp. 3818-3830, ISSN 0032-3861

Su., S.; Jiang, D.D.; Wilkie, C.A. (2004). Novel polymerically-modified clays permit the preparation of intercalated and exfoliated nanocomposites of styrene and its copolymers by melt blending. *Polymer Degradation and Stability,* Vol. 83, No. 2, pp. 333-346, ISSN 0141-3910

Tarantili, P.A.; Mitsakaki, A.N; and Petoussi M.A. (2010). Processing and properties of engineering plastics recycled from waste electrical and electronic equipment (WEEE). *Polymer Degradation and Stability,* Vol. 95, No. 3, pp. 405-410, ISSN 0141-3910

Utracki, L.A. (2004). *Clay-containing polymeric nanocomposites,* vol. 1, Chap. 2, Rapra Technology, ISBN 1-85957-437-8, UK

Vaia, R.A.; Ishii, H.; Giannelis, E.P. (1993). Synthesis and properties of two-dimensional nanostructures by direct intercalation of polymer melts in layered silicates. *Chemistry of Materials,* Vol. 5, No. 12, (December 1993), (March 2010), pp. 1694-1696, ISSN 0897-4756

Valdes, S.S.; Quintanilla M.L.; Vargas E.R.; Rodriguez F.M.; Rodriguez, J.G. (2006). Effect of Ionomeric Compatibilizer on Clay Dispersion in Polyethylene/Clay Nanocomposites. *Macromolecular Materials Engineering,* Vol. 291, No. 2, pp. 128-136, ISSN 1439-2054

Wang, S.; Hu, Y.; Song, L.; Wang, Z.; Chen, Z.; Fan W. (2002). Preparation and thermal properties of ABS/montmorillonite nanocomposite. *Polymer Degradation and Stability,* Vol. 77, No. 3, pp. 423-426, ISSN 0141-3910

Wang, S.; Hu, Y.; Wang, Z.; Yong, T.; Chen, Z.; Fan W. (2003). Synthesis and characterization of polycarbonate/ABS/montmorillonite nanocomposites. *Polymer Degradation and Stability,* Vol. 80, No. 1, (April 2003), pp. 157-161, ISSN 0141-3910

Wang, S.; Hu, Y.; Zong, R.; Tang, Y.; Chen, Z.; Fan, W. (2004a). Preparation and characterization of flame retardant ABS/montmorillonite nanocomposite. *Journal Applied Clay Science,* Vol. 25, No 1-2, pp. 49-55. ISSN: 0169-1317

Wang, S.; Hu, Y.; Song, L.; Liu, J.; Chen, Z.; Fan, W. (2004b). Study on the Dynamic Self-Organization of Montmorillonite in Two Phases. *Journal of Applied Polymer Science,* Vol. 91, No. 3, pp. 1457-1462, ISSN 1097-4628

Wilkie, C.A.; Zheng. X. (2003). Nanocomposites based on poly (ε-caprolactone) (PCL)/clay hybrid: polystyrene, high impact polystyrene, ABS, polypropylene and polyethylene. *Polymer Degradation and Stability,* Vol. 82, No. 3, (December 2003), pp. 441-450, ISSN 0141-3910

Wu, D.; Wu, L.; Zhang, M.; Wu, L. (2007). Effect of epoxy resin on rheology of polycarbonate/clay nanocomposites. *European Polymer Journal,* Vol. 43, No. 5, (May 2007), pp.1635-1644, ISSN 0014-3057

Xiao, J.; Hu, Y.; Lu H.; Cai, Y.; Chen, Z.; Fan, W. (2007). Effect of Order of Mixing on Morphology and Thermal Properties of the Compatibilized PBT and ABS Alloys/OMT Nanocomposites. *Journal of Applied Polymer Science,* Vol. 104, No. 4, (May 2007), pp. 2130-2139, ISSN 1097-4628

Yeh, J.-M. ; Chen, C.-L.; Huang, C.C.; Chang, F.-C.; Chen, S.-C.; Su, P.-L.; Kuo, C.-C.; Hsu, J.-T.; Chen, B.; Yu, Y.-H. (2006). Effect of organoclay on the thermal stability, mechanical strength, and surface wettability of injection-molded ABS–clay nanocomposite materials prepared by melt intercalation. *Journal of Applied Polymer Science,* Vol. 99, No. 4, (February 2006), pp. 1576-1582, ISSN 1097-4628

Yoon, P.J.; Hunter, D.L.; Paul, D.R. (2003a). Polycarbonate nanocomposites. Part 1. Effect of organocaly structure on the morphology and properties. *Polymer,* Vol. 44, No. 18, (August 2003), pp. 5323-5339, ISSN 0032-3861

Yoon, P.J.; Hunter, D.L.; Paul, D.R. (2003b). Polycarbonate nanocomposites: Part 2. Degradation and color formation. *Polymer,* Vol. 44, No. 18, (August 2003), pp. 5341-5354, ISSN 0032-3861

Yoo, Y.; Choi, K.-Y.; Lee J.H. (2004). Polycarbonate/Montmorillonite Nanocomposites Prepared by Microwave-Aided Solid State Polymerization. *Macromolecular Chemistry and Physics,* Vol. 205, No. 14, pp.1863-1868, ISSN: 1521-3935

Zanetti, M.; Camino, G.; Mulhaupt, R. (2001). Combustion behaviour of EVA/fluorohectorite nanocomposites. *Polymer Degradation and Stability,* Vol. 74, No. 3, (December 2001), pp. 413-417, ISSN 0141-3910

Zhang, J.; Jiang, D.D.; Wilkie, C.A. (2006a). Fire properties of styrenic polymer–clay nanocomposites based on an oligomerically-modified clay. *Polymer Degradation and Stability,* Vol. 91, No. 2, (February 2006), pp. 358-366, ISSN 0141-3910

Zhang, J.; Jiang, D.D.; Wang, D.; Wilkie, C.A. (2006b). Styrenic polymer nanocomposites based on an oligomerically-modified clay with high inorganic content. *Polymer Degradation and Stability,* Vol. 91, No. 11, (November 2006), pp. 2665-2674, ISSN 0141-3910

Zong, R.; Hu, Y.; Wang, S.; Song, L. (2004). Thermogravimetric evaluation of PC/ABS/montmorillonite nanocomposite. *Polymer Degradation and Stability,* Vol. 80, No.1, (March 2004), pp. 423-428, ISSN 0141-3910

Zong, R.; Hu, Y.; Liu, N.; Wang, S.; Liao, G. (2005). Evaluation of the thermal degradation of PC/ABS/montmorillonite nanocomposites. *Polymers for Advanced Technologies,* Vol. 16, No. 10, (October 2005), pp. 725-731, ISSN 1099-1581

8

Sintered Materials Based on Copper and Alumina Powders Synthesized by a Novel Method

Marija Korać[1], Željko Kamberović[2], Zoran Anđić[3] and Mirjana Filipović[2]
[1]Innovation center of the Faculty of Technology and Metallurgy in Belgrade,
[2]Faculty of Technology and Metallurgy, University of Belgrade,
[2]Scientific Research Center, Užice,
Serbia

1. Introduction

The intensification in research of nanostructure materials that occurred in recent years was mainly due to their striking potential, i.e. mechanical and physical properties significantly improved compared to the conventional grain materials (Moriarty, 2001; Ristić, 2003).
Nano-structured materials rank in the group of ultra fine, metastable structures containing a high concentration of defects (point defects, dislocations) and boundaries (grain boundaries, interphase boundaries, *etc.*). These materials are structurally different from crystals and amorphous forms because of the fact that grain boundaries and interphases represent a specific state of the solid matter, since the atoms on boundaries are subjected to a periodical potential field of the crystal from both sides of the boundary (Koch, 1999).
The synthesis of powders represents the starting and crucial stage in the production of sintered metal materials with required properties. Considering that the starting structure undergoes certain changes during further processing, but remains essentially preserved in the structure of the final product (Ristić, 2003; Motta et al. 2004), there is an increased necessity for a great number of methods for producing powders.
Copper is widely exploited in industry because of its high electrical and thermal conductivity, even though it possesses low strength especially at elevated temperatures. In order to overcome this problem it can be strengthened by using finely dispersed particles of stable oxides like alumina, titania, yttria etc. Copper-based composite materials are widely applied in the field of electronics and electrical engineering as highly conductive materials for operation at elevated temperatures, as electrodes for resistance spot and seam welding, different contact materials, various switches, thermal and electric conductors, microwave tubes, etc (Lee & Kim, 2004).
Introduction of fine dispersed particles into matrix of base-metal has considerable strengthening effect and nano particles of oxides are especially suitable. Due to their hardness, stability at elevated temperatures and insolubility in copper they represent obstacles for dislocation, grain and sub-grain boundary movement increasing mechanical properties of these materials with insignificant effect on thermal and electrical conductivity (Naser et al., 1997; Trojanova et al., 1999, Tian et al. 2006). Significant reinforcing effects can

be kept even at elevated temperatures. For such reinforcement nano-particles of oxides are suitable (Karwan-Baczewska, 2005).

A very important aspect of dispersion strenghtening is introduction of low volume fraction of dispersed oxide particles into the volume of the base-metal, a uniform distribution of oxide particles and their fine dispersion especially in nanometer scale (Ristic & Milosevic, 2006).

Alumina was known for its exceptional properties such as high melting point, high hardness, excellent thermal stability and chemical inertness. Also, addition of alumina particles can increase the temperature of recrystallization by pinning grain and sub-grain boundaries of copper matrix and blocking the movement of dislocations highly improving strength at elevated temperatures (Plascencia & Utigard, 2005; Lianga et al., 2004). The usual amount of alumina used for dispersion strengthening is from 0.5-5.0 wt.% (Jena et al., 2004), but significant results regarding particle size could be achieved with even higher amounts ranging up to 50 wt.% alumina (Brocchi et al., 2004).

Because these oxides are not soluble in copper P/M techniques must be used instead of conventional ingot metallurgy. Oxide dispersion strengthened (ODS) copper can be successfully synthesized by following methods: highly energetic reactive milling (Ahn et al., 1996), precipitation from solution (sol-gel (Ruys & Mai, 1999), hydrothermal synthesis (Byrappaa & Adschirib, 2007), electrochemical synthesis (Yuana et al., 2007)), internal oxidation (Afshar & Simichi, 2008) etc. The production of powders by the thermo-chemical method is not a new procedure, but in recent years, due to the development of contemporary materials with pre-set properties, there has been an extended interest in this method, especially for the production of nanostructured powders (Wu et al., 2001; Jena et al. 2001; Lee et al, 2001).

Route for synthesis of composite powders based on copper and alumina presented in this research, may be regarded as new for materials in this system, even though phases of this process have been previously investigated by the authors (Korać et al., 2005; Anđić et al., 2006; Anđić et al., 2007a; Anđić, 2007b; Korać et al., 2008a, Korać et al., 2008b) and also previous attempts have been made for application of similar process in system Cu-Ag-Al_2O_3, when a three-component system was produced by mechanically alloying the thermo-chemically synthesized Cu-Al_2O_3 and Cu-Ag powder (Anđić et al., 2006).

Objective of this research was to investigate the possibility of copper strengthening with fine dispersed alumina particles by application of new synthesis method and production of sintered materials with properties suitable for material application as contact material.

After sintering compacts were thermo-mechanically treated in order to simulate real manufacturing process.

Verification of obtained results is performed by tribological investigations.

2. Experimental

Generally synthesis process included two separate routes. The first included synthesis of Cu-Al_2O_3 with 50 wt. % of Al_2O_3 (from now on Cu-50Al_2O_3) by thermo-chemical route. The second part of the synthesis process was mechanical alloying of atomized copper powder with powder of Cu-Al_2O_3 previously synthesized.

Starting raw materials for powder synthesis by thermo-chemical route were soluble salts, nitrates of copper and aluminum of p.a. quality, dissolved in distilled water (50wt.% solution) in suitable ratio for final powder to contain 50wt.% of Al_2O_3 in structure.

Copper powder used in mechanical alloying stage was produced by atomization with particle size 95% <45μm. In previous research (Korać, 2005) electrolytic copper was used for mechanical alloying. Electrolytic copper was replaced with atomized, due to the fact that its dendrite structure prevents formation of homogenous structure of final material, even after prolonged milling time.

Stages of synthesis process included the following processes:

1. Spray drying of nitrate solution using Spray Dryer Büchi B-290 Advanced (Okuyama & Lenggoro, 2003) with inlet/outlet temperature 190/143°C and solution flow rate was 10% of pump power,
2. Dried nitrates were subjected to heat treatment in a laboratory electroresistant tube furnace (Korać et al., 2009) in air at 900°C for 1h in order to form oxides of copper and aluminium,
3. Oxides were then reduced in the same furnace in hydrogen atmosphere (flow rate 20L/h at 350°C for 1h) in order to obtain metallic copper. Previously formed stable Al_2O_3 remained unchanged during reduction (Korać et al., 2009),
4. Mechanical alloying was the next step in the synthesis process. Atomized copper powder was mechanically alloyed with produced composite powder with 50wt.% Al_2O_3 in a ceramic ball mill, with dimensions ø180×160mm. Milling media were corundum balls (≥99% Al_2O_3 with 30mm in diameter) in order to prevent contamination of material; ball to powder ratio was 1:30. Optimal milling time was fixed at 5h for mill rotation of 300min^{-1}. Quantities of added copper powder were adjusted so that the final amount of alumina Al_2O_3 in composite powder would contain 1, 1.5 and 2 wt. %.

Flow-sheet of applied synthesis process is presented in Fig. 1.

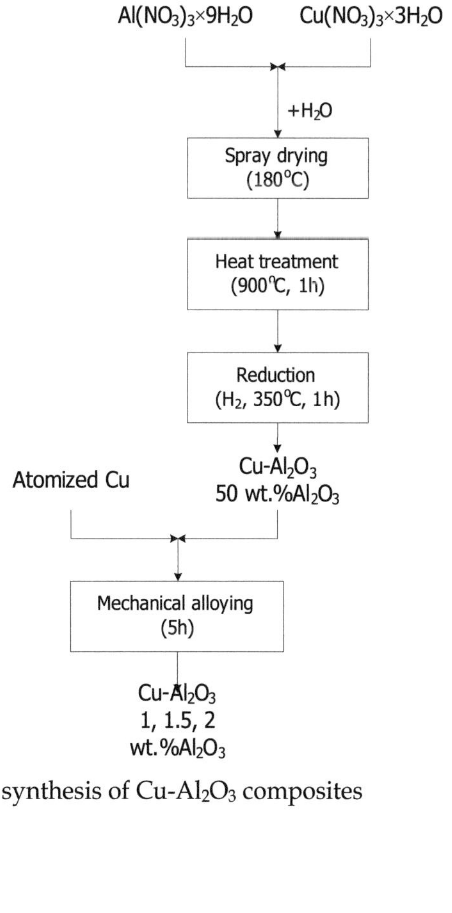

Fig. 1. Flow-sheet for the synthesis of $Cu\text{-}Al_2O_3$ composites

Solution concentration of 50% enabled production of higher amounts of fine particles. If solution concentration was increased coarse particles would be produced, which could disable fine coating of copper particles in later stage of mechanical alloying. On the other hand if solution was too diluted, finer particles would be obtained, but disadvantage is that in this case synthesis process would last longer, with higher equipment engagement, i.e. lower possibilities for industrial application.

Method of choice for acquiring of precursor powder was spray drying. In recent years this method is engaged in large share for production of micron and sub-micron powders (McCandlish et al., 1994). Advantages of this drying method are control of size, morphology and particle composition, as well as significant possibility of industrial application from aspect of price-productivity ratio (Okuyama & Lenggoro, 2003; Iskandar et al., 2003). Also, application of this drying method in Cu-Al_2O_3 system showed very good results in previous research (Anđić, 2006; Korać; 2007).

Temperatures of heat treatment and reduction were previously optimized for this system (Anđić, 2006), as well as milling condition (Korać, 2005).

After mechanical alloying obtained powders were compacted by a uniaxial pressing in order to minimize density gradient trough sample. Used tool had dimensions 8×32mm and height 3mm, and applied compacting pressure was 500 kN.

Sintering of samples was performed in hydrogen atmosphere in isothermal conditions at five different temperatures in the range from 725-925°C for 15 to 120 min.

After consolidation of the obtained powders, the compacted samples were uniaxially compressed by cold rolling, reduction degree of 15 and 30%. In order to determine the stability at higher temperatures, the rolled samples were annealed at temperature of 800°C for one hour in the hydrogen atmosphere.

Produced powders were characterized by X-ray diffraction analysis (XRD). XRD was performed using APD 2000-Ital structures with CuKα radiation, 2θ=0-100°.

Characterization of sintered samples (referred as composite in the following text) included Scanning Electron Microscopy (SEM), Energy Dispersive Spectrometry (EDS), HRF hardness measurements and electrical conductivity.

SEM analysis was performed on JEOL JSM-T20 on polished samples subsequently etched with 40 vol.% HNO_3 solution. EDS analysis was conducted on unpolished composites using Oxford system attached to JEOL SEM JSM-5800.

Ames Portable Hardness Tester was employed for hardness measurements using 1/16" ball with applied load of 60kg. For electrical conductivity measurement Foerster Sigma Test 2.069 operating at 120 kHz and with 8mm electrode diameter, was used.

Values of hardness and electrical conductivity represent the mean value of at least six measurements conducted on the same composite.

Wear testing was performed by the method of Taber in accordance with Guide to friction, wear and erosion testing (ASTM standard MNL 56, 2007). Basic information on test conditions are:

- diameter of the grinding plate - 640mm,
- diameter of the grinding path - 265mm,
- calculated speed rate of the grinding path - 832.5mm
- granulation of Corundum Al_2O_3 Ø100μm.

3. Results & discussion

It was expected that by the proposed synthesis method nano-composite Cu-Al_2O_3 produced by thermo-chemical route would form homogenous compact layer around atomized copper powder particles during mechanical alloying process. This way a ductile, highly conductive core with high strength shell layer of composite, will be achieved. In later stages this microstructure would provide dislocation blocking and prevent grain boundary motion as well as an increase of recrystallization temperature.

3.1 Powder characterization

SEM analysis of Cu-50Al_2O_3 presented in Figure 2 shows that powders are flake like with average particle size of 200-400nm, which have partially agglomerated due to their small size and high surface energy. Average size of present agglomerates is 1-5μm.

Previous results on thermo-chemical synthesis of Cu-Al_2O_3 powders (Korać, 2005) indicates that presented particles in fact represent very fine agglomerates, and that actual size of synthesized powder is in range 20-50nm.

X-ray diffraction pattern of powder containing 50 wt.% Al_2O_3 is presented in Figure 3. Identified peaks correspond to Cu and $CuAl_2O_4$. $CuAl_2O_4$ phase represents the metastable equilibria, developed in microstructure during powder synthesis process, heat treatment and reduction.

Thermo-chemically synthesized powder was used for mechanical alloying of electrolytic copper, SEM micrograph presented in Figure 4.

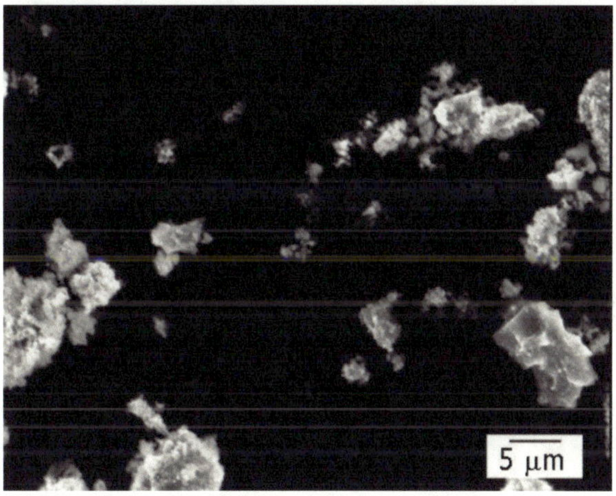

Fig. 2. SEM micrograph of Cu-50Al_2O_3 powder

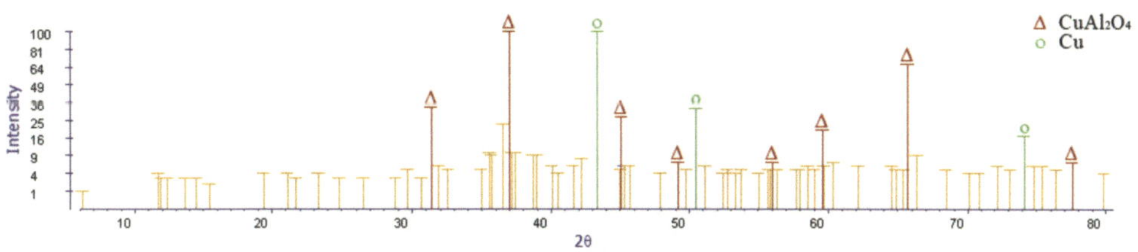

Fig. 3. X-ray diffraction of Cu-50Al_2O_3 powder

Used copper powder particles are mostly spherical shaped with smooth surface. This particle morphology enables achievement of desired structure with highly conductive core with homogenous shell layer around it.

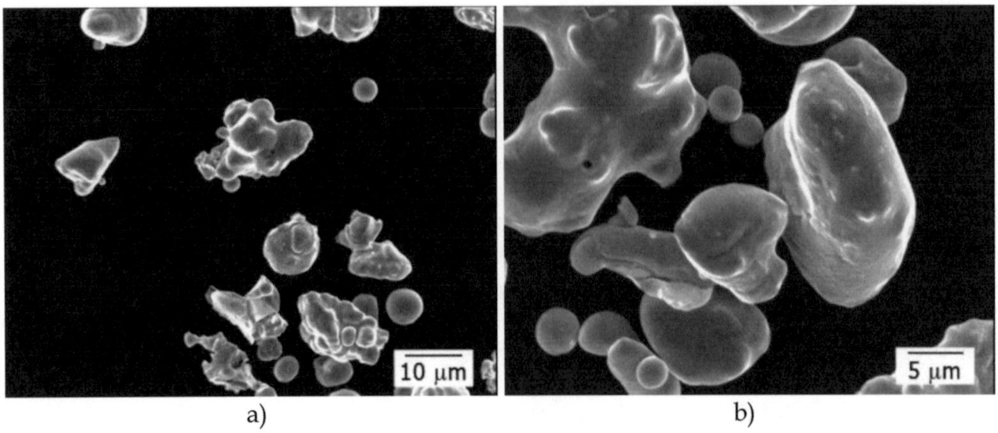

Fig. 4. SEM micrographs of electrolytic copper powder

After 5h of milling of powders, all three compositions showed structure characteristic for mechanical alloying (Figures 5-7).

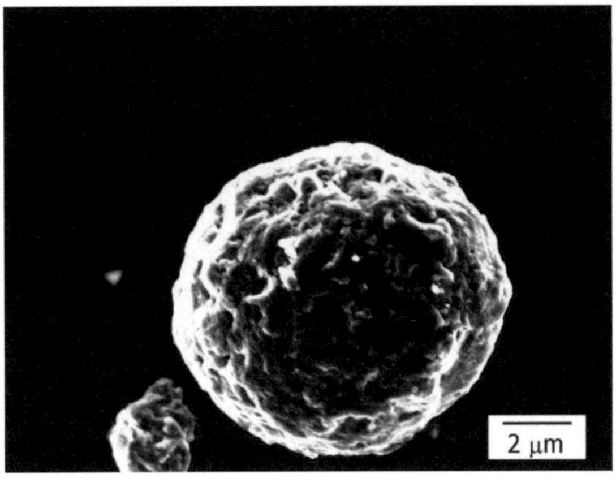

Fig. 5. SEM micrograph of composite powder containing 1 wt.% of Al_2O_3

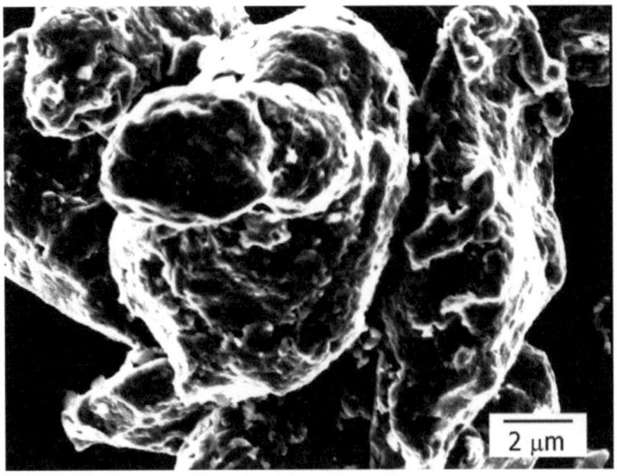

Fig. 6. SEM micrograph of composite powder containing 1.5 wt.% of Al_2O_3

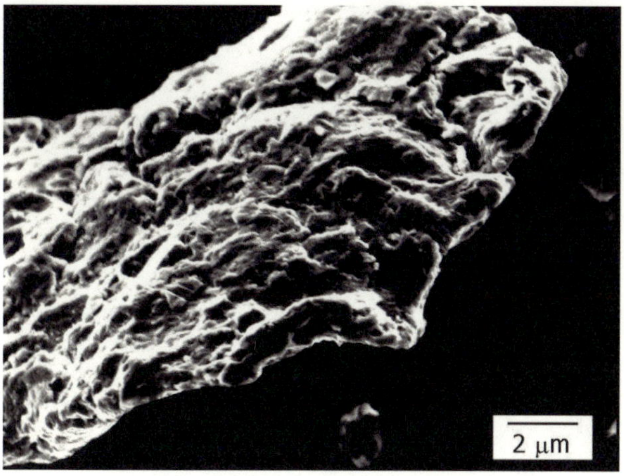

Fig. 7. SEM micrograph of composite powder containing 2 wt.% of Al_2O_3

From these Figures it is obvious that with increased Al_2O_3 content increases particle size. Average particle size of composite powders containing 1% Al_2O_3 was 6μm, 1.5% 8μm and with 2%Al_2O_3 around 12μm. Also, these Figures show that throughout the process of mechanical alloying starting copper particles got reduced in size. Deformation strengthening induced by consecutive plastic deformations by milling media, makes previously ductile particles brittle, further leading to their breakage due to fatigue.
X-ray diffraction patterns (Figure 8.) of powders containing 1, 1.5 and 2 wt. % Al_2O_3 show only peaks corresponding to copper, due to the small amount of Al_2O_3 in system.

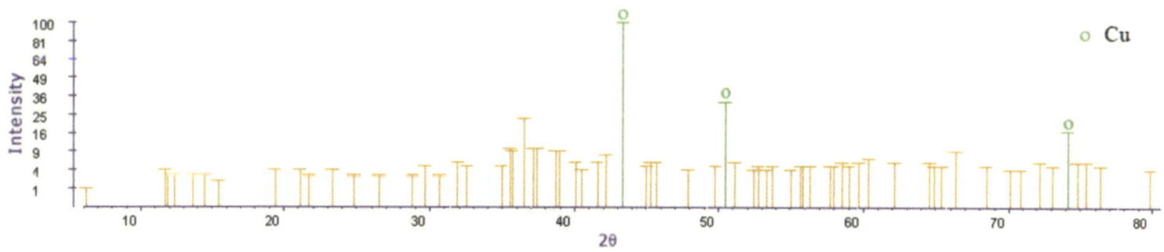

Fig. 8. XRD pattern of Cu-$2Al_2O_3$

3.2 Compacts characterization

SEM analysis of compacted powders after mechanical alloying containing 1, 1.5 and 2 % of Al_2O_3 sintered at 875°C for 1h is presented in Figures 9-11.
From the presented SEM analysis it is confirmed that with increase of alumina content grain size increases from 1μm for Cu-$1Al_2O_3$ up to 15-20μm for Cu-$2Al_2O_3$. Also, sintered compacts exhibit annealing twins and sub-grain boundaries for all three compositions. Twinning might have occurred during the high temperature sintering stage.
Conditions for twins formation are achieved when large number of obstacles is formed in crystal, blocking dislocation movement. Dislocations are accumulated on obstacles causing increase of internal strain in local areas, which together with external strain induce twins formation. Presence of twins indicated lower mobility of dislocations, in other words stabilization of dislocation structure, which is primary condition for improvement of mechanical properties of dispersion strengthened materials.

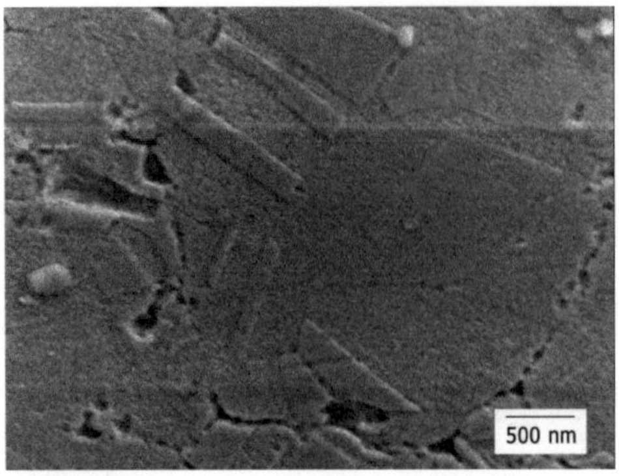

Fig. 9. SEM micrograph of Cu-1Al_2O_3

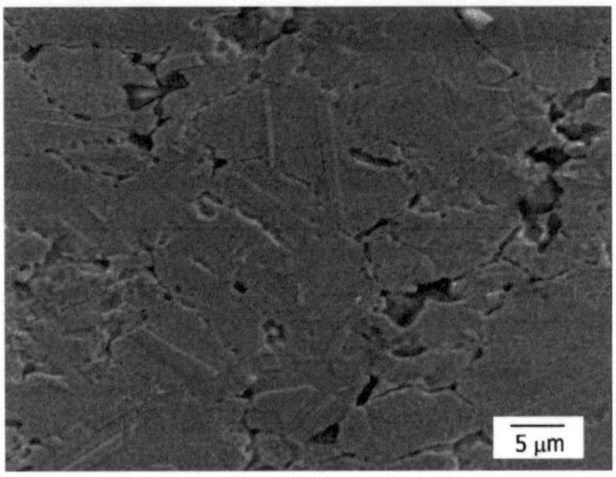

Fig. 10. SEM micrograph of Cu-1.5Al_2O_3

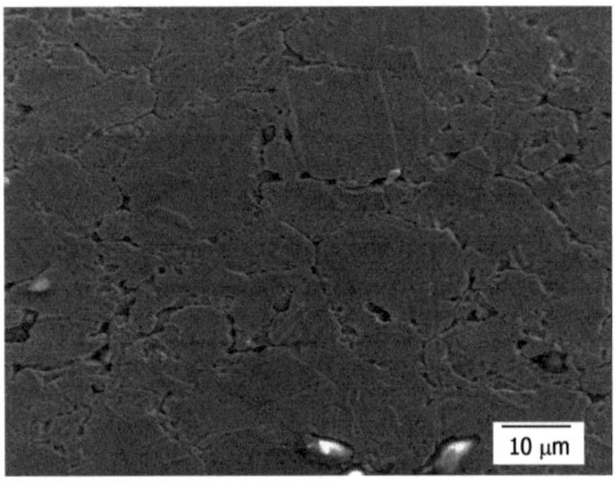

Fig. 11. SEM micrograph of Cu-2Al_2O_3

Size of sub-grains similar to ones presented in Figures 9-11 are expected to be preserved under the influence of temperature as a result of dispersed Al_2O_3 particles blocking sub-grain boundaries and increasing recrystallization temperature, rendering these materials suitable for exploitation at elevated temperatures.

Also, during the period of sintering, formation of a third phase is expected (Lianga et al., 2004), due to the thermodynamically possible eutectic reaction of (Cu+Cu_2O) with Al_2O_3 at the contact surface. Through this reaction eutectic tends to expand and reacts with Al_2O_3

forming $Cu_xAl_yO_z$ phase, compatible with both phases at the interface (Yoshino & Shibata, 1992; Jena et al., 2001). The formed third phase influences the dislocation structure resulting in improvement of mechanical properties, whereas good electrical conductivity is retained. According to literature (Yi et al., 1999) chemical formula of this compound might be derived as $CuAlO_2$ or $CuAlO_4$.
Presence of this phase in the structure and especially at the interface could impede crack propagation and result in higher interface fracture energy (Reimanis et al., 1997).
Figure 12 presents EDS analysis of produced composites with 1, 1.5 and 2 wt. % Al_2O_3.

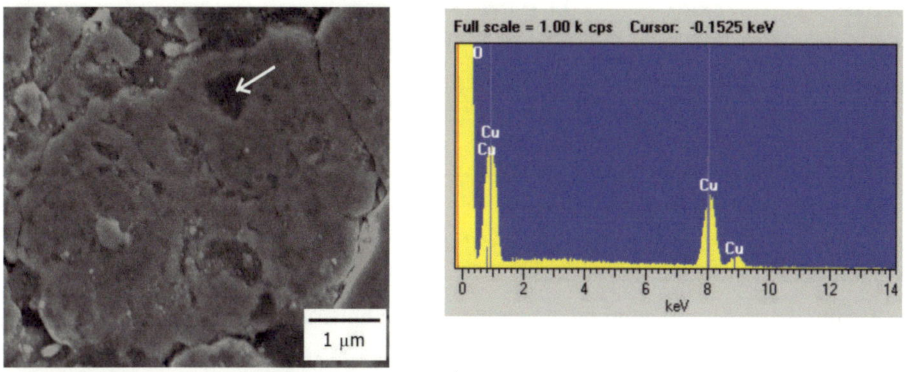

a)

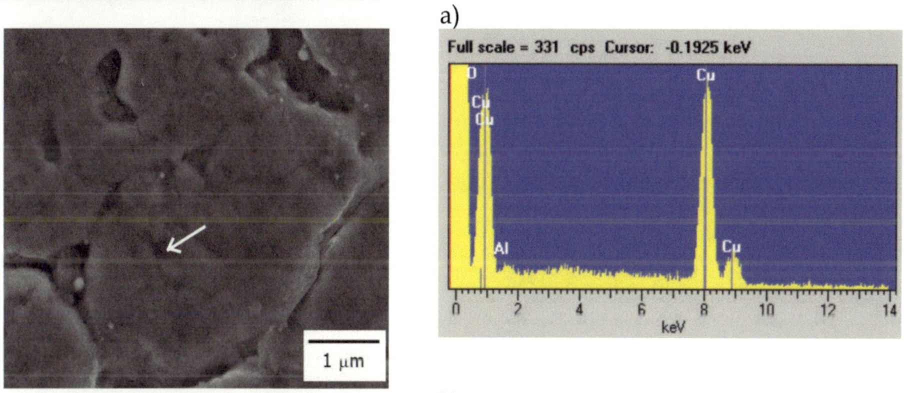

b)

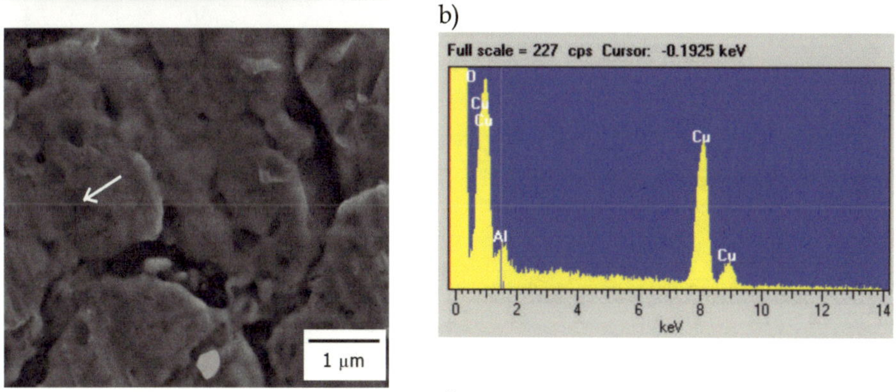

c)

Fig. 12. EDS analysis of marked spots in SEM micrographs of Cu-Al_2O_3 composite with a) 1; b) 1.5 and c) 2wt.% Al_2O_3

Presented EDS results show that in the composites with Cu-1Al_2O_3 aluminum was not detected in the structure, indicating that the whole quantity of Al_2O_3 is incorporated in the copper matrix. With increase of Al_2O_3 content, aluminum is detected which suggests occurrence of segregations in the structure and formation of aluminum-rich regions. These results suggest that the new synthesis method is more suitable for composite materials containing maximum 1wt. % Al_2O_3. This amount could be fully incorporated in the base-metal matrix playing a role as an obstacle to dislocations, grain and sub-grain boundaries movement. However, this role is diminished with increase of Al_2O_3 content by formation of aluminum-rich regions. Furthermore, Figure 12 shows better compaction for sintered compacts with lower Al_2O_3 content, which is supported by porosity measurements presented in Table 1.

Sample	*Min, μm*	*Max, μm*	*Mean, μm*	*Std Error*	*Std Dev.*	*Vv, %*
Cu-1Al_2O_3	0,23	2,41	1,21	0,06	0,43	7,638
Cu-1.5Al_2O_3	0,18	4,46	1,34	0,10	0,72	7,835
Cu-2Al_2O_3	0,45	4,32	1,41	0,11	0,70	8,561

Table 1. Statistical data for the porosity measurements for the samples with different Al_2O_3 content sintered at 875 ° C/1h

These results are also confirmed by measurements of hardness and electrical conductivity as presented in following tables and figures.
Results of HRF hardness measurements are presented in Table 2.

T, °C / t, min	725	775	825	875	925
Cu-1Al_2O_3					
15	52	49	50	52	42
30	52	54	50	50	52
60	53	46	48	50	46
90	48	52	44	45	40
120	47	46	46	52	44
Cu-1.5Al_2O_3					
15	42	42	40	42	40
30	48	44	40	46	38
60	44	44	42	40	42
90	42	43	39	42	38
120	44	39	40	39	43
Cu-2Al_2O_3					
15	42	29	22	26	19
30	26	27	12	20	19
60	30	24	22	26	20
90	24	26	15	20	16
120	30	26	13	23	22

Table 2. HRF hardness of sintered Cu-Al_2O_3 composites

Results show that with increase of temperature and time of sintering, HRF values slightly and unevenly decrease. Increase of Al_2O_3 content has significant effect on hardness. It was expected that increase of Al_2O_3 content in structure would result with an increase of

hardness, but the results show an opposite trend. With increase of Al_2O_3 from 1 to 1.5 wt.% there is a slight decrease in HRF values, while increase up to 2 wt.% Al_2O_3 provokes a rather significant decrease in HRF. Possible reasons for this behavior is ascribed to a positive effect on dispersion strengthening of the copper matrix which occurred when amount of Al_2O_3 dispersoide is up to 1 wt.%, whereas further increase of Al_2O_3 content has a negative effect on hardness as previously postulated by EDS analysis (Fig. 12). This problem supposedly may be solved using hot extrusion process.
Electrical conductivity measurements of composites with different Al_2O_3 content at various sintering conditions are presented in Table 3.

t, min \ T, °C	725	775	825	875	925
			Cu-1Al_2O_3		
15	54,88	59,90	56,83	55,05	55,19
30	53,94	55,39	56,48	57,16	56,06
60	55,44	54,61	56,28	56,69	55,51
90	58,00	55,13	54,54	55,25	54,81
120	56,37	54,24	55,15	56,43	55,87
			Cu-1.5Al_2O_3		
15	42,76	42,49	43,51	43,86	43,81
30	45,07	42,52	42,51	44,02	47,81
60	40,94	43,27	45,48	49,59	49,88
90	39,86	41,87	42,63	46,72	48,08
120	43,19	39,11	42,76	45,80	48,80
			Cu-2Al_2O_3		
15	32,66	32,54	30,71	30,35	31,25
30	29,25	31,27	29,92	31,13	40,70
60	31,93	30,89	31,72	36,94	39,99
90	31,35	31,72	31,76	37,61	39,35
120	30,53	31,66	31,66	33,43	41,21

Table 3. Electrical conductivity of sintered Cu-Al_2O_3 composites, %IACS

With increase of Al_2O_3 content in composite materials values of electrical conductivity decreases.
Values of electrical conductivity are over 55% IACS for compacts with 1wt% Al_2O_3, which correspond to a dispersion strengthened copper alloys intended for application at elevated temperatures. The limitation value for application is 50% IACS (Grant et al., 1984).
Measured values of harness, as well as electrical conductivity are highly dependant on Al_2O_3 content and not so much on sintering conditions.

3.3 Characterization of samples after thermo-mechanical treatment

After consolidation of the obtained powders, the compacted samples were uniaxially compressed by cold rolling, reduction degree up to 30%. In order to determine the stability at higher temperatures, the rolled samples were annealed at temperature of 800°C in one hour in the hydrogen atmosphere.
Cold rolling of samples was performed in two steps: first 15% of size reduction and second of 30%. During cold rolling samples with 1.5 and 2 % of Al_2O_3 were degraded, due to inhomogeneous structure and distribution of Al_2O_3, as presented in Figure 12 and increased

porosity with increase of Al_2O_3 (Table 1). Insufficient packing caused formation of cracks in the middle of the sample and disabled sample structure to endure forces of cold plastic deformation.

Considering the presented research results, shown also in (Korać et al., 2010), which suggest that the best combination of mechanical and electrical properties was achieved in the systems with 1wt.% Al_2O_3 sintered at 875°C/1h, so further experimental results will be related to these compacts only.

SEM analysis of samples after two stages of size reduction and after heat treatment (HT) is presented in Figure 13.

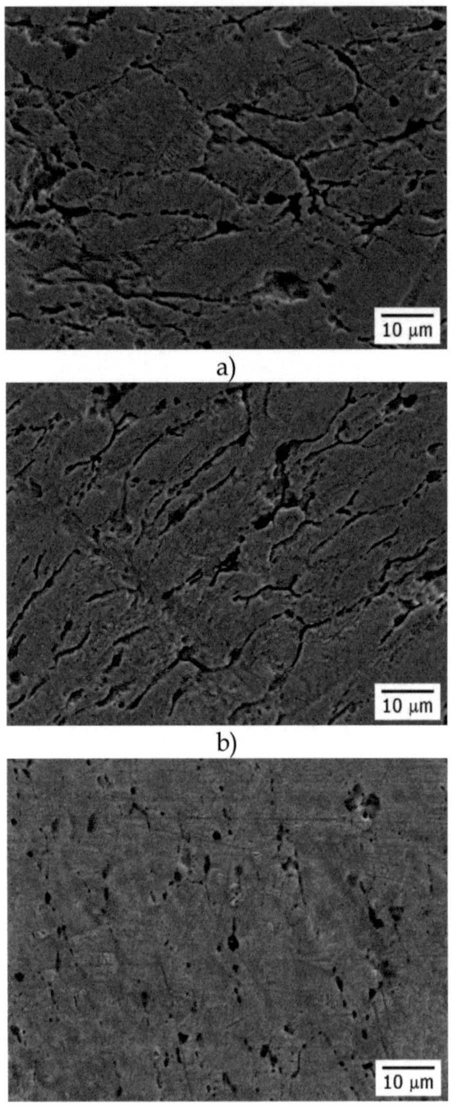

c)

Fig. 13. SEM micrographs of compacts after cold rolling and heat treatment a) size reduction 15%, b) size reduction 30% and c) heat treated

Statistical data on porosity of samples after cold rolling and heat treatment are presented in Table 4. Data for sintered samples is also incorporated in this Table for comparison.

Sample	*Min, µm*	*Max, µm*	*Mean, µm*	*Std Error*	*Std Dev.*	*Vv, %*
sintered	0,23	2,41	1,21	0,06	0,43	7,638
15%	0,36	2,84	0,91	0,07	0,490	7,01
30%	0,29	2,14	0,69	0,04	0,337	5,52
HT	0,16	1,56	0,55	0,03	0,29	4,507

Table 4. Statistical data for the porosity measurements for the samples with different Al_2O_3 content sintered at 875 ° C/1h after cold rolling and heat treatment

In the structure of the samples subjected to cold plastic deformation performed by rolling, porosity is also relatively evenly distributed, with a clearly visible porosity of the directed orientation.

The comparative analysis of the porosity testing results of the samples subjected to cold plastic deformation by rolling and the sintered samples indicates the significant porosity in the sintered samples both from the aspect of the pore size, and also from the aspect of their volume share. In the cold deformed samples by rolling, it is clearly noticeable the accumulation of the pores at grain boundaries, which indicates that the pores in the moment of contact with the grain edge during its movement have been too large to be absorbed by the grain edge and disappeared in it, which has resulted in stopping the grain edge to move and its growth.

Rolling can change the distribution and the size of secondary phase, as well as to reduce the number of pores. After rolling, the particles of Al_2O_3 become more dispersive, while the relative density increases (Korać et al., 2008).

The porosity in the structure of the samples subjected to heat treatment after cold plastic deformation is relatively evenly distributed. The pores are arbitrarily oriented, and the shape of the pores is irregular. Based on relative distribution of porosity and statistical data, the size of pores is in the range from 160nm to 1.56 µm with medium diameter of 550nm. The volume share of porosity is 4.5%.

The comparative analysis of the results of the porosity tests on the sintered samples, the samples subjected to cold plastic deformation by rolling and the HT samples points to significant porosity of the sintered samples and the samples subjected to cold plastic deformation by rolling in relation to the annealed samples as from the aspect of the pore size, so from the aspect of their volume share.

Results of hardness measurements in function size reduction and heat treatment for selected sintering temperature of 875°C are presented in Fig. 14. and for electrical conductivity in Fig.15.

From Fig.14 it could be noticed that with increase of size reduction hardness values are increasing up to the level of 90 HRF indicating stabilization of dislocation structure. Also, curves of dependence HRF-time are stabilizing. Stabilization of curve could be the result of structure compaction and reduction of pore volume fraction in structure. After heat treatment harness values decrease up to the value of 55 HRF. Change of curve slope could be detected for sintering time of 60min. After 30% reduction at 60min slope is changing into plateau. Heat treated samples exhibit slight decrease of hardness values after 60min.

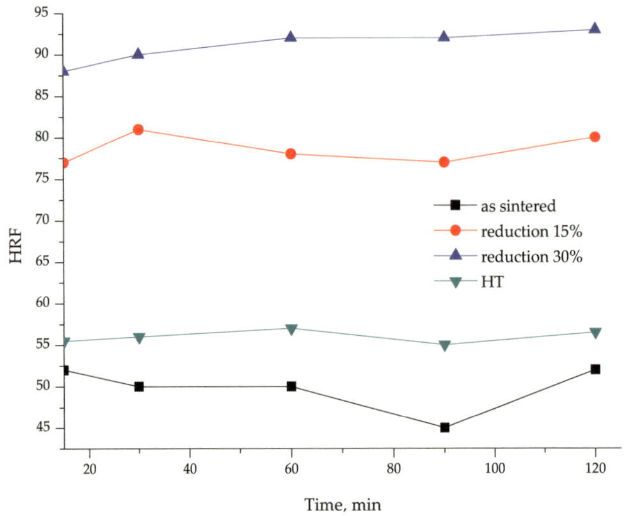

Fig. 14. Dependence of HRF hardness from size reduction and heat treatment for selected sintering temperature of 875°C

Values of electrical conductivity for as sintered samples are over 55% IACS, which correspond to dispersion strengthened copper alloys intended for application at elevated temperatures. With increase of size reduction electrical conductivity is increasing up to the 59.5 %IACS. After heat treatment electrical conductivity continues to increase up to 60.75%IACS for sintering time of 60min, when plateau is reached. After 60 min of sintering hardness and electrical conductivity both reached constant values, indicating completed stabilization of structure.

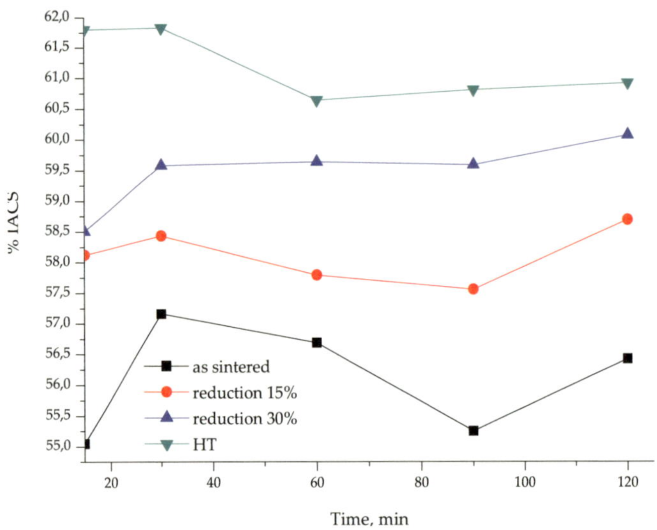

Fig. 15. Dependence of electrical conductivity from size reduction and heat treatment for selected sintering temperature of 875°C

Aiming to evaluate the stability of structure for the sintered samples and each phase of the thermo-mechanical treatment process, the examinations on tribological properties were performed.

It is well known that the HT metals with the surface centered cubic crystal structure have significantly higher resistance to erosive wear in comparison to metals with the spatially centered cubic crystal structure of similar hardness. Cold rolling prior to wear testing does not bring significant improvement in resistance to erosive wear. Also, reducing the speed rate of the erosive wear is achieved with an increase in the index of the deformation hardening. Increasing in the index of the deformation strengthening effects on increasing the value of critical voltage that is required to starting with localize the deformation in the process of wear (Liu et al. 2007).

Tribological tests have shown that the best resistance to wear have the samples after heat treatment (Fig. 16), which can be explained by the fact that after annealing recrystalization and characteristic growth of the grain do not occur, due to block in movement of the grain boundaries homogeneously distribution of alumina and formation of the third phase, identified in (Jena et al., 2004; Jena et al., 2001; Lee & Kim, 2004, Anđić, 2007, Korać, 2009). This is confirmed by the microstructural examinations of the analyzed samples.

Based on the results of quantitative analysis and tribological examinations, it can be concluded that the effect of porosity has a significant influence on the tribological properties, that is, as the volume share of porosity reduces as tribological properties improve.

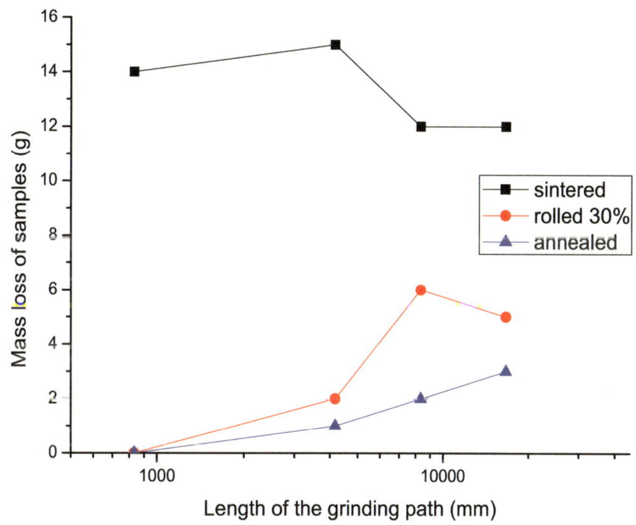

Fig. 16. Wear resistance of the sintered (875°C/60min), rolled and annealed sample

The significant influence on the tribological properties has multiple strengthening mechanisms of the analyzed systems schematically presented in Figure 17.

In the observed system, strengthening the base of copper is achieved by dispersion strengthening, due to the dispersion of fine particles of Al_2O_3 in the base, strengthening by grain boundaries, because of the third phase appearance, as well as the deformation strengthening and strengthening by annealing. During mechanical alloying the coating copper particle with the particles of the composite 50% Al_2O_3 occurs, which is built into its surface. Along with the process of sintering occurs the formation of the compact structure and formation of the third phase on the grain boundary, which causes strengthening on the grain boundaries. Due to the plastic deformation occurs the deformation strengthening, and after heat treatment occurs the strengthening by annealing, which is confirmed by the results of the tribological examinations.

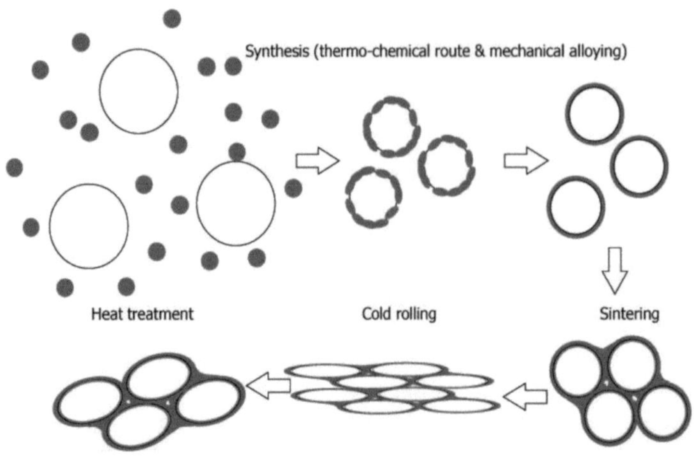

Fig. 17. Proposed multiple strengthening mechanisms

4. Conclusion

Synthesis of nanocomposite Cu-Al_2O_3 powder suitable for obtaining multiple strengthened systems could be successfully performed by combination of the thermo-chemical procedure and the mechanical alloying. To optimize the whole scope of properties that are expected from these materials, the combination of thermo-chemical procedures and the procedure of mechanical alloying is a completely new approach to the synthesis of powders. Thus obtained powders allow obtaining the final product with the excellent effects of strengthening.

Presented results show possibility of usage composite powders based on copper and alumina synthesized by novel method for production of compacts with suitable properties for exploitation at elevated temperatures, i.e. increased hardness with electrical conductivity appropriate for oxide dispersion strengthened copper alloys intended for application at elevated temperatures. Achieved values after thermo-mechanical treatment for hardness are 57HRF and for electrical conductivity 61%IACS.

Limitation for dispersoide content is 1wt. %, due to the decomposition of compacts with higher Al_2O_3 content during mechanical processing. Increased Al_2O_3 content could be achieved by using hot extrusion process.

5. Acknowledgment

Presented research is part of projects TR19032 and TR34033 financially supported by Ministry of Science and Technological Development, Republic of Serbia.

6. References

Afshar A. & Simchi A. (2008), Abnormal grain growth in alumina dispersion-strengthened copper produced by an internal oxidation process, Scripta Materialia, 58 (11), pp. 966-969

Ahn J-H., Song I-H. & Hahn Y-D. (1996), Cu-Based Cermets Prepared By Mechanical Alloying, Materials Transactions JIM, 37 (4), pp. 733-737

Anđić Z. (2007), Ph. D. Thesis, Faculty of Technology and Metallurgy, University of Belgrade (in Serbian)

Anđić Z., Korać M., Kamberović Ž., Vujović A. & Tasić M. (2007), Analysis of the Properties of a Cu-Al_2O_3 Sintered System based on Ultra Fine and Nanocomposite Powders, Science of Sintering, 39(2), pp. 145-152

Anđić Z., Korać M., Tasić M., Raić K. & Kamberović Ž. (2006), The synthesis of ultra fine and nanocomposite powders based on copper, silver and alumina, Kovove materialy, 44(3), pp. 145-150

ASTM standard (2007), MNL 56 - Guide to friction, wear and erosion testing

Brocchi E.A., Motta M.S., Solorzano I.G., Jena P.K. & Moura F.J. (2004), Alternative chemical-based synthesis routes and characterization of nano-scale particles, Materials Science and Engineering B, 112 (2-3), pp. 200-205

Byrappaa K. & Adschirib T. (2007), Hydrothermal technology for nanotechnology, Progress in Crystal Growth and Characterization of Materials, 53 (2), pp. 117-166

Grant J.N., Lee A. & Lou M. (1984), in: Proceedings:"High Conductivity Copper and Aluminium Alloys", Eds. E. Ling and P.W. Taubenblat, The Metallurgical Society of AIME, California, pp. 103–111.

Iskandar F., Gradon L. & Okuyama K. (2003), Control of the morphology of nanostructured particles prepared by the spray drying of a nanoparticle sol, Journal of Colloid and Interface Science 265 (2), pp. 296–303

Jena P. K., Brocchi E. A. & Motta M. S. (2001), Characterization of Cu–Al_2O_3 nano-scale composites synthesized by in situ reduction, Materials Science and Engineering C, 15 (1-2), pp. 175-177

Jena P. K., Brocchi E. A. & Motta M. S. (2001), In-situ formation of Cu- Al_2O_3 nano-scale composites by chemical routes and studies on their microstructures, Materials Science and Engineering A, 313 (1-2), pp. 180-186

Jena P. K., Brocchi E. A., Solórzano I. G. & Motta M. S. (2004), Identification of a third phase in Cu–Al_2O_3 nanocomposites prepared by chemical routes, Materials Science and Engineering A, 371 (1-2), pp. 72-78

Karwan-Baczewska J., Gotman I., Gutmanas E.Y. & Shapiro M. (2005), Small particles with better contacts make nanocomposites kings of conductivity, Metal Powder Report 60 (6), pp. 28-34

Koch C., Bulk Behavior of Nanostructured Materials (1999), Chapter 6, Nanostructure Science and Technology A Worldwide Study, National Science and Technology Council (NSTC), USA, pp. 93-112

Korać M. (2005), Master Thesis, Faculty of Technology and Metallurgy, University of Belgrade (in Serbian)

Korać M. (2009), Ph. D. Thesis, Faculty of Technology and Metallurgy, University of Belgrade (in Serbian)

Korać M., Kamberović Ž. & Filipović M. (2008), Determination of Al_2O_3 particle size in Cu-Al_2O_3 nanocomposite materials using UV spectrophotometry, Metalurgija, 14 (4), pp. 279-284 (in Serbian)

Korać M., Kamberović Ž., Anđić Z. & Filipović M. (2010), Sintered materials based copper and alumina powders synthesised by novel method, Science of Sintering, 42(1), pp. 81-90

Korać M., Kamberović Ž., Tasić M. and Gavrilovski M., Nanocomposite powders for new contact materials based on copper and alumina Chemical Industry & Chemical Engineering Quarterly 14(4) (2008), 215-218

Lee D. W. & Kim B. K. (2004), Nanostructured Cu- Al_2O_3 composite produced by thermochemical process for electrode application, Materials Letters, 58 (3-4), pp. 378-383

Lee D. W., Ha G. H.& Kim B. K. (2001), Synthesis of Cu-Al_2O_3 nano composite powder, Scripta Materialia 44 (8-9), pp. 2137-2140

Liang S., Fan Z., Xua L. & Fang L. (2004), Kinetic analysis on Al_2O_3/Cu composite prepared by mechanical activation and internal oxidation , Composites Part A: Applied Science and Manufacturing, 35 (12), pp. 1441-1446

Liu X-B., Jia C-C., Chen X-H. & Gai G-S. (2007), Microstructures and properties of 1.0%Al_2O_3/Cu composite treated by rolling, The Transactions of Nonferrous Metals Society of China 17, Special issuse International Conference of Nonferrous Materials (ICNFM) Part IIB, pp. 626-629

McCandlish L. E., Kear B. H. & Bhatia S. J. (1994), Spray conversion process for the production of nanophase composite powders, United States Patent, 5.352.269

Moriarty P. (2001), Nanostructured materials, Reports on Progress in Physics, 64 (3), pp. 297–381

Motta M.S., Brocchi E.A., Solórzano I.G. & Jena P.K. (2004), Multidisciplinary Microscopy Research and Education, FORMATEX, pp. 215-223.

Naser J., Ferkel H. & Riehemann W. (1997), Grain stabilization of copper with nanoscaled Al2O3-powder, Material Science and Engineering A, 234-236, pp. 470-473

Okuyama K. & Lenggoro I. W. (2003), Preparation of nanoparticles via spray route, Chemical Engineering Science, 58 (3-6), pp. 537-547

Plascencia G. & Utigard T.A. (2005), High temperature oxidation mechanism of dilute copper aluminium alloys, Corrosion Science 47 (5), pp. 1149-1163

Reimanis I., Trumble K., Rogers K. & Dalglrish B. (1997), Influence of Cu_2O and $CuAlO_2$ Interphases on Crack Propagation at Cu/-Al2O3 Interfaces, Journal of the American Ceramic Society, 80 (2), pp. 424-432

Ristić M., Fundamental Problems of the Science of Materials, TF Čačak and Serbian Academy of Science and Arts, Čačak, 2003, p. 98

Ristić M.M. & Milošević S.Đ. (2006), Frenkel's Theory of Sintering, Science of Sintering, 38 (1), pp. 7-11

Ruys A.J. & Mai Y-W. (1999), The nanoparticle-coating process: a potential sol-gel route to homogeneous nanocomposites, Materials Science and Engineering: A, 265 (1-2), pp. 202-207

Tian B., Liua P., Songa K., Lia Y., Liua Y., Rena F. & Sua J. (2006), Microstructure and properties at elevated temperature of a nano-Al_2O_3 particles dispersion-strengthened copper base composite, Materials Science and Engineering: A, 435-436, pp. 705-710

Trojanová Z., Ferkel H., Luká P., Naser J. & Riehemann W. (1999), Thermal stability of copper reinforced by nanoscaled and microscaled alumina particles investigated by internal friction, Scripta Materialia, 40 (9), pp. 1063-1069

Wu Y., Zhang Y., Huang X. & Guo J. (2001), Preparation of platelike nano alpha alumina particles, Ceramic International, 27 (3), pp. 265-268

Yi S., Trumble K.P. & Gaskell D.R. (1999), Thermodynamic analysis of aluminate stability in the eutectic bonding of copper with alumina, Acta Materialia 47 (11), pp. 3221-3226

Yoshino Y. & Shibata T. (1992), Structure and Bond Strength of a Copper–Alumina Interface, Journal of the American Ceramic Society 75 (10), pp. 2756-2760

Yuana G-Q., Jiang H-F., Lina C. & Liaoa S-J. (2007), Shape- and size-controlled electrochemical synthesis of cupric oxide nanocrystals, Journal of Crystal Growth, 303 (2), pp. 400-406

Part 2

New Materials with Unique Properties

9

Optical Nonlinearities of Colloidal Metal Quantum Dot - Glass Composites for Nanophotonics

Binita Ghosh and Purushottam Chakraborty
Saha Institute of Nuclear Physics,
India

1. Introduction

Since the inception of lasers, nonlinear optics has been a rapidly growing field of research in recent decades. It is based on the study of effects and phenomena related to the interaction of intense coherent light radiation with matter. In other words, nonlinear optics (NLO) is the branch of optics that describes the behaviour of light in nonlinear media, that is, the media in which the dielectric polarization **P** responds nonlinearly to the electric field **E** of the light. This nonlinearity is typically and only observed at very high light intensities (values of the electric field comparable to interatomic electric fields, typically 10^8 V/m) such as those provided by pulsed lasers. Such high powers of laser beams made it possible, for the first time, to observe that the effect of light on a medium can indeed change its optical properties, e.g. refractive index or absorption. When this happens, the light itself also gets affected by this change in a non-linear way; for example, the non-linear response of the material can convert the laser light into new colours, both harmonics of the optical frequency and sum and difference frequencies. Typically, only a laser light is sufficiently intense to modify the optical properties of a material system. In fact, the beginning of the field of nonlinear optics is taken to be the discovery of second-harmonic generation by Fraken *et al.* in 1961 [1], shortly after the demonstration of the first working laser by Maiman in 1960 [2].

The nonlinear effect comes essentially from 'quantum confinement' effect. In an unconfined (bulk) semiconductor, an electron-hole pair is typically bound within a characteristic length, called the 'exciton Bohr radius'. This is estimated by replacing the positively charged atomic core with the hole in the Bohr formula. If the electron and hole are constrained further, then properties of the semiconductor change. Besides confinement in all three dimensions i.e. 'Quantum Dot' - other quantum confined semiconductors include (1) quantum wires, which confine electrons or holes in two spatial dimensions and allow free propagation in the third (2) quantum wells which confine electrons or holes in one dimension and allow free propagation in two dimensions.

Nonlinear optical effect in semiconductor quantum dots was observed in the minuscule crystals of semiconductor material composed of various compounds of chemicals such as cadmium, zinc, tellurium, selenium, sulphur, etc. of sizes less than 500 nanometers. These semiconductor nanoparticles or 'quantum dots' have been found to react to electricity or light by emitting their own light across the visible range of wavelengths from 470 to 730 nm.

The applications of these quantum dots include solid state lighting, with the potential to render incandescent bulbs obsolete in the future and the ability to paint lighting systems onto materials. They are used in the manufacture of medical sensors and solar cells. Thin film deposits of quantum dots on solar cell substrates substantially increase their voltage output by fluorescing the light before it is captured.

An immediate optical fe ature of colloidal quantum dots is their coloration. Such type of coloration due to embedded nanoscale metallic particles within glasses was noted in many stained glass windows in the Medieval era shape of the nanoparticles themselves', followed by Figure 1. These rich colors in stained glass like the metallic sheen associated with naturally embedded nanoparticles in many ceramics, were appreciated and highly valued by artisans, patrons and laymen alike. Stained glass artisans sometimes treasured small vials of materials that we know were metallic oxides obtained from special mines and handed down within their families with careful instructions on how to work with them. When the size of the material particle is reduced to nanoscale, optical properties- particularly colors, can be dramatically affected. In such cases, the wavelength of light is very close to the size of the particles themselves, which causes the way that color is reflected, scattered or absorbed to be dependent on the size and shape of the nanoparticles themselves.

Fig. 1. Stained glass window of Chartres Cathedral

While the material which makes up a quantum dot defines its intrinsic energy signature, the nanocrystal's quantum confined size is more significant at energies near the band gap. Thus quantum dots of the same material, but with different sizes, can emit light of different colours. The physical reason is the 'quantum confinement' effect, which will be discussed later. The larger the dot, the redder (lower energy) is its fluorescence spectrum. Conversely, smaller dots emit bluer (higher energy) light. The coloration is directly related to the energy levels of the quantum dot. Quantitatively speaking, the bandgap energy that determines the energy (and hence colour) of the fluorescent light is inversely proportional to the size of the quantum dot. Larger quantum dots have more energy levels which are also more closely spaced. This allows the quantum dot to absorb photons containing less energy, i.e., those closer to the red end of the spectrum. Recent articles in nanotechnology and in other journals have begun to suggest that the shape of the quantum dot may be a factor in the coloration as well, but as yet not enough information is available. Furthermore, it was shown that the lifetime of fluorescence is determined by the size of the quantum dot. Larger dots have more closely spaced energy levels in which the electron-hole pair can be trapped causing larger dots to show a longer lifetime. As with any crystalline semiconductor, a quantum dot's electronic wave functions extend over the crystal lattice. Similar to a molecule, a quantum dot has both a quantized energy spectrum and a quantized density of electronic states near the edge of the band gap.

A currently active area of research in non-linear optics is concerned with all-optical devices that are designed to switch and process light signals without converting them into electronic form, thus eliminating the electronic "bottleneck" in the speed of electronics used for switching, routing and signal processing. These devices are essentially based on the non-linear optical properties of materials, i.e. changes of the refractive index caused by an intense optical beam.

When a neutral atom is placed in an electric field **E**, the positively charged core and the negatively charged electron cloud surrounding it are both influenced by the electric field. The nucleus is pushed in the direction of the field and the electrons in the opposite direction. In principle, if the field is large enough it can pull the atom apart completely, 'ionizing' it. With less extreme fields, however, equilibrium is soon established, for if the center of the electron cloud does not coincide with the nucleus, leaving the atom 'polarized' with positive charge shifted slightly one way, and negative the other. The atom now has a tiny dipole moment which points in the same direction as **E**. Typically, the polarization **P** (dipole moment per unit volume) is directly proportional to the field, provided E is not too strong. That is

$$P = \chi E \tag{1}$$

where χ is the dielectric susceptibility of the medium. The value of χ depends on the microscopic structure of the substance and also on the external electric field. Materials obeying the above equation are known as linear dielectrics. Linearity means that the displacement of the electrons is proportional to the acting force. Only under this condition the electrons oscillate sinusoidally, with amplitude proportional to the acting force (apart from a phase shift).

The above stated model of the elastically bound electron is too simple and is only an approximation, valid for small amplitudes. The deviations become significant at high oscillation energy. If this energy exceeds the value of the binding energy, the electron is set free. At this level the electron oscillates non-harmonically. Higher frequencies appear in the displacement and in the polarization **P**. In this case, the linear equation between polarization and the field amplitude is no longer valid. The polarization is a complicated function of the field amplitude. It contains not only the linear term of **E**, but also higher order terms. This is traditionally introduced by noting that the dielectric susceptibility can be expanded in a Taylor series in terms of the oscillating optical field. In an isotropic medium the general relation between the polarization **P** (dipole moment per unit volume), induced by the applied electric field **E**, is expressible as a Taylor expansion involving only the magnitude, because the direction of the polarization coincides with that of the field, namely

$$P = \left(\chi^{(1)}E + \chi^{(2)}E^2 + \chi^{(3)}E^3 + ...\right) \quad (2)$$

In this expansion $\chi^{(1)}$ is the normal or linear dielectric susceptibility of the medium and is related to the refractive index η by $\eta^2 = 1+ \chi^{(1)}$. It contributes to all the phenomena associated with linear optics, such as reflection, refraction, interference, etc. It is generally much larger than the nonlinear coefficients $\chi^{(2)}$, $\chi^{(3)}$ and so forth. In general, $\chi^{(n)}$, the complex dielectric susceptibilities, are tensors of rank (n+1) and are related to the microscopic (electronic and nuclear) structure of the material. The even-order nonlinear optical interactions can occur only in non-centrosymmetric crystals, that is, in crystals that do not display inversion symmetry [3]. Since liquids, gases, amorphous solids (such as glass) have inversion symmetry, $\chi^{(2)}$ vanishes identically for such media and consequently they cannot produce second-order nonlinear optical interactions. On the other hand, odd-order nonlinear optical interactions can occur both for centrosymmetric and non-centrosymmetric media. Glasses possess macroscopic inversion symmetry, thereby allowing only odd-order optical nonlinearities. The third-order nonlinear optical phenomena prevail inherently in glasses because of the deficiency of long-range periodicity unlike crystals. Nonlinearity in glasses is in general quite small, but may be enhanced by several orders of magnitude by introducing metal nanoclusters in glass matrices.

The nonlinear response of the nanocluster-glass composites arises from saturable absorption at the excitonic levels. This effect drastically modifies the quantum states of the electrons and their interaction with applied optical fields; referred to as *quantum confinement.* Secondly, when the size of the nanoclusters is much smaller than the wavelength (λ) of the applied optical field (i.e. for clusters with diameters less than $\lambda/20$), the electric field that acts on and polarizes the free charges of these clusters can be vastly different from the macroscopic field outside the metal clusters in the surrounding medium. This polarization, in turn, modifies the dielectric constant of the composite medium. This effect is called *dielectric or classical confinement.* Both quantum and dielectric confinement effects alter the susceptibilities of the composite, thereby modifying the linear and nonlinear refractive indices and the absorption coefficients.

Owing to the difference in dielectric constants between the nanoclusters and the surrounding host material, local field effects arising from dielectric confinement strongly influence the optical properties and can produce major changes in optical response. Therefore, it is important to design a material which has large $\chi^{(3)}$ at the wavelength of the limited light source available and metal - doped glasses have significant possibility in this direction. The large optical nonlinearity of metal-doped glasses is considered mainly due to the local field enhancement effects on the metal microcrystallites under 'surface plasmon resonance (SPR)' condition, which mainly depends on the dielectric functions of metals and the glasses.

2. Nonlinear optical phenomena: physical origin

The equation for the polarization **P** links the material properties with Maxwell's equations for the propagation of electromagnetic waves. It does not give any indication as to the physical mechanisms which give rise to the optical nonlinearities. The origin of the nonlinear susceptibility can be viewed classically as the response of an electron driven by an electromagnetic field in an anharmonic potential well, resulting from the interatomic electric field **E** in the crystalline solid. When a light wave propagates through an optical medium, the oscillating electromagnetic field exerts a polarizing force on all the electrons comprising the medium. Because the inner electrons of the atoms are tightly bound to the nuclei, the major polarizing effect is exerted on the outer or valence electrons. The interatomic field is of the order of 10^{10} V/m. For driving optical fields **E** much weaker than that, the polarization response is essentially linear. For an intense optical field (comparable to or greater than the inter-atomic field) enough to drive the electron beyond the quadratic minimum of the interatomic potential, the response becomes increasingly nonlinear. Based on the classical atomic model of non-linearity (considering the motion of a one-dimensional anharmonic electron oscillator with damping, driven by a laser field), the ratio between the magnitudes of non-linear polarizations of successive orders can be expressed as

$$\frac{P^{(n+1)}}{P^{n}} \approx \frac{e|E|}{m\omega_0^2 r} \approx \frac{e|E|}{e|E_{at}|}$$

where ω_0 is the oscillation frequency of the electron and r is the radius of the equilibrium orbital of the electron ($m\omega_0^2 r$ is actually the expression for the linear force which is equal to $e|E_{at}|$, where E_{at} is the intra-atomic electric field at a distance equal to Bohr radius a_0). Now, $E_{at} = \frac{e}{a_0^2} = 5.15\times10^{11} V/m$. Following the expression for energy flux density of an electromagnetic wave $\left(=\varepsilon c E^2\right)$, the above value of E_{at} corresponds to a laser power of $7\times10^{20} Watt/m^2$, where ε (dielectric constant of vacuum) = 8.8×10^{-12} and c (speed of light) = 3×10^{8}, expressed in SI units.

At low light intensities, the optical fields are a small fraction of the molecules' total internal fields (~5×10^{11} Vm^{-1}). At high photon intensities, the induced fields become a significant fraction of the total. This added internal field substantially modifies the motions of the charges within the molecule and alters the nature of the polarization within the materials. Also, this added internal field "anharmonically" modifies the light wave's propagation, leading to remarkable effects when they occur coherently, constructively. That is when a material's properties (e.g. the interaction length for a process and the strength of the perturbation) are well-matched (i.e. phase-matched) to the laser wavelength, second- and third-order non-linear optical effects can drastically change the colour content or manner of propagation of light. Therefore, the non-linear phenomenon is evidenced by the changes in optical properties as the intensity of light is increased.

We thus expect that under conditions of non-resonant excitation the second-order susceptibility $\chi^{(2)}$ will be of the order of $\chi^{(1)}/E_{at}$ [4]. For condensed matter $\chi^{(1)}$ is of the order of unity, and hence we expect that $\chi^{(2)}$ will be of the order of $1/E_{at}$, or

$$\chi^{(2)} \cong 5\times10^{-8} cm/statvolt = 5\times10^{-8}\left(cm^3/erg\right)^{1/2} = 5\times10^{-8} esu$$

Similarly, we expect $\chi^{(3)}$ to be of the order of $\chi^{(1)}/E_{at}^2$, which for condensed matter is of the order of

$$\chi^{(3)} \cong 3\times10^{-15} cm^2 / statvolt^2 = 3\times10^{-15} cm^3 / erg = 3\times10^{-15} esu$$

These predictions are in fact quite accurate, as one can see by comparing these values with the actual measured values of $\chi^{(2)}$ and $\chi^{(3)}$. For certain purposes, it is useful to express the second- and third-order susceptibilities in terms of fundamental physical constants. Noting that the number density N of the condensed matter is of the order $(a_0)^{-3}$, we find that

$$\chi^{(2)} \cong h^4 / 16\pi^4 m^2 e^5 \text{ and } \chi^{(3)} \cong h^8 / 2^8 \pi^8 m^4 e^{10} \cdot$$

The most common procedure for describing nonlinear optical phenomena is based on expressing the polarization **P**(t) in terms of the applied electric field strength **E**(t) as we have done in Eq (2). The reason why the polarization plays the key role in the description of nonlinear optical phenomena is that a time-varying polarization can act as the source of new components of the electromagnetic field. The wave equation in nonlinear optical media often has the form [4]

$$\nabla^2 E - \frac{\eta^2}{c^2}\frac{\partial^2 E}{\partial t^2} = \frac{4\pi}{c^2}\frac{\partial^2 P^{NL}}{\partial t^2} \tag{3}$$

where η is the usual linear refractive index and c is the speed of light in vacuum. We can interpret this expression as an inhomogeneous wave equation in which the polarization $\mathbf{P}^{NL}$ associated with the nonlinear response drives the electric field **E**. This equation expresses the fact that, whenever $\partial^2 P^{NL} / \partial t^2$ is non-zero, charges are being accelerated, and according to Larmor's theorem from electromagnetism, accelerated charges generate electromagnetic radiation. It should be noted that the power series expansion, expressed by Eq. (2), need not necessarily converge. In such circumstances the relationship between the material response and the applied electric field amplitude must be expressed using different procedures.

The physical origin of nonlinear optical phenomena can be categorized as either structural or compositional [5]. Here 'structural' refers to light induced structure changes, such as a change of electronic density, average interatomic distances, molecular orientation, phase transition, etc. These phenomena belong to the intrinsic category. The 'compositional' refers to light-induced chemical composition changes such as molecular dissociation, polymerization etc. Intrinsic nonlinearity violates the principle of superposition arising from a nonlinear response of the individual molecule unit cell to the fields of two or more light waves, while extrinsic nonlinearity is related to changes in the composition of the medium that results from the absorption or emission of light. In either type of nonlinearity, the optical properties of the medium depend on the intensity of light and the order of nonlinearity can be classified according to the power of the intensity involved.

Colloidal suspension of metal nanoclusters in centro-symmetric dielectric materials (such as glasses) has recently attracted immense interest in materials science for their potential photonic applications. Various types of glasses are very attractive materials for these applications, because of their relatively fast response time. Furthermore, glass-based nanocomposites are, in general, expected to play an important role as materials for various nanotechnology applications, due to low cost, ease of processing, high durability, resistance and high transparency, as well as the possibility of tailoring the behavior of glass-based structures. Since the first attempt of Faraday to explain the nature of the colour induced in glasses by small metallic precipitates, many studies have been dedicated to the properties of metal nanocluster composite glass. In general, the physical properties of these systems change dramatically in the transition from atom to molecule to cluster to solids, where the cluster regime is characterized by the confinement effects, causing thereby these systems particularly interesting. In the last decade, metal nanoclusters embedded in insulating matrices, have received increasing interest due to their peculiar optical, magnetic and catalytic properties when the size becomes comparable to or less than the electronic mean free path. The nonlinear susceptibility $\chi^{(n)}$ of any material is intimately related to its electronic structure. Therefore, it is of utmost importance to understand the relationship between the nonlinearity and electronic structure and thereby the chemical structure of the material so that a predictive capability can be developed for designing device grade materials.

3. Optical properties of metal nanoparticles

Metals contain free electrons which are responsible for their electrical conduction. If a spatial variation of the density of freely moving electrons with respect to the background of positive ions is created, the restoring forces are formed that tend to pull the electrons back to their initial position. Since the electrons have inertia, their distribution overshoots the equilibrium position creating a new charge distribution and with it the restoring forces. The electron charge density thus continues an oscillating motion. Such a system where oppositely charged particles can move freely with respect to each other is called plasma and their oscillatory motion the plasma oscillation. Quantization of the plasma oscillations leads to quasi-particles or the plasmons. Localized surface plasmons or 'surface polaritions' are collective electron charge oscillations in metallic nanoparticles that are excited by light. They exhibit enhanced near-field amplitude at the resonance wavelength. This field is highly localized at the nanoparticle and decays rapidly away from the nanoparticle/dieletric interface into the dielectric background, though far-field scattering by the particle is also enhanced by the resonance. Light intensity enhancement is a very important aspect of localized SPRs and the localization means that these SPRs have very high spatial resolution, limited only by the size of nanoparticles. Figure 2 shows the Schematic of plasma oscillation for a metallic sphere, indicating displacement of the conduction electron charge cloud relative to the nuclei.

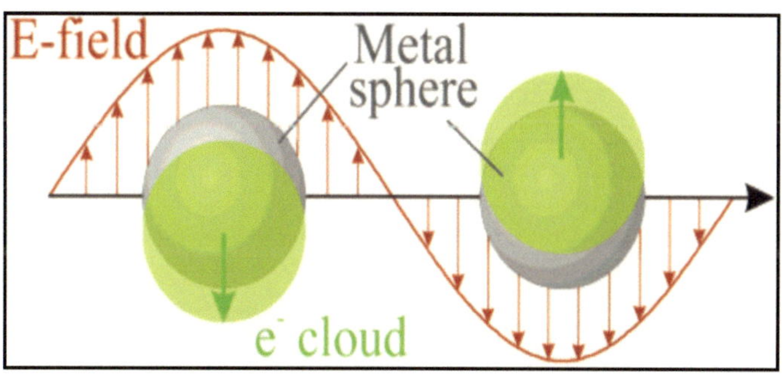

Fig. 2. Schematic of plasma oscillation for a sphere, showing displacement of the conduction electron charge cloud relative to the nuclei.

Assuming the electron charge density ρ to be a continuous function of position and using classical electrodynamics, an equation of motion for $\rho(x,t)$ can be written as

$$\frac{\partial^2 \rho(x,t)}{\partial t^2} = -\frac{4\pi}{m}\frac{\Delta N}{\Delta V} e^2 \rho(x,t) \tag{4}$$

where ΔN is the number of electrons in volume element ΔV while other quantities have their usual meanings. Eq. (4) represents simple harmonic motion with a bulk plasma oscillation frequency

$$\omega_p = \sqrt{\frac{4\pi n e^2}{m}} \tag{5}$$

where $n = \Delta N / \Delta V$ is the average electron density.

For a metal (complex dielectric function $\varepsilon(\omega)$) nanoparticles of radius R embedded in a host of dielectric constant ε_d, the plasmon response in quasi-static approximation ($R < \lambda$, wavelength of light) is essentially dipolar in nature. The dipole plasmon response of the metal nanoparticles is defined by its polarizability

$$\alpha = 4\pi\varepsilon_0 R^3 \left| \frac{\varepsilon(\omega) - \varepsilon_d}{\varepsilon(\omega) + 2\varepsilon_d} \right| \tag{6}$$

The resonance will be observed when the real part $\varepsilon'(\omega)$ of the complex dielectric function $\varepsilon(\omega)$ satisfies the condition

$$\varepsilon'(\omega) = -2\varepsilon_d$$

and its imaginary part $\varepsilon''(\omega) << 1$. $\varepsilon'(\omega)$ is negative when the frequency of electromagnetic wave ω is below the bulk plasmon frequency ω_p. In Drude model, $\varepsilon'(\omega)$ and $\varepsilon''(\omega)$ can be approximated in terms of ω_p and plasma relaxation frequency Γ as

$$\varepsilon'(\omega) = 1 - \frac{\omega_p^2}{\omega^2 + \Gamma^2} \tag{7}$$

$$\varepsilon''(\omega) = 1 - \frac{\omega_p^2 \Gamma}{\omega(\omega^2 + \Gamma^2)} \tag{8}$$

Eqs. (7) and (8) yield the Drude resonance frequency as

$$\omega_D = \frac{\omega_p}{\sqrt{2\varepsilon_d + 1}} \tag{9}$$

The plasma frequency ω_p is scale-invariant but the Drude free electron response varies with geometric shape of the particle. Drude free electron model does not yield the surface plasmon frequency for metal nanoparticles of interest correctly. This is mainly because of the additional electric susceptibility introduced by interband transitions between the outermost *d*- and *s*- orbitals at optical frequencies. These electronic excitations couple strongly with the free electron response yielding large changes in $\varepsilon'(\omega)$ and $\varepsilon''(\omega)$. In addition, these can produce significant damping mechanism for plasmon decay which will reduce the quality of the optical resonance. In case of silver nanoparticles, these interband transitions are responsible for the frequency difference between the theoretically predicted Drude response at ω_D = 5.6 eV and the observed surface plasmon resonance at ω_{sp} = 3.6 eV. More accurate model for the optical properties of metal nanoparticles can be obtained using Mie theory based on Maxwell's equations. Results of Mie theory clearly predict the experimental optical spectra of metal nanoparticles. This theory has also been used to estimate the variations in the electrodynamic response due to the size and anisotropy. It can also account for the optical response of metal nanoparticles beyond the quasi-static limit $(2R/\lambda>0.05)$, whose plasmon resonances are red-shifted by phase retardation and broadened by radiation damping. Higher-order plasmon modes can also become significant. For example, quadrupole resonance in silver and gold nanoparticles can be observed in optical extinction spectra at sizes ≥ 60 nm. For spherical metal clusters, Mie plasmons considering $\varepsilon_d=1$ are shown to occur at frequency [6]

$$\omega_l=\omega_p\sqrt{\frac{l}{2l+1}} \tag{10}$$

where l is the angular momentum of plasmon mode. Using a single photon of appropriate energy one can excite only $l=1$ dipole plasmon oscillations in a metallic cluster.
The dipole surface plasmons are responsible for the formation of giant resonances in photo-absorption spectra of metal clusters [7-10]. They also play an important role in inelastic collisions of charged particles with metal clusters [11-13]. It was also demonstrated that collective excitations make a significant contribution to the electron energy loss spectrum in the region of the surface plasmon resonance. One-photon absorption cross-section in the plasmon approximation is shown to be [14]

$$\sigma_1=\frac{4\pi Ne^2}{mc}\frac{\omega^2\Gamma_1}{(\omega_1^2-\omega^2)^2+\omega^2\Gamma_1^2} \tag{11}$$

where Γ_1 is Landau damping. Connerade *et al.* [13] have also shown that the two-photon absorption can occur for $l=2$ i.e. quadratic mode and is not allowed for $l=1$ plasmon mode (dipolar plasmon). Two-photon absorption cross-section is given by [13]

$$\sigma_2=\left(\frac{4\pi Ne^2}{mc}\right)^2\frac{A^2\hbar}{2\pi\omega_1 N\Delta R^2}\frac{\omega^2}{\omega_1^2}\frac{1}{\left((\omega-\omega_1)^2+\frac{\Gamma_1^2}{4}\right)}\frac{\Gamma_2}{\left((\omega_2-2\omega)^2+\frac{\Gamma_2^2}{4}\right)} \tag{12}$$

where $A\approx 2.79$ and ΔR is the width of the domain in the vicinity of the cluster surface within which plasmon excitations take place. Note that unlike σ_1, σ_2 depends on $\hbar$. It is because $l=1$ plasmon oscillations are purely classical effect, while σ_2 arises from the interaction between dipole and quadrupole plasmon modes and the two-photon absorption profile differs substantially from the one-photon case. Connerade *et al.* [13] have also shown that, in addition to dipole plasmon excitations multipole plasmons (quadrapole, octupole, etc) contribute to the multiphoton excitation process, which results in a significant difference of plasmon resonance profiles between the cross-sections for multiphoton and single-photon absorption. The results were obtained within a theoretical model for the multiphoton excitation of a jellium cluster. This model is applicable to metal clusters.

In general, when an electromagnetic wave propagates in the composite medium, it does not resolve the individual scattering centers, and the medium appears as a homogeneous medium characterized by an effective complex dielectric function. If we consider a composite consisting of small particles occupying a relative volume fraction $p<<1$ in a dielectric host, the complex dielectric constant of the metal cluster, which is a function of the frequency of the incident electromagnetic field by $\varepsilon_m(\omega)=\varepsilon_1(\omega)+i\varepsilon_2(\omega)$. The optical response of the metal nanocluster glass composite systems may be approached following two classes of models, namely, (i) discrete island model, in which each scattering center is considered, and (ii) effective medium theory, in which one scattering center is considered, the rest of the surrounding medium being averaged into a homogeneous medium. Considering the first kind of approach, Maxwell-Garnet [15] considered that the metal clusters embedded in a host matrix of dielectric constant ε_h are polarized by an optical field. The Lorentz local-field relationships show that the effective dielectric constant ε_{eff} of such a composite medium is given by

$$\frac{\varepsilon_{eff}-\varepsilon_h}{\varepsilon_{eff}+2\varepsilon_h}=p\frac{\varepsilon_m-\varepsilon_h}{\varepsilon_m+2\varepsilon_h} \tag{13}$$

For small volume fractions, Eq. (13) may be expanded to the first order in the volume fraction. The optical absorption coefficient is related to the imaginary part of ε_{eff}. Mie was the first to derive an exact description of the optical absorption and elastic scattering by a collection of metal spherical clusters suspended in a transparent medium. By adding the contribution of single clusters, the theory accounts correctly for many experimental cases, provided that interaction effects among the clusters as well as multiple scattering effects are missing. This is concerned, however, with a large class of inhomogeneous composite glasses, where the cluster density is anyhow sufficiently low to allow a description within this regime. The optical absorption coefficient α of a collection of uniform spheres, very small compared to the wavelength λ of light and embedded in a medium of refractive index η_d, is well described by the Mie scattering theory. The general expression for the extinction coefficient, which includes both absorption and scattering, is given by [16]

$$\alpha=p\frac{18\pi n_d{}^3}{\lambda}\frac{\varepsilon_2}{\left(\varepsilon_1+2n_d{}^2\right)^2+\varepsilon_2^2} \tag{14}$$

The extinction coefficient in expression (14) has a maximum at λ for which the condition $(\varepsilon_1+2n_d{}^2)=0$ is satisfied. This is due to the absorption resonance known as the surface plasmon resonance (SPR). The UV-Visible spectrum of a Cu^+ - implanted glass sample shows such a typical SPR band at 589 nm (Fig. 3).

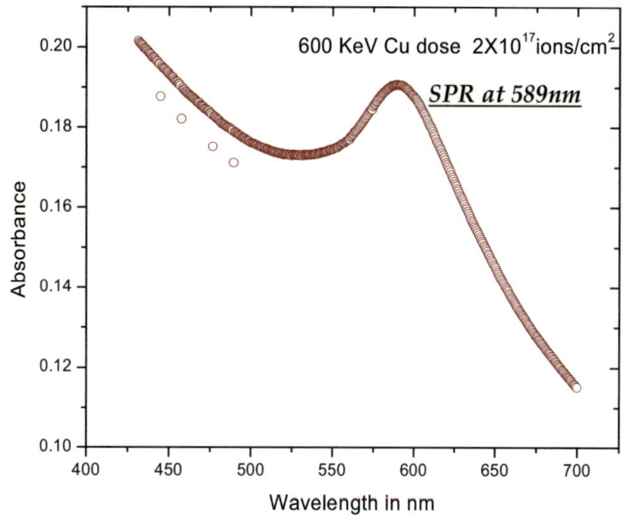

Fig. 3. Optical absorption band for copper nanoclusters in glass

Plasmons are the quanta associated with longitudinal waves propagating in matter through the collective motion of conduction band electrons. Maxwell's equations tell us that an interface between a dielectric (e.g. silica glass) and a metal (e.g. Ag or Au) can support a surface plasmon (SP). An SP is a coherent electron oscillation that propagates along the interface together with an electromagnetic wave. These unique interface waves result from the special dispersion characteristics (dependence of dielectric constant on frequency) of metals. What distinguishes SPs from 'regular' photons is that they have a much smaller wavelength at the same frequency. For example, a HeNe laser, whose free-space emission wavelength is 633 nm, can excite an SP at a Si/Ag interface with a wavelength of only 70 nm. When the laser frequency is tuned very close to the SP resonance, SP wavelengths in the nanometer range can be achieved. The short-wavelength SPs enable the fabrication of nanoscale optical integrated circuits, in which light can be guided, split, filtered, and even amplified using plasmonic integrated circuits that are smaller than the optical wavelength. The position, width and shape of the SPR band are determined by the metal dielectric function, also on the size, shape and concentration of the particles and on the surrounding dielectric medium. The position of the resonant absorption band changes with the cluster size, but remains more or less unchanged for the clusters with less than 10 nm diameter, as was predicted by Mie in his scattering theory. Moreover, quantum corrections to the classical absorption become significant as the cluster dimension approaches 1 nm diameter [17]. Equation (14) gives a band of Lorentzian shape: if the band is narrow, its width w at half maximum absorption (FWHM), i.e. bandwidth, is λ_m^2 / λ_a and is given by

$$w = \frac{\lambda_m^2}{\lambda_a}$$

$$= \left(\varepsilon_0 + 2n_d^{\,2}\right) c / 2\sigma \qquad (15)$$

The d.c. conductivity, σ, is given by

$$\sigma = \frac{N_e e^2 R}{m u_F} \tag{16}$$

where R is the particle radius, $u_F = \left(\frac{2E_F}{m}\right)^{1/2}$ is the electron velocity at the Fermi energy $E_F = \frac{h^2}{2m}(3n8\pi)^{2/3}$ and N_e is the number of electrons per unit volume. Therefore, the experimental result of constant band position is reasonable, because λ_m does not depend upon the conductivity, σ, which is determined by the mean free path of the electrons and therefore, upon the particle size. Alternatively, Equation (15) shows that the dependence of bandwidth upon particle size, i.e. $w \propto 1/R$, as obtained from Equations (15) and (16), results from the changing mean free path of the electrons. Figure (4) gives a comparison between the experimentally measured absorption band for silver particles of about 10 nm diameter, suspended in glass, and the calculated one from Equation (14), using the measured λ_m and bandwidth [18]. The agreement between these curves confirms that the absorption results from the free electrons in the silver particles. The slight increase in absorption at lower wavelengths probably results from the ultraviolet irradiation of the cerium present in the glass matrix [19].

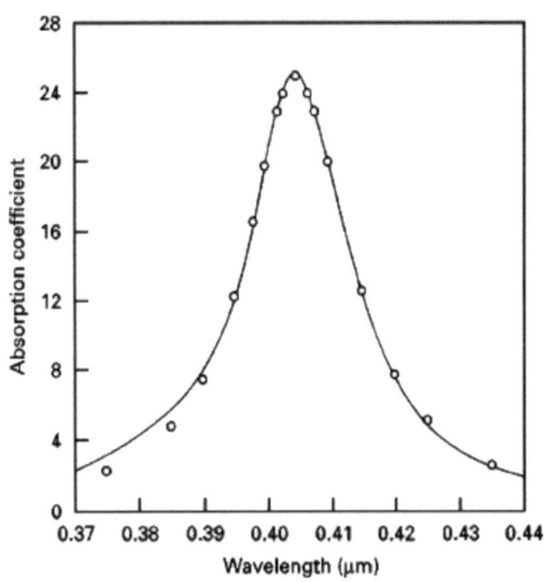

Fig. 4. (o) Measured and (___) calculated absorption bands for silver nanoparticles of about 10 nm diameter, suspended in glass

Eq. (14) predicts a shift of the resonance to longer wavelengths and increasing half-widths for the particles of radius R ≥ 10 nm. Figure (5) shows the optical extinction per unit Ag concentration as a function of wavelength for various values of colloidal radius.

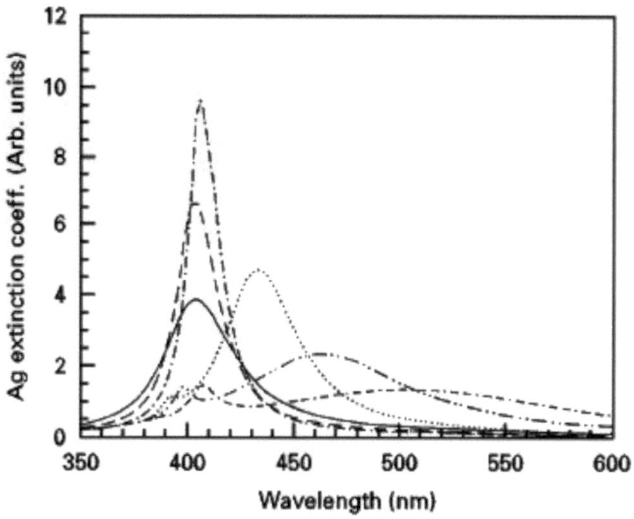

Fig. 5. Mie extinction coefficient of silver colloids of various radii as a function of wavelength of the incident light. R (nm): (__) 2.5, (_ _) 5, (_ .) 10, (......) 25,(_ .. _) 35, (_ _ . _ _) 45.

The average radius of metal spheres, small compared to the wavelength of light, can be approximately estimated from the resonance optical absorption spectrum as per the Mie scattering formula.

$$r_{metal} = \frac{v_F}{\Delta\omega_{1/2}} = \frac{v_F}{2\pi c_0\left(\frac{\Delta\lambda}{\lambda_p^2}\right)} \tag{17}$$

Where v_F − Fermi velocity and $\Delta\omega_{1/2}$ is the full width at half maximum (FWHM) of the SPR absorption when plotted as a function of angular frequency ω, c_0 is the speed of light in vacuum, λ_p is the wavelength where absorption peak appears, and $\Delta\lambda$ gives the bandwidth. This expression [20], however, provides a very crude estimate of the metal particle dimension as it is reliable only when the metal colloids are strictly spherical. It is also difficult to estimate the particle dimensions from the optical absorption spectra as the main features of the SPR bands (i.e wavelength, position and FWHM) depend on size, size distribution, shape, lattice parameters and filling factors of the metal particles.

As the dielectric constant is intensity dependent, the corresponding refractive index is also intensity dependent. The intensity dependent refractive index η is usually expressed as

$$\eta = \eta_0 + \eta_2 I \tag{18}$$

where η_0 is the linear refractive index of the material and is a function of $\chi^{(1)}$ [21]. I is the intensity of light averaged over a period and η_2 is the nonlinear refractive index of the material. The total absorption, α, under saturation can be expressed as [22]:

$$\alpha = \alpha_0 + \beta I \tag{19}$$

β is the coefficient characterizing the nonlinear absorption and is known as the ' two-photon absorption coefficient'. For all-optical devices, the nonlinear refractive index and the two-photon absorption play critical roles. The nonlinear refractive index and the two-photon absorption coefficient are related to the real and imaginary parts of $\chi^{(3)}$ according to the following equations:

$$\eta_2 = 12\frac{\pi \mathrm{Re}\left[\chi^{(3)}\right]}{\eta_0} \tag{20}$$

$$\beta = 96\frac{\pi^2 \omega \mathrm{Im}\left[\chi^{(3)}\right]}{\eta_0^2 c^2} \tag{21}$$

Eqs. (20) and (21) are valid in the case of negligible linear absoption, i.e. when $\alpha_0/2kn_0 << 1$, where $k=2\pi/\lambda$.

4. Ion beam synthesis of metal - glass nanocomposites

Among different possible synthesis processing, ion-beam based techniques proved to be very suitable in synthesizing metal nanocluster glass composites. Moreover, the composition of the clusters can be varied easily by sequential ion implantation in the matrix of two different elements whose energy and dose can be tailored so as to maximize the overlap between the implanted species and to control their local relative concentration. Ion implantation has been shown to produce high-density metal colloids in glasses and other materials. The high-precipitate volume fraction and small size of nanoclusters in glasses lead to the generation of third-order susceptibility much greater than those for metal-doped solids. This has stimulated interest in the use of ion implantation to make nonlinear optical materials. An attractive feature of ion implantation to form these nanoclusters, compared to the classical technique of melt-glass fabrication process, i.e. mixing selected metal powder with molten glass, is that the linear and nonlinear properties occur in a well-defined space in an optical device, and by using focused ion beams, point quantum confinement may be accomplished. Although several other methods, such as ion-exchange [23], sol-gel [24], electrolytic colouration, etc., also exist for introducing the metal into the insulating substrates, the ion implantation has the following advantages of being a generally applicable process; (a) it can be performed at an ambient temperature, (b) it has no side diffusion problem, (c) it offers an accurate control of the total number of ions being added to the target and (4) it offers a predictable depth distribution in the target matrix determined by the incident ion beam energy.

Ion implantation in glassy structures yields in the precipitation of metal colloids at a reasonably higher local concentration because of the large specific volume and more open structure of the glassy state relative to that of the crystalline counterpart. Metal nanoclustres can be easily incorporated within the open spaces present in the glass matrix. Figure (6) shows the schematic structure of SiO_2 in crystalline and glassy form.

Ion implantation of metals into glass has been explored since the last decade as a useful technique to produce nanocomposite materials in which nanometer-sized metal or semiconductor particles are embedded in dielectric matrices. Modifications induced by ion implantation depend on the glass composition as well as on the ion species, and the fluence, energy and the temperature of the process. Both nuclear and electronic processes give rise to

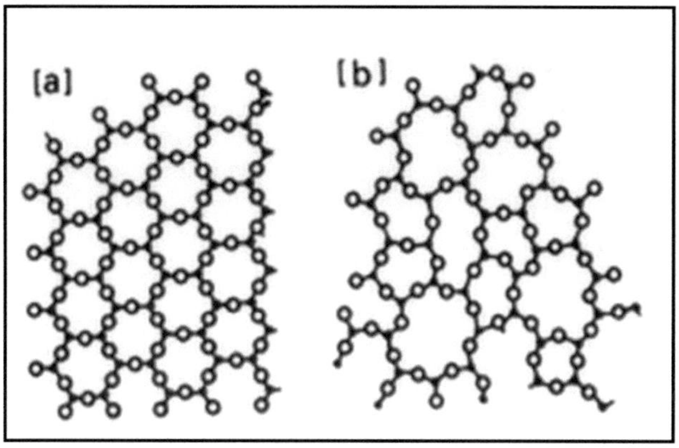

Fig. 6. Schematic 2-D representation of the structure of SiO_2 (a) crystalline (b) glassy form

the structural changes in materials, and several evidences of clusterisation of metal atoms implanted in glass matrices are reported in literatures [25-32]. Alternatively, low-mass ion-beams can be used for promoting cluster aggregation in surface layers of glass previously doped with metal ions by other suitable techniques. High energy ion beam mixing [33] and ion beam assisted deposition [34] have also been successfully exploited for the preparation of clusters-doped silica glass. Figure (7) sketches some of these methods [29]. Furthermore, ion implantation has been used as the first step of combined methodologies that involve other treatments such as thermal annealing in controlled atmosphere, laser or ion irradiation [35-39]. Heat treating the implanted sample reduces the strains and charge imbalances caused by implantation. Moreover, heating increases the diffusion coefficient and the implanted atoms move to the lower energy metallic state of the clusters increasing the localized volume fraction.

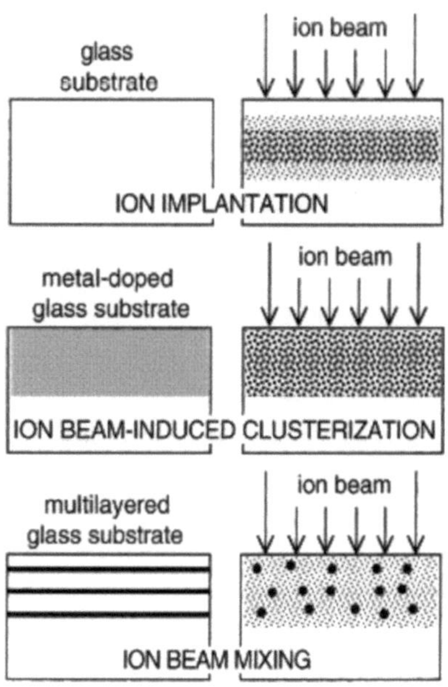

Fig. 7. Ion beam - based methods for promoting metal cluster formation in silica glass

The physical mechanisms governing the cluster formation are presently under debate. Hosono discussed [40] the chemical reactivity of the species involved in the implants on SiO_2 glasses. In the case of weak or negligible chemical interaction, the elements which do not react with the matrix (for example, Ag, Cu, Au...) are considered to directly form metallic precipitates under appropriate conditions. Among those which tend to form bonds with O from the silica network, so forming a defective Si–Si homobond, the cluster formation is predicted when the Gibbs energy for metal oxide formation is greater than that of SiO_2. The comparison of the two energies should be made at the extrapolated effective temperature of 3000 K. The validity of this criterion as a first approximation is good, yet not giving account for the effect of the concentration of the implanted species, nor providing a criterion for fixing the local temperature at which the two energies must be compared. A more general approach is due to Hosono, Matsunami and Imagawa [41, 42]. Starting from the study of the implantation-induced defects in silica, they pointed out the importance of the chemical interaction strength among atoms coming into play in the creation of defects. Although a vast volume of literature is devoted to the nonlinear optical studies in metal nanocluster glass composites, we will discuss here only those works which give meaningful insights into the study of third-order optical nonlinearity; its correlation with electronic and chemical structure and those which offer some useful guidelines for developing potential nonlinear optical materials for photonic applications.

Nanocluster formation in metal-ion implanted silica and glasses has been studied by using different techniques. Out of various metal nanoparticles, silver, copper and gold are extensively studied in semicontinuous thin films, colloids and in different glass matrices for their nonlinear optical properties [43]. Different metal alloy nanoclusters [44, 45] are also studied in this aspect. Optical absorption measurements [46-48] and transmission electron microscopy (TEM) give information on the cluster shape, size and crystalline state. Cross-sectional TEM measurements give the details of the colloid size distribution as a function of depth [48, 49]. Rutherford Backscattering Spectrometry (RBS) and Secondary Ion Mass Spectrometry (SIMS) can be used to obtain the depth distribution of the implanted species. X-ray photoelectron spectroscopy (XPS) and Auger electron spectroscopy (AES) determine the chemical state of the implanted ions. Degenerate four wave mixing (DFWM) [50] and Z-scan [51] methods have been used for measurements of $\chi^{(3)}$ of the metal doped glasses.

Ricard et. al [50] were the first to measure the non-linear properties of such colloids (silver, gold) in bulk silica and other workers [42-55] extended these measurements. The non-linear index η_2 of 160 keV copper-implanted silica at a fluence of $10^{17}\,ions/cm^2$ was determined using the Z-scan method and was found to be around $4\times10^{-14}m^2W^{-1}$ [42]. More recently, the formation of various metal colloids (copper, silver, gold, lead, tin, iron, phosphorus etc.) in silica glasses by ion implantation under various conditions and measurements of $\chi^{(3)}$ values in these cases have been reported [53-56]. The highest value of $\chi^{(3)}$ $\left(3\times10^{-6}e.s.u\right)$ was obtained, so far, for tin implanted $\left(2\times10^{17}\,ions/cm^2 dose\right)$ silica glass [57] at a wavelength of 500 nm; corresponding to surface plasmon resonance (SPR) of tin. Figures (8) and (9) show the TEM images of such microstructures. As shown in these figures, the implanted ions form spherical nanoclusters. The crystallinity of the nanoclusters is evidenced [57] from X-ray diffraction studies, as shown in Fig. 10. All the XRD peaks are identified as those of metallic tin crystallites and relative peak intensities confirm that there is no evidence of preferential orientation of the microcrystallites.

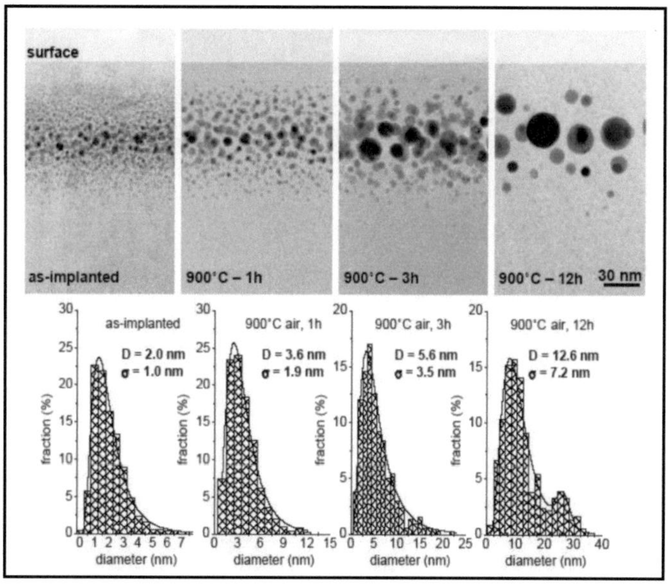

Fig. 8. TEM results on Au-implanted silica samples annealed in air at 900°C for different time intervals: cross-sectional bright-field micrographs with the corresponding histogram of size distribution for the as-implanted and annealed for 1 h, 3 h, 12 h samples.

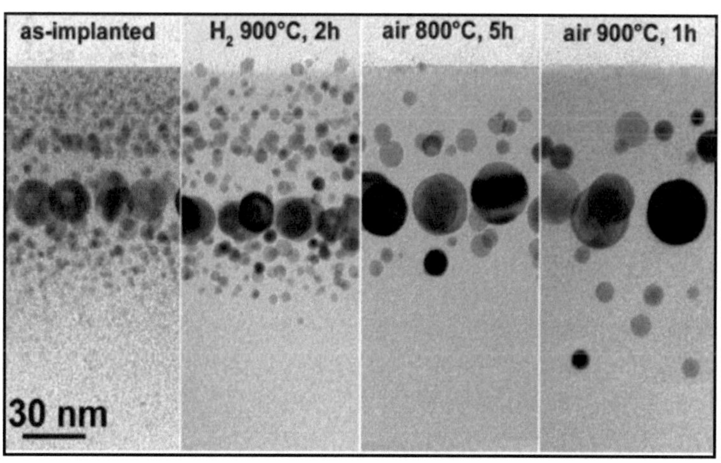

Fig. 9. Cross-sectional bright-field TEM images of sequentially ion implanted silica with Pd (130 keV) and Ag (130 keV) at dose 3×10^{16} ions/cm^2: (a) as implanted; (b) annealed in reducing atmosphere at 9000C for 2 hours; (c) annealed in air at 8000C for 5 hours; (d) annealed in air at 9000C for 1 hours.

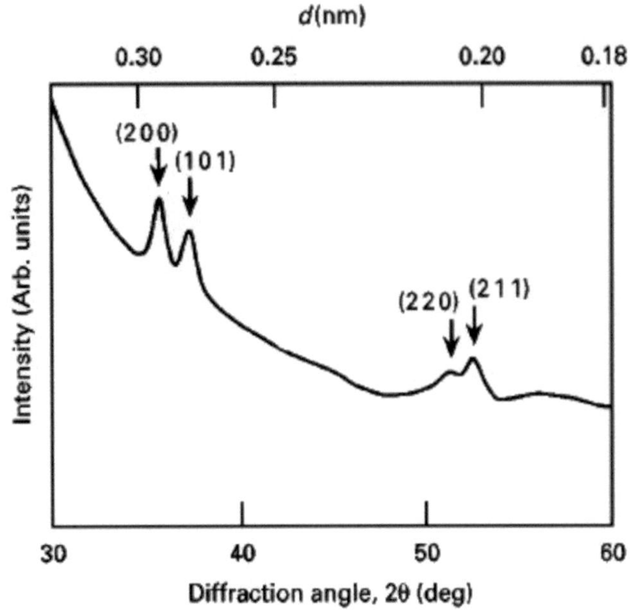

Fig. 10. XRD pattern of Sn^{+} -implanetd silica glass

Another example of changing both the linear and nonlinear optical properties of silica by implanting 2.0 MeV copper, 350 keV tin, 1.5 MeV silver and 3.0 MeV gold have been studied by D. Ila *et. al* [58]. Resonance enhancement of $\chi^{(3)}$ with in-diffusion of gold was exploited for high non-linearity at wavelength above 1000 nm [59]. Another significant work reported [60] of copper nanoparticles prepared by the reduction of copper (II) acetate in water and 2-ethoxyethanol using hydrazine under reflux. The synthesized nanoparticles exhibited a distinct absorption peak in the region 572−582 nm. The nonlinear optical properties of the copper colloids were first measured using the Z-scan technique and $\chi^{(3)}$ values obtained were found to be of the order of 10^{-11}−10^{-12} esu. Recently reported works includes synthesis of metal nanocluster composite glass by Cu ions implanted into silica with dose of 1×10^{17} ions/cm [61, 62]. Third-order nonlinear optical properties of the nanoclusters were measured at 1064 nm and 532 nm excitations using Z-scan technique. Absolute third-order nonlinear susceptibility $\chi^{(3)}$ of this kind of sample was determined to be 2.1×10^{-7} esu at 532 nm and 1.2×10^{-7} esu at 1064 nm. Nonlinear optical properties of Ag–Cu nanoclusters co-doped in a SiO_2 substrate prepared using the sol-gel technique were investigated under the influence of the second harmonic of a high-power nanosecond Nd:YAG laser [63]. Third-order nonlinear susceptibility $\chi^{(3)}$ was determined to be of the order of 10^{-7} esu and nonlinear refractive index was found to be of the order of 10^{-9} cm^2 /Watt. The measured nonlinear refractive indices (~10^{-6} e.s.u.) for these microcrystallite-doped glass composites compare well to that of II-VI semiconductor-doped glasses and are much larger than that of colloidal melt-glasses.

For a metal cluster with a cubic symmetry, the fourth-rank tensor $\chi^{(3)}$ has four independent components. Of these, $\chi^{(3)}{}_{xxxx}$ includes contributions from both thermal and electronic processes. The $\chi^{(3)}{}_{xxyy}$ component, on the other hand, is sensitive only to electronic processes. The enhancement in the electronic component of $\chi^{(3)}$ due to the increased volume fraction

(~30%) of the nanoclusters in the composite is obviously an important consideration from the point of view of non-linear optics, but achieving such a high concentration of dopants is a non-trivial materials problem. A volume fraction of such magnitude, obtainable by ion implantation only, cannot be made in a melt-glass because of the constraints of chemical and thermodynamic equilibrium. This points to one major advantage of ion-implantation as a technique for synthesizing nanocluster-glass composites.

5. Magnitudes and temporal responses of χ(3) in nanocluster - glass composites

Progress in non-linear optical materials is the key to creating functional photonic devices for optical communications, sensing and computing. Understanding the correlation between materials processing and optical nonlinearities is critical to the development of advanced non-linear materials for photonic devices. Optical switching at terahertz frequencies requires nonlinear optical materials with the following properties: picosecond or shorter response times, low-power switching threshold, high thermal stability, low two-photon absorption, wavelength tunability, and a high threshold for laser-induced damage.
There are many materials with high third-order nonlinearities, although only a large value of $\chi^{(3)}$ is not sufficient. The key issue for light-wave technology is to optimize performance of third-order materials in real device geometries under realistic operating conditions. Moreover, the long-term mechanical, thermal and optical stability of potential waveguide devices can only be tested accurately in actual device geometries. Thus the results achieved with metal-nanocluster glass composites made by ion implantation are significant because of their potential for making a high-performance nonlinear material using an established ion-implantation-based waveguide fabrication technology [64]. Figure (11) shows a comparison between a variety of electronic and photonic materials with respect to switching energy and switching speed [42].

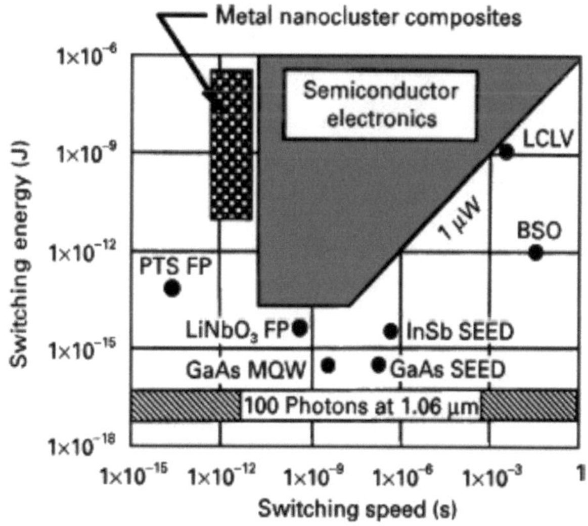

Fig. 11. Comparison of a variety of electronic and photonic materials with respect to switching energy and switching speed

As shown in the hatched area, nanocluster composites of Au : SiO_2 and Cu : SiO_2 have faster switching speeds than any material, except para-toluene sulphonate (PTS), which, however, suffers from long-term photochemical damage. The shaded area in the figure indicates the range of switching energies projected from theoretical estimates for metal nanocluster composites. While no single material exhibits all the characteristics of an ideal optical switch, metal nanocluster composites do seem to offer an attractive combination. However, both for the demonstration of device potential and for studying optical physics in these interesting non-linear materials, it may be advantageous to carry out the non-linear studies in actual waveguide geometries. For example, because two-photon absorption, which is proportional to the imaginary part of $\chi^{(3)}$, arising not only from the clusters but also from the defects created by ion implantation, severely limits all-optical device performances, a fundamental understanding of its origin and relation to size, shape and density of the clusters in the waveguide is crucial. In optical switching devices and waveguide, a key-parameter is the non-linear index of refraction, proportional to the real part of $\chi^{(3)}$. Measurements of the magnitude and the temporal response characteristics of $\chi^{(3)}$ are the most important part of the experiments.

5.1 Degenerate Four Wave Mixing (DFWM)

5.1.1 Principle

When a nonlinear material is illuminated by a pair of coherent waves of equal magnitude traveling in different directions, they result in a phase grating within the material, which in turn can diffract a third wave (not necessarily coherent with the first two) incident on the material into a fourth one. The diffracted phase conjugated wave ids the fourth wave. This process is called 'four-wave mixing' and has been demonstrated in several crystals. When all the four waves involved in the process have the same frequency, it is known as the degenerate four wave mixing (DFWM). The effect can be understood in terms of the third-order nonlinear susceptibility $\chi^{(3)}$. Physically, one may understand this process by considering the individual interactions of the fields within a dielectric medium. The first incident field causes in the dielectric an oscillating polarization, which re-radiates with some phase shift determined by the damping of the individual dipoles; this is just traditional Raleigh scattering described by linear optics. The application of a second field will also drive the polarization of the dielectric, and the interference of the two waves will cause harmonics in the polarization at the sum and difference frequencies. Now, the application of a third field will also drive the polarization at the sum and difference frequencies. This beating with the sum and difference frequencies is what gives rise to the fourth field in four wave mixing. Since each of the beat frequencies produced can also act as new source fields, a bewildering number of interactions and fields may be produced from this basic process.

5.1.2 Theory

The two coherent waves ($E = \frac{1}{2} E_0 \cos \omega t$) with a time variation create an interference pattern or standing wave within the crystal. The pattern consists of regions near the nodes where the local field is essentially zero, and regions near the antinodes where its value is close to $E_0 \cos \omega t$. Retaining terms up to the third order in polarization equation (Eq (2)), and adding a small extra field $\mathbf{E}'$ acting as the third field, we then have, up to the first order in $\mathbf{E}'$:

$$P+\delta P=\chi(E+E')+\chi^{(2)}\left(E^2+2EE'\right)+\chi^{(3)}\left(E^3+3E^2E'\right)$$

Therefore,

$$\delta P=E'\left(\chi+2\chi^{(2)}E+3\chi^{(3)}E^2\right) \tag{22}$$

Now if **E′** is incoherent with E, the ratio $\delta P/_{E'}$ is obtained by averaging the bracketed term in Equation (22) over many periods of ω. The average of $E_0\cos\omega t$ of course is zero; that of $E_0^2\cos^2\omega t$ is $E_0^2/_2$ and so, we have on average,

$$\delta P/_{E'}=\left(\chi+3/_2\chi^{(3)}E_0^2\right) \tag{23}$$

The refractive index seen by the field **E′** is thus given by,

$$\mu_d=\left(1+\delta P/_{\varepsilon_0E'}\right)^{1/2}$$

$$=\left[1+\left(\chi+3/_2\chi^{(3)}E_0^2\right)/_{\varepsilon_0}\right]^{1/2} \tag{24}$$

which depends on the field E_0, and is therefore different at the nodes and the antinodes of the standing wave pattern created by the first two waves. The wave **E′**, therefore, sees the standing wave 'written' into the crystal as a phase grating, which diffracts according to the Bragg's law. Figure (12) shows the forward and backward mixing arrangements of the DFWM setup [50]. In the former case there are two input beams, usually obtained by a mirror placed beyond the nonlinear medium. In both the cases, the nonlinear polarization radiates a signal field at ω in the phase-matched direction. The pump and the probe waves interfere giving rise to both amplitude gratings (resulting from the nonlinear loss due to two-photon absorption) and refractive index or phase gratings (due to the intensity dependent refractive index change), from which the waves scatter coherently. The output intensities of signal and probe are related to the nonlinear absorption which can be related to the nonlinear refractive index.

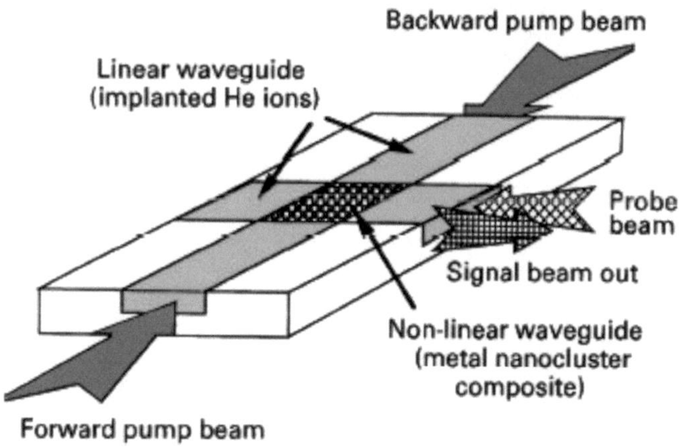

Fig. 12. The forward and backward mixing arrangements of the DFWM set up.

Measurements of the magnitude and the temporal response characteristics of $\chi^{(3)}$ are the most important part of the experiments. One can measure $\chi^{(3)}$ at wavelengths both on and off the SPR by forward DFWM in the geometry of Fig. (12) [30]. In guided-wave geometry, it is possible to use a CW mode locked laser for four-wave mixing studies with counter-propagating pump-beams. In DFWM configuration, three different light fields, E_1, E_2 and E_3, of the same frequency are incident on the non-linear optical material to produce a fourth beam, E_4. It is also called phase conjugation or real-time holography. Two different types of volume holograms, i.e. reflection and transmission holograms can be constructed. In either type, the third beam is used as the reading beam and the fourth one is the result of the diffraction of the hologram, making E_4 conjugate to E_3. The third-order nonlinear optical susceptibility obtained from the above technique is given by [65]

$$\chi^{(3)} = \frac{8c^2 n_d^2 \varepsilon_0 \alpha n^{1/2}}{3\omega I_1 (1-T)} \tag{25}$$

where n_d and α are linear refractive index of the host matrix and absorption coefficient of the nanocluster-glass composite, respectively, both measured at the same frequency, ω, of the optical signal close to the SPR of the composite, I_1 is the pump intensity, c is the velocity of light in vacuum, T is the transmittance of the medium at ω , and n is the diffraction efficiency of the holographic grating. DFWM experiments also give important information on the switching speed. $\chi^{(3)}$ for gold nanoclusters in glass was measured by Magruder *et al.* [66] by DFWM using a mode-locked Q-switched, frequency doubled Nd:YAG laser at a wavelength of 532 nm with a pulse repetition frequency of 10 Hz and a nominal pulse width of 35±5 ps. This wavelength is near the SPR of gold (530 nm). The two forward-going pump-beams and a weak probe-beam were arranged to intersect in the composite-layer containing the metal clusters. The probe was delayed with respect to the two pump beams using a computer-controlled optical delay line. The pump and probe beams interacted coherently via the third-order non-linear susceptibility to produce a phase-conjugated signal, detected in a photomultiplier. Identical measurements were carried out on unimplanted samples; no measurable DFWM signal was observed. Figure (13) shows the intensity of the phase-

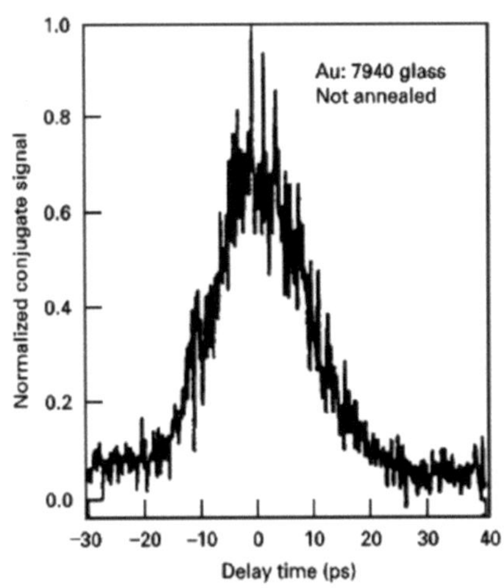

Fig. 13. Intensity of the phase-conjugated beam in a DFWM measurement with cross-polarized pump-probe configuration as a function of pump-probe time in picoseconds.

conjugated signal as a function of pump probe delay time [66]. The symmetric shape of the time spectrum indicates that the third-order response is no longer than that of the pulse width (~ 35 ps). The measured values of $\chi^{(3)}$ were found to be 1.0×10^{-10} e.s.u and 1.7×10^{-10} e.s.u. for this nanocluster composite layer without and with heat treatment, respectively.

5.2 Z- Scan experiment

5.2.1 Principle

Z-Scan technique [51], based on the principles of spatial beam distortions, allows us to determine both sign and magnitude of nonlinear refractive index η_2 and nonlinear absorption coefficient β and it is very sensitive to detect small nonlinear refraction. The excitation source used in the Z-Scan measurement is a mode-locked laser. Each laser pulse has a Gaussian spatial profile and is focused by a converging lens. The sample is placed near the waist of a focused Gaussian beam and is scanned along the direction of propagation of the incident laser beam (Z-axis) of the light. As the sample is translated along Z-direction, it experiences a different incident field at different Z-positions. The light intensity transmitted across the nonlinear material is measured in the far field (FF) as the sample is moved along the direction of the propagation of light, in the open Z-Scan mode of operation. The transmittance of the nonlinear medium is measured through a finite aperture placed in the far field as a function of the sample position (Z). Figure (14) shows the schematics of a Z-scan set up. When the sample moves towards the focus from negative Z, the laser power density intercepted by the sample increases, giving rise to self-focusing [67]: owing to the Gaussian transverse intensity profile of the laser beam, the original plane wavefront gets progressively more disturbed, in a way similar to that imposed by a positive lens (in the case of positive nonlinearity of the material), leading to a self-focusing that shifts the position of the actual focal point. A photodiode detector simultaneously measures the intensity of the transmitted beam through the sample as a function of the position of the sample, giving the Z-Scan curve, the shape of which reveals the presence of any absorptive nonlinearity in the sample.

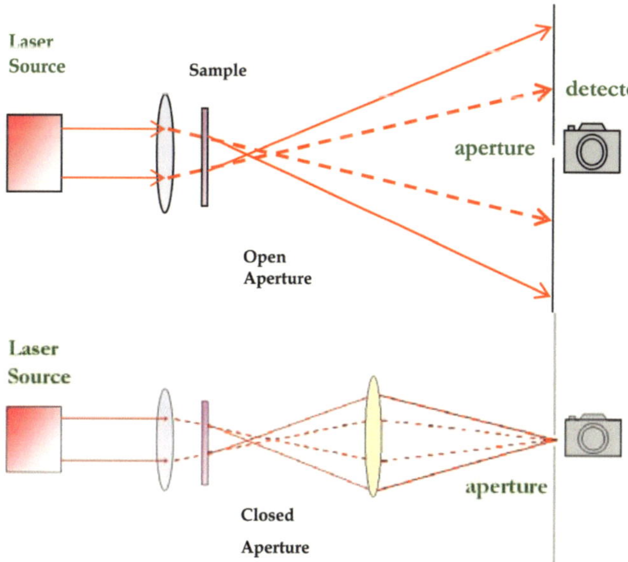

Fig. 14. Schematics of Z-scan experiment

The sign of the nonlinear refraction is readily obtained from a Z-scan signature. An increase in transmittance in the pre-focal region followed by a decrease in the post-focal region (peak-valley) in the Z-scan signature denotes a negative nonlinear refraction, whereas a valley-peak configuration denotes a positive nonlinearity. Removal of the aperture, i.e., collecting all the light on detector, which is referred to as an open-aperture Z-scan, will result in a flat response for a purely refractive nonlinearity. It is because the sensitivity to nonlinear refraction is entirely due to the aperture, and its removal completely eliminates the effect. However, if nonlinear absorption is also present, then it will be reflected as a transmission variation in the open aperture scan. Multi-photon absorption suppresses peak and enhances the valley while saturation of absorption produces the opposite effect in a closed-aperture Z-scan. Thus, apart from the magnitude of the nonlinearity (both real and imaginary parts), this technique provides a direct measurement of the sign of nonlinearity also. The sign of the nonlinearity is an important parameter for practical realization of optical signal processing devices.

5.2.2 Theory

When a matter is subjected to an intense laser radiation, it exhibits intensity-dependent index of refraction given by

$$n(I)=n_0+n_2I$$

That is,

$$n(I)=n_0+\Delta n(I) \tag{26}$$

where n_0 is the linear refractive index and $\Delta n(I)\equiv n_2I$ is the intensity dependent change in the refractive index. n_2 is called the refractive index intensity coefficient and is related to $\chi^{(3)}$ by

$$\chi^{(3)}(esu)=\frac{n_0^2}{4\pi^2}10^{-6}\,n_2$$

where n_2 is expressed in cm^2 / GW.
Consequently, the phase change of the intense laser beam on propagation through a distance L through the nonlinear medium will be

$$\begin{aligned}\Phi&=kn(I)L\\&=K(n_0+\Delta n(I))L\\&=\Phi^L+\Phi^{NL}\end{aligned}$$

In the above expression, the first term is the linear phase shift. The second term is the nonlinear phase shift arising from nonlinear refraction and can be written as

$$\begin{aligned}\Phi^{NL}&=k(\Delta n)L\\&=kn_2IL\end{aligned}$$

It is thus clear that due to effect of n_2, the beam passing through a nonlinear optical medium has an additional phase shift which is intensity dependent. Consequently, after

passing through a nonlinear medium, the beam wavefront is modified due to intensity-dependent phase changes. In Z-scan technique, these nonlinear phase front distortions are translated into a transmission variation of an aperture placed in the far field. Hence the transmittance of the aperture is a measure of the nonlinear phase change or n_2 and consequently, $\chi^{(3)}$.
The electric field of a Gaussian beam having waist radius w_0 and traveling in the +Z direction is represented as

$$E(Z,r,t) = E_0(t)\frac{w_0}{w(Z)}\exp\left(-\frac{r^2}{w^2(Z)}-\frac{ikr^2}{2R(Z)}\right)e^{-i\phi(Z,t)} \tag{27}$$

where

$$w^2(Z) = w_0^2\left(1+\frac{Z^2}{Z_0^2}\right) \tag{28}$$

is the beam cross-section at Z,

$$R(Z) = Z\left(1+\frac{Z_0^2}{Z^2}\right) \tag{29}$$

is the radius of curvature of the wavefront at Z,

$$Z_0 = \frac{\pi w_0^2}{\lambda} \tag{30}$$

is the Rayleigh range,

$$k = \frac{2\pi}{\lambda} \tag{31}$$

is the wave vector, and λ is the laser wavelength, all in free space. $E_0(t)$ denotes the radiation electric field at the focus and contains the temporal envelope of the laser pulse. The exponential term $\exp\{-i\phi(Z,t)\}$ represents the radial phase variations. If the sample length is small enough such that the changes in the beam diameter within the sample due to either diffraction or nonlinear refraction can be neglected, the sample is regarded as thin. Considering a thin sample and using the slowly varying envelope approximation, the wave equation for the phase and intensity can be respectively written as

$$\frac{d\Delta\phi}{dz'} = \Delta n(I)k \tag{32}$$

$$\frac{dI}{dz'} = -\alpha(I)I \tag{33}$$

where z' is the propagation distance inside the sample and $\alpha(I)$, in general, includes the linear and nonlinear absorption terms. For cubic nonlinearity and negligible absorption, Eqs.

(32) and (33) are solved to give the phase shift $\Delta\phi$ at the exit surface of the sample, which simply follows the radial variation of the incident irradiance at a given position of the sample Z :

$$\Delta\phi(Z,r,t)=\Delta\phi_0(Z,t)\exp\left(-\frac{2r^2}{w^2(Z)}\right) \tag{34}$$

with

$$\Delta\phi_0(Z,t)=\frac{\Delta\phi_0(t)}{1+Z^2/Z_0^2} \tag{35}$$

where $\Delta\phi_0(t)$ is the on-axis phase shift at focus and is defined as

$$\Delta\phi_0(t)=k\Delta n_0(t)L_{eff} \tag{36}$$

where

$$L_{eff}=\frac{1-\exp(-\alpha l)}{\alpha} \tag{37}$$

and

$$\Delta n_0=n_2 I_0(t) \tag{38}$$

Here α is the linear absorption coefficient of the sample, l is the sample length, n_2 is the intensity dependent refractive index and $I_0(t)$ is the on-axis irradiance at focus (i.e., at Z = 0). The complex electric field exiting the sample E_e will contain the nonlinear phase distortion and can be represented as

$$E_e(Z,r,t)=E(Z,r,t)\exp(-\alpha L/2)\exp\{i\Delta\phi(Z,r,t)\} \tag{39}$$

Now, applying the Gaussian beam decomposition method, we decompose the complex electric field at the exit plane of the sample given by Eq. (40) into a summation of Gaussian beams through a Taylor series expansion of the nonlinear phase term $\exp\{i\Delta\phi(Z,r,t)\}$. That is,

$$\exp\{i\Delta\phi(Z,r,t)\}=\sum_{m=0}^{\infty}\frac{[i\Delta\phi_0(Z,t)]^m}{m!}\exp\left[-\frac{2mr^2}{w^2(Z)}\right] \tag{40}$$

Now each Gaussian beam can be simply propagated to the aperture plane where they are reassumed to reconstruct the beam. The resultant electric field at the aperture plane is given by

$$E_a(r,t)=E(Z,r=0,t)\exp(-\alpha L/2)\sum_{m=0}^{\infty}\frac{[i\Delta\phi_0(Z,t)]^m}{m!}\frac{w_{m0}}{w_m} \exp\left(-\frac{r^2}{w_m^2}-\frac{ikr^2}{2R_m}+i\theta_m\right) \tag{41}$$

where we have used the following abbreviations:

$$w_{m0}^2 = \frac{w^2(Z)}{2m+1}$$

$$w_m^2 = w_{m0}^2 \left[g^2 + \frac{d^2}{d_m^2} \right]$$

$$g = 1 + \frac{d}{R(Z)}$$

$$d_m = \frac{k w_{m0}^2}{2}$$

$$R_m = d \left[1 - \frac{g}{g^2 + d^2/d_m^2} \right]^{-1}$$

and

$$\theta_m = \tan^{-1} \left[\frac{d/d_m}{g} \right]$$

Here d is the propagation distance from the sample to the aperture. The transmitted power through the aperture is obtained by spatially integrating $E_a(r,t)$ over the aperture radius r_a, giving

$$P_T(t) = c\varepsilon_0 \pi \int_0^{r_a} |E_a(r,t)|^2 \, r \, dr \tag{42}$$

Including the pulse temporal variation, the normalized Z-scan transmittance $T(Z)$ can be calculated as

$$T(Z) = \frac{\int_{-\infty}^{\infty} P_T(t)\, dt}{S \int_{-\infty}^{\infty} P_i(t)\, dt} \tag{43}$$

where

$$P_i(t) = \frac{\pi w_0^2 I_0(t)}{2} \tag{44}$$

is the instantaneous input power and

$$S = 1 - \exp(-2r_a^2 / w_a^2) \tag{45}$$

is the aperture linear transmittance, with w_a denoting the beam radius at the aperture in the linear regime. In case the sample exhibits nonlinear absorption, open aperture scan will reflect that, while closed aperture scan records coupled refractive and absorptive parts. Therefore, one has to deduce the nonlinear absorption coefficient from open aperture scan, and then using closed aperture, the refractive and absorptive nonlinearity can be separated out. For example, if the nonlinear absorption is of two-photon in nature, then the absorption coefficient is given by

$$\alpha(I) = \alpha_0 + \beta I$$

The phase shift and irradiance distribution at the exit surface of the sample as

$$I_e(Z,r,t) = \frac{I(Z,r,t)e^{-\alpha L}}{1+q(Z,r,t)} \tag{46}$$

and

$$\Delta\phi(Z,r,t) = \frac{kn_2}{\beta}\ln[1+q(Z,r,t)] \tag{47}$$

where $q(Z,r,t) = \beta I(Z,r,t)L_{eff}$. Combining Eqs. (46) and (47), the complex electric field at the exit surface of the sample is given as

$$E_e(Z,r,t) = E(Z,r,t)e^{-\alpha L/2}(1+q)^{(ikn_2/\beta - 1/2)} \tag{48}$$

We can see that Eq. (48) reduces to (39) in the limit of no nonlinear absorption (i.e. $\beta = 0$). For $|q| < 1$, following a binomial series expansion in powers of q, Eq. (48) can be expressed as an infinite sum of Gaussian beams similar to the purely refractive case described above as follows:

$$E_e(Z,r,t) = E(Z,r,t)e^{-\alpha L/2}\sum_{m=0}^{\infty}\frac{q(Z,r,t)^m}{m!}\left[\prod_{n=0}(ikn_2/\beta - 1/2 - n + 1)\right] \tag{49}$$

Now we can write the complex field pattern at the aperture plane in the same manner as described in purely refractive case provided the phase term in Eq. (40) is replaced by

$$f_m = \frac{(i\Delta\phi_0(z,t))^m}{m!}\prod_{n=0}^{m}\left(1 + i(2n-1)\frac{\beta}{2kn_2}\right) \tag{50}$$

The normalized transmittance can now be estimated following the same procedure as described in purely absorptive case. It is evident from Eq. (50) that the absorptive and refractive contributions to the far-field beam profile and hence to the Z-scan transmittance are coupled to each other due to the coupling factor $\beta/2kn_2$ which is the ratio of the imaginary to the real parts of the third-order nonlinear susceptibility $\chi^{(3)}$. When the aperture is removed, the Z-scan transmittance is insensitive to the beam distortion and is only a function of the nonlinear absorption. The total transmitted power in that case (S = 1) is obtained by spatially integrating Eq. (46), which yields

$$P(Z,t) = P_i(t)e^{-\alpha L}\frac{\ln[1+q_0(Z,t)]}{q_0(Z,t)} \tag{51}$$

where

$$q_0(Z,t) = \frac{\beta I_0(t)L_{eff}}{1+\frac{Z^2}{Z_0^2}} \tag{52}$$

For a temporally Gaussian pulse, Eq. (51) can be time-integrated to give the normalized energy transmittance as

$$T(Z,S=1) = \frac{1}{\sqrt{\pi}q_0(Z,0)}\int_{-\infty}^{\infty}\ln\left[1+q_0(Z,0)e^{-\tau^2}\right]d\tau \tag{53}$$

For $|q_0| < 1$, this energy transmittance can be expressed in terms of the peak irradiance in a summation form as

$$T(Z,S=1) = \sum_{m=0}^{\infty}\frac{\left[-q_0(Z,0)\right]^m}{(m+1)^{3/2}} \tag{54}$$

Thus, once an open aperture (S = 1) Z-scan is performed, the nonlinear absorption coefficient β can be estimated using Eq. (54). With β known, the Z-scan with aperture in place (S < 1) can be used to extract the nonlinear refractive index coefficient η_2. Similarly in case of other nonlinearities (e.g. saturation of absorption, excited state absorption, free carrier absorption etc.), we can deduce the refractive and absorptive contribution to nonlinearity by modifying the equations for phase shift and beam attenuation, given by Eqs. (33) and (34) respectively, and solving for nonlinear phase shift and field amplitude at the exit surface of the sample.

Sheik *et. al.* [51] have shown that the above tedious calculations can be simplified provided the phase change induced by the nonlinear medium is small. For $|\Delta\phi| < \pi$, the peak-valley difference in transmission in case of purely refractive nonlinearity is given by

$$T_{p-v} = 0.405(1-S)^{0.25}kn_2 I_0 L_{eff} \tag{55}$$

which is valid within 3%. The peak-valley separation along Z-axis in this case is related to Rayleigh range through

$$Z_{p-v} = 1.7\,Z_0 \tag{56}$$

Therefore, we can get quick estimate of nonlinearity by just measuring the peak-valley difference in transmission.

5.2.3 Z-scan simulations

Figure (15) shows open aperture Z-scan simulations for positive and negative β respectively. In case of positive β (two-photon absorption, excited state absorption or free-carrier absorption), we get a valley at the beam waist position of the focused beam while negative β (saturation of absorption) results in a peak at the same position. Thus, the sign of absorptive nonlinearity is readily obtained from the open aperture Z-scan signature. If the sample does not exhibit any nonlinear absorption, then no transmittance variation will be recorded in the open-aperture

Figure (16) shows Z-scan simulation for purely refractive positive and negative nonlinearities corresponding to a phase change $\Delta\phi = \pm 0.5$ and an aperture parameter $S = 0.2$. As explained earlier, negative nonlinear refraction results in increase in transmittance in the pre-focal region followed by decrease in the post-focal region (peak-valley) while positive nonlinear refraction exhibits the opposite behaviour. Also shown is the open aperture simulation in which there is no transmittance variation in case of purely refractive nonlinearity.

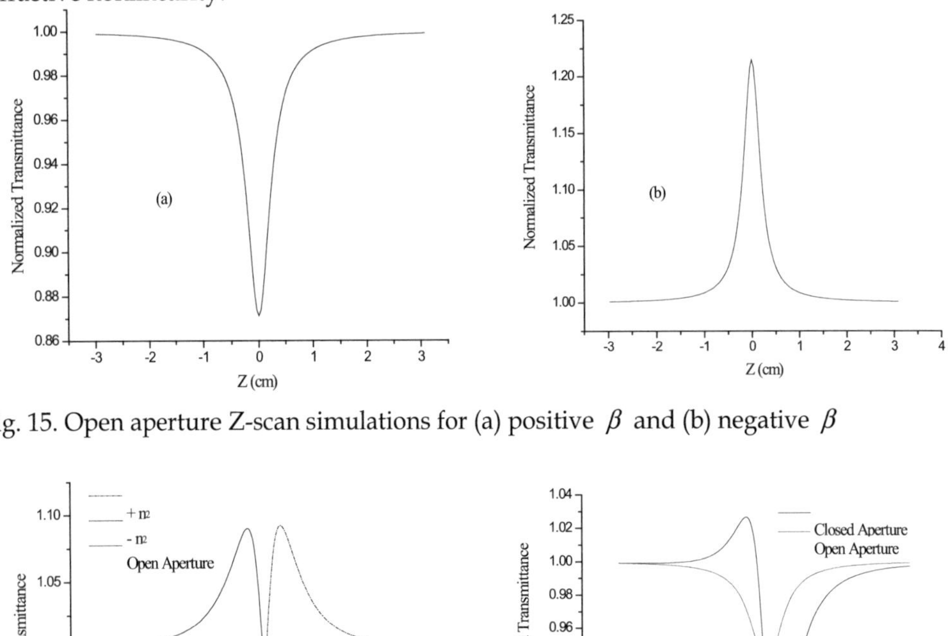

Fig. 15. Open aperture Z-scan simulations for (a) positive β and (b) negative β

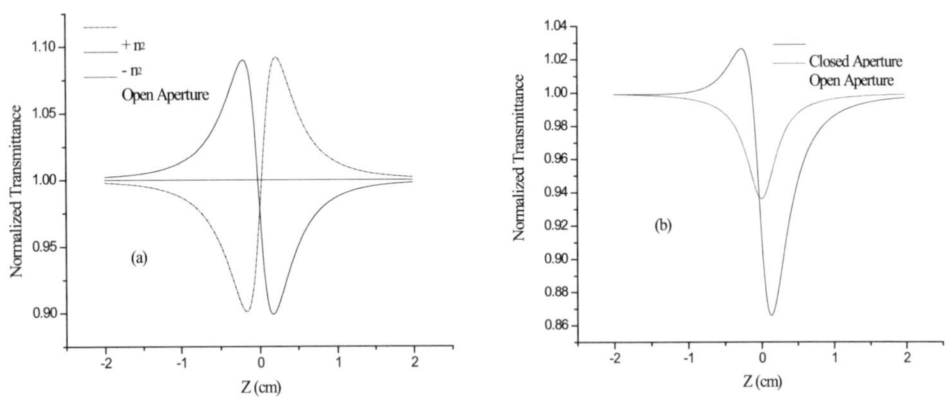

Fig. 16. Z-scan simulations for (a) purely refractive and (b) refractive + absorptive nonlinearity.

When nonlinear absorption is also present, then in the closed aperture signature, the peak gets suppressed while the valley gets enhanced for positive absorptive nonlinearity. The opposite behaviour will be exhibited in case of negative absorptive nonlinearity. Fig. shows simulation for a negative refractive nonlinearity and positive absorptive nonlinearity for a phase change of $\Delta\phi = -0.5$ and two-photon absorption parameter $q_0 = 0.2$.

5.3 Anti-Resonant Ring Interferometric Nonlinear Spectroscopy (ARINS)

5.3.1 Principle

Lee and Hughes (LH) [68] proposed a simple, sensitive, single beam technique based on an anti-resonant ring (Sagnac) interferometer for simultaneously measuring the real and imaginary contributions to optical nonlinearity, and called it Anti-resonant Interferometric Nonlinear Spectroscopy (ARINS). The ARINS technique utilizes the dressing of two unequal-intensity counter-propagating pulsed beams with differential nonlinear phases, which occurs upon traversing the sample. This difference in phase manifests itself in the intensity dependent transmission. Photo-detection of the transmission of the ARINS yields spatially and temporally integrated response. Consequently, it is the pulse energy and not the instantaneous power that is detected. Figure (17) shows the schematic diagram of ARINS setup [69].

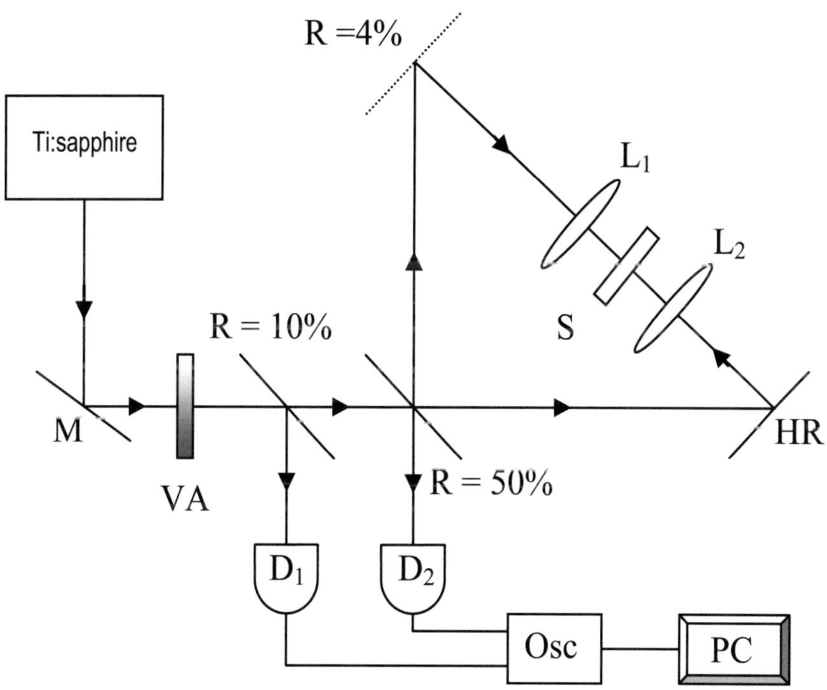

Fig. 17. ARINS experimental setup. M- mirror, L – lenses, HR – high reflectivity mirror, S – sample, VA – variable attenuator D- detectors, Osc – digital oscilloscope, PC – microcomputer

A 50-50 beam splitter divides the incoming pulsed beam into two counter-propagating pulses having a π phase difference. The pulses propagating in the clockwise (CW) direction are reflected by an uncoated flat with 6° wedged rear surface while that propagating in the counterclockwise (CCW) direction are reflected by a high reflectivity mirror. Thus this technique is based on the dressing of two unequal intensity counter-propagating pulses.

These two pulses again recombine at the beam splitter to yield ARINS transmission, $|E_{out}|^2 \propto |E_{cw} + E_{ccw}|^2$, where E_{cw} and E_{ccw} are the optical fields travelling in clockwise and counterclockwise directions respectively. Inside the ring there is a unit magnification telescope comprising of a pair of identical lenses in 2f configuration. One of the lenses focuses the pulses into the sample placed at its focus while the other one re-collimates them. The two counter propagating fields traversing the ring acquires linear as well as intensity dependent nonlinear phase shifts (if the sample exhibits nonlinear response). Since both the fields traverse the same optical path through the ring and encounter the same interactions with the optical elements, linear interactions would affect their amplitude and phase in identical manner. For an exactly 50% beam splitter, in the absence of any nonlinear interactions, the two returning fields will be having the same amplitude and phase difference between them and consequently they will interfere destructively at the beam splitter to yield zero transmission. In this 'balanced' condition all the input power is reflected back to the incident direction. Measurement against this dark background provides the basis for the improved sensitivity essential for measuring relatively weak signals. Any small deviation from the ideal splitting ratio (δ) results in a leakage from the ARINS and is responsible for the background that limits the sensitivity of the measurement.

If the sample under investigation exhibits nonlinear response, then the two unequal-intensity counter-propagating pulses will undergo different phase changes after passing through the sample. Their superposition on the beam splitter will result in the intensity dependent transmission of the ARINS, which is related to the nonlinear response of the sample. However, differential dressing of the two counter propagating pulses with nonlinear phases is possible only when the two do not interact simultaneously in the sample. Else, the same nonlinear phase will be impressed on both by the cross-action phenomenon. Therefore, we prevent the temporal overlap of the two pulses in the sample by spatially offsetting it with respect to the center of the ARINS in such a way that the intense beam (i.e. CCW beam in our geometry) should reach the sample first in order to initiate nonlinear processes. The time difference between the arrivals of the two pulses ($\Delta\tau_{arr}$) determines the nature of the nonlinear optical process that can be studied depending on its response time. Nonlinear processes with decay time longer that $\Delta\tau_{arr}$ do not contribute to the intensity dependent transmission of the ARINS as both the pulses will be affected simultaneously in that case. The delay window thus acts as an ultrafast gate. Hence this technique has the unique ability to filter the nonresonant electronic contributions from integrating (or slow, e.g. thermal) nonlinearities or those arising from long lived (resonant) states and makes it ideal for time-resolved studies and ultrafast gating. If the arrival times of the two pulses in the sample were reversed, one would measure the unfiltered response.

5.3.2 Theory

Equation Chapter 2 Section 1Let us consider a collimated, spatially and temporally coherent Gaussian pulse with electric field amplitude E_0, incident on the beam splitter, where it splits into two counter-propagating beams with electric fields E_{cw} and E_{ccw}. The general form for the electric field of a Gaussian beam is given by

$$E(Z,r,t) = E_0 \frac{w_0}{w(Z)} \exp\left[\frac{-r^2}{w^2(Z)}\right] \exp\left[i\phi(Z)\right] F(t) \tag{57}$$

where Z is the distance of propagation, r is the transverse coordinate, w_0 is the beam waist (Z = 0), $\phi(Z) = \exp\left[i\left(kZ - \tan^{-1}(Z/Z_0)\right)\right]$ is the phase of the Gaussian beam (with pulse duration τ_p and wave vector k) and the function $F(t) = \exp\left[(-2\ln 2)t^2/\tau_p^2\right]$ gives the temporal variation of the pulse. The spot size at a distance Z is $w^2(Z) = w_0^2\left[1 + \left(Z^2/Z_0^2\right)\right]$, where Z_0 is the Rayleigh range. For simplicity, we assume that the Rayleigh range is larger than the sample thickness (thin sample approximation). We ignore the slight difference in the spot sizes for the CW and CCW beams at the lenses that arises because the sample is offset from the center of the ARINS. The intensity ratios in CW and CCW directions are $(1/2 - \delta)$ and $(1/2 + \delta)$ respectively, where, as said earlier, δ represents the small deviation from the ideal splitting ratio of the beam splitter. The electric field at the incident face for the two counter propagating beams can then be expressed as [70]

$$E_{cw} = -\sqrt{1/2 - \delta}\, E_0 \exp\left(-r^2 / w_0^2\right) F(t) \sqrt{R} \tag{58}$$

$$E_{cw} = \sqrt{1/2 + \delta}\, E_0 \exp\left(-r^2 / w_0^2\right) F(t)$$

where R is the reflectivity of the uncoated flat. For the weaker CW beam, $\alpha(I) = \alpha$, the coefficient of linear absorption and $n(I) = n_0$, the linear refractive index. However, for the stronger CCW beam, $\alpha(I) = \alpha + \beta I$ and $n(I) = n_0 + n_2 I$, where β is the effective nonlinear absorption coefficient and n_2 is the nonlinear index of refraction. The electric field of a pulsed Gaussian beam at the exit face of the sample (thickness = L) with a nonlinear absorption and nonlinear refractive index is

$$E_{exit}(r,t) = \frac{E_0(r,t)}{\sqrt{1+q}} \exp(-\alpha L / 2) \exp(-ikn_0 L) \exp\left(-ikn_2 \ln(1+q) / \beta\right) \tag{59}$$

where $E_0(r,t)$ is the incident electric field, $q = \beta I_{in} L_{eff}$, $L_{eff} = \left[1 - \exp(-\alpha L)\right]/\alpha$ is the effective length of the sample, I_0 is the intensity incident on the beam splitter, $I_{in} = (1/2 + \delta) K' I_0 = K I_0$ is the intensity incident on the sample, K' is a constant (<1) accounting for the reflection losses at the sample and lens surfaces while $K = (1/2 + \delta) K'$ is another constant. Using Eqs. (57), (58) and (59), the electric field of the two counter-propagating beams at the corresponding exit faces of the sample can be written as

$$E_{cw}^{exit} = -\sqrt{(1/2-\delta)}\,E_0 \exp\left(\frac{-r^2}{w_0^2}\right)\exp(-\alpha L/2)\exp(-ikn_0 L)F(t)\sqrt{R} \tag{60}$$

$$E_{ccw}^{exit} = \sqrt{(1/2+\delta)}\frac{E_0}{\sqrt{1+q}}\exp\left(\frac{-r^2}{w_0^2}\right)\exp(-\alpha L/2)\exp(-ikn_0 L)\exp\left(-ikn_2 I_{in} L_{eff}\right)F(t) \tag{61}$$

When the two beams arrive again at the beam splitter after one trip round the ring, the electric fields in the transmission branch are given by multiplying Eqs. (60) and (61) by the respective splitting ratios

$$E_{cw}^{t} = -(1/2-\delta)E_0 \frac{w_0}{w(Z)}\exp\left(\frac{-r^2}{w^2(Z)}\right)\exp(-\alpha L/2)\exp(-ikn_0 L)\exp\left[i\phi(Z)\right]F(t)\sqrt{R} \tag{62}$$

$$E_{ccw}^{t} = (1/2+\delta)\frac{E_0}{\sqrt{1+q}}\frac{w_0}{w(Z)}\exp\left(\frac{-r^2}{w^2(Z)}\right)\exp(-\alpha L/2)\exp(-ikn_0 L)\exp\left[i\phi(Z)\right] \times\exp\left(-ikn_2 I_{in} L_{eff}\right)F(t)\sqrt{R} \tag{63}$$

where $w(Z)$ is the spot size at the beam splitter after one round trip. The values of $w(Z)$ and w_0 can be determined experimentally. The ARINS leakage is given by

$$\left|E_{out}(r,t)\right|^2 = \left|E_{cw}^{t}(r,t) + E_{ccw}^{t}(r,t)\right|^2 \tag{64}$$

Substituting Eqs. (62) and (63) in Eq. (64), we get

$$\left|E_{out}\right|^2 = \left[\frac{w_0^2}{w^2(Z)}\exp(-\alpha L)\exp\left(\frac{-2r^2}{w^2(Z)}\right)F^2(t)\right] \times\left[\left(\frac{1}{2}-\delta\right)^2 + \frac{(1/2+\delta)^2}{1+q} + \left(2\delta^2 - \frac{1}{2}\right)\cos\left(kn_2 I_{in} L_{eff}\right)\right]\left|E_0\right|^2 R \tag{65}$$

For $q << 1$, Eq. (65) becomes

$$\left|E_{out}\right|^2 = \left[\frac{w_0^2}{w^2(Z)}\exp(-\alpha L)\exp\left(\frac{-2r^2}{w^2(Z)}\right)F^2(t)\right] \times\left[4\delta^2 + \beta I_{in} L\delta + \left[\left(\frac{\beta I_{in}}{4}\right)^2 + \left(\frac{kn_2 I_{in}}{2}\right)^2\right]L^2\right]\left|E_0\right|^2 R \tag{66}$$

The relevant measured quantity in this experiment is the transmitted pulse energy:

$$W = 2nc\varepsilon_0 \int_{-\infty}^{\infty} \int_0^{\infty} E_{out}^2 \, 2\pi r \, dr \, dt \tag{67}$$

Using Eq. (66), and carrying out the integration in Eq. (67), we get

$$W = 2nc\varepsilon_0 \left(\frac{\pi\sqrt{\pi} w^2(Z) R\tau I_0 \exp(-\alpha L)}{2\sqrt{\ln 2}} \right) \times \left[4\delta^2 + \frac{\beta L \delta I_{in}}{2\sqrt{2}} + \left[\left(\frac{\beta}{4}\right)^2 + \left(\frac{kn_2}{2}\right)^2 \right] \frac{L^2 I_{in}^2}{3\sqrt{3}} \right] \tag{68}$$

Defining

$$I_{out} = 2(\ln 2)^{1/2} W / \pi^{3/2} \tau w^2(Z)$$

which has the dimensions of intensity, Eq. (68) can now be expressed as

$$I_{out} = 2nc\varepsilon_0 R I_0 \exp(-\alpha L) \times \left[4\delta^2 + \frac{\beta L \delta I_{in}}{2\sqrt{2}} + \left[\left(\frac{\beta}{4}\right)^2 + \left(\frac{kn_2}{2}\right)^2 \right] \frac{L^2 I_{in}^2}{3\sqrt{3}} \right] \tag{69}$$

Using $I_{in} = K I_0$ as explained earlier, we get

$$I_{out} = 2nc\varepsilon_0 R \exp(-\alpha L) \times \left[4\delta^2 I_0 + \frac{\beta L \delta K}{2\sqrt{2}} I_0^2 + \left[\left(\frac{\beta}{4}\right)^2 + \left(\frac{kn_2}{2}\right)^2 \right] \frac{L^2 K^2}{3\sqrt{3}} I_0^3 \right] \tag{70}$$

For low intensity levels $q << 1$, the ARINS transmission I_{out} as a function of I_{in} is a cubic polynomial [70]. The coefficients are directly related to the coefficient of nonlinear refractive index, n_2 and to the coefficient of nonlinear absorption, β (which may be due to 'saturation of absorption' or 'two-photon absorption' or 'excited-state absorption'). As stated earlier, δ is related to the linear leakage and limits the sensitivity. For $\delta = 0$, $I_{out} \propto I_{in}^3$, in which case the simultaneous evaluation of n_2 and β becomes difficult. However, a small nonzero δ makes the analysis much simpler as each term of the polynomial can be evaluated directly. If $\delta > 0$ and the samples exhibits two-photon absorption or excited-state absorption (i.e. $\beta > 0$), then all the coefficients of the polynomial become positive and I_{out} will be a continuously increasing function of I_{in}. On the other hand, if the sample exhibits saturation of absorption (i.e. $\beta < 0$) then the sign of β opposes the increase in nonlinear leakage. In this case, the quadratic term in the polynomial becomes negative and I_{out} shows saturation at a relatively low intensity, where the effect of δ is cancelled by the saturable absorption coefficient, β. However, at higher intensities, the cubic term becomes dominant and I_{out} starts increasing continuously. For $\delta < 0$, the curvature of ARINS leakage as a function of I_{in} is reversed for the two processes. Thus, by carefully choosing the value of δ, both the origin of nonlinearity and the values of nonlinear coefficients can be conveniently obtained.

5.3.3 ARINS simulations

The calculated ARINS transmission using Eq. (70) with typical values of $n_2 = 1.8\times10^{-6}\, cm^2 / GW$ and $\beta = 0.378\, cm / GW$ by choosing different values of δ is shown in Figure (18).

The above simulation shows that the ARINS leakage is very sensitive to both sign and magnitude of δ. In fact, it is the combination of both δ and β which determines the sign and magnitude of the quadratic term in Eq. (70). If the sample exhibits nonlinear absorption (not saturation) then $\beta > 0$. In this case, for $\delta > 0$, all the coefficients of the polynomial will be positive and I_{out} will be a continuously increasing function of I_{in}. However, if the sample exhibits saturation of absorption ($\beta < 0$), then the quadratic term in Eq. (70) becomes negative and I_{out} shows saturation at a relatively low intensity where the effect of δ is cancelled by the saturable absorption. Thus the curvature experiences a point of inflexion in this case. At higher intensities, the cubic term in Eq. (70) dominates and therefore I_{out} increases continuously. For $\delta < 0$, the curvature of ARINS leakage as a function of I_{in} is reversed for the two processes.

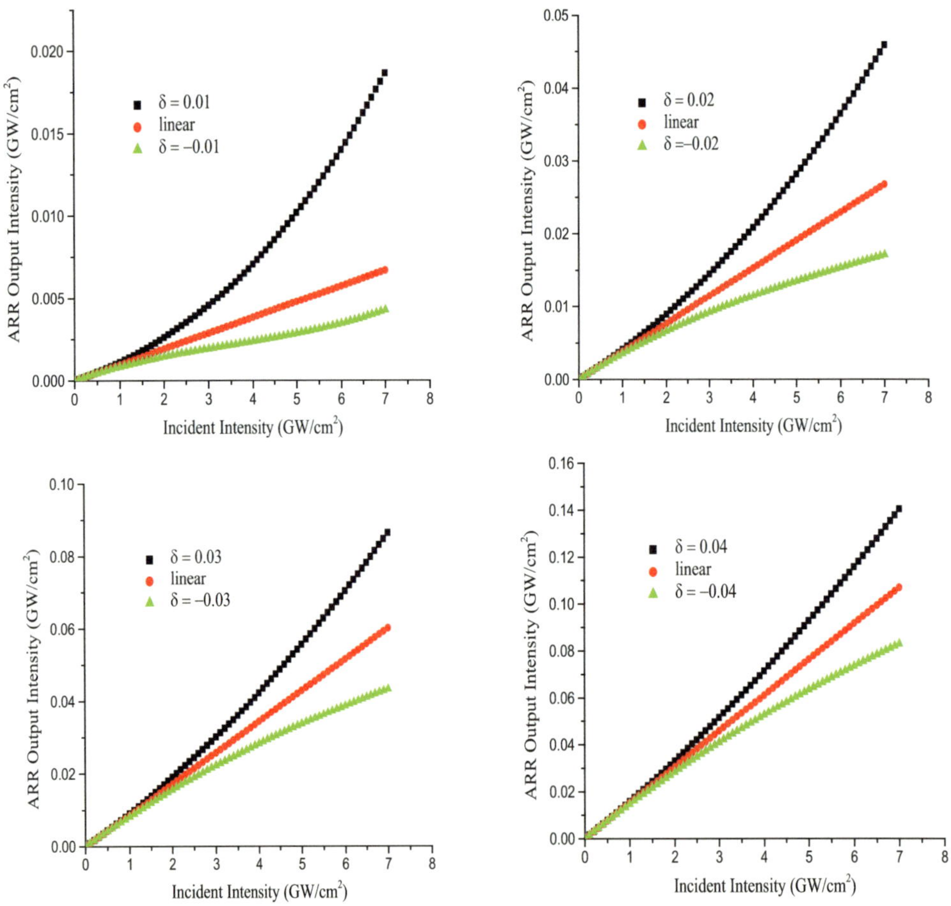

Fig. 18. ARINS simulation for (a) $\delta = \pm 0.01$, (b) $\delta = \pm 0.02$, (c) $\delta = \pm 0.03$ and (d) $\delta = \pm 0.04$

6. Z-Scan and ARINS: a comparative discussion

Since Z-Scan is a simple single-beam technique, it has no difficult alignment other than keeping the beam centered on the aperture. It yields magnitude as well as sign of $\chi^{(3)}$. It is also possible to separate out refractive and absorptive parts (real and imaginary parts) of the nonlinearity by performing simultaneous closed and open aperture scans. However, it gets affected by external noise. Since the sample passes through the beam waist, catastrophic self-focusing can damage the sample, particularly in thin films if significant absorption is present. Since the whole principle is based on the wavefront modification due to nonlinear effect, anomalies like sample scatter and surface inhomogeneity can cause artifact to Z-scan signatures.

On the contrary, both beams in ARINS travel the same path inside the ring and thus, any external jerk will affect both beams identically [71]. Also ARINS is very sensitive and can detect changes of less than 0.005% in the laser amplitude. Hence it can be used to measure nonlinearity of thin films having thickness as low as a few tens of nm. Furthermore, this technique has the unique capability of discriminating against different nonlinear optical processes based on their response times. Any value of δ results in a leakage from the ARR and is responsible for the background that limits the sensitivity of the measurement. Therefore, the alignment of the technique has to be very precise so as to keep δ as low as possible (~ 0.02). This technique does not provide the sign of nonlinearity. It has to be supplemented by other technique like Z-scan to get the sign of β and n_2. Since ARINS technique is based on the time difference between the arrivals of two pulses at the sample, it cannot be used for cw beam.

Keeping in view of the potential advantages of Z-Scan and ARINS techniques over each other, a combination of these two techniques has recently been applied successfully to have very accurate estimates of the third-order optical susceptibilities of ion beam synthesized metal (copper, gold, silver) nanocluster glass composites [72, 73]. The silver ion implanted samples have shown the highest nonlinear susceptibility of ~ 10^{-10} e.s.u at off-resonant condition and ~ 10^{-6} e.s.u at resonant condition, i.e at wavelength near to SPR of silver nanoparticles. The gold and copper implanted samples exhibited nonlinearty of the order 10^{-12} e.s.u. Attempts towards achieving higher orders of $\chi^{(3)}$ of such nanocomposites in the femtosecond time domains are, however, of prime interest for futuristic device technology.

7. Photonic applications: scope and challenges

Nonlinear optics has stirred many phenomena like fabrication of metal nanocluster-based novel nonlinear materials, harmonic generations, optical solitons, parametric amplification, stimulated Raman scattering, self-induced transparency, modulational instability, etc., which find a variety of applications ranging from high data transmission in optical communication, switching, amplifiers, pulse reshaping, pulse compression, tunable lasers to encoded message transmission. Surface plasmon absorption of metal nanoparticles have found large applications for chemical and biological sensing due to sensitive spectral response to local surface environment and ease of monitoring the light signal due to strong light scattering and absorption [74]. Due to the enhanced third-order nonlinear susceptibility of the metal nanoclusters, especially near the SPR frequency, these nanoparticles embedded in dielectric matrices have practical applications in ultrafast all-optical switching devices [75-78]. All optical switches are the base for optical parallel data processing systems and optical time domain multiplexing transmission systems. Nonlinear optical switching devices are based on optical waveguide structures for they provide strong beam confinement in prescribed patterns. In these devices, the intensity of an optical signal is used as a parameter that causes switching between two output channels that perform logic operations.

Today, there are a large number of nonlinear optical materials for specific wavelengths, with various damage thresholds, and with various optical characteristics. The major areas of interest are *photonic metamaterials* (materials having negative refractive index) and *plasmonics*, which bring together the two scientific communities of nanoplasmonics and metamaterials, both of which deal with tailored metal/dielectric and metal/semiconductor nanostructures.

Recently, metal nanocluster glass composites displaying negative refractive index - a property not found in any known naturally occurring materials - have drawn significant scientific interest, underscoring the remarkable potential of metamaterials to facilitate new developments in electromagnetism. Sophisticated structures for such metamaterials have to be prepared by various design technologies. Lots of groups throughout the world are involved in the synthesis of these kinds of materials. Different approaches have been made to synthesize the negative refractive index materials. These include fabricating a multilayer metal-dielectric stack by using focused ion-beam milling [79], printed circuit board technology [80], self assembly approach [81, 82], fabricating regular arrays of metal pillars by high-resolution electron-beam lithography on a glass substrate [83,84], and by sol-gel technique [85]. Regular arrays of gold and silver pillars fabricated by high resolution electron-beam lithography or by ion-implantation, both on glass substrates have been reported to have negative refractive index. By providing access to new realms of material response, metal nanoclusters in glasses have and will continue to impact the fields of physics, materials science, engineering, optics, nanotechnology, and many other branches of science. Gold and silver are the materials of first choice for the realization of a large number of interesting nanoscale devices and structures due to their unique chemical and optical properties. Moreover, nanocomposites made of silver and gold nanowires imbedded in a sol-gel host have been morphologically and optically investigated.

Metallic nanoparticles have distinctly different optical characteristics than SPs at planar interfaces. Nanoparticles show strong optical resonances, again because of their large free-electron density. As a result, a plane wave impinging on a 20 nm diameter Ag particle is strongly 'focused' into the particle, leading to a large electric field density in a 10 nm region around the particle. Ordered arrays of nanoparticles can possess even further enhanced field intensities as a result of plasmon coupling between adjacent particles [86]. By varying nanoparticle shape or geometry, the SP resonance frequency can be tuned over a broad spectral range. For example, Au ellipsoids or silica colloids covered with an Au shell show resonances that coincide with the important telecommunications wavelength band. The ability to achieve locally intense fields has many possible applications, including increasing the efficiency of LEDs, (bio-) sensing, and nanolithography.

Arrays of metal nanoparticles can also be used as miniature optical waveguides. In linear chain arrays of nanoparticles, a plasmon wave propagates by the successive interaction of particles along the chain. The propagation length is small (~100 nm), but may be increased by optimizing particle size and anisotropy. What makes these nanoparticle array waveguides attractive is that they provide confinement of light within ~50 nm along the direction of propagation, a 100-fold concentration compared to dielectric waveguides.

8. References

[1] P. A. Franken, A. E. Hill, C. W. Peters, and G. Weinreich, *Phys. Rev. Lett.* 7, 118 (1961)

[2] T. H. Maiman, R. H. Hoskins, I. J. D'Haenens, C. K. Asawa, and V. Evtuhov, *Phys. Rev.* 123 *1151(1960)*

[3] P. Chakraborty, *Journal of Material Science*, 33 2235 (1998)

[4] Robert W. Boyd, *Nonlinear Optics*, pp 4, Academic Press (2003)

[5] C.N.Ironside, Ultrafast all-optical Switching, *Contemporary Physics*, 34 (1993)

[6] G. Mie, *Ann. Phys.* (Leipz) 25, 377 (1908)

[7] M. Brack, *Rev. Mod. Phys.* 65, 677 (1993)

[8] C. Brechignac and J.P. Connerado, J. Phys. B : *At. Mol. Opt. Phys.* 27, 3795 (1994)

[9] F. Alasia, R.A. Broglia, H.E. Roman, L.I. Serra, G. Colo and J.M. Pacheco, *J. Phys. B : At. Mol. Opt. Phys.* 27, L663-L649 (1994)

[10] M. Madjet, C. Guet and W.R. Johnson, *Phys. Rev. A* 51, 1327 (1995)

[11] A.V. Korol and A.V. Solov'yov, *J. Phys. B : At. Mol. Opt. Phys.* 30,1105 (1996)

[12]W. Ekardt, *Phys. Rev. B* 33, 8803 (1986)

[13] W. Ekardt, *Phys. Rev. B* 36, 4483 (1987)

[14] L.G. Gerchikov, A.N. Ipatov, A.V. Solov'yov and W. Greiner, *J. Phys. B : At. Mol. Opt. Phys.* 31, 3065 (1998)

[15] Jean-Patirick Connerade and Andrey V Solov'yov, *arXiv : physics/02020661v1* (27th Feb 2002)

[16] J.C. Maxwell-Garnett, *Philos. Trans. R. Soc A* 203, 385 (1904)

[17] G.W. Arnold, J.A. Bordes, J. Appl. Phys. 48 ,1488 (1977)

[18] W. C. Huang and J. T. Lue, *Phys. Rev. B* 49, 179 (1994)

[19] *Idem, ibid.* 42 414 (1965)

[20] W. J. Doyle, *Phys. Rev.* 111 1067(1958)

[21] M. Sheik-Bahae, M.Hutchings, D.C. Hagan, Dispersion of bound electronic nonlinear refracting solids, *IEEE J. Quant. Elect.* 1296

[22] Y.Gao, X.Zhang, Y.Li, H. Liu, Y.Wang, Q.Chang, W.Jiao and Y. Song, *Opics Communications* 251, 429, (2005)

[23] G. Battaglin, R. Polloni, G. De Marchi, F. Cavale, F. Gonella, G. Mattei, P. Mazzoldi, A. Quaranta, F. Spizzo, G. De and R.F. Haglund Jr, in '*Proceedings of the International Conference on Fibre Optics and Photonics, Photonics -96*' (Ed: J.P. Raina and P.R. Vaya), Tata McGraw-Hill Publishing Co., New Delhi, pp-36

[24] M. Mennig, M. Schmitt and H. Schmidt, *J. Sol-Gel Sci. Tech.* 8, 1035 (1997)

[25] G.W. Arnold, J. Appl. Phys. 46, 4466 (1975)

[26] F. Gonella, P. Mazzoldi, in *Handbook of Nanostructured Materials and Nanotechnology*, Vol. 4 (Ed: Hari Singh Nalwa) pp 81-158, Academic Press (2000).

[27] E. Cattaruzza, *Nucl. Instr. and Meth. B* 169, 141 (2000)

[28] G. Mattei, *Nucl. Instr. and Meth. B* 191, 323 (2002)

[29] F. Gonella, *Nucl. Instr. and Meth. B* 166, 831(2000)

[30] R.F. Haglund, Jr., L. Yang, R.H. Magruder III, C.W. White, R.A. Zuhr, L.Yang, R. Dorsinville and R.R. Alfano, *Nucl. Instr. and Meth. B* 91, 493 (1994)
[31] N. Skelland and P.D. Townsend, *Nucl. Instr. and Meth. B* 93, 433 (1994)
[32] A. Meldrum, R.F. Haglund, L.A. Boatner and C.W. White, *Adv. Mater.* 13, 1431 (2001)
[33] F. Garrido, F. Caccavale, F. Gonella, A. Quaranta, *J. Eur. Opt. Soc. A: Pure Appl. Opt.* 4, 771 (1995)
[34] F. Garrido, J.C. Dran, L. Thome, C. Meneghini, F. Gonella and A. Quaranta *Nucl. Instr. and Meth B* 115, 561 (1996)
[35] F. Gonella, G. Mattei, P. Mazzoldi, C. Sada, G. Battaglin and E. Cattaruzza, *Appl. Phys. Lett.* 75, 55 (1999)
[36] D. Ila, Z. Wu, C.C. Smith, D.B. Poker, D.K. Hensley, C. Klatt, S. Kalbitzer, *Nucl. Instr. and Meth B* 127/128, 570 (1997)
[37] B. Ghosh, P. Chakraborty, B. Sunderavel and C. Vijayan, *Nucl. Instr.and Meth. B* 266, 1356 (2008)
[38] H. Hosono, Jpn. H. Hosono. *Jpn. J. Appl. Phys.* 32, 3892 (1993)
[39] H. Hosono and N. Matsunami. *Phys. Rev. B* 48, 13469 (1993)
[40] H. Hosono and H. Imagawa. *Nucl. Instr. and Meth. B* 91, 510 (1994)
[41] R. H. Magruder III, J. E. Witting and R. A. Zuhr, *J. Non-Cryst. Solids* 163, 162 (1993).
[42] P. D. Townsend, *Rep. Prog. Phys.* 50, 501 (1987) .
[43] K. Uchida, S. Kaneko, S. Omi, C. Hata, H. Tanji, Y. Asahara, A. J. Ikushima, T. Tokizaki, and A. Nakamura, *J. Opt. Soc. Am. B, Opt. Phys.*, 11 (7), 1236 (1994)
[44] R. Philip, G. Ravindra Kumar, N. Sandhyarani, and T. Pradeep, *Phys. Rev. B, Condens. Matter,* 62 (19), 13160 (2000)
[45] M. Falconieri, G. Salvetti, E. Cattaruzza, F. Gonella, G Mattei, P. Mazzoldi, M. Piovesan, G. Battaglin, and R. Polloni, *Appl. Phys. Lett.*, 73 (3), 288 (1998)
[46] R.H. Magruder, III, R.F. Haglund,Jr. , L. Yang, J.E. Witting and R.A. Zuhr, *J. Appl. Phys.* 76, 708 (1994)
[47] R.L. Zimmerman, D. Ila, E.K. Williams, B. Gasic, A. Elsamadicy, A.L. Evelyn, D.B. Poker, D.K. Hensley, David J. Larkin, , *Nucl. Instr. and Meth. B* 166-167, 892 (2000)
[48] G. Battaglin, E. Cattaruzza, F. Gonella, G. Mattei, P. Mazzoldi, C. Sada and X. Zhang, *Nucl. Instr. and Meth. B* 166, 857 (2000)
[49] L. C. Nistor, J. Van Landuyt, J. D. Barton, D.E. Hole, N.D. Skelland and P.D. Townsend, *J. Non-Cryst. Solids* 162, 217 (1993)
[50] D. Ricard, P. Roussognal and C. Flytzanis, *Opt. Lett.* 10, 511 (1985)
[51] M. Sheik-Bahae, A. A. Said, T. Wei, D.J. Hagan and E.W. Van Stryland, *IEEE J. Quantum Eletcron.* QE-26, 760 (1990)
[52] R. F. Haglund, Jr., L. Yang, R.H. Magruder III, J.E. Witting, K. Becker and R.A. Zuhr, *Opt. Lett*, 18, 373 (1993)
[53] P. Mazzoldi, F. Caccavale, E. Cattaruzza, P. Chakraborty, L. Tramontin, A. Boscolo-Boscoletto, R. Bertoncello, E. Trivillin, G. Battaglin and G. W. Arnold, *Nucl. Instr. and Meth. B* 91, 505 (1994)
[54] P. Mazzoldi, T. Tramontin, A. Boscolo-Boscoletto, G. Battaglin and G. W. Arnold, *ibid.* 80/81, 1192 (1993)

[55] R. H. Magruder III, Do Henderson, S.H. Morgan and A. Zuhr. *J. Non-Cryst. Solids* 152, 258 (1993)
[56] H. Hosono, Y. Y. Suzuku, Y. Abe, Y. L. Lee, K. Oyoshi and S. Tanakar, *ibid.* 142, 287 (1992)
[57] Y. Takeda, T. Hoiki, T. Motohiro and S. Noda, *Nucl. Instr. and Meth. B* 91, 505 (1987)
[58] D. Ila, E.K. Williams, S. Sarkisov, C.C. Smith, D.B. Poker and D.K. Hensley, *Nucl. Instr. and Meth B* 141, 289 (1998)
[59] Y. Menke, M. Ferraris, C. Corbari and J. Fage-Pedersen, *Journal of Non-Crystallien Solids*, 345/346, 366 (2004)
[60] H. H. Huang, F. Q. Yan, Y. M. Kek, C. H. Chew, G. Q. Xu, W. Ji, P. S. Oh and S. H. Tang, *Langmuir*, 13(2), 172 (1997)
[61] Y. H. Wang, Y.M. Wang, J.D. Lu, L.L. Ji, R.G. Zang and R.W. Wang, *Optics Communications* 283(3), 486 (2010)
[62] B. Ghosh, P. Chakraborty, C. Vijayan, S. Mahapatra and P. Mazzoldi, *Materials Letters* 61, 4512(2007)
[63] P. Prem Kiran, G. De and D. Narayana Rao, *IEE Proc.-Circuits Devices Syst.* 150(6), 559 (2003)
[64] R. H. Magruder III, J. E. Witting and R. A. Zuhr, *J. Non-Cryst. Solids* 163, 162 (1993).
[65] R. C. Caro and R. C. Gower, *IEEE J. Quantum Electron.* QE-18, 1375 (1982)
[66] R. H. Magruder III, L. Yang, R. F. Haglund, C. W. White, Lina Yang, R. Dorsinville and R. R. Alfano, *Appl. Phys. ¸Lett.* 62 1730 (1993).
[67] Y. R. Shen, *The principles of Nonlinear Optics*, Wiley, pp 1-563, New York, (1984)
[68] H. W. H. Lee, R.S. Hughes, *Opt. Lett.* 19, 1708 (1994)
[69] Parinda Vasa, B.P.Singh, Praveen Taneja, Pushan Ayub, *Optics Communications* 233, 297 (2007).
[70] M. C. Gabriel, N.A. Whitaker Jr., C.W. Dirk, M.G.Kuzyk and M. Thakur, *Opt. Lett.* 16 1334 (1991)
[71] R. Trebino, C. C. Hayden, *Opt. Lett.* 16, 493(1991)
[72] B. Ghosh and P. Chakraborty, *Nucl. Instrum. Meth B (in press), doi: 10.1016/j.nimb 2010.11.055* (2010).
[73] B. Ghosh, P. Chakraborty, B. P. Singh and T. Kundu, *Applied Surface Science*, 256, 389 (2009)
[74] Y. Xia, N.J. Halas, *MRS Bulletin*, 30, 338 (2005)
[75] N. Pincon, B. Palpant, D. Prot, E. Charron, S. Debrus, *Eur. Phys. J. D* 19, 395 (2002)
[76] H. S. Jun, K. S. Lee, S. H. Yoon, T. S. Lee, I.H. Kim, J. H. Jeong, B. Cheong, D. S. Kim, K. M. Cho, W.M. Kim, *Phys. Stat. Sol. (A)*, 203, 1211 (2006)
[77] M. S. Hu, H. L. Chen, C. H. Shen, L. S. Hong, B. R. Huang, K. H. Chen, L. C. Chen, *Nat. Mater.* 5, 102 (2006)
[78] W. Schrof, S. Rozouvan, E.V. Keuren, J. Schmitt, G. Decher, *Adv. Mater.* 3, 338 (2002)
[79] Jason Valentine *et al, NATURE LETTERS* 455, 376 (2008)
[80] Dongo Kim, Wangju Lee and Jaeick Choi, *Appl Phys A* 97, 461 (2009)
[81] J. F Galisteo *et. al*, *Journal of Optics A: Pure Appl. Opt.* 7, 244 (2005)
[82] Santhi Ani Joseph *et al, Journal of Crystal Growth* 311, 1152 (2009)

[83] A. N. Grigorenko *et. al, Appl. Phys. Lett.* 88, 124103 (2006)
[84] A. Perentes and I. Utke, *Nanotechnology* 16, 273 (2005)
[85] A. M. Chinie, A. Stefan and S. Georgescu, *Romanian Reports in Physics* 57, 412 (2005)
[86] Albert Polman and Harry A. Atwater, *Materials Today,* January 2005, p. 56

10

Structural Characterisation of Kraft Pulp Fibres and Their Nanofibrillated Materials for Biodegradable Composite Applications

Gary Chinga-Carrasco[1], Arttu Miettinen[2], Cris L. Luengo Hendriks[3], E. Kristofer Gamstedt[4] and Markku Kataja[2]

[1]*Paper and Fibre Research Institute (PFI),*
[2]*Department of Physics, University of Jyväskylä,*
[3]*Centre for Image Analysis, Swedish University of Agricultural Sciences,*
[4]*The Ångström Laboratory, Uppsala University,*
[1]*Norway*
[2]*Finland*
[3,4]*Sweden*

1. Introduction

The utilization of wood pulp fibres in composite materials has gained major interest during the last years. One of the major motivations has been the potential of wood pulp fibres and their nanofibrillated derivatives for increasing the mechanical properties of some materials. However, in order to exploit the full potential of wood pulp fibres and cellulose nanofibrils as reinforcement in hydrophilic and hydrophobic matrices, several characteristics of fibres and their interactions with a given matrix need to be understood.

With the increasing capabilities of novel microscopy techniques and computerized image analysis, structural analysis is moving forward from visual and subjective evaluations to automatic quantification. In addition, several microscopy techniques for obtaining 2D and 3D images of a given composite material, including field-emission scanning electron microscopy (FESEM) and X-ray micro-computed tomography (X-μCT), have evolved considerably during the last years. X-μCT is a non-destructive method for obtaining the three-dimensional structure of a physical material sample. It is well suited for structural analysis of complex heterogeneous materials such as paper, biological materials and fibrous composites (Samuelsen *et al.*, 2001; Holmstad *et al.*, 2005; Axelsson, 2008). In addition, FESEM is a powerful technique for assessment of a variety of materials. One of the major advantages of FESEM is its versatility and high-resolution power (Chinga-Carrasco *et al.*, 2011). Structures down to 1-2 nm can thus be visualized and quantified.

In this work we will focus on practical and complementary imaging and image analysis techniques. We will also give a brief introduction to SEM, X-μCT and to 3D image analysis methods, emphasizing topics that are relevant for characterisation of composite materials. Selected case studies of wood pulp fibre-reinforced composite materials and their corresponding microstructure-property relationships will be discussed.

2. Wood pulp fibres and microfibrillated cellulose

Wood pulp fibres (Fig. 1) are regaining interest within several industry sectors. Wood pulp fibres are a natural resource, renewable and biodegradable. This is a major advantage in a world moving towards environmental-friendly products, where major efforts are being made to develop sustainable materials. Considering sustainability and recyclability as major requirements, wood pulp fibres are ideal components for novel composite materials.

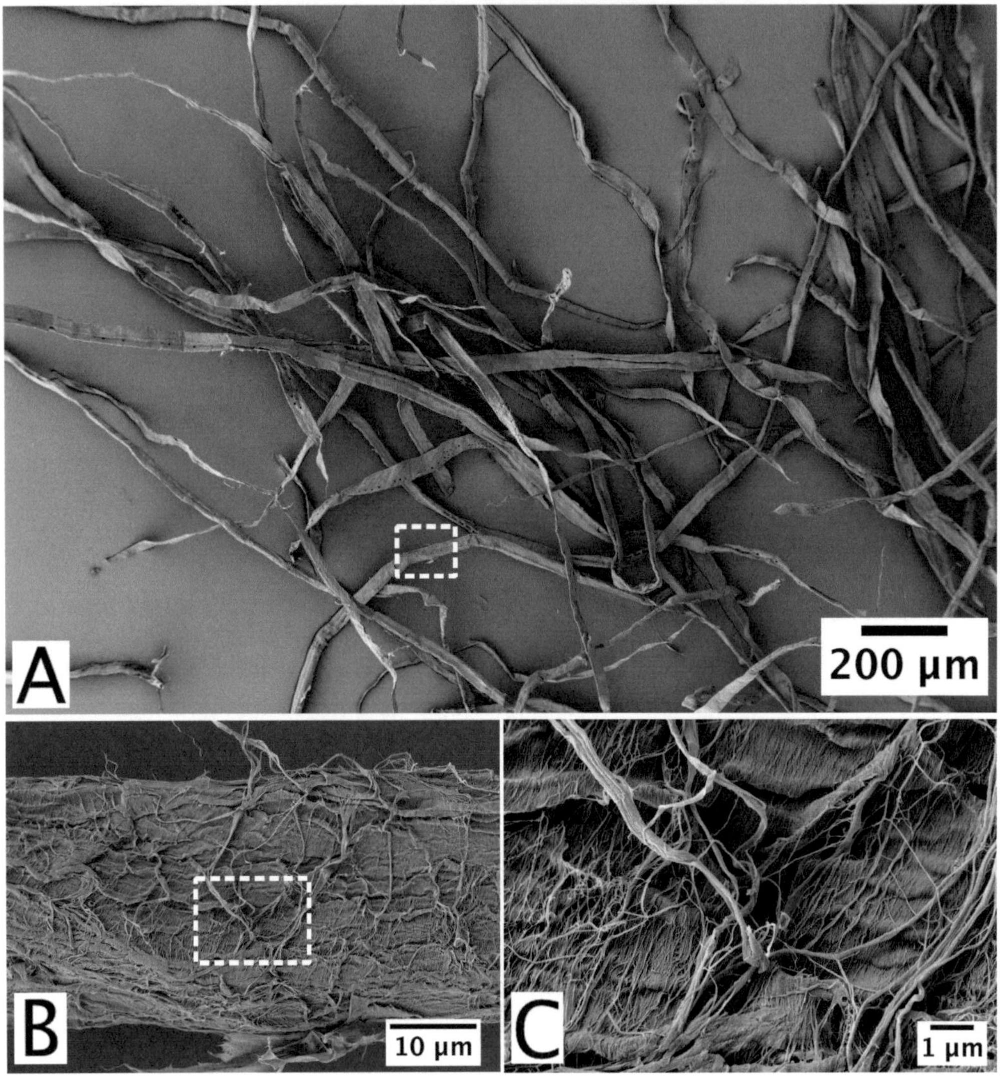

Fig. 1. FESEM of the structure of kraft pulp fibres. (A) Kraft fibres exemplifying their high aspect ratio. (B) A surface structure of a single fibre. (C) The microfibrils composing the surface structure. The dashed rectangles in (A) and (B) correspond to the images in (B) and (C), respectively.

Wood pulp fibres have a relatively high aspect ratio. Typical lengths are between 1 and 3 mm (Fig. 1A). Typical widths are between roughly 10 and 50 µm (Fig. 1B). The wall structure of cellulose fibres is mainly composed of microfibrils, with reported values of diameter in the nanometer-scale (Fig. 1C). The microfibrils are arranged differently in the various layers of a fibre wall structure (see *e.g.* Meier, 1962; Heyn, 1969). The wall of cellulose fibres is roughly composed of a primary wall and 3 secondary wall layers, *i.e.* S1, S2 and S3.
Cellulose fibres can also be disintegrated into their structural nano-components. This approach was introduced in the beginning of the eighties for commercial purposes (Turbak *et al.*, 1983; Herrick *et al.*, 1983). The novel material was denominated microfibrillated cellulose (MFC). The material has also been given a series of different denominations, including nanofibrillated cellulose, nanofibrils, nanofibres and nanocellulose (Abe *et al.*, 2007; Ahola *et al.*, 2008; Mörseburg & Chinga-Carrasco, 2009; Klemm *et al.*, 2010). MFC can be considered a nano-material, provided that the material is composed of a major fraction of individualized nanofibrils (Fig. 2). In this study, nanofibrils are considered the material produced through a homogenization process, having at least one dimension less than 100 nm.

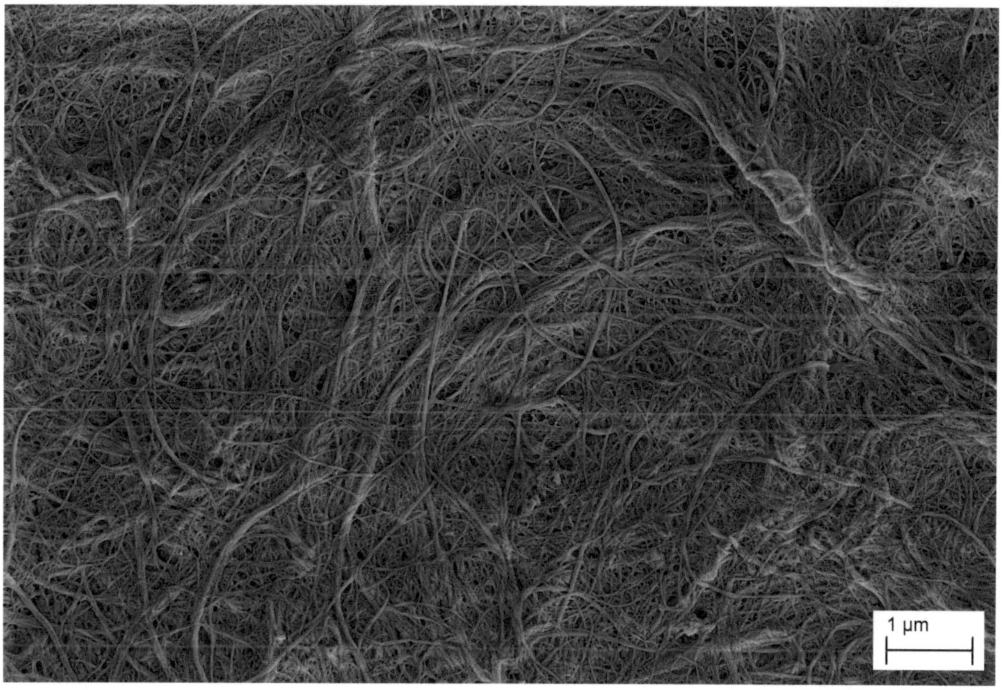

Fig. 2. FESEM image of microfibrillated cellulose, produced from kraft pulp fibres . The material is composed mainly of cellulose nanofibrils.

A series of new approaches have been developed for production of MFC (Saito *et al.*, 2006; Päkkö *et al.*, 2007; Wågberg *et al.*, 2008). The procedure for producing MFC may include mechanical, enzymatic and chemical pre-treatments. Each pre-treatment seems to produce a material with different morphology and different diameter size distribution. Depending on the applied amount of energy, homogenization without pre-treatment may produce a material containing nanofibrils, fibre fragments and poorly fibrillated fibres. On the other side, chemi-mechanical pre-treatments yield a narrow nanofibril diameter size distribution. This has been considered a confirmation of the positive effect that chemi-mechanical pre-treatments have on facilitating the fibrillation of cellulose fibres (see Syverud *et al.*, 2010).

3. Production of fibre-reinforced composites and cellulose nanofibrils

In this chapter, kraft pulp fibres (Fig. 1) will be applied as reinforcement in a fibre-polylactic acid (PLA) composite material and as a source for production of cellulose nanofibrils (Fig. 2).
A homogenization process was applied for producing cellulose nanofibrils (Fig. 2). The kraft pulp fibres were beaten prior to the homogenization. The homogenization was performed with a Rannie 15 type 12.56X homogenizer operated at 1000 bar pressure. The pulp consistency during homogenizing was 1%. The fibrillated material was collected after 3 passes through the homogenizer.
The following procedure was applied for production of fibre-PLA composites. The kraft pulp fibres were pelletized before manufacturing the composites. The applied equipment was a Kahl flat die pelletising press (Kahl 14-175, Reinbek, Germany). For details on the procedure see Nygård *et al.* (2008). Composites of PLA reinforced with 10%, 30% and 40% kraft fibre loadings were produced. The kraft pulp fibres and PLA were blended in a compounding unit (double screw) equipped with an injection moulding unit. Dogbone samples were made by injection moulding.

4. Structural characterisation

Proper structural characterization of cellulose fibres, nanofibrils and their corresponding composite materials requires an adequate utilization of specialized equipment for detailed assessments. In this respect it is most important to be aware of the advantages and limitations of modern microscopy techniques, and apply their complementary capabilities. In this work we emphasize the complementary capabilities of X-μCT for 3D characterization and SEM for complementary assessments at the micro and nano-scales. While X-μCT requires none or minor sample preparation, electron microscopy techniques may require adequate preparation for exposing a given structure. In the following sections, some of the techniques applied for fibre structural characterization will be described.

4.1 Scanning electron microscopy (SEM)

SEM has several modes of operation, from conventional secondary electron imaging (SEI) mode for studying fibre surfaces to specialized field-emission SEM (FESEM) for assessment of nano-structures. Image acquisition from fibre surfaces requires none or minor preparation. The fibre samples may be covered with a conductive metallic layer. Uncoated fibre samples may also be visualized with environmental or low-vacuum SEM. Well-prepared fibre samples reveal structures in the nanometre scale, such as the microfibrils observed in fibre wall structures (Fig. 1C).

4.1.1 Preparation for electron microscopy

One of the principal objectives with preparation techniques is to preserve a given structure in a particular state. Preparation is especially necessary for several electron microscopy techniques.

In the case of wood pulp fibres, dedicated preparation techniques have been developed. This includes *e.g.* freeze-drying, cryofixation and critical point drying (de Silveira *et al.*, 1995; Duchesne and Daniel, 2000). Freeze-drying is relatively simple to perform and has been applied in this study as a step in the preparation of the kraft pulp fibres.

Freeze-drying has facilitated the preparation of single fibres and bundles of fibres for surface structural analysis in SEI mode (Fig. 1; Chinga-Carrasco *et al.*, 2010). In addition, SEM in backscatter electron imaging (BEI) mode has been applied for cross-sectional analysis (Reme *et al.*, 2002; Chinga-Carrasco *et al.*, 2009). The SEM-BEI mode yields contrast based on the local average atomic number of a given structure. SEM-BEI mode requires distortion-free and smooth surfaces of the studied samples. A well-established method consists on i) embedding in epoxy resin, ii) grinding using abrasive papers and iii) polishing with a cloth using a fine diamond paste (Reme *et al.*, 2002). If modern equipment is available, blocks can be prepared quickly and effectively. This is a major advantage, as the cross-sectional structural characteristics of large fibre populations can be quantified (Fig. 3). The quantification of fibre cross-sectional characteristics is of importance in several applications such as; i) verification of fibre development due to different pulping processes, ii) evaluation of pre-treatments (*e.g.* enzymatically, chemically) on the fibre morphology, for homogenization purposes and iii) assessment of the relationship of fibre morphology and composite characteristics.

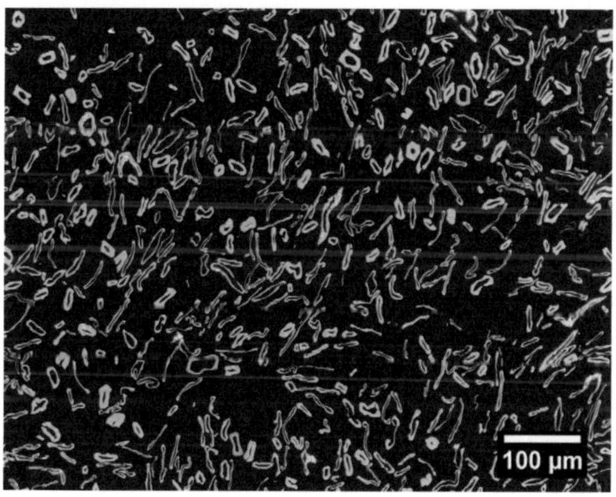

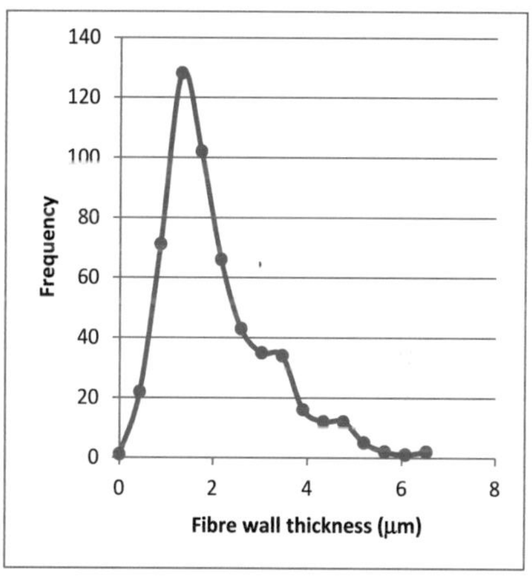

Fig. 3. SEM analysis of fibre cross-sections. (Left) Cross-sectional image acquired in backscattered electron imaging mode. (Right) The fibre wall thickness distribution of the kraft fibres applied in this study. See also Chinga-Carrasco *et al.* (2009).

4.2 X-ray micro-computed tomography (X-µCT)

Tomographic imaging facilities, based on synchrotron radiation and capable of resolution in the micrometre range, have been in use for more than a decade. During the past few years, the techniques have developed rapidly towards higher resolution. Furthermore, table-top tomographic scanners based on x-ray tube have become available. Depending on the techniques used, the resolution of the 3D tomographic images can vary from millimetres down to a few tens of nanometres (Fig. 4). In structural analysis of *e.g.* composite materials, resolution of the order of a micrometre is typically used. Such resolution is generally available also with current commercial tomographic table-top scanners.

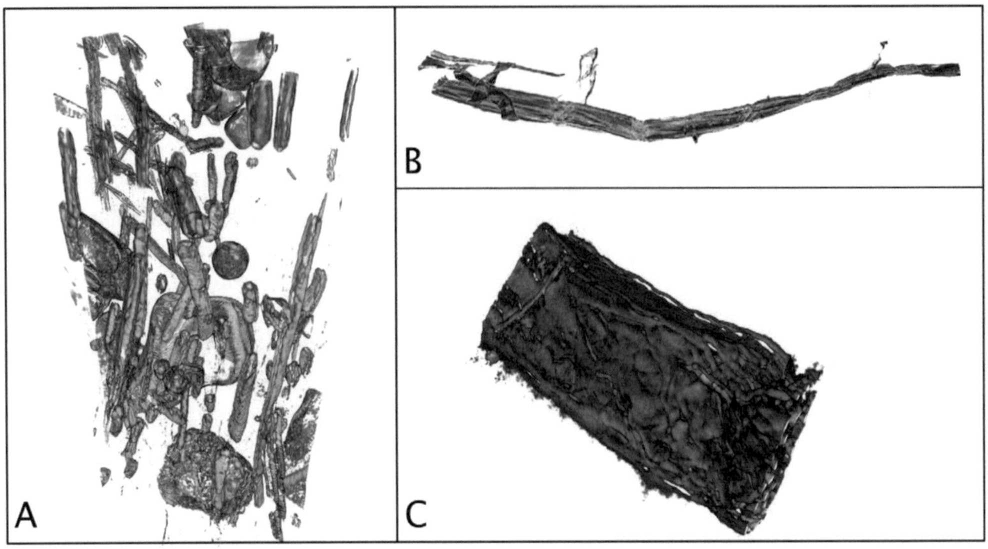

Fig. 4. X-μCT images acquired at different scales. (A) Wood fibre composite material, only fibres and some air bubbles are visible. The height of the sample is about 420 μm. Note that long fibres (Fig. 1A) are cut by the edges of the image. (B) Single wood fibre whose length is about one millimetre. (C) About 150 μm long part of a single wood fibre showing fibrils on the surface of the fibre. The images were acquired at ESRF ID19 synchrotron beamline (A), using XRadia microCT device (B) and using XRadia nanoCT device (C).

The basic components of a typical X-μCT apparatus are the x-ray source, object stage, scintillator plate and CCD camera. Some devices include additional optical systems for x-rays or for visible light. The entire device also contains a computer system for data acquisition and for reconstruction. Tomographic imaging is done by firstly acquiring a series of x-ray images of the object from several directions. Typically, of the order of 1000 individual x-ray images are taken while the object is rotated in small angular steps for 180 degrees around a vertical axis. These two-dimensional projection images are then processed computationally to reconstruct a three-dimensional digital image of the structure. Two different modalities of x-ray tomography are generally used, *i.e.* absorption mode and phase contrast mode. Within the absorption mode tomography, the numerical value associated with each 'voxel' in the reconstructed 3D image represents the local value of x-ray absorption coefficient of the material. For non-crystalline (non-diffractive) materials the absorption coefficient correlates with the density of the material. Consequently, an absorption mode tomographic image of such a material may be interpreted as an approximation of the density distribution within the sample. Interpretation of a phase contrast tomographic image is more complicated. In general, such an image emphasizes regions of high density gradient, *e.g.* intrinsic surfaces between various components in a composite material. In the following we will concentrate on absorption mode X-μCT.

An important advantage of the X-μCT method is that no particular sample preparation is needed. Consequently, material samples can be imaged under various conditions with respect to *e.g.* humidity, state of deformation, *etc.* The same physical sample may also be scanned several times in different conditions and studied with other, complementary methods. In addition to giving a graphical visual view on the structural characteristics of heterogeneous materials, the tomographic images can be utilized to obtain detailed quantitative information. Various segmentation and 3D image analysis methods have been developed for *e.g.* separating various components of the heterogeneous material and analysing them individually for structural properties such as relative volume, spatial distribution, domain size and shape distributions, specific area, connectivity, orientation, *etc.* Combining X-μCT with numerical simulation enables more advanced analysis of also dynamic characteristics such as transport and elastic properties of materials. A noteworthy further advantage of the method is that various analyses of quite different nature can be made based on the same basic data, *i.e.* the 3D digital image of the actual material structure.

4.3 Image processing and analysis

Depending on the particular device for image acquisition, digital images may contain varying amounts of noise, which must be removed before using the images for further analysis. In the case of X-μCT, several algorithms for such filtering have been developed (Jähne, 2002). One of the simplest adaptive filters is the variance weighted mean filter (Gonzalez & Woods, 2002), based on the assumption that the local variance of grey value is higher near edges of internal structures or domains than in the bulk far away from the edges. Another common noise removal method is the bilateral filter which is based on replacing the value of each pixel with a weighted average of values of the surrounding pixels (Tomasi & Manduchi, 1998). The weight function is a Gaussian function that depends on the Euclidean distance and on the grey value distance from the centre pixel. In addition, the non-linear SUSAN (Smallest Univalue Segment Assimilating Nucleus) filter (Smith and Brady, 1997), which is similar to the bilateral filter, has proven to be suitable for noise-removal in X-μCT images of wood fibre composites (Axelsson, 2009).

Successful noise removal facilitates many straightforward analyses on digital images. For example, the fibre content in a composite material can be estimated simply by segmenting the image into fibre pixels and background pixels based on a suitable threshold value, and by calculating the ratio of the number of fibre pixels to the number of all pixels. In the case of X-μCT analysis, this requires that the values of the x-ray absorption coefficient (density) of the fibres and the matrix are sufficiently distinct such that a successful thresholding becomes possible.

As mentioned above, various characteristics of fibre-reinforced composites can be quantified, including fibre spatial distribution, fibre orientation and volume fraction. In addition, the fibre and nanofibril lengths are of major importance as these characteristics affect some critical mechanical properties of a given composite material. Methods for quantification of fibre and nanofibril lengths are described in the following sections.

4.3.1 Quantification of fibre length distribution

The fibre length distribution of a composite material is a basic quantity that is often needed for microstructural modelling or for optimization of processing parameters in the manufacturing phase. Finding the fibre length distribution of a material from a three-dimensional image is a nontrivial problem. In this section we present a two-phase algorithm for such analysis.

Before proceeding further, let us consider the problem of measurement of fibre length from a finite volume such as an X-μCT image. The measurement will be affected by the edges of the volume, which cut long fibres into shorter segments (see Fig. 4A and Fig. 6). Additionally, the probability of a randomly positioned fibre being in the volume is related to the length of the fibre and to the size of the finite volume. Thus, the fibre length distribution measured from a finite volume and the true fibre length distribution in the sample material do not coincide. This effect can be corrected for by utilizing a linear measurement model of the fibre length distribution. This indicates the relation

$$m = \alpha n + \epsilon ,$$

where m is the fibre length distribution measured from the image, n is the actual fibre length distribution in the sample material, ϵ is a random error vector and α is a matrix that describes the measurement process. The correction is performed by solving the equation for n using, *e.g.* Tikhonov regularization. For details on the matrix α and the solution process, see Miettinen *et al.* (2011).

The method for measuring fibre length is based on a granulometry approach. The granulometry is a well-established tool to characterise texture by the size of its components (Matheron, 1975; Soille, 2003). The granulometry is, in essence, a multi-scale version of the mathematical morphology operations of opening or closing. It applies an opening (or closing) at many different scales, and summarises the result of each operation by summing the grey values of the output image. When applied to a properly prepared image, and with careful normalisation, it yields an estimate of the size distribution of the objects in an image (Luengo Hendriks, 2004). The size of an object can be defined by a suitably chosen opening (or closing) operation. Most commonly, isotropic openings are used. These use the object's width as its size. Another common possibility is the area opening, which uses the object's volume. We propose to use the path opening (Luengo Hendriks, 2010), which selects on the object's length. Applying granulometry with path openings on suitable images with fibres thus yields an estimate of the fibre length.

4.3.2 Quantification of fibril length

Proper homogenisation of kraft pulp fibres yields fibrillated materials, which are composed of a major fraction of cellulose nanofibrils (Fig. 2). The diameters of cellulose nanofibrils are in the nanometre scale (< 100 nm). The quantification of nanofibril diameters has been applied for evaluating the fibrillation of a given pulp and how this is affected by a given pre-treatment before homogenisation (see *e.g.* Chinga-Carrasco *et al.*, 2011). In addition to the nanofibril diameter as a structural characteristic, the quantification of the corresponding nanofibril length is most important. However, quantification of the nanofibril length is demanding. This is due to the morphology of the nanofibrils, which may have diameters and lengths in the nanometre and micrometre-scale, respectively. Visualization of nanofibrils requires high-resolution and large field of view. In order to fulfil these requirements a proposed approach is to acquire several adjacent high-resolution images with a FESEM and stitch them digitally (Fig. 5A). The image exemplified in Fig. 5 is composed of various adjacent and stitched images. The images were then filtered with a Fast Fourier Transform (FFT) bandpass filter to preserve the structures corresponding to the fibrillated materials.

SEM images may be difficult to analyse automatically. The contrast in SEM-SEI images is obtained, not through differences in surface properties as in optical imaging, but through orientation of the surface and shadowing effects. This means that the brightness of a pixel is not related to material properties, but to local geometry in the sample. In the acquired images, pixels corresponding to the nanofibrils have slight differences with respect to the intensity of the background. Human vision is particularly adept at interpreting this type of image, but it is deceptively tricky to teach a program to do the same. This is the so-called "shape from shading" problem, well studied in the computer vision literature (Zhang, 1999). However, applying such computer vision techniques to SEM images yields unreliable results (Czepkowshi, 1996).

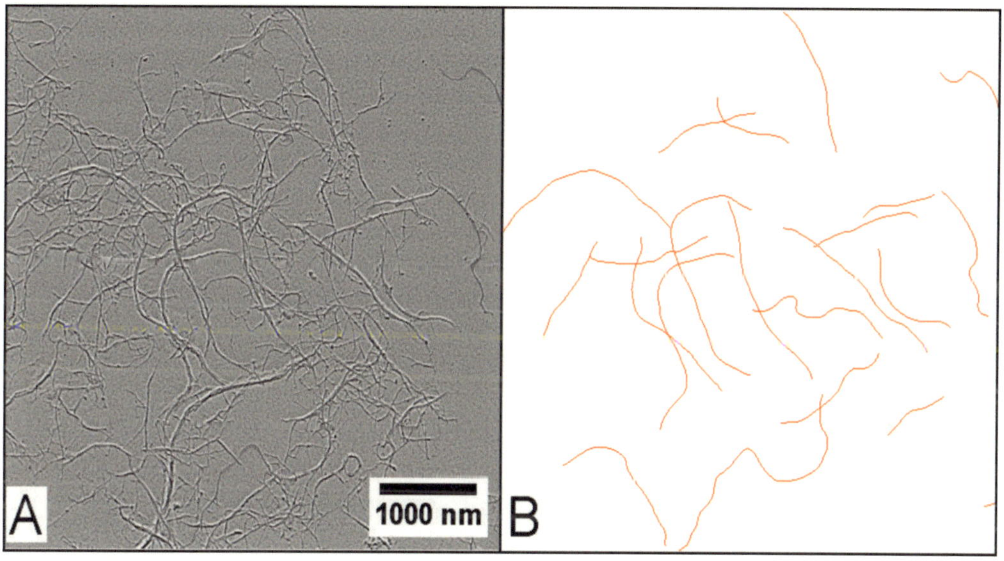

Fig. 5. Quantification of nanofibril lengths based on FESEM images. (A) Digital image acquired in SEI mode. The image has been enhanced for better visualization. (B) The corresponding processed image with paths drawn on some nanofibrils.

In this chapter we propose a semi-automated approach to measuring nanofibrils with high accuracy. Through a simple user interface, the analyst clicks on the two ends of a fibril. The software can then compute an optimal path between those two positions, and draw the path over the image. The analyst can then refine the path through additional clicks. This may be necessary only in specific situations, such as with strongly bent fibrils and fibrils that intersect at multiple locations. Once the fibril is found, an accurate length measurement can be calculated.

The following procedure was applied for detecting the nanofibrils exemplified in Fig. 5. First, we inverted the grey values in the image, such that the shadow regions were represented by high grey values. Next, we determined the optimal path between two points. A path is a sequence of neighbouring pixels, such that the first and last pixels in the path are the points selected by the analyst. We defined the optimal path as the path over which the integrated grey value were minimal. That is, if a path deviated from the fibril and went across a shadow region, it would yield a larger grey value integral than a path that nicely followed the fibril. Optimal path finding is a well-studied problem, and can be solved with a famous algorithm due to Dijkstra. In the image analysis world, the grey-weighted distance transform (GWDT) is the best-known version of the Dijkstra algorithm (Verbeek and Verwer, 1990). We applied the GWDT to compute the grey-weighted distances to one of the two selected points, and then, starting at the other point, found the path of steepest descent, which always takes one back to the first point. This path necessarily corresponds to the optimal path according to our criterion. When corrections were necessary, the analyst added control points. In this case, we found the optimal path between subsequent points, and appended these paths to obtain a trace of the whole fibril. This path was thus composed of a dense set of points, one pixel apart, and represented by the coordinates of these points (Fig. 5).

Measuring the length of the path can be accomplished by first smoothing the path. The sequence of x and y coordinates was filtered using a Gaussian low-pass filter with an appropriate sigma (σ) value. We found that $\sigma = 3$ yielded good results and this value was applied for smoothing the path. Next, the Euclidean distance between subsequent points was calculated and summed. Summing the Euclidean distance in this way overestimates the length if the path has not been smoothed (Vossepoel and Smeulders, 1982). Finally, the length in pixels can be converted to a physical length by multiplying with the pixel size.

In this section we have described a method that may be used for quantification of nanofibril length, although the approach is demanding. The approach will be improved, considering (i) adequate procedures for sample preparation of well-dispersed nanofibrils, (ii) alternative methods for image acquisition, *e.g.* transmission electron microscopy and (iii) alternative and improved methods for nanofibril tracking and quantification.

5. The effect of fibres and nanofibrils on the mechanical properties of composite materials

All of the above examples of fibre and fibril length are of great importance for the mechanical properties of the composites or networks composed of fibres or fibrils. Structural composites for load-carrying applications are based on long fibres acting as reinforcement, whereas high-volume composites with particle 'reinforcement' are preferred to reduce costs by using cheaper filler or to improve processability. For new cellulose-based materials with improved mechanical properties, it is therefore of interest to control the length of the reinforcement. Microscopic techniques to quantify fibre length distribution are therefore of practical interest.

Essentially it is the aspect ratio of a cylindrical fibre, *i.e.* its length to diameter ratio, which controls the average load in the fibre. This is perhaps best illustrated by an oriented fibre embedded in a softer matrix, which is subjected to a far-field stress in the direction of the fibre. Since load is transferred from the matrix to the fibre at the interface by shear stresses, there will be ineffective lengths at the fibre ends, where build-up in fibre tensile stress takes place. In parallel, processing methods should be developed to manufacture larger quantities at lower cost, while maintaining proper control of the microstructure to assure improved mechanical properties. This balance between processability and performance is an omnipresent challenge in development of materials of this kind. Hence, compromises between processability and performance have to be considered. A particular feature of wood pulp fibres is that they are easily dispersed in water and that they form strong bonds to each other upon drying, which is not the case for other typical composite fibres, such as glass or carbon fibres. This behaviour could potentially be utilized as a processing advantage, using wet forming techniques and water suspensions in mixing the composite constituents. Experience in the paper making community could be used to manufacture novel composites, in combination with processing techniques presently used to manufacture polymer matrix composites.

Wood-plastic composites (WPCs), based primarily on extruded wood flour filled polyolefins, have a large market as building components in North America. The aspect ratio of the wood reinforcement is typically low (below 5). The resulting mechanical properties are therefore limited, since the wood particles may act as stress raisers rather than load-carrying fibres. The aspect ratio is a critical microstructural parameter that controls stiffness and strength, which should be increased for WPCs (Fowler *et al.*, 2006). The pivotal role of fibre length is perhaps best illustrated by the concept of ineffective length introduced by Cox (1953). For an embedded fibre oriented along the direction of load, there are short distances close to the two fibre ends that do not carry full load, since stress is transferred by interfacial shear from the fibre ends. The effect of fibre length or aspect ratio on mechanical properties is large for small values, but then reaches an asymptotic value. Above a certain value of the aspect ratio, say ~ 20, further increase might not be worth the slight improvement in stiffness and strength (Nordin, 2004). The desired increase in fibre length, compared to WPCs, has prompted use of wood pulp fibre reinforcement, where the wood tracheids could have an aspect ratio of 60 or more (McHenry & Stachurski, 2003). Although the pulp fibres have sufficiently high aspect ratios, conventional melt processing of wood-fibre reinforced polymers usually degrades the fibre length to the extent that it adversely affects the mechanical properties (Almedar *et al.*, 2008). This motivates the development of the above-mentioned techniques to quantify fibre length for wood-fibre composites. Fibre length is known to affect several important engineering properties of composites. Several models, with experimental support, have been developed to link the fibre length to some key engineering properties:

- Stiffness, according to shear-lag model of Cox (1953)
- Tensile strength, according to the model Fukuda & Chou (1981), although fibre dispersion may overshadow the fibre-length effects since wood fibres tend to aggregate on drying
- Fracture toughness, according to the model of Piggott (1970)
- Dimensional stability on moisture uptake, according to the model by Almgren *et al.* (2010), where increased stiffness is shown to constrain swelling

All of these engineering properties are considered in materials selection and design, depending on the application at hand. It is not only the fibre length and its distribution that affects these properties. Obviously, the fibre orientation distribution and fibre content are of great importance (Fig. 6). These parameters are however relatively easier to characterize by image analysis. Efforts to develop methods for fibre length characterization are therefore most important.

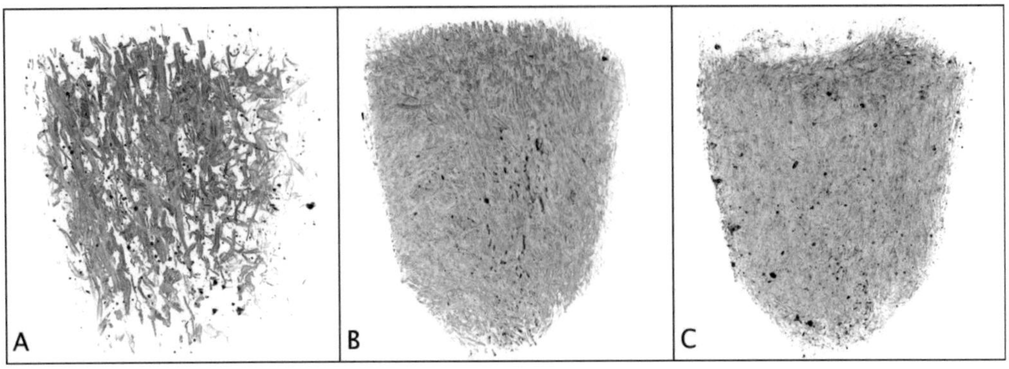

Fig. 6. X-μCT images of kraft fibre-PLA composites, consisting of 10 weight-% (A), 30 weight-% (B) and 40 weight-% of fibre (C). The images were acquired with XRadia microCT device.

In this chapter we have described a method based on granulometry for quantification of fibre length in X-μCT 3D images (see section 4.3.1). The method appears robust and yields reasonable results. The quantification of fibre length in the samples presented in Fig. 6 yields values of length below 200 μm. The fibre length distributions are relatively broad, ranging from small (<10 μm) to relatively long fibre fragments (100-200 μm). It is worth to mention that the kraft fibres had an average fibre length of 1690 μm before composite manufacturing (measured with a FiberMaster device). This clearly indicates that the process applied for manufacturing the composite material has shortened the fibre length, which reduces the corresponding mechanical strength (for details see Miettinen *et al.*, 2011).

As discussed above, the geometrical variability of natural wood fibres imposes challenges on the development of robust methods for quantification of fibre length in composite materials. Such quantification is demanding and requires tailor-made algorithms for a given application. There is thus scope for scientific research work in this domain, especially within non-destructive quantification of fibre morphology in composite materials.

An additional example of the application of X-μCT in the study of the properties of wood fibre reinforced composite materials is shown in Fig. 7. In this particular case, the entire test specimens (dogbones) of the material were first imaged using X-μCT to find the distribution of possible defects and irregularities. Mechanical tests were then performed to determine the stress-strain curve and tensile strength. After tensile testing, the fractured regions were imaged again using both X-μCT (Fig. 7) and SEM (Fig. 8). Comparison of the X-μCT images taken before and after the tensile test allowed identifying the critical structural features leading to fracture (Fig. 7). It appears that relatively large fibre agglomerates seem to have initiated the failure and crack propagation. The agglomerates, which are poorly dispersed fibres in the PLA matrix, were confirmed by SEM images (Fig. 8). Fractography based on SEM is a relatively common technique for failure analysis (see *e.g.* McCoy, 1994; Hein, 2001). The fracture area clearly showed the occurrence of local areas with higher concentration of fibres (Fig. 8C), compared to neighbouring regions (Fig. 8B). The case-study presented in this section exemplifies the complementary capabilities of two modern techniques for structural analyses, *i.e.* X-μCT and (FE)SEM.

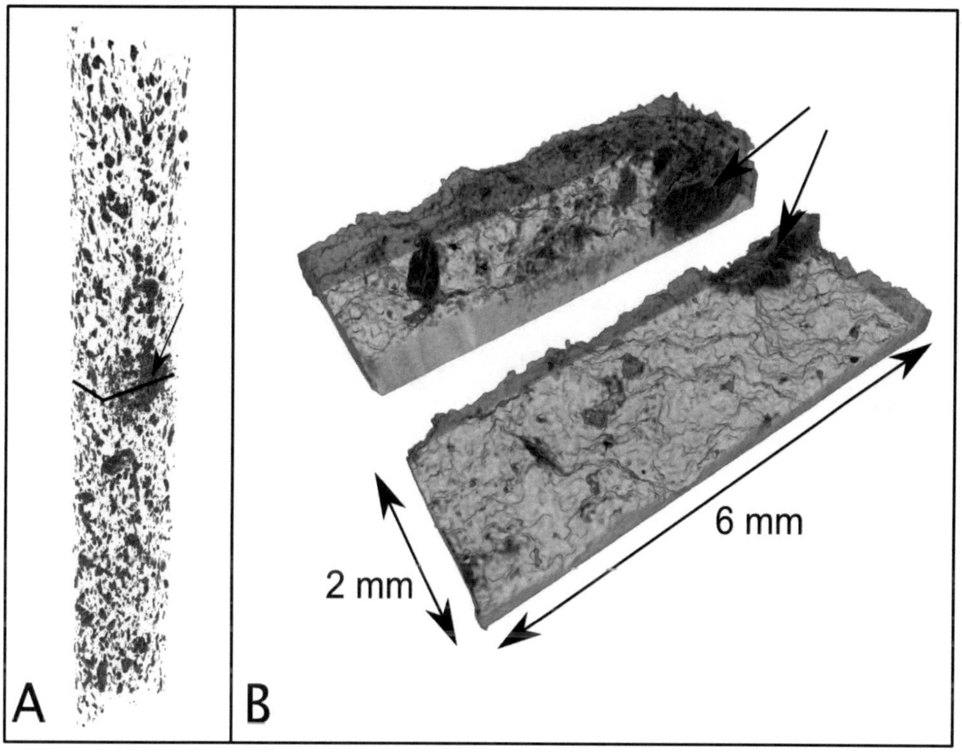

Fig. 7. X-μCT images of a kraft fibre-PLA composite sample (dogbone) of approximate dimensions 60 mm × 6 mm × 2 mm. The images have been acquired with a SkyScan 1172 tabletop tomography device. (A) Fibre bundles inside the intact specimen. (B) Fracture surfaces after tensile testing. The arrows in (A) and (B) indicate the critical defect (dark area) that caused the failure of the structure. The black line in A) indicate the approximate position of the fracture.

These structure-property relations essentially apply also to the corresponding nanofibrils, made from kraft pulp fibres. The main difference is the dimension and the specific surface area. The aspect ratio depends on the fibrillation procedure, although it is possible to produce sufficiently long nanofibrils that they can be considered continuous, *i.e.* of indefinite length from a load-carrying perspective (Henriksson *et al.*, 2007). The high surface area to volume ratio of nanofibrils, compared to wood fibres, results in an enhanced ability to form a tight network with strong interfibrillar bonds and significantly improved mechanical properties (Henriksson *et al.*, 2008). On the other hand, the tendency of nanofibrils to aggregate makes it more difficult to process and produce nanofibril-based materials in larger quantities.

Fig. 8. SEM-SEI imaging of a fracture area. (A) Fracture area after a tensile testing of a kraft fibre-reinforced PLA composite. Facture areas corresponding to the areas marked with white (B) and black (C) dashed rectangles in (A). Note the differences with respect to the fibre spatial distributions in the two fracture areas.

Concerning the reinforcement itself, the stiffness is considerably higher for nanofibrils than wood fibres, since the reinforcing cellulose is aligned in the direction of the nanofibrils. For wood fibres, the cellulose microfibrils have chiral orientation in the cell wall with an inclination angle given by the microfibril angle (MFA). The lumen and softer wood polymers like lignin and hemicellulose also contribute to the lower stiffness of the wood fibres. Cheng *et al.* (2009) have estimated the stiffness of single fibrils to 84 ± 23 GPa by three-point bending in an atomic force microscope, compared with tensile stiffness values 10-25 GPa of various wood pulp fibres (Leopold, 1966). Based on the same arguments, the strength of nanofibrils should be considerably higher than that of wood fibres, although the former is extremely difficult to test, due to the miniscule dimensions of cellulose nanofibrils. Furthermore, nanofibrils are relatively uniform and have fewer defects compared to the corresponding wood fibres. This property contributes to an increased strength, provided that the fibrillation process does not induce damage and degradation of the crystalline nanostructure.

Cellulose nanofibrils have thus a great potential as reinforcement, although the processing challenges should be overcome in order to upscale the production. The small dimensions of nanofibrils (diameter 3.5-100 nm) make them almost impossible to be distinguished in presently utilized X-μCT techniques. Hence, high-resolution electron microscopy must generally be used to estimate the dimensions of a given nanofibril material. However, with further development of tomographic and microscopic characterization techniques, the nanostructure-property relations could be addressed from a mechanistic viewpoint, similarly to the microstructure-property relations of wood-fibre composites.

The long-term goal is to conceive useful methods to quantify microstructural parameters of wood-fibre composites that affect relevant engineering properties. Additionally, models that link these microstructural parameters to the mechanical properties should be developed. Some methods can be borrowed from work on short-fibre composites based on glass or carbon fibres, although the special characteristics of wood fibres should be taken into account. Such characteristics include; (i) wood fibres are generally not straight and uniform, (ii) wood fibre properties depend on the moisture content, (iii) wood fibres may form strong inter-fibre hydrogen bonds and thus create networks, and (iv) the mechanical properties of wood fibres depend on their corresponding microfibril angle.

6. Conclusions

Given the importance of wood pulp fibres in several renewable applications it has been considered most important to develop and demonstrate adequate methods for characterising their complex morphology and assess how fibre structures are affected by specific processes. In this respect we have focused on practical and complementary imaging (X-μCT and FESEM) and the corresponding image analysis techniques necessary to quantify a given structure.

We have demonstrated that X-μCT is most appropriate for exploring non-destructively the surface and bulk structures of fibre-reinforced composites. As a case study, we described a novel method for measuring the fibre length distribution in a kraft fibre-PLA composite. The results revealed a significant reduction of the fibre length from >1500 μm to <200 μm, which could impose a limitation in the proper application of wood fibres as reinforcement in biodegradable composites. In addition, we have demonstrated how strength properties of a given composite may be reduced due to defects caused by relatively large fibre agglomerates. This new insight has been possible due to the non-destructive capabilities of X-μCT, which facilitates the visualization of a composite specimen before and after mechanical testing. The effect of other key microstructural parameters, such as fibre length, orientation and relative content on mechanical properties have also been discussed.

This chapter has also emphasized the applicability of various scanning electron microscopy (SEM) techniques. SEM analyses can be performed in backscatter electron imaging and secondary electron imaging modes for quantification of fibre cross-sectional dimensions and assessment of fracture areas, respectively. In addition, the resolution power of modern field-emission SEM (FESEM) expands and complements structural studies of wood fibre structures and their nanofibrillated materials. A potential method for quantification of cellulose nanofibril length has been described. Such methods are most important for quality control of nanofibrillated materials and for expanding our understanding of structure-property relationships of fibre/nanofibril-reinforced composites.

7. Acknowledgement

The WoodWisdom-Net WoodFibre3D project is thanked for funding part of this work.

8. References

Abe, K.; Iwamoto, S. & Yano, H. (2007). Obtaining cellulose nanofibers with a uniform width of 15 nm from wood. *Biomacromolecules*, Vol. 8, pp. 3276-3278.

Ahola, S.; Salmi, J.; Johansson, L.-S.; Laine, J. & Österberg, M. (2008). Model films from native cellulose nanofibrils. Preparation, swelling, and surface interactions. *Biomacromolecules*, Vol. 9, pp. 1273-1282.

Alemdar, A.; Zhang, H.; Sain, M.; Cescutti, G. & Müssig, J. (2008). Determination of Fiber Size Distributions of Injection Moulded Polypropylene/Natural Fibers Using X-ray Microtomography. *Advanced Engineering Materials*, Vol. 10, Nos. 1-2, pp. 126-130.

Almgren, K.M.; Gamstedt, E.K. & Varna, J. (2010). Contribution of Wood Fiber Hygroexpansion to Moisture Induced Thickness Swelling of Composite Plates. *Polymer Composites*, Vol. 31, No. 5, pp. 762-771.

Axelsson M. (2008). Estimating 3D fibre orientation in volume images. In Proc. Of 19th Int. Conf. Pattern Recognition, Tampa, Florida, USA.

Axelsson, M. (2009). Image Analysis for Volumetric Characterisation of Microstructure. PhD thesis, Swedish University of Agricultural Sciences, Uppsala, Sweden.

Cheng, Q.; Wang, S.; Harper, D. (2009). Effects of Process and Source on Elastic Modulus of Single Cellulose Fibrils Evaluated by AFM. *Composites Part A*, Vol. 40, No. 5, pp. 583-588.

Chinga-Carrasco, G.; Lenes, M.; Johnsen, P.O. & Hult, E.-L. (2009). Computer- assisted scanning electron microscopy of wood pulp fibres: dimensions and spatial distributions in a polypropylene composite. Micron, Vol. 40, No. 7, pp. 761-768.

Chinga-Carrasco, G.; Johnsen, P.O. & Øyaas, K. (2010). Structural quantification of wood fibres surfaces - morphological effects of pulping and enzymatic treatment. *Micron*, Vol. 41, No. 6, pp. 648-659.

Chinga-Carrasco, G.; Yu, Y. & Diserud, O. (2011). Quantitative electron microscopy of cellulose nanofibril structures from Eucalyptus and Pinus radiata kraft pulp fibres. *Microscopy and microanalysis*. In Press, Vol. 17, No. 4.

Cox, H.L. (1952). The Elasticity and Strength of Paper and Other Fibrous Materials, *British Journal of Applied Physics*, Vol. 3, No. 2, pp. 72-79.

Czepkowshi, T.; Sówko, W. (1996). Some limitations of surface profile reconstruction in scanning electron microscopy. *Scanning*, Vol. 18, No. 6, pp. 433-446.

de Silveira, G.; Forsberg, P. & Conners, T.E. (1995). Scanning electron microscopy: a tool for the analysis of wood pulp fibres and paper. *In Surface analysis of paper*. Ed. Terrance E. Conners, Sujit Banerjee, Boca Raton, CRC Press.

Duchesne, I. & Daniel, G. (2000). Changes in surface ultrastructure of Norway spruce fibres during kraft pulping - visualization by field emission-SEM. *Nordic Pulp Paper Resources Journal*, Vol. 15, No. 1, pp. 54-61.

Fowler, P.A.; Hughes, J.M. & Elias, R.M. (2006). Biocomposites: Technology, Environmental Credentials and Market Forces, *Journal of the Science of Food and Agriculture*, Vol. 86, No. 12, pp. 1781-1789.

Fukuda, H. & Chou, T.-W. (1981). A Probabilistic Theory for the Strength of Short Fibre Composites. *Journal of Materials Science*, Vol. 16, No. 4, pp. 1088-1096,

Gonzalez, R. C., Woods, R. E. (2002). *Digital Image Processing* (Second edition), Prentice Hall, ISBN 0-130-94650-8, USA.

Hein, L.R.O. (2001) Quantitative fractography by digital image processing: NIH Image macro tools for stereo pair analysis and 3-D reconstruction. *Journal Microscopy*, Vol. 204, pp. 17- 28.

Henriksson, M.; Henriksson, G.; Berglund, L.A. & Lindström, T. (2007). An Environmentally Friendly Method for Enzyme-Assisted Preparation of Microfibrillated Cellulose (MFC) Nanofibers. *European Polymer Journal*, Vol. 43, No. 8, pp. 3434-3441.

Henriksson, M.; Berglund, L.A.; Isaksson, P.; Lindström, T. & Nishino, T. (2008). Cellulose Nanopaper Structures of High Toughness. *Biomacromolecules*, Vol. 9, No. 6, pp. 1579-1585.

Herrick, F.W.; Casebier, R.L.; Hamilton, J.K. & Sandberg, K.R. (1983). Microfibrillated Cellulose: Morphology and accessibility. *Journal Applied Polymer Science, Applied Polymer Symposium*, Vol. 37, pp. 797-813.

Heyn, A.N. (1969). The elementary fibril and supermolecular structure of cellulose in soft wood fiber. *Journal Ultrastructure research*, Vol. 26, pp. 52-68.

Holmstad, R.; Gregersen, Ø.W.; Aaltosalmi, U.; Kataja, M.; Koponen, A.; Goel, A. & Ramaswamy, S. (2005). Comparison of 3D structural characteristics of high and low resolution X-ray microtomographic images of paper. *Nordic Pulp and Paper Research Journal*, Vol. 20, No. 3, pp. 283-288.

Jähne, B. (2002). *In Digital Image Processing*. Springer, Berlin.

Klemm, D.; Kramer, F.; Moritz, S.; Lindström, T.; Ankerfors, M.; Gray, D. & Dorris, A. (2010). Nanocellulose: A new Family of Nature - Based Materials. *Green nanomaterials*, DOI: 10.1002/anie.200.

Leopold, B. (1966). Effect of Pulp Processing on Individual Fibre Strength, *TAPPI*, Vol. 49, No. 7, pp. 315-318.

Luengo Hendriks, C.L. (2004). *Structure Characterization Using Mathematical Morphology*. PhD Thesis, Delft University of Technology, The Netherlands.

Luengo Hendriks, C.L. (2010). Constrained and Dimensionality-Independent Path Openings. *IEEE Transactions on Image Processing*, Vol. 19, No. 6, pp. 1587-1595.

Matheron, G. (1975). *Random Sets and Integral Geometry*. Wiley, New York.

McCoy, R.A. (1994). SEM Fractography and Failure Analysis of Nonmetallic Material. *Journal of failure analysis and prevention*, Vol. 4, No. 6, pp. 58-64.

McHenry, E. & Stachurski, Z.H. (2003). Composite Materials Based on Wood and Nylon Fibre. *Composites Part A.*, Vol. 34, No. 2, pp. 171-181.

Meier, H. (1962). Chemical and morphological aspects of the fine structure of wood. *Pure applied chemistry*, Vol. 5, pp. 37-52.

Miettinen, A.; Luengo Hendriks, C.; Chinga-Carrasco, G.; Gamstedt, E.K. & Kataja, M. (2011). A non-destructive X-ray microtomography approach for measuring fibre length in short-fibre composites. Submitted for publication.

Mörseburg, K. & Chinga-Carrasco G. (2009). Assessing the combined benefits of clay and nanofibrillated cellulose in layered TMP-based sheets. Cellulose, Vol. 16, No. 5, pp. 795-806.

Nordin, L.-O. (2004). *Wood Fiber Composites: From Processing and Structure to Mechanical Performance*, PhD thesis, Luleå University of Technology, ISSN 1402-1544

Nygård, P.; Tanem, B.S.; Karlsen, T.; Brachet, P. & Leinsvang, B. (2008). Extrusion-based wood fibre-PP composites: Wood powder and pelletized wood fibres - a comparative study. *Composites Science and Technology*, Vol. 68, No. 15-16, pp. 3418-3424.

Piggott, M.R. (1970). Theoretical Estimation of Fracture Toughness of Fibrous Composites. *Journal of Materials Science*, Vol. 5, No. 8, pp. 669-675.

Pääkkö, M.; Ankefors, M.; Kosonen, H.; Nykänen, A.; Ahola, S.; Österberg, M.; Ruokolainen, J.; Laine, J.; Larsson, P.T.; Ikkala, O. & Lindström, T. (2007). Enzymatic hydrolysis combined with mechanical shearing and high-pressure homogenization for nanoscale cellulose fibrils and strong gels. *Biomacromolecules,* Vol. 8, 1934-1941.

Reme, P.A.; Johnsen, P.O. & Helle, T. (2002). Assessment of fibre transverse dimensions using SEM and image analysis. *Journal of Pulp and Paper Science,* Vol. 28, No. 4, pp. 122-128.

Saito, T.; Nishiyama, Y.; Putaux, J.L.; Vignon, M. & Isogai, A. (2006). Homogeneous Suspensions of Individualized Microfibrils from TEMPO-Catalyzed Oxidation of Native Cellulose. *Biomacromolecules,* Vol. 7, No. 6, pp. 1687-1691.

Samuelsen, E.J.; Helle, T.; Houen, P.-J.; Gregersen, Ø.W. & Raven, C. (2001). Three dimensional imaging of paper by use of synchrotron X-ray microtomography. *Journal of Pulp and Paper Science,* Vol. 27, No. 2, pp. 50-53.

Smith, S.M. & Brady, J.M. (1997). SUSAN - a new approach to low level image processing. *International Journal of Computer Vision,* Vol. 23, No. 1, pp. 45-78.

Soille, P. (2003). *Morphological Image Analysis*, 2nd ed. Springer, Berlin.

Syverud, K.; Chinga-Carrasco, G.; Toledo, J. & Toledo, P. (2010). A comparative study of *Eucalyptus* and *Pinus radiata* pulp fibres as raw materials for production of cellulose nanofibrils. *Carbohydrate Polymers*, Vol. 84, No. 3, pp. 1033-1038.

Tomasi, C. & Manduchi, R. (1998). Bilateral filtering for gray and color images. *Proceedings of the sixth IEEE International Conference on Computer Vision*, ISBN 81-7319-221-9, India, January 1998.

Turbak, A.F.; Snyder, F.W. & Sandberg, K.R. (1983). Microfibrillated cellulose, a new cellulose product: properties, uses, and commercial potential. *Journal of Applied Polymer Science, Applied Polymer Symposium*, Vol. 37, pp. 815–827.

Verbeek, P.W.; Verwer, B.J.H. (1990). Shading from shape, the eikonal equation solved by grey-weighted distance transform. *Pattern Recognition Letters* Vol. 11, pp. 681-690.

Vossepoel, A.M.; Smeulders, A.W.M. (1982). Vector code probability and metrication error in the representation of straight lines of finite length. *Computer Graphics and Image Processing,* Vol. 20, No. 4, pp. 347-364.

Wågberg, L.; Decher, G.; Norgren, M.; Lindström, T.; Ankerfors, M. & Axnas, K. (2008). The build-up of polyelectrolyte multilayers of microfibrillated cellulose (MFC) and cationic polyelectrolytes. *Langmuir,* Vol. 24, pp. 784-795.

Zhang, R.; Tsai, P.-S.; Cryer, J.E.; Shah, M. (1999). Shape from Shading: A Survey. *IEEE Transactions on Image Processing,* Vol. 21, No. 8, pp. 690-706.

11

Properties of Nanofillers in Polymer

Damien M. Marquis, Éric Guillaume* and Carine Chivas-Joly
Laboratoire national de métrologie et d'essais (LNE)
France

1. Introduction

Polymer nanocomposites have existed for decades, as carbon black, pyrogenic silica and diatomite were used as additives in polymers. Nevertheless, their characterizations and the effect of properties induced by the nanometric scale of fillers was not fully understood at these times. The real starting point, corresponding to an understanding of the action of these fillers, is generally considered as corresponding to the first papers on a polyamide-6 filled with nanoclays published by Usuki et al. (1993) and Okada A. (1995), from Toyota R&D. Both these papers called it "*Hybrid*" material. Rapidly, research increased, and the first use of the term "*nanocomposites*" appeared in 1994 [Lan & Pinnavaia (1994), Lan et al. (1995), Giannelis (1996)]. After these pioneers, a lot of researches started on various fillers. The demand for continual improvement in the performances of thermoplastic and thermoset polymer materials has led to the emergence of these new technologies. Nanofillers lists increased within years (nanoclays, nano-oxides, carbon nanotubes, POSS , etc.), as well as the matrix in which they are used and interactions with traditional fillers. Nowadays, the development of polymer nanocomposites is one of the most active area of development of nanomaterials. The properties imparted by the nanoparticles are various and focus particularly on strengthening the electrical conduction and barrier properties to temperature, gases and liquids as well as the possible improvement of fire behaviour. As a method which consists of reinforcing polymer chains at the molecular scale in the same way than the fibres at the macroscopic scale, nanocomposites [Biron (2004), Gloaguen & Lefevre (2007)] represent the new generation of two-phased materials, associating a basic matrix to nanofillers inserted between polymer chains. Nanofillers can significantly improve or adjust the different properties of the materials into which they are incorporated, such as optical, electrical, mechanical, thermal properties or fire-retardant properties, sometimes in synergy with conventional fillers. The properties of composite materials can be significantly impacted by the mixture ratio between the organic matrix and the nanofillers.

2. Classification of nanocomposites

2.1 Terminology

Terminology issues were solved recently with standardization. A standardization committee, ISO TC229 "*Nanotechnologies*", started in 2005, and a joint working group with IEC 113 "Nanotechnology Standardization for Electrical and Electronic Products and Systems" considered terminology and nomenclature issues. This group produced a set of documents

*Corresponding author

limiting confusion in definition of nano-objects [ISO/TS27687 (2008), ISO/TS11360 (2010), ISO/TS88004 (2011)].

Nanocomposites are considered as one family of nanomaterials, where a nano-object is

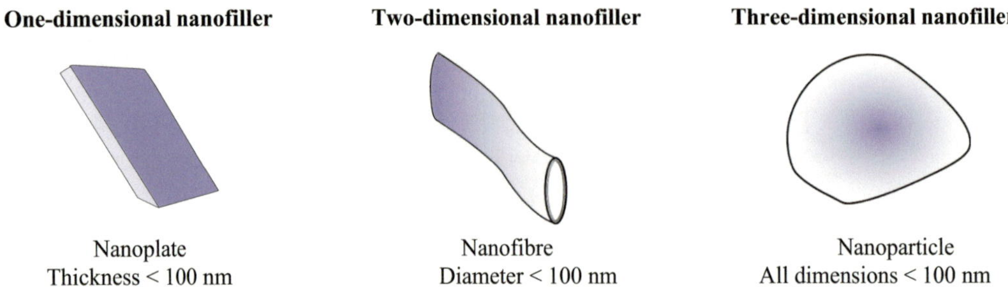

Fig. 1. Nano-objects used for nanocomposites, as defined in ISO/TS27687 (2008)

dispersed into a matrix or a phase. The other families are mainly nanostructured materials in surface, multi-layer or volume. A nanocomposite is a multiphase solid material where one of the phases has one, two or three dimensions of less than 100 nm [Ajayan et al. (2003)]. The nanocomposites can be distributed according to the nanofillers, as classified in reference [ISO/TS27687 (2008)] and highlighted in Fig. 1:

- One-dimensional nanofiller: in the form of plates, laminas and/or shells;
- Two-dimensional nanofiller: nanotubes and nanofibres which diameter is lower than 0.1 μm;
- Three-dimensional nanofiller: isodimensional nanoparticles such as nanometric silica beads.

Nanofillers are introduced in polymer at rates from 1% to 10% (in mass). They are incorporated in addition to traditional fillers and additives, and eventually traditional reinforcement fibres such as glass, carbon or aramide fibres. The diverse nanofillers used are described in the following paragraphs.

2.2 Nanoclays

Nanoplate fillers can be natural or synthetic clays, as well as phosphates of transition metals. The most widely used reinforcement is clay due to its natural abundance and its very high form factor. Clay-based nanocomposites generate an overall improvement in physical performances. The most widely used ones are the phyllosilicates (smectites). They have a shell-shaped crystalline structure with nanometric thickness. Clays are classified according to their crystalline structures and also to the quantity and position of the ions within the elementary mesh. The elementary or primitive mesh is the simplest atomic geometric pattern, which is enough for duplicating the crystalline network, by repeating itself indefinitely in the three directions. Table 1 presents the various natural and synthetic nanoclays available and used as fillers in polymers. The most common usage concerns organomodified Montmorillonite (MMT), a natural phyllosilicate extracted from Bentonite. Raw formula of Montmorillonite is $(Na,Ca)_{0,3}(Al,Mg)_2Si_4O_{10}(OH)_2,nH_2O$.

2.3 Nano-oxides

Titanium dioxide: Rutile is commonly used in polymers as a white pigment, as particles from 200 to 300 nm. The nanoparticle of titanium dioxide is not Rutile, but Anastase, and both are

Family		Groupe	Formula
Phyllosilicates	TO(1:1)	Kaolinite	The reference plate is formed from a tetrahedral plate T and a octahedral plate O. The thickness of the layer is about 0,7 *nm*. Kaolinite $Al_4\ Si_4\ O_{10}.(OH)_8$
	TOT(2:1)	Smectite,(Talc, Mica, Mommorillonite), Sepiolite	Two tetrahedral plates T in both sides of an octahedral plate O form the reference plate. The thickness of the layer is about 1 nm. The group includes many minerals that are major constituents of clays.
	TOT:O(2:1:1)	Chlorite, Bentonite, Saponite	The reference plate is formed of three plates TOT and another isolated O plate. The thickness of the layer is about 1,4 *nm*. Chlorite di-tri $Al_2Mg_3Si_4O_{10}(OH)_8$
Polysilicate	Natural	Kenyaite, Magadiite, Kanemite, Ilerite, Silhydrite, Zeolite.	Magadiite ($Na_2Si_{14}O_{29}H_2O$)
	Synthetic	FluoroHectorite, Zeolite	
Double lamellar hydroxide	Synthetic	Hydrotalcite	Hydrotalcites: $(Mg_6Al_2(OH)_{16})(CO_3^{2-})4H_2O$

Table 1. Nanoclays identification.

tetragonal crystal forms of titanium dioxide. Anastase is commonly used as spherical particles with diameter around 20 *nm*. This nanoparticle has photocatalytic properties. Anatase can be converted by hydrothermal synthesis into titanium nanotubes [Mogilevsky et al. (2008)]. These nanotubes have an outer diameter of 10 to 20 *nm*, an inner diameter of 5 to 8 *nm* and a length of 1 μm. *Nanoalumina*: Alumina particles are used as fillers in a wide range of size, from 20 *nm* to micrometric sizes. They are made of spherical crystal particles of Al_2O_3. Nanoparticles of alumina are frequently used as inert fillers in polymers, but highlight catalytic properties in some conditions. *Nano-antimony-Tin oxide* (ATO): The raw formula of this nanoparticle is Sb_2O_5/SnO_2. This is a tetragonal crystal particle of a diameter of around 15 *nm*, commonly used as flame retardant. *Nanosilica*: Nanosilica correspond to a large family of nanoparticles from various origins. The most commonly used is a natural one, called diatomite. This filler comes from the skeleton of a unicellular alguea (the diatomea) forming sedimentary layers. It is constituted from ultrafine particles of 750 *nm*. Two families of widely used synthetic nanosilicas are pyrogenic silica, forming particles from 5 to 100 *nm*, and silica fume, forming particles of about 100 *nm*. Precipited silica historically used in polymers is not a nanoparticle, as its diameter is between 1 and 10 μm when micronized.

2.4 Carbon nanotubes

Carbon nanotubes (CNT) were discovered by [Oberlin et al. (1976), Endo et al. (1976)], without application, and then rediscovered by [Iijima (1991)]. The fibre could present a nanometric diameter and length of some orders of magnitude in comparison with its diameter. In general, three kinds of carbon nanotubes are considered (Fig. 2):

- Single-wall carbon nanotubes (SWCNT). They present a diameter between 1 and 2 *nm*;

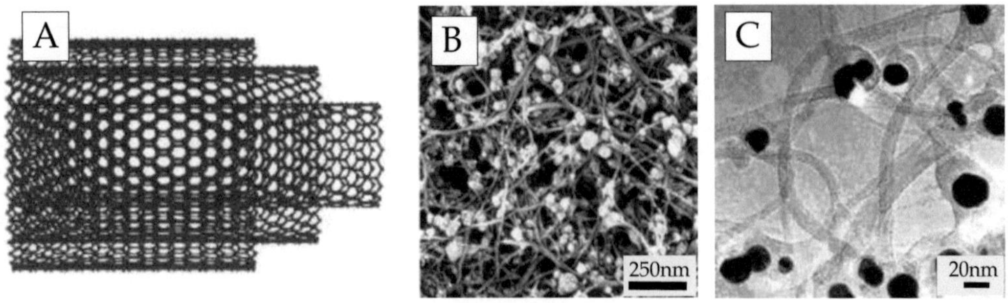

Fig. 2. Carbon nanotubes. a) Multi-Wall carbon nanotube (MWCNT). Idealized view; b) SEM image of nanotubes obtained by arc with 30 to 40 wt% catalytic residues (width of picture corresponds to 2 μm); c) TEM image showing the catalytic residues within nanotubes with 25 wt% catalyst residue.

- Double-wall carbon nanotubes (DWCNT). Diameter is between 2 and 4 nm;
- Multi-wall carbon nanotubes (MWCNT). They present a diameter between 4 and 150 nm.

These nanotubes present a theoretical range of properties incredible (Young modulus up to 1 TPa, heat conductivity of 3000 $W.m^{-1}.K^{-1}$, electric conductivity of 107 $S.m^{-1}$, etc.), but considering perfect nanotubes individually makes no sense. They nevertheless provide a wide range of new properties when used in nanocomposites, depending on their purity and dispersion in the matrix.
Carbon nanotubes are produced by two possible ways: a catalytic chemical vapour decomposition process at medium temperatures (600-1000$^\circ C$) and an electric discharge (arc) process under helium at high temperature (3000 to 4000$^\circ C$). Both processes produce a mix between SWCNT, DWCNT and MWCNT, with surface defaults (e.g. some pentagonal cycles in place of aromatic rings), and present important catalytic residues. Another important parameter is the chirality of the CNT (i.e. the direction and deviation in the "roll" process of CNT), only when they are studied individually, as the processes generate a mix of all possible chiralities.

2.5 Other nanofillers

Metallic nanoparticles: Most common metallic nanoparticles are nanosilver, nanozinc and nanogold fillers. These particles have a catalytic behaviour, which leads to antibacterial properties at surface. Their electrical and magnetic properties are also used for nanocomposites. *Silsesquioxanes*: Silsesquioxanes are big synthetic cage macromolecules used as nanofillers in polymers. Their raw formula has been found to be $R - SiO_{3/2}$ with R typically being alkyl or organo-functional groups. The high three-dimensional symmetry and nanometer size make silsesquioxanes building blocks for nanocomposites. The diversity of possible functional groups along with their controlled orientation in $3 - D$ space allows for highly tailored nanometer-by-nanometer construction in all three dimensions. An example of a silsesquioxane structure is given in Fig. 3.a). The most common Silsesquioxanes found are polyhedral oligosilsesquioxanes (POSS), octasilsesquioxanes of the $[ROSiO_{1.5}]_8$ type and its polymeric derivatives, and $[R - phenylSiO_{1.5}]_8$ type and its polymeric derivatives. They are used to make structured $3 - D$ nanocomposites for thin films, monoliths, and fibre reinforced composites. At present time, their expensive price limits their usage in polymers. *Boehmite*: Boehmite is a mineral aluminium hydroxide orthorhombic dipyramidal cristal used as a

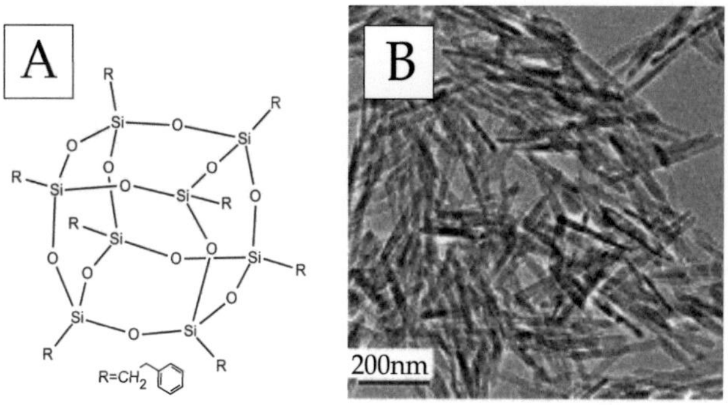

Fig. 3. a) Silsesquioxanes ; b) Boehmites.

nanofiller in nanocomposites. It could be obtained naturally or from chemical synthesis from saturated solutions. Its raw formula is $\gamma - AlO(OH)$. A picture of boehmite nanorods is given in Fig. 3.b.

3. Production of nanocomposites

3.1 Functionalization and dispersion of the nanofiller

As stated previously, two of the governing parameters for the properties of nanocomposites are the dispersion of the filler and its interaction with the matrix. Both are influenced by the usage of a chemical treatment at the surface of the filler. This surface treatment is possible for every kind of nanofillers. The use of nanofillers in thermoplastics and thermoset resins [Suh et al. (2000)] requires pre-treatment to improve the interfacial interactions between the filler and the matrix. The introduction of appropriate organic links between the matrix and the filler by functionalization improves the intercalation of polymer chains between the layers.

3.1.1 Nanoclays

MMT is extracted from the bentonite by a process of sedimentation in water [Kornmann et al. (1998)], and then activated by sodium ions (Na^+) to improve clays swelling. It is then possible to insert organic cations, such as silanes [Fu & Qutubuddin (2004)], ammonium bromides [Kornmann et al. (1998)] or ammonium chlorides such as vinylbenzyloctadecyldimethyl ammonium chloride (VOAC) or vinylbenzyldodecyldimethyl ammonium chloride (VDAC) [Chen et al. (2007a)]. These ions reduce the surface energy of clay, making it more organophilic, thus facilitating access between the layers of monomers and polymers with different polarities. Nowadays, pre-modified MMT are often used and commercially available (Closite®, Garamite®, etc). The incorporation of montmorillonite-layered silicates in a thermoplastic melt needs the exchangeable cations or alkaline earth from the clay to be replaced by organophilic cations. The organophilic cations used are commonly quaternary ammoniums. Fig. 4.a presents commonly used cation for polyethylene and propylene nanocomposites. Fig. 4.b presents commonly used cation for polyamide nanocomposites. The apophyllite (Phyllosilicate of formula $KFCa_4$ $[Si_8O_{20}]$, $8H_2O$) can be functionalized by syllilation reaction [Dean et al. (2007)]. The apophyllite functionalized ($n-$Octylsiloxy apophyllite) can be introduced into polyester or vinylester resins. Dean et al. (2007) studied the influence of the implementation of nanoclay composite/epoxy on the improvement of composite properties, using various dispersal methods (mechanical or ultrasonic methods) and various concentrations of nanoclay (mass fraction 0, 1, 2.5 and 5%). The author

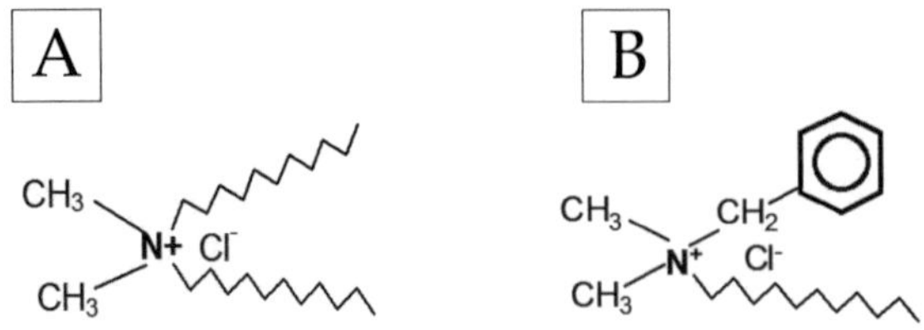

Fig. 4. Example of quaternary ammonium used for exfoliation in thermoplastics nanocomposite production. a) Nanowire 5; b) Nanowire 9

highlighted a strong influence of this parameter. The objective is to obtain an interaction between the nanofiller and the matrix at the nanometric scale. The intercalated structure is the common objective when dealing with nanoclays, as shown in Fig. 5 for example. Three

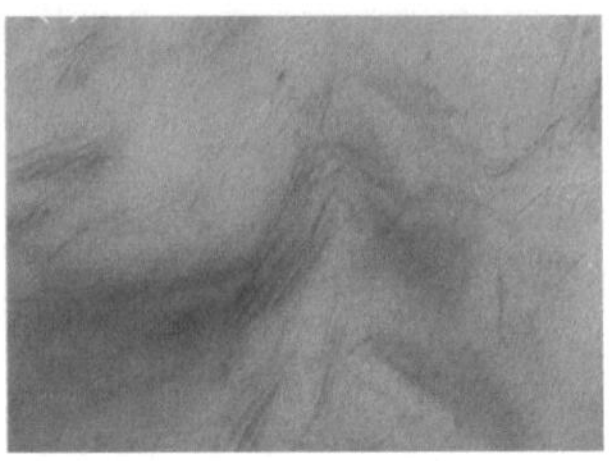

Fig. 5. Dispersion of modified nanoclay into poly ε-caprolactone (PCL), from Labidi et al. (2010). Width of the picture is 250 nm

cases are possible, which affect considerably the properties of the material:

1. The nanofillers are not mixed with the matrix, but form regions. The material is then a microcomposite.
2. The polymer matrix intercalates into the filler, between plates.
3. The fillers are totally exfoliated and dispersed

3.1.2 Nano-oxides

The use of metal oxides as nanofillers requires surface functionalization of these nanoparticles. Guo et al. (2007) and Guo et al. (2008) have reported the functionalization of Fe_2O_3 and other oxides by methacryloxypropyl trimethoxisilane di-functional in tetrahydrofurane. The objective with nano-oxides is to obtain a maximized exfoliation and full dispersion into the matrix.

3.1.3 Carbon nanotubes

The objective of carbon nanotubes is to maximize the dispersion into the matrix. The dispersion of CNTs in a thermoset matrix type can be improved without modifying the surface of CNTs and thanks to high shear rates [Song & Youn (2005)]. The dispersion can also be improved by functionalizing the surface of CNTs. These modifications improve the interfacial interactions with the matrix. Three ways are possible:

3.1.3.1 Chemical functionalization (covalent bounds):

Various techniques exist for the chemical functionalization of carbon nanotubes:

- Functionalization by exposure to ultraviolet rays and ozone [Sham & Kim (2006), Li et al. (2008)] to introduce carboxylic acid groups on the surface. This UV/O_3 treatment improves the dispersion and interfacial adhesion of CNTs to the epoxy matrix.
- Functionalization with amines [Fiedler et al. (2006), Shen et al. (2007), Chen et al. (2008)]: The amine can form covalent bonds with the epoxy matrix. The first reaction requires the formation of carboxylic acid binding on the nanotube. This carboxylation can be carried out by adding an acid (H_2SO_4/HNO_3) [Shen et al. (2007)] or amino dendrimers (APNM-O) [Sun et al. (46)].
- Functionalization with silanes [Ma et al. (2007)]: CNTs are first oxidized by exposure to UV in the presence of ozone, then reduced in a hybrid aluminum-lithium solution followed by a silanization. This functionalization with silanes improves the dispersion of CNTs in matrix just like epoxy improves thermal and mechanical properties but decreases thermal conductivity.
- Functionalization of bridges by adding poly(acryloyl chloride) (PACl) [Zou et al. (2008)]: PACl functionalized CNT react with epoxy monomers. CNTs thus grafted epoxy (epoxy-CNT) are mixed with an epoxy resin. This technique improves the thermal and mechanical properties while working at very low rates of CNT (0.1-1.0 wt%).
- Functionalization by an epoxy group [Chen et al. (2006)]. Multiwall CNT (MWCNT) are functionalized monomer diglycidylether of bisphenol A (DGEBA). The monomer is then added to the functionalized MWCNT and the addition of a tertiary amine results in the anionic polymerization of this monomer and the creation of covalent bonds between the resin matrix and the MWCNT-epoxy.

3.1.3.2 Physical functionalization:

The dispersion of CNTs in a thermoset matrix can be improved by non-covalent physical treatments, which have the advantage of not damaging the structure of the CNT, or introduce defects.

- Functionalization by adding a surfactant [Fiedler et al. (2006), Fan et al. (2004), Geng et al. (2008)]: physical adsorption of surfactant on the surface of CNTs decreases its surface tension, which prevents the formation of aggregates. Polyoxyethylene octyl phenyl ether (or Triton $X-100^{®}$) is frequently used as a non-ionic surfactant for MWCNT. The hydrophobic part of the surfactant is absorbed by the CNT, while the hydrophilic part forms hydrogen bonds with the matrix.
- Functionalization by metal ions: The functionalization of CNTs could be performed with Ag+ by a mixing process in the presence of ammonium bicarbonate [Ma et al. (2007), Zhao et al. (2008b)]. It could also be done coating CNTs with nickel [Zhao et al. (2008b)], silver nanowire [Zhao et al. (2008a)]or iron [Zhao et al. (2009)] encapsulated inside the nanotube.

3.1.3.3 Surface deposition:

The dispersion can also be improved by plasma deposition of a very thin polymer film (2-7 *nm*) on the inner and outer surfaces of CNTs [Shi et al. (2003), Lee et al. (2007)]. The dispersant may also be a block copolymer [Cho et al. (2008)].

3.2 Processes of nanocomposite production

3.2.1 Nanoclays and nano-oxides

For nanoclay-based composites, three main ways are used for the synthesis of nanocomposite:

- In-situ: monomer is introduced between clay plates, then polymerized in-situ.
- Direct: polymerization and incorporation into clay plates are performed together simultaneously. It could be performed using temperature as the initiator for thermoplastics, or by catalyze during the polymerization process (solution-assisted)
- Co-precipitation: clays and precursors are dispersed together in a solution, then co-precipited by a catalytic way.

The important aspect to control during the process is then the homogeneity of the dispersion. If nanoparticles are aggregated inside the material, it could lead to the formation of a microcomposite and nanoscale interactions could not be obtained. A large majority of authors working at laboratory scale with nanocomposites (especially thermosets) use the same manufacturing process based on the direct way, described by the company Ashland [Twardowska & Dammann (2005), Twardowska et al. (2005)] as follows:

1. Chains of polymer (commonly polyesters or vinylesters) are mixed with styrene and nanofillers previously treated and functionalized (in order to improve exfoliation).
2. The mixture is mechanically stirred at a high shear rate to improve the intercalation, and then placed under vacuum to remove air bubbles. To improve the dispersion of the filler in the resin, a step of cold ultrasound sonication is used.
3. At room temperature, an initiator (methyl ethyl ketone peroxide, MEKP) and an accelerator (cobalt-octate) are added for the crosslinking step. The high viscosity mixture is mixed, then again placed under vacuum to remove air bubbles before being poured into a mold previously covered with a coating limiting the membership of the mixture (paraffin and polyvinyl acetate, Teflon). The mold is rotated so as to avoid sedimentation of particles in the bottom of the mold. Crosslinking is obtained through successive stages of cooking and post-baking pan (called "*cure*" and "*post treatment*").

Some variations could also be found in the relevant literature [Suh et al. (2000)]. Patent [Nichols & Chou (1999)] concerning a large number of resins and inorganic materials in the form of plates mention also a classical in-situ polymerization method.

3.2.2 Carbon nanotubes

Nanocomposites containing carbon nanotubes are made by different processes, depending on whether or not prepreg (fibre mat) is used. The various nanocomposites that do not use pre-impregnated fibre processes are high-speed mechanical stirring [Xu et al. (2004)] and 3-roll milling process [Seyhan et al. (2007a), Seyhan et al. (2007b)]. Nevertheless, one of the main applications for nanocomposites containing carbon nanotubes is the casting of a large object subjected to high stresses. These materials often contain glass or carbon fibres reinforcing type and layout requires special techniques such as Wet/Hand lay-up process [Chandradass et al. (2007)], Resin Transfer Molding (RTM)[Schubel P.J. (2006), Wichmann et al. (2006), Fan et al. (2004)], or Vacuum Assisted Resin Transfer Molding (VARTM) [Hsiao & Gangireddy (2008), Sadeghian et al. (2006), Zhu et al. (2007)].

Layered hybrid materials (laminates) in thermoset matrix containing carbon nanotubes and reinforcing glass fibres are manufactured in 2 stages: CNTs are separated and dispersed in the mixture containing the polymer (by 3-roll milling process or ultrasound). In a second step, the suspension polymer / CNT is added to a mat of fibre (prepreg). Processes refered previously could then be used. Depending on the matrix, viscosity of the mix and process used, the incorporation of CNTs could be performed up to 2% in mass.

4. Properties of nanocomposites

4.1 Modification of mechanical properties of thermoset composites

4.1.1 Nanoclays

Application to resin concrete: Nanoclay, treated and untreated montmorillonite were incorporated to strengthen a resin concrete (11% polyester resin and 89% of mineral fillers (sand and limestone)) [Jo et al. (2008)]. Only the treated montmorillonite increases the strength and modulus in compression of concrete. An increase of 20% of the mechanical properties is obtained for a rate of nanofillers between 5 and 7%. Without surface treatment, performances are comparable to those lacking of nanofiller. Exfoliation is supposed to explain the differences.

Application to polyester resins: [Bauer et al. (2008)] shows that the strengthening of a polyester resin by montmorillonite nanofillers treated by grafting silane functions led to a sharp increase in mechanical properties and in particular the energy propagation of rupture. Value is doubled compared to the virgin resin, with a proportion of nanofillers 1.5% by volume. The mixing time between treated montmorillonite and hydroxypropyl acrylate (reactive solvent can be used instead of styrene crosslinked unsaturated polyester resins) may affect the mechanical properties of nanocomposite, depending on the treatment of montmorillonite [Kim et al. (2004)]. Nanoclays can also be used to reduce the shrinkage of polyester resins [Shubel et al. (2006)]. The withdrawal is a major drawback of polyester resins, because it causes dimensional changes and distortion of composite parts. The withdrawal of a neighbor virgin resin is lowered from 7.5% to 5.8% with only the addition of 1 wt% clay nanofiller. The breaking strength of a nanocomposite with 4% nanofillers is increased by 108% and Young's modulus of 53%. In contrast, an anti-withdrawal conventionally used such as polyvinyl alcohol (PVA) should be used in higher proportions (about 30%). This causes a decrease in breaking strength and flexural modulus of 24% and 32% respectively. *Application to vinylester resins*: The incorporation of a 4% nanoclay in a vinylester resin slightly increases the modulus of the nanocomposite, but decreases the elongation at break compared to the virgin resin [Chen et al. (2007a)]. Further work on the hybrid resins modified vinylester/epoxy with two different types of clay nanofillers [Karger-Kocsis et al. (2003)] showed that the fracture energy of nanocomposites containing 5 wt% nanofillers was doubled compared to that of pure resin. Chen et al. (2007a) also highlights the importance of surface functions of nanofillers on the final properties of composites. *Application to Epoxy resins*: The breaking strength and elasticity modulus in compression of epoxy resins can be increased by nanoclay (montmorillonite with surface treatment). The wear resistance is also changed (to a 30% increase) but depends on the rate and morphology of nanoparticles. Depending on the rate of nanofillers, it is possible to obtain either a lubricating effect of the surface which protects the nanocomposite or an increase of mass loss by friction [Lin (2007)]. Regarding previous paragraphs, it appears that the mechanical properties of polymers can be greatly enhanced by nanoclays, depending on two parameters:

- The surface treatment of nanofillers is essential to increase the filler/matrix adhesion, otherwise there is no change in the properties of the nanocomposite.

- The dispersion of nanofillers is another governing parameter. It requires considerable energy: vigorous mixing or ultrasonic dispersion

In these circumstances, it appears that for optimum rate of 5% of nanofillers, the characteristics to break can be doubled. Regarding the tensile properties, the results are contradictory depending on the author and resin, with a small increase in elastic modulus and decrease in elongation at break or a 53% increase in elastic modulus and increase of 108% elongation at break.

4.1.2 Nano-oxides

Application to polyester resins: For highly cross linked polyester resins, if alumina nanoparticles (15 nm) are not surface treated, the properties of the nanocomposite are lower than the resin alone [Zang et al. (2007)]. Cons by grafting aminosilanes functions to increase adhesion nanofiller-matrix can increase by 100% the K1C stress intensity factor of a composite for a loading rate of 4.5%. The mechanical properties of unsaturated polyester resin are considerably improved by the incorporation of 4% TiO_2 nanoparticles with an average grain size of 27 nm, only if nanoparticles have a strong adhesion due to surface defects [Xiao et al. (2002)]. The presence of titanium dioxide increases the stress at rupture in traction of 47% and Young's modulus of 22%. It leads to an increase of flexural strength at break of 173% and modulus of elasticity flexion of 22%. The impact strength is increased of about 60% of it. Reference [65] also highlights the breaking strength and modulus of elasticity in compression of epoxy resins can be increased by TiO_2 nanoparticle. *Application to Epoxy resins*: The abrasion resistance of epoxy resins is generally improved by fillers such as graphite, polytetrafluoroethylene (PTFE) or short carbon fibres. A comparison between conventional fillers and / or the incorporation of nanoparticles of TiO_2 indicates that for a content of 5% in volume, the wear resistance of a nanocomposite containing nanoparticles of TiO_2 is higher than that of resin containing the same amount of PTFE (particle size 4 μm) [Zang et al. (2007)]. However the best wear resistance of epoxy resins is achieved by a combination of traditional fillers (15% graphite and 15% short carbon fibres) with TiO_2 nanoparticles (5% by volume). As for nanoclays, dispersion and surface treatments are the governing parameters for the effect of nano-oxides on mechanical properties. Various studies have shown that treated nano-alumina (4.5% by volume) can double the K_{1C} of virgin polyester resins. Nanoparticles of TiO_2 (4% in volume) increase significantly the characteristics of flexural and tensile and impact strength of virgin polyester and the abrasion resistance of epoxy resins.

4.1.3 Carbon nanotubes and nanofibres

Work on strengthening a vinylester resin by carbon nanofibres for rates from 0 to 25 wt% [Xu et al. (2004)] features show that flexural modulus and strength decrease with the rate of nanofibres. It should be noted however that the bending tests were conducted only on nanocomposites with nanofillers that have received no surface treatment and therefore with a low fibre-matrix interaction. Xu et al. (2004) also deals with enhancing resins and vinyl ester hybrid resins, vinyl ester/vinyl ester and epoxy-urethane by MWCNT to a rate of 2%. Toughness (stress intensity factor and fracture energy) was slightly increased (+27% and +35%) for an expense ratio of 1% and decreased for higher rates. An important point is that only the outer wall of the MWCNT is coupled with the matrix and therefore the effectiveness of capacity decreases with the number of walls or layers (Walls). Seyhan et al. (2007b) presents the reinforcement of polyester resins by carbon nanotubes, using Double-Wall Carbon Nanotubes (DWCNT) and Multi-Wall Carbon Nanotubes untreated and treated (grafted amine functions). Nanocomposites comprising a rate of 0.5% of treated nanotubes have tensile properties slightly higher than that of virgin resin, and exhibit an increase of 5% to 17% of the tensile strength. Nanocomposites containing untreated fillers have the same characteristics as pure resin. Carbon nanofibres have also been used to improve resistance to delamination of glass-fibre reinforced polyester composite [Sadeghian et al. (2006)]. The critical energy for delamination crack growth in mode 1 is increased by 100% to a rate of 1% nanofibres. These studies highlight the need and difficulties of the dispersion of nanotubes in the matrix. Using a three-roll mixer gives better results than commonly used methods such as mixture by stirring or ultrasonic dispersion. The rate of these nanofillers for optimum effect is around 1% and improvements in tensile properties and fracture resistance obtained are low compared to the unique characteristics of these materials. As a known effect, there is the resistance to delamination of a glass-reinforced polyester, which was doubled by carbon nanofibres. Note that these nanofillers must be surface treated so as to improve a characteristic of resins. The dispersion of nanofillers is difficult, critical and requires high shear rates (three-roll mixer).

4.2 Modification of thermophysical properties of composites

All the nanocomposites studied hereafter exhibit anisotropic thermal conductivity. The methods of preparation of composites often direct, intentionally or not, the nanofillers within the matrix. Conductivity measured in the direction of orientation of the nanofillers is therefore higher than the one measured in the transverse direction. The majority of studies concern carbon nanotubes in polymers. For Carbon nanotubes, Vergne (2007) explains that many parameters influence the thermal conductivity of such composites: the contact resistance, the nature, dispersion and orientation of carbon nanotubes, the impurities in the formulation of thermoset resins or synthesis of CNT, the dispersants, nanotube bundles, metal catalysts and carbonaceous residues.

4.2.1 Nanoclays

Very few data are found in the litterature related to thermal properties of nanocomposites using nanoclays. Few data corresponding to micrometric clay-based fillers could be found by Al-Malah & Abu-Jdayil (2007) and Abot et al. (2003). They show that increasing the amount of charge (between 25 and 60 wt%) increases the thermal conductivity of the composite obtained. Al-Malah & Abu-Jdayil (2007) justifies this behaviour with the highest thermal conductivity of mineral fillers (38 $W.m^{-1}.K^{-1}$ for Al_2O_3 and 2.7 $W.m^{-1}.K^{-1}$ for SiO_2) compared to resins like polyester (0, 1 $W.m^{-1}.K^{-1}$). As a consequence, it remains probably true for smaller fillers. Haque et al. (2002) present the thermogravimetric analysis and DMA of a composite epoxy/glass fibre with 1 to 2wt.% of nanoclays (montmorillonite). The incorporation of nanoparticles increases the glass transition temperature by a few degrees and the decomposition temperature of about 20°C.

4.2.2 Carbon nanotubes

Decomposition temperature: The addition of MWCNT in epoxy resin have no significant influence on the decomposition temperature of the composite. Zhou et al. (2008) has made measurements of decomposition by thermogravimetric analysis under nitrogen at 10°$C.min^{-1}$ on specimens of epoxy containing 0 to 0.4 wt% MWCNT. Carbon nanotubes of a diameter of 30-50 nm and 3 μm long and the matrix polymer are mixed using a high intensity ultrasound method. All samples start to decompose around 340°C, and are completely decomposed at

460°C whatever the concentration of MWCNT is. *Thermal conductivity*: Works on epoxy/CNT nanocomposites all indicate an increase in thermal conductivity. Biercuk et al. (2002) asserts that epoxy with a charge of % SWCNT correspond to a thermal conductivity increased of almost 125% at room temperature. The influence of mass fraction (ranging from 0.05 to 3.0%) of carbon nanotubes MWCNTs on the thermal conductivity of composite epoxy/MWCNT was studied by Evseeva & Tanaeva (2008) from -150°C to 150°C. The introduction of 0.1 to 1.0% of MWCNTs increases the thermal conductivity of epoxy resin by about 40%. A larger amount of nanotubes is decreasing thermal conductivity, probably due to poor dispersion of nanotubes at high concentrations. Indeed, the specific surface load increases when the size of particles or clusters of particles decreases. However, more specific surface loads are high and their effectiveness on the performance of the nanocomposite is great. *Coefficient of thermal expansion*: Gonnet (2004) studied the variations in coefficient of thermal expansion and glass transition of an epoxy/SWCNT nanocomposite with 1 and 5 wt.% of carbon nanotube. The coefficient of thermal expansion of the composite before the glass transition is lower than average than the epoxy resin alone. It is lower for a mass fraction of nanotubes for a 5% mass fraction of 1%. The coefficient of thermal expansion of the composite after the glass transition is higher than that of the epoxy resin alone. The glass transition temperature of the composite is lower than the resin alone. The glass transition temperature of 5 wt.% nanocomposite is lower than for the 1 wt.%. Dos Santos et al. (2008) noted that the expansion coefficient of epoxy resin remains constant with or without the addition of 0.5 wt% MWCNT, whether before or after the glass transition. Similarly, Yuen et al. (2007) notes that the expansion coefficient of a composite epoxy/MWCNT is not affected if the mass fraction of MWCNT is less than 0.5%. However, its glass transition temperature increases from 167°C for pure epoxy to 189°C for the nanocomposite. According to the literature, the thermophysical properties of these composites vary with the nature of nanofillers, their mass fraction, morphology and dispersion. The thermal conductivity of nanocomposites using thermoset resins always increases with the addition of nanofillers and depends strongly on the nature of the charges, and their concentration and quality of their dispersion. The conclusions are highly variable from one publication to another. The thermal response of these nanocomposites is different from that of similar composites containing fillers of larger size, heat transfer to macroscopic scales and sub-micron being different. The coefficient of thermal expansion, glass transition and decomposition temperature of nanocomposites using thermoset resins decrease with the addition of nanofillers for some authors, while others have almost found the reverse. Given the large number of parameters to which the thermophysical properties of nanocomposite appear sensitive, especially about their implementation, it is difficult to predict a priori the thermal behaviour of such composites without a preliminary study on a case-by-case basis. These studies could examine the evolution of their thermal properties based on influence parameters such as nature, rate, and dispersion of fillers, and look for possible correlations.

4.3 Modification of optical and radiative properties of composites

Amorphous polymers are transparent to the visible range [Trotignon et al. (1989)]. The addition of pigments and fillers increases the light scattering and opacifies the material. Thermosets are generally uncharged amorphous and therefore transparent. In the infrared, the absorption spectra are characteristic of the polymer (molecular vibration modes). The conventional thermoset generally charged have a high emissivity and are opaque. If the polymer is opaque to the considered wavelength and whether the contents nanofillers are small, it is very likely that those of the polymer mainly influence the radiative properties of

the composite. Indeed, in this case the only important properties are the surface properties (reflectivity, thermal emissivity) and the surfaces of nanofillers present on the surface are very low compared to that of the polymer. The effect will be totally different for nanostructured objects, which could behave as metamaterials.
It is possible to maintain the transparent properties of the polymer as the levels of nanofillers are small but sufficient to significantly improve the mechanical properties and heat resistance of the nanocomposite [Bharadwaj et al. (2002)]. For semi-transparent material, transmittance is more likely to decrease with an increasing nanofiller content [Bharadwaj et al. (2002), Thompson et al. (2003)]. Radiation intensity phenomena can be obtained using semiconductor particles, especially nanocrystals.

4.3.1 Nanoclays

According to Bharadwaj et al. (2002), the influence of nanoclay plates (aluminosilicates) depends on the size and shape of particles. The decrease of transmittance is due to scattering and absorption from plates. Authors suggest that transparency is even increased when the dispersion of plates is important.
According to Chrismann et al. (2010), the incorporation of "Organo-modified layered silicates" (OMLS) particles is limited in transparent materials due to the diffusion induced by the excessively large flat surfaces of the plates relative to the thicknesses.

4.3.2 Carbon nanotubes

Because of their highly anisotropic nature, carbon nanotubes are only incorporated in thin films to preserve the transparency of the polymer [Chrismann et al. (2010)] to modify optical properties. The incorporation of 0.5% SWNT (single wall nanotube) in a PMMA film of 20 μm in thickness gives a transmission factor of 46% for visible wavelength of 500 nm.

4.3.3 Nano-oxides

The use of "classic" nano-oxide particles (silica, alumina, zirconia) appears promising for the reinforcement of transparent polymers. In general, these are quasi-spherical nanoparticles that are used in transparent materials to prevent the diffusion phenomena [Chrismann et al. (2010)]. It is possible to produce transparent composite with refractive index or variations index chosen. The use of these materials is limited to the production of special optical components (lenses, optical fibres, waveguides, etc).
Chen et al. (2007c) present the characterization of a nanocomposite polyurethane/TiO_2 emissivity. Emissivities measured for the band 8-14 μm are: 0.945 for the raw polyurethane, 0.925 for the pure TiO_2 and 0.538 for the composite polyurethane/TiO_2. The TiO_2 content in the composite is not clearly stated. It seems that it is possible to generate composite materials with an emissivity, which is considerably lower than the emissivity of individual components. According to the authors, the change in emissivity is due to interfacial phenomena that alter the molecular vibration modes. Lin et al. (2004) present the results of measurements of spectral emissivity on a nanocomposite film polyimide/$BaTiO_3$. Emissivities averages for the spectral band 8-14 μm are respectively 0.703, 0.574 and 0.605 for mass concentrations of particulate $BaTiO_3$: 0%, 14.7% and 28.7%. Emissivity variations are due to interfacial interactions. The spectral variations (depending on the wavelength) measured relative to a nanocomposite with 14.7% and 28.7% of $BaTiO_3$ are similar to those of unfilled polymer. Thompson et al. (2003)] announced that the emissivity of a nanocomposite polyimide/indium-tin-oxide is higher than the polymer alone.

4.4 Modification of electrical and magnetic properties of composites

4.4.1 Nano-oxides

Nanocomposite materials consisting of nano-oxide and a conductive matrix paved the way for new applications such as smart windows, new toner, copier, conductive paints and rechargeable batteries [Maeda & Armes (1994), Butterworth et al. (1995), Jarjayes et al. (1995), MacDiamid (1993)]. A first class of materials corresponds to the association of π-conjugated polymers with permanent magnetic nanoparticles (Fe_3O_4, $CoFe_2O_4$, $NiFe_2O_4$). These magnetic nanoparticles exhibit superparamagnetic properties. Each nanoparticle behaves as a nano-magnet. The collective behaviour of magnetic nanoparticles can be greatly affected by their ability to aggregate. Thus, optimal performance can be achieved by dispersing these nanoparticles in a non-magnetic polymer matrix. Nanocomposites containing ferrite and a polyamide (PA) or polypyrole (Ppy) matrix present both interesting electric and magnetic properties [Turcu et al. (2006)]. This allows consideration of multiple applications such as electromagnetic interference shielding or microwave absorbing materials. Other works evoke the dispersion of TiO_2 nanoparticles [Nabid et al. (2008)] or SiO_2 [Dutta & De (2007)] matrices to add conductive or dielectric optical properties to nanocomposites.

4.4.2 Carbon nanotubes

The CNTs are used in π-conjugated polymer matrices to improve significantly the electrical and mechanical properties, and these composites are often denoted NT-CP (Carbon Nanotube - Conducting Polymer) in the literature. These materials exhibit specific properties related to electronic interactions between fillers and the matrix. The electrical properties are enhanced and thus give rise to all sorts of original applications: Schottky diodes, capacitors, high performance electrodes in rechargeable batteries [Chen et al. (2002), Hughes et al. (2002)], nanowires for optoelectronic applications, organic light emitting diodes (OLEDs). Polyacrylonitrile (PAN) [Cochet et al. (2001)], polypyrole (PPy) [Chen et al. (2002), Hughes et al. (2002), Baibarac et al. (2003)], polythiophene [Philip et al. (2004)] and polyacetylene [Tchmutin et al. (2003)] based nanocomposites are commonly found in research and industrial applications nowadays. Their electrical properties are dependent on the synthesis methods, as it influences interactions between CNT and polymer. Nanocomposite PAN/CNTs exhibit electrical conduction properties that can be described simply as either parallel conductors when the interaction is Van der Waals forces or charge transfer when there are covalent bonds. There is no percolation in these materials. Some studies have concluded that improving the electrical properties of the NT-CP was due to the doping of polymer by nanotubes [Lee et al. (2001), Zengin et al. (2002)].

4.4.3 Metallic nanoparticles

Metallic nanoparticles (NPs) correspond to crystallite metal (Fe, Ag, Au, etc) of spherical shape. The incorporation of such nanoparticles in a conductive polymer matrix allows increasing the electrical conductivity of the composite in a synergical way. The applications are diverse and of interest for many areas: photovoltaic cells, memory devices, anticorrosion coatings, supercapacitors, sensors or catalysts. Thus, a composite of platinum nanoparticles (Pt) or palladium dispersed in a matrix of PPy shows a remarkable electrocatalytic activity. A nanofiller Fe in a polystyrene matrix (insulant) or PPy (conductor) provides a magnetic composite insulator or conductor. These materials are used as electromagnetic shielding for electronic components. A number of studies are devoted to the use of nanoparticles of noble metals (Au, Ag, Pt) often associated with PPy or PAN [Chen et al. (2007b), Bose & Rajeshwar

(1992)]. The objective is to develop nanocomposites for opto-electronics, optical filters or absorbers in different frequency bands by exploiting the properties of nonlinear optical NPs, photonic crystals, materials with high electrochemical properties, and biological or chemical sensors.

4.4.4 Metal-CNT systems

Functionalization of carbon nanotubes by metal ions is also proposed to form wave-absorbing materials. Thus, the CNTs are coated with silver ions [Zhao et al. (2008a)] or nickel [Zhao et al. (2008b)]. Silver nanowire [Zhao et al. (2008b)] or iron [Shi et al. (2003)] could also be encapsulated inside the nanotube. Such epoxy CNT-modified material is capable of absorbing in microwaves range [Zhao et al. (2008b)].

4.5 Modification of gas diffusion properties

The gas and solvent diffusion properties of polymers could be modified using nanofillers, especially when nanoplates are used [Al-Malah & Abu-Jdayil (2007)]. This parameter is essential in some applications such as food packaging. The structure conferred to the polymer when nanoplates are used induces an increase in the distance to be traveled by gases between plates. This distance is qualified by the "*tortuosity*" (deviousness) factor (Fig. 6). It is especially present with nanocomposites using nanoclays, and depends on dispersion, diffusion, exfoliation and the orientation of the plates. Improved barrier properties of gas transfer is related to the parallel orientation of organo-modified nanoclay plates, and therefore requires a high degree of exfoliation. This property will lead to the creation of new food packaging materials, as oxygen diffusion is one of the governing parameters of food conservation.

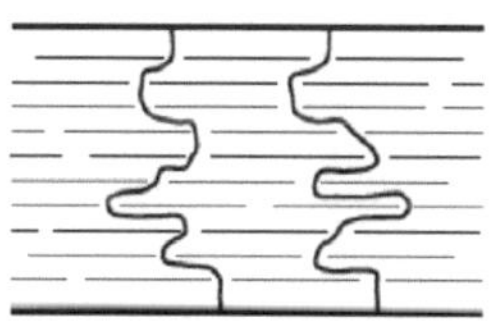

Fig. 6. Tortuosity for gas diffusion properties of lamellar nanocomposite

4.6 Modification of fire behaviour of composites

Materials used to meet performance criteria against fire involve the use of additives to improve their behaviour. To meet the specifications in terms of choice of materials, new systems to improve the fire behaviour of polymers based on the introduction of nanofillers can become an alternative to traditional ways. Since the first studies on organomodified clays [Giannelis (1996), LeBaron et al. (1999)], various studies have highlighted the interest of various types of nanoparticles (nano-oxides [LeBaron et al. (1999)], polyorganosesquioxanes [Finaa et al. (2006)], carbon nanotubes [Peeterbroech et al. (2007)]) or natural (organo-modified phyllosilicates, delaminated talc, kaolin submicron [Le Bras et al. (2005)]). A synergy between classical flame retardant and nanofillers may be interesting as it may decrease the amount of flame retardant and thereby reduce the loss on other properties (including mechanical). Extensive details on fire behaviour of nanocomposites could be found in Morgan & Wilkie (2007) and Morgan (2009). Gilman (1999) and (2000) has demonstrated that the nanoclays particles brought a significant improvement in the flammability properties of the polymer within which these were dispersed. This enhanced fire resistance is attributable to the formation of a compact and superficial layer of clay shells during the exposure to flames, which thus act as a thermal barrier. The fire behaviour of the matrix will obviously be dependent upon the way the nanocomposites fillers are dispersed within such matrix. Fillers should be evenly dispersed to confer flame-retardant properties. The mechanism of action according to which the nanoparticles confer flame-retardant properties is attributable to the accumulation of inorganic material upon the composite surface together with the formation of char. This shield will reduce the material and heat transfers between the flame and the polymer. It then reduces the radiative flux directed towards the substrate by diffusing the incident radiation towards the gaseous phase and as a result, limits the degradation rate of the material [Gilman (2000)]. The organophile montmorillonite clays generate improvements in the mechanical properties without losing impact strength. An even or perfect dispersion of clay, i.e. an exfoliated structure, will show an important increase in the stiffness of the composite material for low load rates. The synergy existing between the traditional reinforcements, such as glass fibres and nanoclays is of particular interest. The association of glass fibres and montmorillonite produces a much more rigid material with a larger temperature range than the traditional composite material that has the same reinforcement rate. The effective flame retardancy is obtained by combining the nanoparticles effect with conventional flame-retardants such as metal hydroxides, phosphorous or halogenated compounds. Their mode of action is based on physical mechanisms and physico-chemical properties [Le Bras et al. (2005)]. The large majority of nanocomposites defined for their fire behaviour properties use nanoclays. The barrier effect of layered nanoparticles (nano-clay) corresponds to two effects:

- Limiting the transfer of volatile fuel gas flow associated with the decomposition of polymer,
- Limiting diffusion of oxygen inside the material.

The formation of a particulate barrier is related to the mobility of particles toward the surface exposed to the external heat source as well as the removal of polymer decomposition. The incorporation of nanoparticles also tends to generate an increase in viscosity, which limits the heat and mass transfers in material. The presence time increased in the condensed phase decomposition products of the polymer facilitates dehydrogenation reactions catalyzed by the large area of nano-particles, leading to the formation of condensed structures charred (char) that grow on the surface of residual material. Furthermore, incorporation of nanotubes in EVA also showed a greater presence of free radicals in the combustion residue. The way of thermal decomposition of the polymer tends to be modified in the presence of nanoparticles, with a slowed emission of volatile fuel, even in the absence of conventional flame retardant. However, a fraction of the nanoparticles could be released into the atmosphere during combustion [Laachachi (2005), Chivas-Joly et al. (2010)]. The use of double-layered hydroxides is an alternative to the use of cationic clays for making epoxy-based composites with good fire properties [Ibeh et al. (2006)]. The cone calorimeter results on vinylesters containing clays (Cloisite 15 A) or POSS highlighted a reduction of maximum heat release rate and increase of time to ignition, but no modification on total heat released [Chigwada et al. (2005)]. A combination with a phosphorus-based flame retardant promotes a greater improvement. Table 2 presents results obtained by various authors, using nanoclays in thermoplastic and thermoset polymers. Common compositions between classical flame-retardants and nanoparticules that have a synergy effect are:

	Nanofiller used	Polymer matrix	Effect on fire behaviour on cone calorimeter test (in comparison with polymer alone)
Chigwada et al. (2005)	Montmorillonite (Cloisite 15 A)	Vinylester, Bisphenol A/epoxy,	Decrease of maximum heat release rate
	Montmorillonite (1%, 3% and 5%)	PS, HIPS, ABS, PE, PP	Decrease of maximum heat release rate
	Montmorillonite (1% and 2%)	PA6	Increase of ignition time, Decrease of maximum heat release rate
Kandola (2008)	Montmorillonite (Cloisite 25 A) (5%)	Polyester	Increase of ignition time, Decrease of maximum heat release rate
		Polyester + APP	Increase of ignition timeDecrease of maximum heat release rate Increase of smoke released
Gianelli & Camino (2006)	Montmorillonite (5%)	Polyester	Decrease of maximum heat release rate
Hussain et al. (2004)	Montmorillonite	Phosphorous-modified epoxy resin	Decrease of maximum heat release rate

Table 2. Effects of nanofillers on fire behaviour of various nanocomposites.

- Organo-modified nanoclays in combination with metallic hydroxides, e.g. aluminium trihydrate (ATH) or magnesium dehydrate (MDH);
- Organo-modified nanoclays in combination with phosphorous compounds, to form intumescent structure, e.g. Ammonium polyphosphate (APP);
- Nano-hydroxides and nano-oxides in combination with phosphorous compounds (APP, phosphinates);
- Carbon nanotubes in combination with metallic hydroxides.

Pyrogenic silica and nanoalumina produce synergistic effects in combination with ammonium polyphosphate [Cinausero (2009)]. The use of treated silicas (octylsilane) enhances the performance of the system at constant composition flame retardant. Improving the fire behaviour of polymer materials using nanoparticles is related to the existence of barrier effects, but also catalytic process and capture of radicals. The incorporation of nanoparticles can advantageously be combined with the reactive flame-retardants.

5. Perspectives

Nanoscale dimensions can increase significantly the physical interactions, physico-chemical and chemical interfaces in materials. The morphologies obtained for the nanocomposites and the ability to modify the interfaces are essential to maximize the properties. For all charges, surface treatment and mixing are key points dictating the performance of nanomaterials. The variety of combinations between nanofillers, matrix and classical additives allow wide possibilities of material specifications improvements: mechanical, thermal, optical, electrical properties as well as reaction to fire.

The improvement in the quality of fillers (e.g. less catalyse residues in carbon nanotubes) and progress in production processes will lead to an increasing diffusion of such nanocomposites in a wide range of applications with strong requirements.

6. References

Abot, J., Yasmin, A. & Daniel, I. (2003). Mechanical and thermoviscoelastic behavior of clay/epoxy nanocomposites, *Materials Research Society Symposium*, pp. 167–172.

Ajayan, P., Schadler, L. & Braun, P. (2003). *Nanocomposite science and technology*, Ed. Wiley.

Al-Malah, K. & Abu-Jdayil, B. (2007). Clay-based heat insulator composites: thermal and water retention properties, *Applied Clay Science* 37: 90–96.

Baibarac, M., Baltog, I., Lefrant, S., Mevellec, J. & Chauvet, O. M. (2003). Polyaniline and carbon nanotubes based composites containing whole units and fragments of nanotubes, *Chemistry of Materials* 15(4149-4156): 21.

Bauer, M., Kahle, O., Landeck, S., Uhlig, C. & Wurzel, R. (2008). High performance composites using nanotechnology, *Advanced Materials Research* 32: 149–152.

Bharadwaj, R., Mehrabi, A., Hamilton, C., Trujillo, C., Murga, M., Fan, R., Chavira, A. & Thompson, A. (2002). Structure-property relationships in cross-linked polyester clay nanocomposites, *Polymer* 43(3699-3705).

Biercuk, M., Llaguno, M. & Hyun, J. (2002). Carbon nanotube composites for thermal management, *Applied Physics Letters* 80(15): 2767–2769.

Biron, M. (2004). *Thermosets and composites, Technical information for plastics users,*, Elsevier Ltd.

Bose, C. & Rajeshwar, K. (1992). Efficient electrocatalyst assemblies for proton and oxygen reduction: the electrosynthesis and characterization of polypyrrole films containing nanodispersed platinum particles, *Journal of Electroanalytical Chemistry* 333(235-256): 1–2.

Butterworth, M., Corradi, R., Johal, J., Lascelles, S., S., M. & S.P., A. (1995). Zeta potential measurements on conducting polymer-inorganic oxide nanocomposite particles, *Journal of Colloid And Interface Science* 174(510-517): 2.

Chandradass, J., Ramesh Kumar, M. & Velmurugan, R. (2007). Effect of nanoclay addition on vibration properties of glass fibre reinforced vinyl ester composites, *Materials Letters* 61: 4385.

Chen, C., Yebassa, D. & Raghavan, D. (2007a). Synthesis, characterization, and mechanical properties evaluation of thermally stable apophyllite vinyl ester nanocomposites, *Polymers for Advanced Technologies* 18: 574–581.

Chen, J., Huang, Z., Wang, D., Yang, S., Li, W., Wen, J. & Ren, Z. (2002). Electrochemical synthesis of polypyrrole films over each of well-aligned carbon nanotubes., *Synthetic Metals* 125(289-294): 3.

Chen, J., Zhou, Y., Nan, Q., Sun, Y., Ye, X. & Wang, Z. (2007c). Synthesis, characterization and infrared emissivity study of polyurethane/tio_2 nanocomposites, *Applied Surface Sciences* 253: 9154–9158.

Chen, W., Auad, M., Williams, J. & Nutt, S. (2006). Improving the dispersion and flexural strength of multiwalled carbon nanotubes-stiff epoxy composites through ï£¡-hydroxyester surface functionalization coupled with the anionic homopolymerization of the epoxy matrix, *European Polymer Journal* 42: 2765.

Chen, W., Li, C., Chen, P. & Sun, C. (2007b). Electrosynthesis and characterization of polypyrrole/au nanocomposite, *Electrochimica Acta* 52(8): 2845–2849.

Chen, X., Wang, J., Lin, M., Zhong, W., Feng, T., Chen, X., Chen, J. & F., X. (2008). Mechanical and thermal properties of epoxy nanocomposites reinforced with amino-functionalized multi-walled carbon nanotubes, *Materials Science and Engineering: A* 492: 236.

Chigwada, G., Jash, P., Jiang, D. & Wilkie, C. (2005). Fire retardancy of vinyl ester nanocomposites : synergy with phosphorus-based fire retardants, *Polymer degradation and stability* 89: 85–100.

Chivas-Joly, C., Guillaume, E., Ducourtieux, S., Saragoza, L., Lopez-Cuesta, J., Longuet, C., Duplantier, S., Bertrand, J., Calogine, D., Minisini, B. & Gensous, F. (2010). Influence of carbon nanotubes on fire behavior and on decomposition products of thermoplastic polymers., *INTERFLAM*, Nottingham.

Cho, J., Daniel, I. & Dikin, D. (2008). Effects of block copolymer dispersant and nanotube length on reinforcement of carbon/epoxy composites, *Composite: Part A* 39: 1844.

Chrismann, A., Longuet, C. & Lopez-Cuesta, J. (2010). Transparent polymer nanocomposites: A new class of functional materials, *in* J. T. ed (ed.), *Nanomaterials and Surface Engineering*.

Cinausero, N. (2009). *Etude de la dégradation thermique et de la réaction au feu de nanocomposites ï£¡ matrice PMMA et PS.*, PhD Thesis, University of Montpellier 2 (France).

Cochet, M., Maser, W., Benito, A., Callejas, M., Martï£¡nez, M., Benoit, J., Schreiber, J. & Chauvet, O. (2001). Synthesis of a new polyaniline/nanotube composite: "in-situ" polymerisation and charge transfer through site-selective interaction, *Chemical Communications* 21(16): 1450–1451.

Dean, K., Krstina, J., Tian, W. & Varley, R. (2007). Effect of ultrasonic dispersion methods on thermal and mechanical properties of organoclay epoxy nanocomposites, *Macromolecular Materials and Engineering* 292: 415–427.

Dos Santos, A., Leite, T. & Furtado, C. (2008). Morphology, thermal expansion, and electrical conductivity of multiwalled carbon nanotube/epoxy composites, *Journal of Applied Polymer Science* 2: 979–986.

Dutta, K. & De, S. (2007) Electrical conductivity and dielectric properties of sio2 nanoparticles dispersed in conducting polymer matrix, *Journal of Nanoparticle Research* 9(4): 631–638.

Endo, M., Koyama, T. & Hishiyama, Y. (1976). Structure improvement of carbon fibers prepared from benzene., *Japanese Journal of Applied Physic* 15(11): 2073–2076.

Evseeva, L. & Tanaeva, S. (2008). Thermal conductivity of micro and nanostructural epoxy composites at low temperatures, *Mechanics of composite materials* 44(1): 87–92.

Fan, Z., Hsiao, K. & Advani, S. (2004). Experimental investigation of dispersion during flow of multi-walled carbon nanotube/polymer suspension in fibrous porous media, *Carbon* 42: 871.

Fiedler, B., Gojny, F., Wichmann, M., Nolte, M. & Schulte, K. (2006). Fundamental aspects of nano-reinforced composites, *Composite Science and Technology* 66: 3115.

Finaa, A., Abbenhuisb, H., Tabuania, D. & G., C. (2006). Metal functionalized poss as fire retardants in polypropylene, *Polymer Degradation and Stability* 91: 2275–2281.

Fu, X. & Qutubuddin, S. (2004). Synthesis of unsaturated polyester-clay nanocomposites using reactive organoclays, *Polymer Engineering and Science* 44: 345.

Geng, Y., Liu, M., Li, J., Shi, X. & Kim, J. (2008). Effects of surfactant treatment on mechanical and electrical properties on cnt/epoxy nanocomposites, *Composite : Part A* 39: 1876.

Gianelli, W. & Camino, G. (2006). Fire behaviour of polyester-clay nanocomposites, *Fire and Materials* 30: 333–341.

Giannelis, E. (1996). Polymer layered silicate nanocomposites, *Advanced Materials* 8: 29.

Gilman, J. (1999). Flammability and thermal stability studies of polymer layered silicate (clay) nanocomposites, applied clay science, 15(1): 31–49.

Gilman, J. (2000). Flammability properties of polymer layered silicate nanocomposites polypropylene and polystyrene nanocomposites, *Chemistry of Material* 12: 1866–1873.

Gloaguen, J. & Lefevre, J. (2007). Nanocomposites polymers/silicates en feuillets, *Techniques de l'ingénieur* AM 5205: 19.

Gonnet, P. (2004). *Thermal conductivity and coefficients of thermal expansion of SWNTS/Epoxy nanocomposites*, PhD Thesis, University of Florida (USA).

Guo, Z., Lei, K., Li, Y., Ng, H., Prikhodko, S. & Hahn, H. (2008). Fabrication and characterization of iron oxide nanoparticles reinforced vinyl-ester resin nanocomposites, *Composites Science and Technology* 88: 1513.

Guo, Z., Wei, S., Shedd, B., Scaffaro, R., Pereira, T. & Hahn, H. (2007). Particle surface engineering effect on the mechanical, optical and photoluminescent properties of zno/vinyl-ester resin nanocomposites, *Journal of Materials Chemistry* 17: 800.

Haque, A., Hossain, F. & Dean, D. (2002). s_2-glass/vinyl ester polymer nanocomposites : manufacturing, structures, thermal and mechanical properties, *17th Annual Technical Conference, American Society for Composites*, West Lafayette, Indiana, pp. 1–9.

Hsiao, K. & Gangireddy, S. (2008). Investigation on the spring-in phenomenon of carbon nanofiller-glass fiber/polyester composites manufactured with vacuum assisted resin transfer molding, *Composite: Part A* 39: 834.

Hughes, M., Shaffer, M., Renouf, A., Singh, C., Chen, G., Fray, D. & Windle, A. (2002). Electrochemical capacitance of nanocomposite films formed by coating aligned arrays of carbon nanotubes with polypyrrole., *Advanced Materials* 14(5): 382–385.

Hussain, M., Varley, R., Mathys, Z., Cheng, Y. & G.P., S. (2004). Effect of organo-phosphorus and nano-clay materials on the thermal and fire performance of epoxy resins, *Journal of applied polymer science* 91: 1233–1253.

Ibeh, C., Hui, D., Shivakumar, K. & Rajapakse, Y. (2006). Literature survey of current state of research on nanocomposite and multifunctional materials.
URL: *http://www2.pittstate.edu*

Iijima, S. (1991). Helical microtubules of graphitic carbon, *Nature* 354(6348): 56–58.

ISO/TS11360 (2010). *Nanotechnologies - Methodology for the classification and categorization of nanomaterials.*

ISO/TS27687 (2008). *Nanotechnologies - Terminology and definitions for nano-objects – Nanoparticle, nanofibre and nanoplate.*

ISO/TS88004 (2011). *Parts 1 to 7 -(under development) Nanotechnologies - Vocabulary.*

Jarjayes, O., Fries, P. & C., B. (1995). New nanocomposites of polypyrrole including $\gamma - fe_2o_3$ particles: electrical and magnetic characterizations., *Synthetic Metals* 69(1-3): 343–344.

Jo, B., Park, S. & Kim, D. (2008). Mechanical properties of nano-mmt reinforced polymer composite and polymer concrete., *Construction and building Materials* 22: 14–20.

Kandola, B. (2008). Fire retardation of polymeric composites by physical/chemical modification of the resin matrix, *Internation conference in Composite in fire CIF 5*, Newcastle.

Karger-Kocsis, J., Gryshuk, O., Frohlich, J. & Mulhaupt, R. (2003). Interpenetrating vinylester/epoxy resins modified with organophilic layered silicates, *Composites Science and Technologies* 63: 2045–2054.

Kim, H., Oh, D.H., L. H. & Min, K. (2004). Effect of reactive diluent on properties of unsatured polyester/montmorillonite nanocomposites, *Journal Of Applied Polymer Science* 92: 238–242.

Kornmann, X., Berglund, L. & Sterte, J. (1998). Nanocomposites based on montmorillonite and unsaturated polyester, *Polymer Engineering and Science* 28: 1351.

Laachachi, A. (2005). *Polyméthacrylate de méthyle (PMMA) : Développement de nouveaux systï£¡mes retardateurs de flamme ï£¡ base de nanocharges minï£¡rales. Recherche de synergies avec des montmorillonites et des composï£¡s phosphorï£¡s.*, PhD Thesis, University of Metz (France).

Labidi, S., Azema, N., Perrin, D. & Lopez-Cuesta, J. (2010). Organo-modified montmorillonite/poly(ε-caprolactone) nanocomposites prepared by melt intercalation in a twin-screw extruder., *Polymer Degradation and Stability* 95(3): 382–388.

Lan, T., Kaviratna, P. & Pinnavia, T. (1995). Mechanism of clay tactoid exfoliation in epoxy-clay nanocomposites., *Chemistry of Materials* 7(11): 2144–2150.

Lan, T. & Pinnavaia, T. (1994). Clay-reinforced epoxy nanocomposites, *Chemistry of Materials* 6: 2216–2219.

Le Bras, M., Wilkie, C., Bourbigot, S., Duquesne, S. & Jama, C. (2005). New applications of mineral fillers,: the use of mineral fillers in micro and nano-composites, *Fire Retardancy of Polymers*, The Royal Society of Chemistry.

LeBaron, P., Wang, Z. & Pinnavaia, T. (1999). Polymer layered silicates nanocomposites : an overview, *Applied Clay Science* 15: 11–29.

Lee, H., Mall, S., He, P., Shi, D., Narasimhadevara, S., Yeo-Heung, Y., Shanov, V. & Schulz, M. (2007). Characterization of carbon nanotube/nanofiber-reinforced polymer composites using an instrumented indentation technique, *Composite: Part B* 38: 58.

Lee, S., Katayama, T. andKajii, H., Araki, H. & Yoshino, K. (2001). Electrical and optical properties of conducting polymer-c_{60}-carbon nanotube system., *Synthetic Metals* 121(1-3): 1591–1592.

Li, J., Wong, P. & Kim, J. (2008). Hybrid nanocomposites containing carbon nanotubes and graphite nanoplatelets, *Materials Science and Engineering: A* 660: 483–484.

Lin, B., Liu, H., Zhang, S. & Yuan, C. (2004). Structure and infrared emissivity of silicon-containting polyimide/$batio_2$ nanocomposite films, *Journal of Solid state chemistry* 177: 3849–3852.

Lin, J. (2007). Compression and wear behaviour of composites filled with various nanoparticles, *Composites: Part B* 38: 79–85.

Ma, P., Kim, J. & Tang, B. (2007). Effects of silane functionalization on the properties of carbon nanotube/epoxy nanocomposites, *Composite Science and Technology* 67: 2965.

MacDiamid, A. (1993). Conducting polymers, science and technology, *Second Brazilian polymer conference*, Brazil: Plenum Publ. Corp.

Maeda, S. & Armes, S. (1994). Preparation and characterisation of novel polypyrrole-silica colloidal nanocomposites, *Journal of Materials Chemistry* 4: 935.

Mogilevsky, G., Chen, Q., Kleinhammes, A. & Wu, Y. (2008). The structure of multilayered titania nanotubes based on delaminated anatase, *Chemical Physics Letters* 460(4-6): 517–520.

Morgan, A. (2009). Polymer nanocomposites flammability and flame-retardancy, *in* L. Kin-Tak, F. Hussain & L. K. (eds), *Nano and biocomposites*, CRC Press, p. 408.

Morgan, A. & Wilkie, C. (2007). *Flame Retardant Polymer Nanocomposites*, Edt Wiley-Interscience.

Nabid, M., Golbabaee, M., Moghaddam, A., Dinarvand, R. & Sedghi, R. (2008). Polyaniline/tio2 nanocomposite: Enzymatic synthesis and electrochemical properties., *International Journal of Electrochemical Science* 3: 1117.

Nichols, K. & Chou, C. (1999). Polymer composite comprising a inorganic layered material and a polymer matrix and a method for its preparation, from dow chemical, *US Patent* 5 952: 093.

Oberlin, A., Endo, M. & Koyama, T. (1976). Filamentous growth of carbon through benzene decomposition, *Journal of Crystal Growth* 32(3): 335–349.

Okada A., Usuki A., . (1995). The chemistry of polymer-clay hybrids, *Materials Science and Engineering C* 3: 109–115.

Peeterbroech, S., L., B., Alexandre, M., Bnagy, J., Viville, P., Lazzaroni, R. & Dubois, P. (2007). The influence of the matrix polarity on the morphology and properties of ethylene vinyl acetate copolymers-carbon nanotube nanocomposites, *Composites Science and Technology* 67: 1659–1665.

Philip, B., Xie, J., Chandrasekhar, A., Abraham, J. & Varadan, V. (2004). A novel nanocomposite from multiwalled carbon nanotubes functionalized with a conducting polymer, *Smart Materials and Structures* 13(2): 295–298.

Sadeghian, R., Gangireddy, S., Minaaie, B. & Hsiao, K. (2006). Manufactoring carbon nanofibers toughned polyester/glass fiber composites using vacuum assisted resin transfer molding for enhancing the mode-1 delamination resistance, *Composites Part A* 37: 1787–1795.

Schubel P.J., Johnson M.S., W. N. R. C. (2006). Characterisation of thermoset laminates for cosmetic automotive applications: Part iii - shrinkage control via nanoscale reinforcement, *Composite: Part A* 37: 1757.

Seyhan, A., Gojny, F., Tanoglu, M. & Schulte, K. (2007a). Critical aspects related to processing of carbon nanotube/unsaturated thermoset polyester nanocomposites, *European Polymer Journal* 43: 374–379.

Seyhan, A., Gojny, F., Tanoglu, M. & Schulte, K. (2007b). Rheological and dynamic-mechanical behavior of carbon nanotube/vinyl ester-polyester suspensions and their nanocomposites, *European Polymer Journal* 43: 2836.

Sham, M. & Kim, J. (2006). Surface functionalities of multi-wall carbon nanotubes after uv/ozone and teta treatments, *Carbon* 44: 768.

Shen, J., Huang, W., Wu, L., Hu, Y. & Ye, M. (2007). The reinforcement role of different amino-functionalized multi-walled carbon nanotubes in epoxy nanocomposites, *Composite Science and Technology* 67: 3041.

Shi, D., Lian, J., He, P., Wang, L., Xiao, F. & Yang, L. (2003). Plasma coating of carbon nanofibers for enhanced dispersion and interfacial bonding in polymer composites, *Applied Physics Letters* 83: 5301.

Shubel, P., Johson, M., Warrior, N. & Rudd, C. (2006). Characterisation of thermoset laminates for cosmetic automotive applications: Part 3- shrinkage control via nanoscale reinforcement, *Composites: Part A* 37: 1757–1772.

Song, Y. & Youn, J. (2005). Influence of dispersion states of carbon nanotubes on physical properties of epoxy nanocomposites, *Carbon* 43: 1378.

Suh, D., Lim, Y. & Park, O. (2000). The properties and formation mechanism of unsaturated polyester-layered silicate nanocomposite depending on the fabrication methods, *Polymer* 41: 8557.

Sun, L., Warren, G., O'Reilly, J., Everett, W., Lee, S., Davis, D., Lagoudas, D. & Sue, H. (46). Mechanical properties of surface-functionalized swcnt/epoxy composites, *Carbon* 2008: 320.

Tchmutin, I., Ponomarenko, A., Krinichnaya, E., Kozub, G. & Efinov, O. (2003). Electrical properties of composites based on conjugated polymers and conductive fillers, *Carbon* 41(7): 1391–1395.

Thompson, C., Herring, H., Gates, T. & Connel, J. (2003). Preparation and characterization of metal oxide/polyimide nanocomposites, *Composites Science and Technology* 63: 1591–1598.

Trotignon, J., Verdu, J., Piperaud, M. & Dobraczinski. (1989). *Prï£¡cis de matiï£¡res plastiques*, 4 edt, Nathan.

Turcu, R., Bica, D., Vekas, L., Aldea, N., Macovei, D., Nan, A., Pana, O., Marinica, O., Grecu, R. & Pop, C. (2006). Synthesis and characterization of nanostructured polypyrrole-magnetic particles hybrid material, *Romanian Reports in Physics* 58: 359.

Twardowska, H. & Dammann, L. (2005). hermosetting inorganic clay nanodispersions and their use, from ashland inc, *US Patent 6* 887: 931.

Twardowska, H., Singh, R. & Dammann, L. (2005). Thermosetting inorganic clay nanodispersions and their use, from ashland inc, *US Patent 6* 841: 607.

Usuki, A., Kojima, Y., Kawasumi, M., Okada, A., Fukushima, Y., Kurauchi, T. & Kamigaito, O. (1993). Synthesis of nylon 6-clay hybrid, *Journal of Materials Research* 8: 1179–1184.

Vergne, B. (2007). *Mise en forme de composites NanoTubes de Carbone/Alumine et modélisation de leur conductivité thermique*, PhD Thesis, University of Limoges (France).

Wichmann, M., Sumfleth, J., Gojny, F., Quaresimin, M., Fiedler, B. & Schulte, K. (2006). Glass-fibre-reinforced composites with enhanced mechanical and electrical properties-benfits and limitations of a nanoparticle modified matrix, *Engineering Fracture Mechanics* 73: 2346.

Xiao, Y., Wang, X., Yang, X. & Lu, L. (2002). Nanometre-sized tio_2 as applied to the modification of unsatured polyester resin, *Material Chemistry and Physics* 77: 609–611.

Xu, J., Donohoe, J. & U., P. J. (2004). Preparation, electrical and mechanical properties of vapor grown carbon fiber (vgcf)/vinyl ester composites, *Composite: Part A* 35: 693.

Yuen, S., Ma, C., Wu, H. & Kuan, H. (2007). Preparation and thermal, electrical, and morphological properties of multiwalled carbon nanotube and epoxy composites, *Journal of Applied Polymer Science* 103: 1272–1278.

Zang, Z., Breidt, C., Chang, L., Haupert, F. & Friedrich, K. (2007). Enhancement of the wear resistance of epoxy: short carbon fibre, graphite, ptfe and nano-tio_2, *Composites: Part A* 35: 1385–1392.

Zengin, H., Zhou, W., Jin, J., Czerw, R., Smith Jr., D., Echegoyen, L., Carroll, D., Foulger, S. & Ballato, J. (2002). Carbon nanotube doped polyaniline, *Advanced Materials* 14(20): 1480–1483.

Zhao, D., Li, X. & Shen, Z. (2008a). Electromagnetic and microwave absorbing properties of multi-walled carbon nanotubes filled with ag nanowires, *Materials Science and Engineering: B* 150: 105.

Zhao, D., Li, X. & Shen, Z. (2008b). Microwave absorbing property and complex permittivity and permeability of epoxy composites containing ni-coated and ag filled carbon nanotubes, *Composite Science and Technology* 68: 2902.

Zhao, D., Li, X. & Shen, Z. (2009). Preparation and electromagnetic and microwave absorbing properties of fe-filled carbon nanotubes, *Journal of Alloys and Compounds* 471(1-2): 457–460.

Zhou, Y., Wu, P., Cheng, Z. & Ingram, J. (2008). Improvement in electrical, thermal and mechanical properties of epoxy by filling carbon nanotube, *Express polymer letters* 2(1): 40–48.

Zhu, J., Imam, A., Crane, R., Lozano, K., Khabashesku, V. & Barrera, E. (2007). Processing a glass fiber reinforced vinyl ester composite with nanotube enhancement of interlaminar shear strength, *Composite Science and Technology* 67: 1509.

Zou, W., Liu, Y., Yang, X., Li, H. & Zhang, C. (2008). Functionalization of mwnts using polyacryloyl chloride and the properties of cnt-epoxy matrix nanocomposites, *Composite Science and Technology* 68: 3259.

Part 3

Applications of New Materials

12

Novel Electroceramic: Polymer Composites - Preparation, Properties and Applications

Padmaja Guggilla and Ashok K. Batra
Alabama A&M University,
USA

1. Introduction

Engineers consistently demand new material systems, vital in emerging novel technological applications. Thus, this demand dictates that material scientists develop new material systems. The modern electronic devices and systems require diverse and specific functional properties in materials; which cannot be met in single-phase materials (Batra et al., 2008). Composite technology, where a novel functional material is fabricated by combining two or more chemically different materials or phases, for example, ceramics and polymers in an ordered manner or just mixing, is playing an important starring role. In recent decades, a large number of ceramic-polymer electronic composites have been introduced for medical, telecommunication and microelectronics applications, and devices ranging from micro-mechanical systems (MEMS; Bio-MEMS) through sensors and actuators (Taya, 2008). The composites have a unique blend of polymeric properties such as mechanical flexibility, high strength, design flexibility and formability, and low cost, with the high electro-active functional properties of ceramic materials. In these materials, it is, thus, possible to tailor, physical, electronic and mechanical properties catering a variety of applications. As a result, the composite as a whole is described using a set of microstructural characteristics, for example, connectivity, volume fractions of each component, spatial distribution of the components, percolation threshold and other parameters. Thus, the response of an electronic composite (electroceramic-polymer) to an external excitation (electric field, temperature, stress, etc.) depends upon the response of individual phases, their interfaces as well as the type of connectivity. As a result, an Electronic Composite can be broadly be described as exhibiting electromagnetic, thermal, and/or mechanical behavior while maintaining structural integrity (Taya, 2008).
With a recent progress in nano- science and technology, there is an increasing interest in polymer nano-composites both in scientific and in engineering applications (Kochary et al., 2008; Guggilla et al., 2009; Sebastian et al., 2010), therefore, a brief account of their electronic properties and applications will discussed. In the light of many technologically important applications in this field, the proposed chapter will be exhaustive and complete in all respects for naive to seasoned scientists.

2. Composite materials with piezoelectric, ferroelectric and pyroelectric functionalities

All crystals can be categorized into 32 different classes. These classes are point groups divided by using the following symmetry elements: (1) Center of symmetry, (2) axes of rotation, (3) mirror planes, and (4) several combinations of them. The 32 point groups are subdivisions of 7 basic crystal systems that are, in order of ascending symmetry, triclinic, monoclinic, orthorhombic, tetragonal, rhombohedral (trigonal), hexagonal, and cubic. Of the 21 classes of the 32 point groups that are noncentrosymmetric, which is a necessary condition for piezoelectricity to exist, 20 of them are piezoelectric. Of the 20 piezoelectric crystal classes, 10 crystals have pyroelectric properties. Within a given temperature range, this group of materials is permanently polarized. Compared to the general piezoelectric polarization produced under stress, the pyroelectric polarization is developed spontaneously and kept as permanent dipoles in the structure. As this polarization varies with temperature, the response is termed as pyroelectricity. The ferroelectric group is a subgroup of the spontaneously polarized pyroelectric crystals. Though the polarization of ferroelectric is similar to the polarization of pyroelectric, there is a difference in the two polarizations because the ferroelectric polarization is reversible by an external applied electric field, provided that the applied field is less than the dielectric breakdown of the materials. Therefore, materials that can be defined as ferroelectrics must have two characteristics: (1) The presence of spontaneous polarization and (2) reversibility of the polarization under an electric field. A ferroelectric material is therefore pyroelectric, piezoelectric, and noncentrosymmetric. It follows that, not all piezoelectric materials are pyroelectric, and not all pyroelectric materials are ferroelectric material as depicted in Figure 1. The Table 1 lists the important ferroelectric/pyroelectric materials being investigated and/or are being used in the various devices (Lal et al., 1993). Figure 1 show some important ferroelectric materials and their uses in various devices based on their unique piezoelectric, pyroelectric, ferroelectric, electro-optic, acousto-optic, and dielectric properties. The most commonly used materials for pyroelectric applications include triglycine sulfate (TGS); lead (Pb) zirconate titanate, $PbZrTiO_3$ (PZT); lead titanate, $PbTiO_3$ (PT); barium titanate, $BaTiO_3$(BT); lithium (Li) tantalate, $LiTaO_3$ (LT); and polymers: poly (vinylidene fluoride) (PVDF) and co-polymers including P (VDF-TrFE) and P (VDF-TFE).

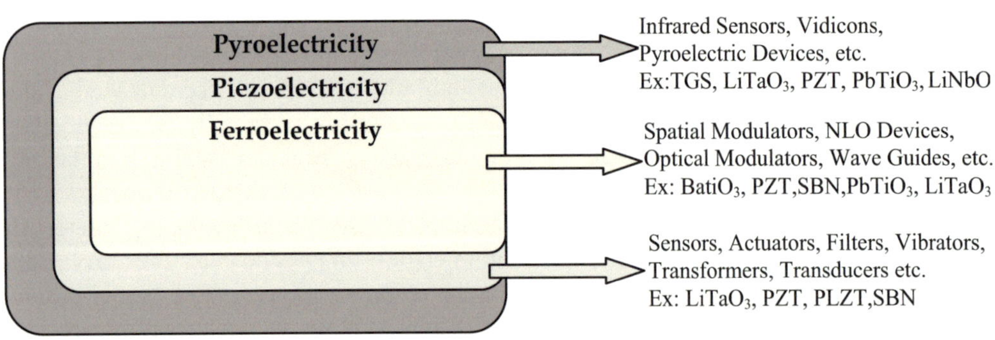

Fig. 1. The relationship between pyro, piezo, and ferroelectricity

Ferroelectrics' Family	Ferroelectric Material	Chemical Formula	Abbreviation
Perovskite type	Barium titanate	$BaTiO_3$	BT
	Lead zirconate titanate ceramics)	$PbZr_xTi_{1-x}O_3$	PZT
	Barium Strontium titanate	$Ba_xSr_{1-x}TiO_3$	BST
Lithium niobate	Lithium Niobate	$LiNbO_3$	LN
	Lithium Tantalate	$LiTaO_3$	LT
Barium Type	Barium Strontium Niobate	$Ba_{5x}Sr_{5(1-x)}Nb_{10}O_{30}$	BSN
TGS type	Triglycine Sulfate	$(NH_2CH_2COOH)_3 \cdot H_2SO_4$	TGS
Polymer	Poly (vinyldene Fluoride)	-	PVDF
	Poly (vinylidene Fluoride-trifluoroethylene)	-	P(VDF-TrFE)

Table 1. List of important pyroelectric materials Investigated/being Investigated for use in various devices

2.1 Physics and chemistry of composite materials

There are thousands of materials available for use in engineering applications. Most materials fall into one of three classes that are based on the atomic bonding forces such as metallic, ionic, covalent, and Van der walls etc. of a particular material; these three classifications are metallic, ceramic and polymeric. Additionally, different materials can be combined to create a composite material. Within each of these classifications, materials are often further organized into groups based on their chemical composition or certain physical or mechanical properties. Composite materials are often grouped by the types of materials combined or the way the materials are arranged together. Figure 2 clearly demonstrates a list of some of the commonly classification of materials within these four general groups of materials (Taya, 2008).

2.2 Constituents of composite materials

Composite materials are the materials made up of two or more materials combined together in such a way that the constituent materials are easily distinguished. The composite materials consist of basically three phases as discussed below and presented in Figure 3.

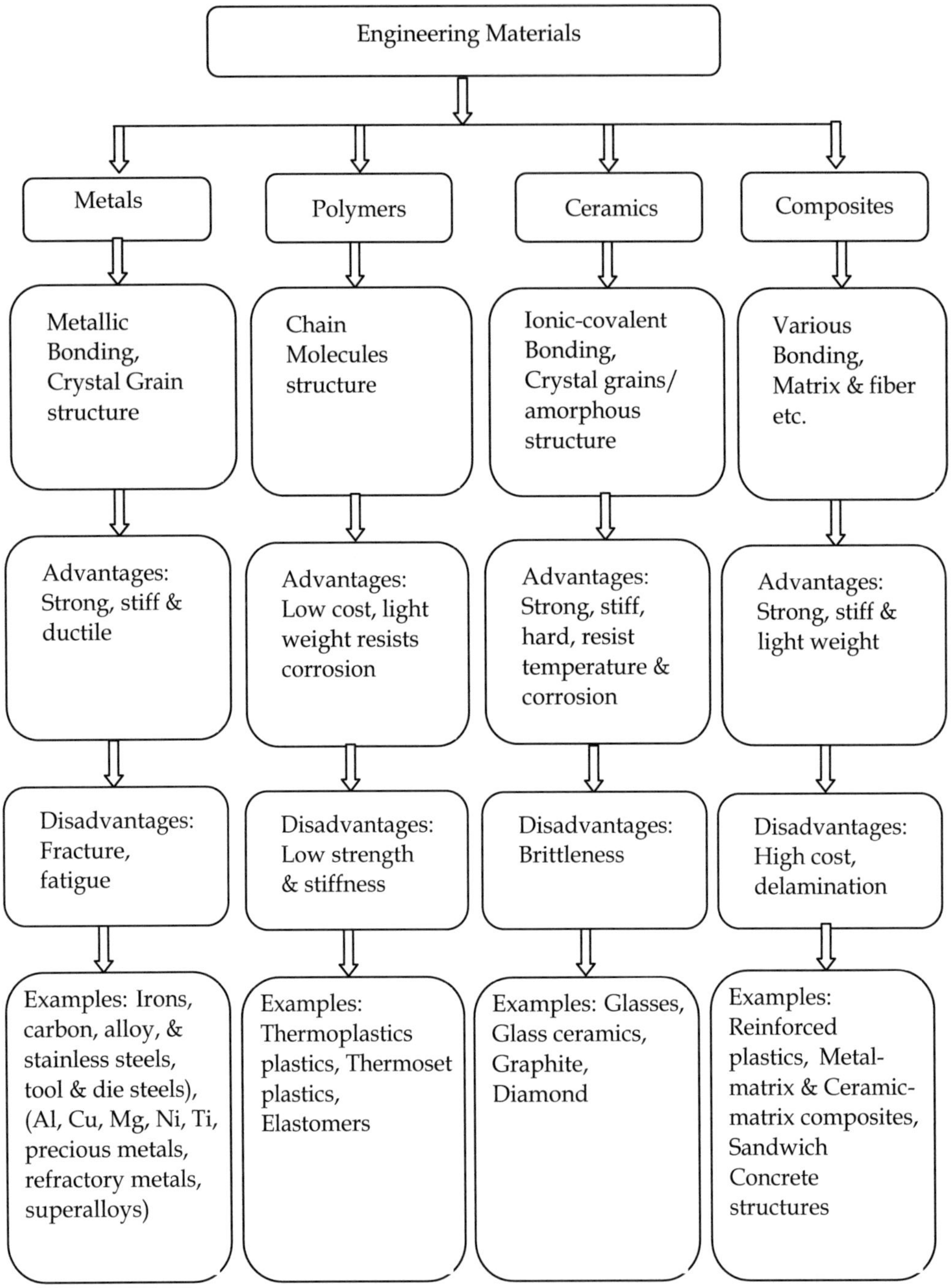

Fig. 2. General characteristics of the major classes of engineering materials

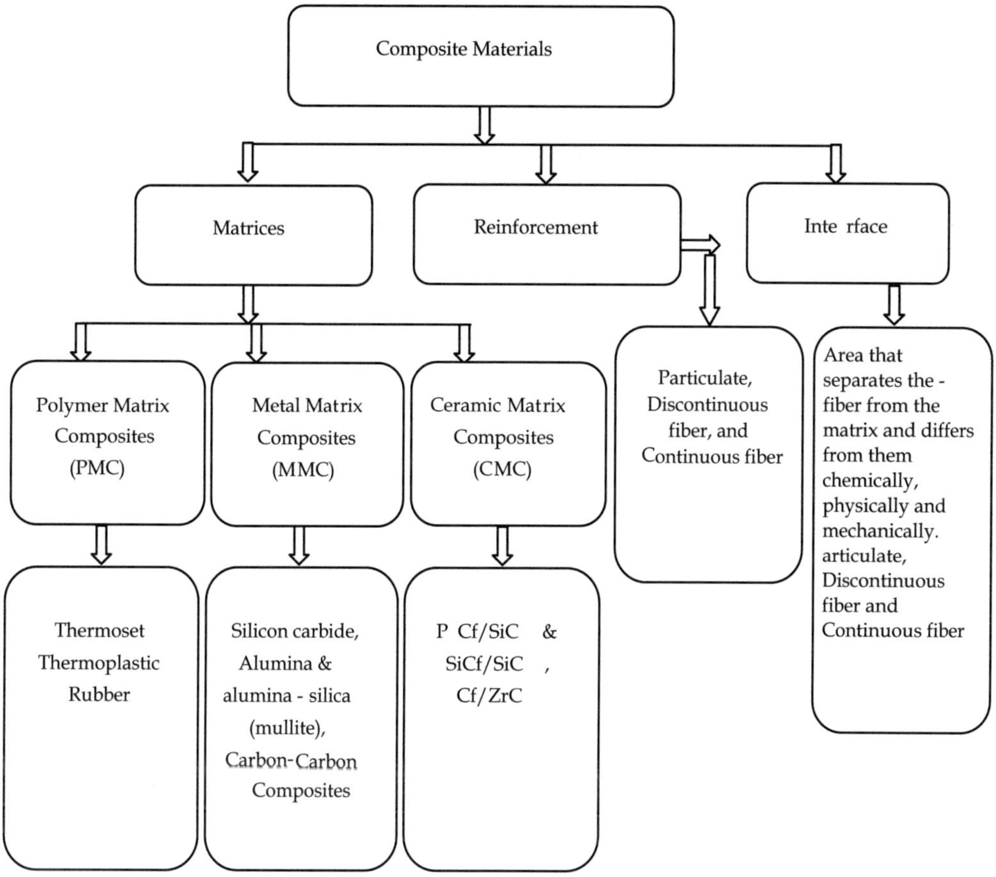

Fig. 3. Composite materials and types of constituents

2.2.1 Matrix phase

It is a continuous phase or the primary phase. It holds the dispersed phase and shares a load with it. It is made up of metals, ceramics or polymers depending on the tuple of composite in investigation.

2.2.2 Dispersed (reinforcing) phase

It is the second phase (or phases) which is imbedded in the matrix in a continuous/ discontinuous form. Dispersed phase is usually stronger than the matrix, therefore it is sometimes called reinforcing phase in case of structural composites. This reinforcement is a strong, stiff (functional) integral component which is incorporated into matrix to achieve desired properties or functionalities, basically it means desired property enhancement. It can be fiber or particles of any shape and size including nanoparticles as well.

2.2.3 Interface

This is the zone across which the matrix phase and reinforcing phases interact (chemical, physical, mechanical, electrical and others). This region in most composite materials has a finite thickness because of diffusion and/or chemical reactions between the fiber and the matrix.

A multiphase material formed from a combination of materials which differ in composition or form, remain bonded together, and retain their identities and properties. Composites maintain an interface between components and act in concert to provide improved specific or synergistic characteristics not obtainable by any of the original components acting alone. Composites include: (1) fibrous (composed of fibers, and usually in a matrix), (2) laminar (layers of materials), (3) particulate (composed of particles or flakes, usually in a matrix), and (4) hybrid (combinations of any of the above) (Mallick, 1997).

A few common types of composites as listed above are presented in the following Figure 4.

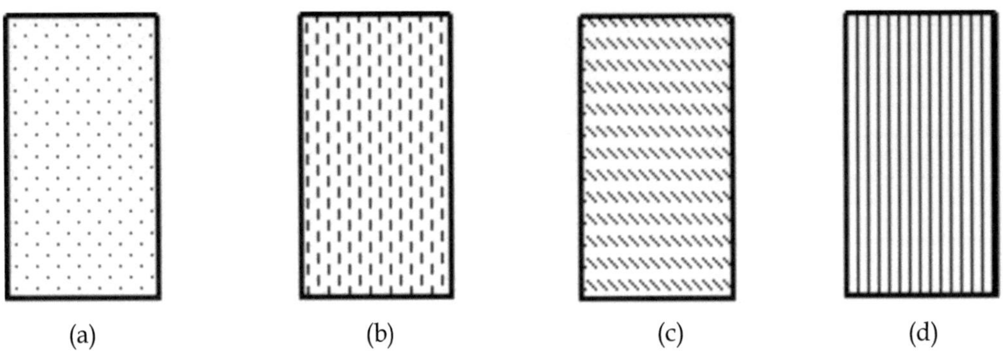

Fig. 4. (a) Particulate and random (b) Discontinuous fibers and unidirectional (c) Discontinuous fibers and random (d) Continuous fibers and unidirectional

3. Theory of pyroelectric detector

A pyroelectric detector, based on conversion of photons to phonons to electrons, as shown in Figure 5, is a capacitor whose spontaneous polarization vector is oriented normal to the plane of the electrodes. Incident radiation absorbed by the pyroelectric material is converted into heat, resulting in a temperature variation (dT) and thus, the magnitude of the spontaneous polarization. Changes in polarization alter the surface charge of the electrodes, and to keep neutrality, charges are expelled from the surface which results in a pyroelectric current in the external circuit. The pyroelectric current depends on the temperature change with time. Therefore, pyroelectric devices are considered to be 'AC' coupled device. Pyroelectric current (Ip) is proportional to Area (A) and rate of change of temperature (dT/dt) of the detecting element, one can write:

$$Ip = p.A\frac{dT}{dt} \tag{1}$$

where p is the pyroelectric coefficient. Determining the electrical response of a pyroelectric detector requires analysis of the thermal and electrical circuits and optical parameters. The processes that take place during radiation detection are illustrated in Figure 5.

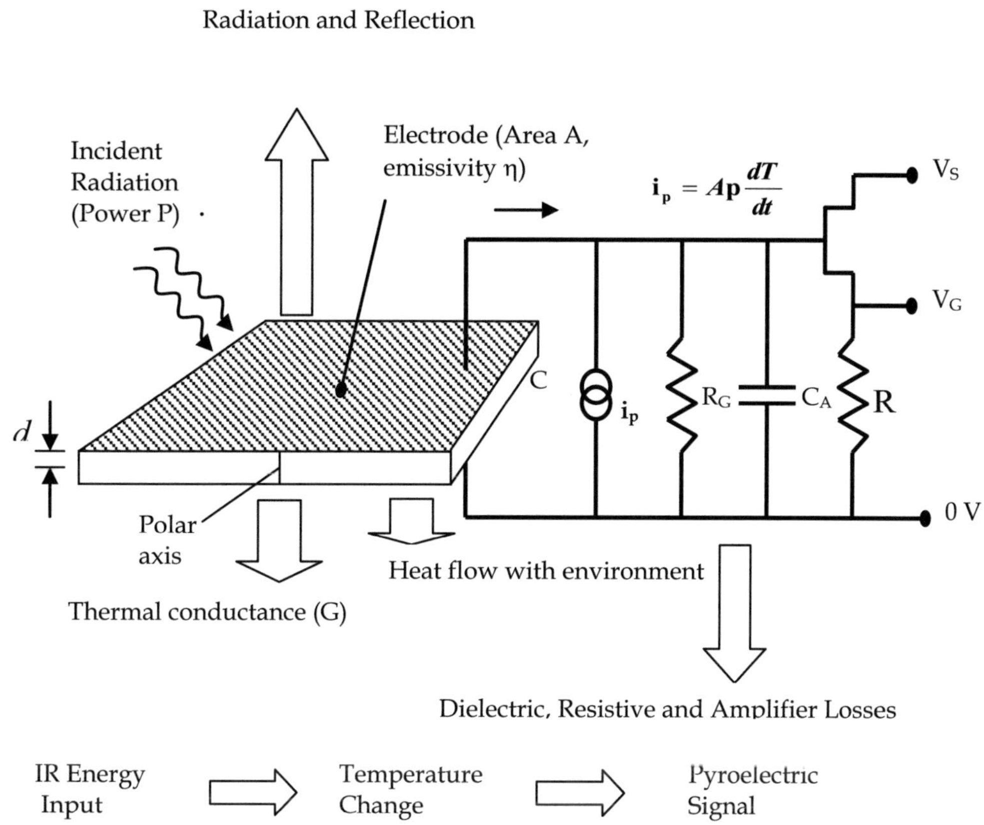

Fig. 5. A Schematic diagram of a pyroelectric detector and detection processes

4. Materials figure-of-merits for infrared detectors

The important properties to look for in sensors are low dielectric constant and loss, high pyroelectric coefficient, and low specific heat. However, important figure-of- merits (F) (Sidney et al., 2000; Lal et al., 1993 and Whatmore et al., 2001) are,

$$F_I = p / c' \qquad \text{for high current detectivity,} \tag{2}$$

$$F_V = p / c'\varepsilon' \qquad \text{for high voltage responsivity,} \tag{3}$$

$$F_D = p / \sqrt{(c'\varepsilon'')} \qquad \text{for high detectivity,} \tag{4}$$

Where p is the pyroelectric coefficient, c' is the specific heat of the element, ε' is the dielectric constant and ε'' is the imaginary part of dielectric constant (dielectric loss).
Bauer et al., 1991 has developed following the figure-of-merits, when the pyroelectric element is placed on a substrate that is acting as a heat sink i.e. whose thermal conductivity is infinite where k is the thermal conductivity of the pyroelectric element.

$$F_I = p/k \qquad \text{for high current detectivity} \qquad (5)$$

$$F_V = p/k\varepsilon' \qquad \text{for high voltage responsivity} \qquad (6)$$

$$F_D = p/\sqrt{(k\varepsilon'')} \qquad \text{for high detectivity.} \qquad (7)$$

5. Novel ceramic: polymer matrix composites

The search for new material systems for piezoelectric and pyroelectric infrared detecting device applications has led to the design of 'ferroelectric: polymer' composites. Recent studies on ceramic-polymer based pyroelectric composites show potential usefulness via large area, lightweight, enhanced strength, and flexible infrared sensing behavior. Thus, composites based on pyroelectric ceramic particles embedded in polymer possess hybrid properties derived from individual components. These hybrid properties include large pyroelectric coefficients of ceramic material and excellent mechanical strength, formability and robustness of the polymer, eventually useful for infrared detectors. Infrared radiation sensors can be generally divided into two classes: i) photon detectors, in which the radiation absorption process directly produces a measurable effect, e.g., generation of photoelectrons or charge carrier pairs in a photoconductor; and ii) thermal detectors, in which absorbed radiation is converted first to heat, which subsequently produces a measurable effect. Pyroelectrics, along with thermocouples, thermopiles, thermistors, and bolometers, belong to the later class. Pyroelectric infrared (PIR) detecting devices have the following advantages over the photon infrared sensors: sensitivity in very large spectral bandwidth limited only by the ability of the sensor to absorb the incident radiation; sensitive in very wide temperature range without the need of cooling; low power requirements; relatively fast response; generally low cost materials; cannot be detected being a passive device; temperature range of operation can be changed by the variation of the amount of the constituents (such as Lead zirconate titanate, Potassium tantalate niobate and others); and suitable for space applications because of light weight consuming less power having no bulky cooling equipment.

The diphasic polymer composites consist of ceramic particles embedded in the polymer matrix. The properties of these composites depend on the following factors: (i) properties of its constituents; (ii) volume fraction of each constituent; (iii) polarizability of particles; and (iv) nature of inter-connecting these particles. Composite materials comprising of ferroelectric material particles embedded in polymer materials with different connectivities have generated great interest among known pyroelectric infrared detecting materials such as lead titanate (PT); lead zirconate titanate (PZT); barium titanate (BT) ; and triglycine sulfate (TGS) having polymer matrix host poly-vinylidene fluoride (PVDF) or polyvinylidene fluoride-trifluoroethylene [P(VDF-TrFE)], and virgin pyroelectric materials (Dias and Das-Gupta, 1996 and Aggarwal et al., 2010). It is worth mentioning that novel pyroelectric ceramics are defined as nano-ceramics as well.

5.1 Pyroelectric composite materials for un-cooled infrared detectors

5.1.1 Importance of pyroelectric composites

Zook and Liu, 1976 proposed that pyroelectric material with a figure-of-merit, $(p/\sqrt{\varepsilon})$, significantly higher than presently known ferroelectric materials is unlikely to be found. This was based on the fact that relationship:

$$P_o = \left(k\varepsilon_o C_c / v\right) \tag{8}$$

where P_o is the polarization, ε_o is the permittivity of free space, C_c is the Curie constant, k is the Boltzmann's constant, and v is the volume per unit dipole, which has been derived by using Devonshire's two-level dipolar effective field model and the Aizu-Lines lattice dynamic effective field model. C_c and v are materials dependent, so in order to maximize polarization one has to maximize Cc or/and minimize the dipole volume v. As of now, the Curie constant of known ferroelectrics are in the range of $2x10^5$ to $5x10^5$K and minimum polarizable volume seems to have a minimum of approximately $50A^{o3}$, therefore, finding materials with higher polarization seems unlikely. Increase in performance may be possible by working on 0-3 composites. The figure-of-merit (for high voltage responsivity) of the pyroelectric detector is proportional to pyroelectric coefficient and inversely proportional to dielectric constant of pyroelectric material. To decrease the dielectric constant, efforts have also been made by forming a matrix-void composite. In the forming process only particles are introduced in the matrix so as to decrease the dielectric constants and hence obtain an increase in the figure-of-merit is obtained. Similarly 0-3 connectivity polymer-ceramic composite shall give attractive results.

5.2 Theoretical modeling of electronic composites

A number of models have been proposed or used to predict the dielectric constants, piezoelectric and pyroelectric coefficients of mixture of two or more components (Dias and Das-Gupta, 1996; Sidney and Das-Gupta, 2000; Kohler et al., 1998; Tressler et al., 1999). A summary of dielectric constants and pyroelectric coefficients mixtures formulas is described below. Most of the formulas for mixture give expression in terms of dielectric constants of phases, their volume fractions, and depolarization factors of inclusions. A simple case where the mixture rule for calculation of permittivity, using a model of a capacitance consisting of different homogenous dielectrics connected in series or parallel is shown in Figure 6a & 6b respectively along with formulas of permittivity (Newnham et al., 1978 and Tressler et al., 1999). However, in case of one phase consisting of particles dispersed in other matrix phase, the calculations become more complicated because the field distortion caused by a polarized particle depends on its orientation and shape with respect to the applied field. Thus, different formulas have been proposed, and important ones are listed below.

Maxwell-Garnett formula:

$$\varepsilon_{eff} = \varepsilon_1 \frac{\varepsilon_1 v_1 (1 - A) + \varepsilon_2 (v_2 + A v_1)}{\varepsilon_1 + A v_1 (\varepsilon_2 - n\varepsilon_1)} \tag{11}$$

where v_1 and v_2 are the volume fractions of the matrix and inclusions respectively. The depolarization factor of the inclusions in the direction perpendicular to the capacitance plates is A. ε_1 and ε_2 are the dielectric constant of matrix and inclusions respectively. ε_{eff} is the dielectric constant of the composite. In equation (11), the dielectric properties are calculated from electric-field average taken over both the component of the mixture. Brugemann assumed that in Maxwell-Garnett formula holds in a diluted limit when an infinitesimal amount of inclusions are added in the mixture, which leads to following equation (12), obtained by solving the differential equation solved with proper initial conditions.

Brugemann formula:

$$\frac{\varepsilon_{eff} - \varepsilon_2}{\varepsilon_1 - \varepsilon_2}\left(\frac{\varepsilon_1}{\varepsilon_{eff}}\right)^A = v_1 \tag{12}$$

Bottcher introduced a model, in which particles of both the phases are dispersed in an effective medium with dielectric constant ε_{eff}. Assuming that the average dipole field due to particles vanishes, the Bottcher formula is obtained. Bottcher formula:

$$\varepsilon_{eff} = \varepsilon_1 + (\varepsilon_2 - \varepsilon_1)v_2 \frac{\varepsilon_{eff}}{\varepsilon_{eff} + (\varepsilon_2 - \varepsilon_{eff})A} \tag{13}$$

Looyenga formula:

$$\varepsilon_{eff}^{1-2A} = v_1\varepsilon_1^{1-2A} + v_2\varepsilon_2^{1-2A} \tag{14}$$

In Loovenga's model, it was assumed that the composite behaves similarly as a system containing the same overall composition, but composed of spheres that itself are heterogeneous and having a slight different composition.
In the following formulas φ is the volume fraction of phase 2, ε_{eff} is the dielectric constant of composite.
Logarithmic formula:

$$\log\varepsilon_{eff} = (1-\varphi)\varepsilon_1 - \varphi\log\varepsilon_2 \tag{15}$$

Wagner formula:

$$\varepsilon_{eff} = \varepsilon_1 \frac{2\varepsilon_1 + \varepsilon_2 + 2\varphi(\varepsilon_2 - \varepsilon_1)}{2\varepsilon_1 + \varepsilon_2 - \varphi(\varepsilon_2 - \varepsilon_1)} \tag{16}$$

Landauer formula:

$$(1-\varphi)\frac{\varepsilon_{eff} - \varepsilon_1}{2\varepsilon_{eff} - \varepsilon_1} + \varphi\frac{\varepsilon_{eff} - \varepsilon_2}{2\varepsilon_{eff} + \varepsilon_2} = 0 \tag{17}$$

Banno formula:

$$\varepsilon_{eff} = \frac{a^2(a+(1-a)n)^2\varepsilon_1\varepsilon_2}{a\varepsilon_2 + (1-a)n\varepsilon_1} + (1-a^2(a+((1-a)n))\varepsilon_2 \qquad \varphi = a^3 \tag{18}$$

Pletto Formula:

$$\varepsilon_{eff} = (1-\varphi)\frac{\varphi\varepsilon_1 A^2 + (1-\varphi)\varepsilon_2}{(1+\varphi(A-1))^2} + \phi\frac{\varphi\varepsilon_1 + (1-\varphi)\varepsilon_2 B^2}{(\varphi + (1-\varphi)B)^3} \tag{19}$$

where

$$A = \frac{3\varepsilon_2}{\varepsilon_1 + 2\varepsilon_2}$$

$$B = \frac{3\varepsilon_1}{2\varepsilon_1 + \varepsilon_2}$$

5.3 Design of composites - connectivity models

For fabrication of a composite, properties of the components, amount of each phase present, and how they are interconnected viz connectivity, are important. Newnham et al., 1978 and Tressler et al., 1999 proposed the concept of connectivity. Any phase in a mixture can be self-connected in zero, one, two, and three dimensions. For example, inclusions dispersed in a polymer host material shall have connectivity 0 while host polymer shall have connectivity 3. Thus, we can say composite with connectivity 0-3 or 0-3 composites. In a two phase composite system, there can be ten different connectivities, which are 0-0, 0-1, 0-2, 0-3, 1-1, 1-2, 1-3, 2-2, 2-3, and 3-3. In this format, first digit denotes the connectivity of inclusions and second digit denotes the host. Generally, the host is a polymer in case of polymer composites. A few connectivities are presented in Figure 7 using a cube as a building block along with real examples. Based on the above concept, in 0-3 connectivity composite, there is random distribution of active particulates in a 3D host polymer matrix.

6. Fabrication of polymer-ceramic composites

The 0-3 connectivity composites are easy to fabricate, which allows for commercial production of these composites in a cost effective manner (Nalwa, 1995). The composites which have shown some promising results and are prepared by simple composite fabrication routes are described in this section. As a result, most composites are fabricated as an active pyroelectric and polymer based diphasic samples. Polymer component can be polar or non-polar polymers. The second more popular approach is using the sol gel

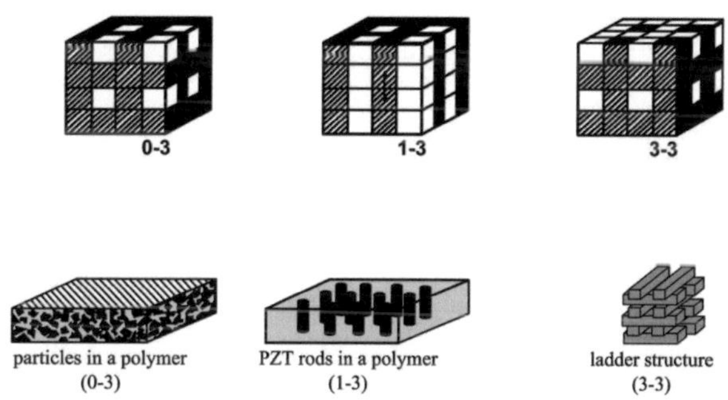

Fig. 7. Connectivity Models (Newnham et al., 1978)

synthesis route. Though the glass-ceramics have also shown noticeable pyroelectric sensing elements, this section doesn't include specific descriptions section on such materials. The 0-3 connectivity composite may be prepared by mixing the ceramic particles in a hot rolling mill with softened thermoplastic polymer, and thin films of composites then be produced by high-pressure casting at softening temperature of the polymer. With a thermo-set polymer such as epoxy, the mixing can be made at room temperature with the right proportion of the resin, hardener and the ceramic powder (Nalwa, 1995 and Yamazaki and Tayama, 1981).
In solvent casting, first a polymer is dissolved in a suitable solvent and then electro-active ceramic powder is added and mixed/dispersed. A mixture so obtained is kept in a suitable container for solvent to evaporate. The film so obtained is hot pressed at crystallization temperature of the polymer. The problems are faced during the mixing operation, which are linked to a poor distribution of the ceramic inclusions, poor adhesion of component phases

or air bubbles in the composites. An agglomeration of ceramic particles is also present in the solution, which may be overcome by lowering the polymer viscosity via heating during the ultrasonic mixing process. For the fabrication of 0-3 connectivity composites films using spin coating technique. A typical example for the preparation of P (VDF-TrFE): PZT composite, first of all, a suitable amount of polymer, P (VDF-TrFE) is dissolved in methyl-ethyl-ketone (MEK) to form a solution (PMix). A requisite amount of nano-ceramic (PZT) powder is then added and the mixture is ultrasonically agitated for several hours to break-up the agglomerates and to disperse the ceramic powder uniformly in the copolymer solution. With this composite solution (nPMix), a thin film can be deposited on conducting electroded substrate using a spin coating technique. The steps involved in fabricating films by spin coating are illustrated in Figure 9. The film is then annealed for 2-3 hours in air at 130 °C for the present case and top electrode is deposited for testing (Abdullah and Das-Gupta, 1988 & 1990; Abdullah, 1989). The flow chart illustrating various steps involved in fabricating composite films are shown in Figure 8.

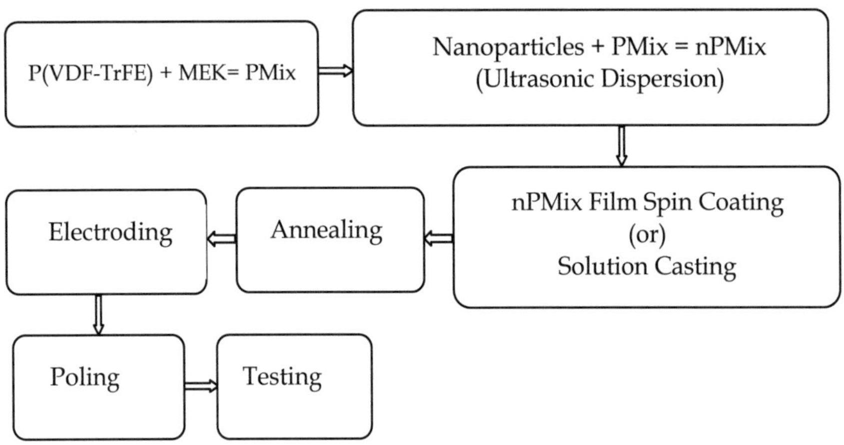

Fig. 8. A Flow chart for fabrication of thin-film of composite

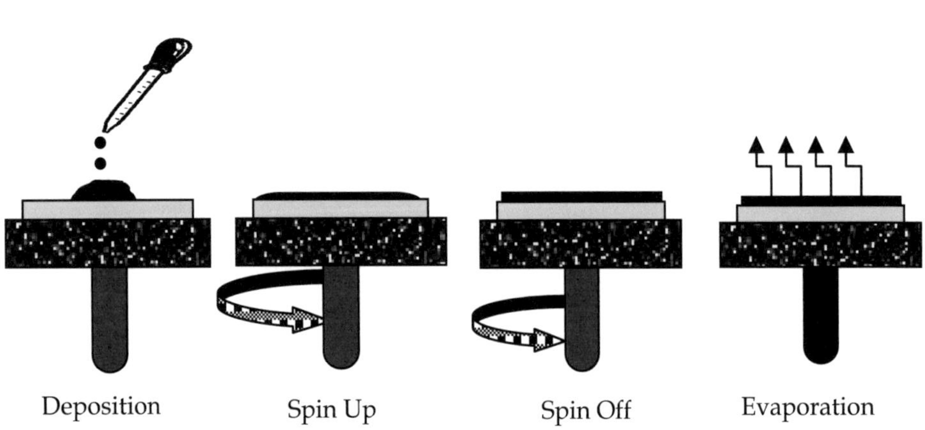

Fig. 9. Spin Coating Technique Processing Steps

7. Poling of composites

To develop pyroelectric-/electro-activities, these composites are initially subjected to an external electric field. This process is called poling. The composites can be poled (i.e., polarized) by following methods namely: thermal, electric field, corona, electronic-beam poling; plasma, and hysteresis poling (Sessler, 1994).

In thermal poling a high DC step voltage of appropriate magnitude (1–10 MV/m) is applied at an elevated temperature for an extended period of time (from a few minutes to a few days) in vacuum or silicon oil bath. The sample is then cooled down to room temperature with the polarizing DC voltage on. This way, polar alignment is stabilized. The ceramic content shall be polarized provided the applied field is greater than the coercive field of the ceramic. In some cases for poling polymers, film is stretched while the corona discharge is on. This produces uniform orientation of domains and large piezoelectric coefficients (Sessler, 1994; Das-Gupta and Doughty, 1986; Kaure et al., 1991).

In field poling, an applied field is up to 10 MV/m and poling time is usually a few seconds. After poling, the sample is shorted to remove the charges.

In corona poling (Sessler, 1994), a non-metalized face of the sample is exposed to Corona discharge between a needle (Dias and Das-Gupta, 1996), biased at 10 to 15kV, and the electroded rear face of the sample is grounded. This way one side electroded sample is charged on the unelectroded face by means of a corona current produced at a corona point (Dias and Das-Gupta, 1996; Sessler, 1994). The samples are exposed for about a few seconds to minutes at room temperature or elevated temperature. The charges deposited by corona discharge set up a field in the sample and cause alignment of the dipoles to occur.

In electron beam poling, the field generated by injected electrons polarizes the sample. One side of the sample is electroded only. A non-metalized sample is irradiated with electrons. The injected electrons are trapped, after drift, in a thin layer. Thus, electric field between the electrons and their counter charges on the rear electrodes polarizes the sample.

In hysteresis poling, an electric field of a low frequency (0.1Hz) is applied to the sample at room temperature.

Ploss et al., 2000, Ploss et al., 2001 and Chan et al., 1999 have demonstrated a poling technique for ferroelectric composites of ceramic particles like lead titanate or lead zirconate titanate in poly (vinylidene fluoride-trifluoride) or poly (Vinylidene fluoride-trifluoroethylene). A copolymer, in the matrix can be poled parallel or in anti-parallel directions. When the matrix and inclusions are poled in opposite directions, the pyroelectric response will be reduced, but the piezoelectric response will be reinforced. On the other hand, if two phases are polarized in parallel, the pyroelectric response reinforces while piezoelectric activity partially cancels, thereby reducing vibration-induced electrical noise in pyroelectric sensors. The composite films cited above, a dc field of the order of 50 MV/m for 1 hour at 115°C (above Curie temperature of the polymer) is used to polarize the ceramic phase. When cooling to room temperature, the electric field is kept on, so that a polymer phase is polarized in the same direction as the ceramic phase. The composite is reheated to 95°C and poled under dc field of 50MV/m for another two hours to polarize the polymer phase completely.

8. Properties and infrared detector performance parameters of important pyroelectric: polymer composites

8.1 Micro- and nano-pyroelectric polymer composites

The dielectric and pyroelectric coefficient, on TGS-PVDF composites with different proportion of TGS have been investigated (Wang et al., 1991). It was found that the variation of dielectric and pyroelectric coefficients with temperature for composites were as reported for TGS single crystals. Furthermore, it was reported that with 50 wt% of TGS particles in the composite, the figure of merit is the largest. It was also shown that D* value of detectors made out of TGS composite reaches up to (5-7)×10^7cm $Hz^{1/2}$ W^{-1}. Wang et al., (Wang et al., 1993) investigated the 45 micron to 75 micron TGS particles of different volume fraction (up to 80 vol. fraction), dispersed in organic solvent for the preparation of films using solution-casting technique. It was observed that the pyroelectric coefficient increased to 90 μC/m²K and figure of merit (p/ε) to 3.3 with 80 % volume of TGS particles. A novel technique has been used in the fabrication of ATGS-PVDF oriented film (Changshui et al., 1998). A high electric field was applied during preparation of the film. It was also reported that PVDF: ATGS (L-alanine doped TGS) that with highest field grains get oriented to b- axis when field is increased perpendicular to b-axis. With a field of 10 kV/cm, better results for dielectric constant ~10 and pyroelectric coefficient ~ 30 μC/m²K at about 30 °C have been reported. Recently Yang et al., 2006 performed an extensive and noteworthy study on TGS: P (VDF-TrFE) composites with various volume fractions (0.05 to 0.43) of TGS embedded in P (VDF-TrFE). The pyroelectric coefficient varied from 32μC/m²K to 102μC/m²K. The dielectric constant increased from 9.66 to 12.27 while the dielectric loss decreased from 0.021 to 0.008. The two phases of samples were polled in same direction, in which pyroelectric coefficient reinforced while the piezoelectric contribution partially cancels out. The low piezoelectric activity in pyroelectric composite is an asset as it reduces the vibration induced noise. TGS: P (VDF-TrFE) is a good candidate for sensing element in pyroelectric infrared detecting devices.

The detailed study on the pyroelectric and piezoelectric properties of PT-P (VDF-TrFE) composites gave the maximum value of pyroelectric coefficient, with 54 % volume fraction of the particles (Chan et al., 1998). Pyroelectric coefficient and dielectric constant were 40.7 μC/m²K and 57.3 respectively which gives a value of 0.71 for figure-of-merit. However, maximum value of figure-of-merit (0.92) for composite having 49 % vol. fractions was obtained when only ceramic phase was poled. However, when both the phases were poled then pyroelectric coefficient and dielectric constant were 69.2 μC/m²K and 55 respectively which gives the value of figure-of-merit of 1.24. The pyroelectric properties of $PbTiO_3$-P(VDF-Tree) 0-3 nano-composites films fabricated using with various volume fraction of ceramic of particles of about 70 nm size have been investigated by Chen et al., 1998. They could fabricate films with only up to volume fraction of 12% due to problem of agglomeration. The films were prepared by spin coating method on aluminum-coated glass. All the figure-of-merits (F_I, F_V and F_D) reported by them showed an increase with increase of volume fraction of lead titanate ceramic particles. The results obtained for pyroelectric properties are shown in Figure 9. It was concluded that pyroelectric coefficient of the composite with 12% volume fraction of PT was 40% higher than that of the polymer. A D* = 1.2×10^7 cm $Hz^{-1/2}W^{-1}$ has been achieved in composite films of [($Pb_{0.8}Ca_{0.2}$) TiO_3:P(VDF-TrFE)] prepared using appropriate amount of ceramic particles having size less than 100 nm (Zhang et al., 2000). The values obtained for dielectric and pyroelectric coefficients by Zhang et al., 2001 for nano-crystalline calcium (0.11%), modified lead titanate (PCLT) are presented in Table 2. F_V and F_D exhibit maxima around 0.11 volume fraction of PCLT nanoparticles. The figure-of-merits were 35% higher than poled copolymer. A noteworthy study was performed on thin film deposited on silicon substrate, consisting of 12 vol. % of nano-sized lanthanum and calcium modified lead titanate embedded in P (VDF-TrFE) 70/30 matrix to form pyroelectric sensors with three different configurations. The maximum specific detectivity (D*) 1.3×10^{7}cm $Hz^{1/2}/W$ (at 1 kHz), 2.11×10^{7}cm $Hz^{1/2}/W$ (at about 300Hz) and 2.8×10^{7}cm $Hz^{1/2}/W$ (between 5 and 100Hz) respectively, was reported. Authors suggested that sensors fabricated with PCLT/P (VDF-TrFE) nano-composites have potential to be use in silicon based pyroelectric sensors (Zhang et al., 1999). The characteristics of 8 x 1 array fabricated using modified lead titanate powder prepared by sol-gel method dispersed in polymer, P (VDF-TrFE) were studied by Zhang et al.1999, Loiacono and Dougherty, 1978. The amount of powder used was 12 vol. %. The film of thickness of 12 μm was prepared by spin-coating method on glass substrate. After poling the film, it was removed from the glass substrate and bonded of a silicon chip with readout electronic circuitry to form 8x1 integrated linear arrays. The variation of specific detectivity (D*) with frequency is shown in Figure 9. The maximum value of D*obtained in this array was 1.45×10^{7}cm $Hz^{1/2}W^{-1}$. However this value is 1-2 orders lower than found in single crystalline materials. Lead magnesium niobate-lead titanate (abbreviated PMN-PT) with 30% mol PT: P (VDF-TrFE) 70/30 composite has been investigated by Lam and Chan, 2005. With increasing KTN content, the ferroelectric, piezoelectric and pyroelectric properties of the composite improve comparative to the PMNZT/P(VDF-TrFE) biphasic composite with the same PMNZT volume fraction due to increasing poling degree of the PMNZT particles (Xia et al. 2008).

Hilczer et al., 2002 have performed an important study of pyroelectric and dielectric dispersion response of P $(Z_{0.5}T_{0.5})O_3$-P ($VDF_{0.5}$-$TrFE_{0.5}$) nanocomposite film samples of 100-200μm thick having powder of $P(Z_{0.5}T_{0.5})O_3$ crystallite size ~30nm. Composites ($PZT_{0.32}PVDF_{0.68}$, $PZT_{0.24}[P(VDF_{0.5}\text{-}TrFE_{0.5})_{0.76}]$ displayed an excellent pyroelectricity. It was observed by dielectric dispersion studies that the role of polymer heterogeneity at crystalline level is important in PZT-PVDF nanocomposites in the temperature range from 315K to 395K. In PZT- P ($VDF_{0.5}$-$TrFE_{0.5}$) composites, this becomes important above ~350K due to effect of the ferroelectric-paraelectric phase transition. It was reported that the variation of dielectric constant is temperature and frequency independent between ~275K to 310K. Due to excellent figure-of-merits values for these composite, it was proposed that the sensor operating temperature could be in the range of 290K to 310K. Recently authors of this article performed an extensive study on P (VDF-TrFE): modified PZT composites films prepared by solution casting technique and also with spin-coating method. The results are presented in Table 2 (Batra et al., 2003). Recently thick films of 0-3 composites of lead zirconate titanate ceramic and polyvinylidene - trifluorethylene copolymers have been produced by spin-coating on gold coated silicon wafer (Dietze et al., 2007). A pyroelectric coefficient of 92 $\mu C/m^2K$ has been measured for composites with 20 % volume ceramic. For the composites with 10 and 20 vol. % ceramic, voltage and current figures-of-merit are higher than PZT thick films. A most important disadvantage with the use of PVDF as sensor material is the requirement of a high poling field. PVDF and lithium tantalate (LT) composites are investigated for their pyroelectric and dielectric properties as infrared detector. The choice of LT was decided by the fact that LT has much lower poling field requirement and larger pyroelectric coefficient, with comparable dielectric constant to PVDF. Studies show that LT/PVDF nanocomposite can prove to be a better alternative to both single crystal $LiTaO_3$ and pure PVDF for pyroelectric sensor applications. (Satapathy et al., 2009)

Lithium tantalite [($LiTaO_3$), LT] ceramic particles have been incorporated into a polyvinylidene fluoride-trifluoroethylene [P (VDF-TrFE) 70/30 mol%] copolymer matrix to form composite films. The films were prepared using solvent casting method with the LT powder homogeneously dispersed in the P (VDF-TrFE) copolymer matrix with various volume fractions. Electrical properties such as the dielectric constant, dielectric loss, and pyroelectric coefficient have been measured as a function of temperature as well as frequency. In addition, materials' figures-of-merit have also been calculated to assess their use in infrared detectors. The results show that the fabricated lithium tantalite: polyvinylidene fluoride-trifluoroethylene composite films have a good potential for uncooled infrared sensor applications operating at moderate temperatures (Batra et al., 2009).

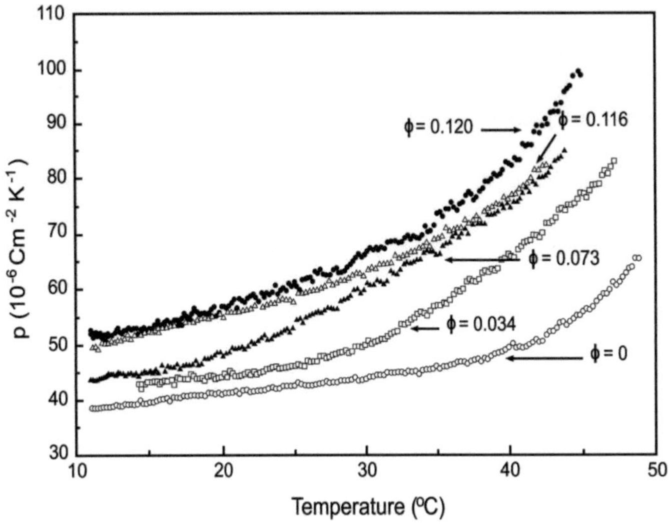

Fig. 9. Pyroelectric coefficients of the composite films with various volume fraction of ceramic (ϕ) lead titanate as function of temperature (Chen et al., 1998).

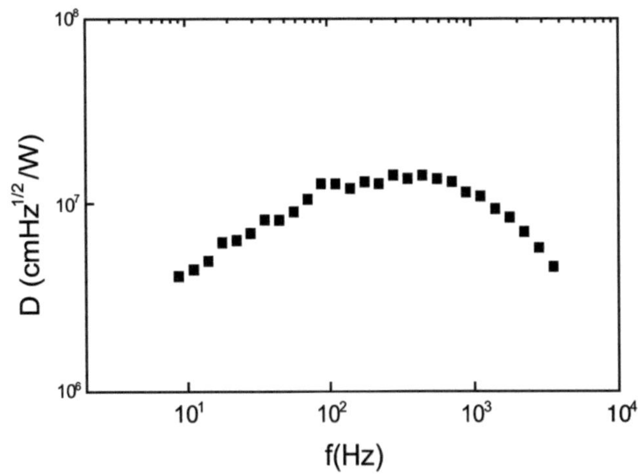

Fig. 10. Specific detectivity of an element in an array (Zhang et al., 1999).

Sample	Fi = p (μC m^{-2}K^{-1})	ε′	$F_V = p/\varepsilon'$ (μC m^{-2}K^{-1})	References
P(VDF-TrFE) (Commercial Sample)	34	16.1	2.1	Chan et al., 1998
PVDF (Commercial Sample)	38	5.3	5.7	Dias and Das-Gupta, 1996
P(VDF-TrFE)70/30 (Commercial Sample)	41.2	11.0	3.7	Dias and Das-Gupta, 1996
PT(vol 62%):PVDF Used dispersoids ceramic grains and pressed into thin films of 30-70 μm range	130	54	2.4	Dias and Das-Gupta, 1996
PLZT(26..27wt%) :P(VDF-TrFE Used dispersoids ceramic grains and Solution Casting method for thick films of range 20 μm	16.7	25	0.66	Guggilla, 2007
PLZT (12..33wt%): nano P(VDF-TrFE) Used dispersoids ceramic grains and Solution Casting method for thick films of range 20 μm	9.0	14.12	0.64	Guggilla, 2007
PZT (vol. 50%): PVDF	90	10	0.11	Dias and Das-Gupta, 1996
PZT (vol.50%): P(VDF-TrFE)	39	118	0.33	Lang and Das-Gupta, 2000
TGS:PVDF	30	10.0	3.0	Kobune, 1997
BT: PVC	0.08	3.7	0.02	Dias and Das-Gupta, 1996
PLZT (10.57wt%) :P(VDF-TrFE) Used dispersoids ceramic grains and Solution Casting method for thick films of range 30 μm	11.07	33.15	.33	Guggilla, 2007

Table 2. Material characteristics and Figure-of-Merits of 0-3 Composites

8.2 Metal-nanoparticles reinforced pyroelectric polymer composites

As both Lithium tantalite (($LiTaO_3$), LT) and polyvinylidene fluoride-trifluoroethylene (P (VDF-TrFE)) have excellent ferroelectric and pyroelectric properties, LT ceramic particles have been incorporated into P (VDF-TrFE) copolymer matrix with 70/30 mol% to form 0-3 composite films. Due to the low relative permittivity of LT the pyro-activity and performance of infrared detector will be increased [4]. However, pyroelectric particles in the 0-3 composites cannot be fully poled due to the screening effect of the polymer matrix i.e. important properties of the composite are reduced. Sakamoto et al [5] indicated that PZT/PU composite doped with graphite particles improved the poling behavior of PZT phase and intern improved pyroelectric and piezoelectric properties. Oltean et al [6] studied some electrical properties of metallic iron reinforced polymeric composite materials subjected to stress state: electric field, and temperature variation and mechanical load. In order to increase the conductivity of these LT: P (VDF-TrFE) samples it is proposed to embed the silver nano particles into this system.

Films of LT: P (VDF-TrFE) composites with 2% volume fraction of LT powder and with Ag nanoparticles particles have been fabricated using 'solution cast' method. This technique is a very useful and inexpensive technique for manufacturing composite pyroelectric sensor because composite films can be fabricated with less energy, time and effort as compared to ceramic and single crystal fabrication. To assess their use as pyroelectric infrared detectors, various figures-of-merit of composite films have been calculated and compared with pure P (VDF-TrFE) film fabricated with the described techniques in the previous section.

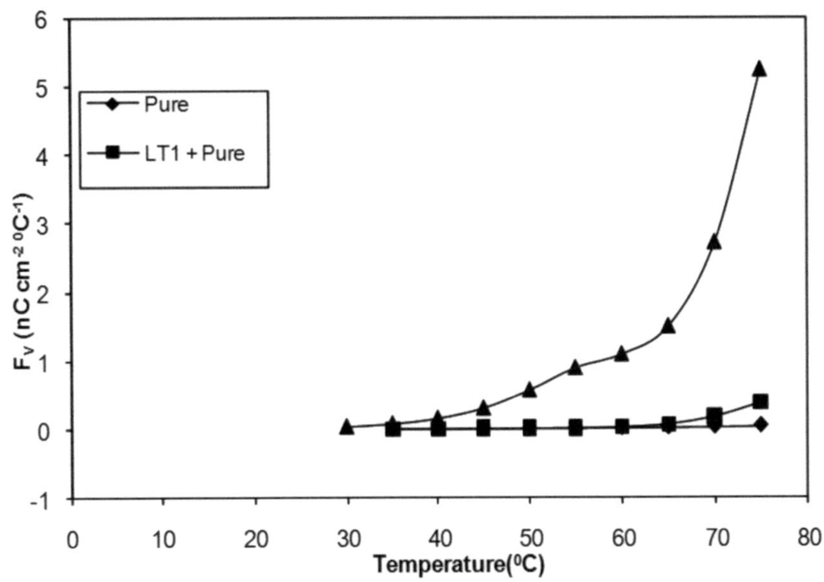

Fig. 11. Temperature Dependence of Figure-of-Merit, F_V, for Pure PVTF, PLT1 and PLT1Ag films

The pyroelectric performance figures-of-merit, the voltage responsivity increases with the volume fraction of LT and with addition of silver nanoparticles in the polymer matrix as shown in figure 11. Preliminary data on Figures-of-Merit of composite films are higher than P (VDF-TrFE) film poled under similar conditions. Based on the preliminary results obtained, LT: P (VDF-TrFE) and LT: P (VDF-TrFE) +Ag films are attractive for use in infrared sensing elements especially where low level applications and curved surface detectors are needed.

9. Advantages of micro- and nano-pyroelectric composites

Some of the advantages of the Micro and nano-pyroelectric composites are greater tensile and flexural strength for the same dimension of polymer part, reduced weight for the same performance increased dimensional stability, improved gas barrier properties for the same film thickness and higher chemical resistance.

10. Applications of pyroelectric detectors

These uncooled pyroelectric infrared detectors are used in many applications and few of them are: Air Quality Monitor; Atmospheric Temperature Measurement; Earth Position Sensor; Earth Resources; Engine Analysis; Fire Alarm; Gas Analyzer; Glass Processing; Horizon Sensor; Infrared Detection; Infrared Spectrometer; Interferometer; Interplanetary Probe; Intrusion Detector; Laser Detection; Laser Power Control; Liquid Fuel Analysis; Meteorology; Plasma Analysis; Plastic Processing; Pollution Detection; Position Sensor; Radiometer; Reflectance Measurements; Remote Sensing; Sky Radiance; Solar Cell Studies; Human Sensors; Forest Fire Detection; Pyroelectric Vidicons; Ear Thermometer; Terahertz Detection, Biomedical Imaging; Satellite-Based IR Detection at 90°K; and Optical Wave Guide Studies; X-ray Detector; Microwave Detector; facial recognition; traffic control; vision testing; law enforcement and border patrol. Recent applications are their use in Waste Energy Harvesting for Micro-electric Generators.

11. Conclusions

Composites of polymers blended with ferroelectric ceramics have been well studied; design, processing and characterization (from viewpoint of their applications). Ferroelectric ceramic: polymer composites with 0-3 connectivity pattern owe their popularity to easy fabrication procedure that allows for mass production at a relatively low cost. Special reference has been made to composites made with modified lead titanate and lead zirconate titanate. These ceramics have relatively high pyroelectric properties, which gives composites substantial advantage with additional strength and flexibility of the polymer. One additional advantage of this class of ceramic–polymer composite is that when the ceramic and polymer (P (VDF-TrFE)) are poled in the same direction, the pyroelectricity of two phases reinforces while their piezoelectricity partially cancels. It minimizes the micro-phony effect arising from vibration-induced electrical noise. It is evident that higher current responsively (F_I) and detectivity (F_D) figure-of-merit can be obtained in composite materials fabricated with modified PT/PZT electro-ceramic and co-polymer (P (VDF-TrFE)) in right proportion.

However, only fabrication of detector with these composites shall prove their worthiness in real world. The note-worthy results have been reported by Zhang et al., 2000; a maximum value of D* 1.45 X10^7 cm $Hz^{-1/2}W^{-1}$ at 280Hz. for an 8x1 integrated linear pyroelectric array using PCLT/P(VDF-TrFE) composite. The composite detectors are useful for low-level applications. These detectors can be used and most importantly, where parameters like

flexibility, high strength, and large area of the infrared detector are required, e.g. cavity-shaped detector.
According to Dias and Das-Gupta, 1996, the composites however have the advantage of being easier to polarize in thicker self-supporting samples, preventing thus the need for a substrate. It can be concluded that there is a strong potential of 0-3 composites use in for low-level applications, including integration with semiconductor processing. Other technique of enhancing the pyroelectric performance such as using bimorph structure (Alexe and Pintilie, 1995) for composites can give higher voltage responsivity. However, the pyroelectric elements used in general have to be thin for the reasons of low thermal mass (and hence the high ΔT change), most papers included in the current discussions have not been the basis of such selection or requirements. The aim of this chapter has been to overview the advantages of making composites and to study the gains in the pyroelectric figures of merit or the detectivity of the sensing elements. By having such guideline studies the further advances in designing and fabrication of composite sensing elements of desirable thickness or dimensions can be made as the future composite fabrication goals.

12. Acknowledgements

Authors gratefully acknowledge the partial support of the present work by NSF HBCU RISE project # 0927644. Special thanks to Drs. M. E. Edwards, R. B. Lal, M. D. Aggarwal for keen interest in our work.

13. References

Abdullah, M. J., and Das-Gupta, D. K., (1988). *Ferroelectrics,* 87 312

Abdullah, M. J., (1989) A Study of Electro-active Properties of Polymer/ceramic Composites, University of Wales - Si ECS (Bangor)

Abdullah, M. J., and Das-Gupta, D. K., (1990). *IEEE Trans. Electr. Insnrl.,* Vol. 25(3) 605.

Aggarwal, M. D., Batra, A. K., Guggilla, P., Edwards, M. E., Penn, B. G., and Currie, J. R. (2010). Pyroelectric Materials for Uncooled Infrared Detectors: Processing, Properties and Applications. *NASA Technical Memorandum,* 216373 1-71

Alexe M. and Pintilie, L. (1995). Analysis of the pyroelectric bimorph as radiation detector, *Infrared Physics & Technology,* 1 pp (949-954)

Batra, A. K., Aggarwal, M. D., Lal, R. B. (2003). Growth and characterization of doped DTGS crystals for infrared sensing devices. *Materials Letters* 57 pp (3943-3948)

Batra, A. K., Aggarwal, M. D., Edwards, M. and Bhalla, A. S. (2008). Present Status of Polymer: Ceramic Composites for Pyroelectric Infrared Detectors. *Ferroelectrics,* 366 84-121

Batra, A. K., Corda, J., Guggilla, P., Aggarwal, M. D., Edwards, M. E. (2009). Electrical properties of Silver Nanoparticles Reinforced $LiTaO_3$:P(VDF-TrFE) Composite Films. *SPIE Proceedings Vol. 7419;* Infrared Systems and Photoelectronic Technology IV, pp(721313-2-721313-9), San Francisco, CA August 2009

Dietze, M., Krause, J., Solterbeck, C. H., and Es-Souni, M. (2007) Thick film polymer-ceramic composites for pyroelectric applications. *J. Appl. Physics* 101 (054113-1-054113-7)

Bauer, S., and Ploss, B. (1991). *Sensor and Actuator A* (25-27) 417

Chan, H. L. W., Chan, W. K., Zhang, Y., and Choy, C. L. (1998). Pyroelectric and piezoelectric properties of lead titanate/polyvinylidene fluoride-trifluoroethylene 0-3 composites. *IEEE Trans. Dielec. and Elec Ins.*, 5(4) pp(505-512)

Chan, H. L. W., Ng, P. K. L., and Choy, C. L. (1999). *Appl. Phys. Lett.*, 74(20) 3029

Chang, J., Batra, A. K., and Lal, R.B. (1996). Growth and properties of urea-doped triglycine sulfate (UrTGS) crystals. *Journal of Crystal Growth* 158 pp(284-288)

Changshui, F., Qingwu, W., and Hongsheng, Z. (1998). Preparation and Pyroelectric properties of oriented composite ATGS-PVDF film. *J. Korean Phys. Soc.*, 32 pp(S1843-S1845)

Chen, Y., Chan, H. L. W., and Choy, C. L. (1998). Pyroelectric properties of PbTiO3/P(VDF-TrFE) 0–3 nano-composite films. *Thin Sold. Films* 323 pp(270-274)

Das-Gupta, D. K., Doughty, K. J., (1986). *Appl. Phys. Lett.*, 27 4601

Dias, C. J., and Das-Gupta, D. K., (1996). *IEEE Trans. Die. Elect. Ins.* 3(5) 706 and references therein

Dietze, M., Krause, J., Sollerbeck, C. H., Es-Souni, M. (2007). Thick film polymer-ceramic composites for pyroelectric applications. *JAP* (101) pp (054113-1-054113-7)

Dieguez, E., Plaza, J. L., Aggarwal, M. D., Batra, A. K. (2010). Czocharlski Growth of Photo-refractive Crystals In: *Handbook of Crystal Growth*, Dhanaraj, G.; Byrappa, K.; Prasad, V.; Dudley, M., pp (559-598) Springer, 978-3-540-74182-4, USA

F. Kochary, M. D. Aggarwal , A. K. Batra, R. Hawrami , D. Lianos , A. Burger: Growth and electrical characterization of the lead magnesium niobate-lead titanate (PMN-PT) single crystals for piezoelectric devices. J. Mater Sci: Mater Electron., 19 (2008)1058-1063

Guggilla, P., Batra, A. K., Currie, J. R., Aggarwal, M. D., Lal, R. B. (2006). Pyroelectric Ceramics for Infrared Detection Applications. *Materials Letters* 60 pp (1937-1942)

Guggilla, P., (2007) Studies on pyroelectric materials for infrared sensor application. Ph.D. Dissertation, Alabama A & M University, Normal, Alabama, U.S.A.

Guggilla, P., Batra, A.K., Edwards, M. E. (2009). Electrical Characterization of $LiTaO_3$: P (VDF-TrFE) Composites. *J. Materials Science* **44**, pp (5469-5474)

Hilczer, B., Kulek, J., Markiewicz, E. and Kosec, M. (2002). Dielectric relaxation in ferroelectric PZT-PVDF nanocomposites. *Journal of Non-Crystalline Solids*, 5, pp (167–173)

Kaure, T., Nath, R., and Perlman, M. M. (1991). *J. Phys. D: Applied Physics*, 24 1848

Kobune, M., Mineshige, A., Fujii, S., and Maeda, Y. (1997). *Jpn. J. Appl. Phys.*, 36 5976

Kohler, R., Padmini, P., Gerlach, G., Hofmana, G., Bruchhaus, R. (1998). *Int. Ferroelectrics*, 22 383

Krakovsky, I., and Myroshnychenko, V. (2002). *J. Appl. Phys.*, 992 (11) 6743 and references therein

Lal, R. B., and Batra, A. K., (1993). Growth and properties of triglycine (TGS) sulfate crystals Review. *Ferroelectrics* 142 pp (51-82)

Lal, R. B., and Batra, A. K., (1996). Growth of crystals for infrared detectors and second harmonic generation devices. Ceramic Transactions, vol.60, American Ceramic Society, OH

Lam, K. H., and Chan, H. L. W. Composites Science & Technology, 65(2005) 1107

Lee, M. H., Guo, R., and Bhalla, A. S. (1998). Pyroelectric Sensors, *Journal of Electroceramics*, 2 pp (229-242)

Loiacono, G. M., and Dougherty, J. P. (1978). Final Technical Report (contract no DAAK70-77-C-0098) submitted to Night Vision and Electro-optics Laboratories, Fort Belvoir, Virginia

Mallick, P.K. (1997).Constituents In: Composites Engineering Handbook (Materials Engineering, ISBN: 978-0824793043 CRC Press, USA

Nalwa, H. S. (1995). Ferroelectric Polymers, ed.page 598 Marcel Dekker, NY

Newnham, R. E., Skinner, D. P., Cross, L. E. (1978). *Mat. Res. Bull.*, 13 523

Ploss, B., Ng, W. Y., Chan, H. L., Ploss. B., and Choy, C. L. (2001). *Composite science and technology*, 61 957

Ploss, B., Shin, F. G., Chan, H. L. W., and Choy, C. L. (2000). *Appl. Phys. Lett.*, 76(19) 2776

R. E. Newnham, D. P. Skinner, L. E. Cross:_Connectivity and piezoelectric-pyroelectric composites, Mat. Res. Bull., 13 (1978) 525-536

Satapathy, S., Gupta, P. K., and Varma, K. B. R. (2009). Enhancement of nonvolatile polarization and pyroelectric sensitivity in lithium tantalate (LT)/poly (vinylidene fluoride) (PVDF) nanocomposite. J. Phys. D: Appl. Phys. 42 055402 (6pp)

Sessler, G. M. (1994). Ferroelctric Polymer and Ceramic-Polymer Composites, *Key Engn. Mat.*, vol. 92-93 249

Sidney B. L., and Das-Gupta, D. K., (2000). *Ferroelectrics Review*, 2 pp (217-223)

Taya, M. (2008). Introduction In: *Electronic Composites*, pp (1-19), Cambridge University Press, ISBN-13: 9780521057318, UK

Tressler, J. F., Alkoy, S., Dogan, A., Newnham, R. E. (1999). *Composites (A)*, (30) 477

Wang, M., Fang, C. S., and Zhuo, H. S. (1991). Study on the pyroelectric properties of TGS-PVDF composites, *Ferroelectrics* 118 pp (191-197)

Wang, Y., Zhong, W., and Zang, (1993). P. Pyroelectric properties of ferroelectric-polymer composite, *J. App. Phys.*, 74(1) pp (512-524)

Whatmore, R. W., and Watton. (2001) R. Pyroelectric Materials and Devices Infrared Detectors and Emitters: Materials and Devices. *Kluwer Academic Publishers*,

Xia, Y. F., Ming, Z. D., Wei, D. Z., Yuan, C. Z., and Lin, J. S. (2008). The influence of ferroelectric KTN particles on electric properties of 0–3 ferroelectric composites. J. Phys. D: Appl. Phys. 41 (2008) 055408 (5pp)

Yamada, T., Ueda, T., and Kitayama, T. (1982). J. Phys. Phys. 53 (4) 4328

Yamazaki, H. and Tayama, T. K. (1981). Ferroelectrics, 33 147

Yang, Y., Chan, H. L. W., and Choy, C. L. (2006). Properties of triglycine sulfate/poly (vinylidene fluoride-trifluoroethylene) 0-3 composites, *J. Mat. Sci.*, 41 pp (251-258)

Zhang, Q. Q, Chan, H. L. W., and Choy, C. L. (1999). Dielectric and pyroelectric properties of P (VDF-TrFE) and PCLT–P (VDF-TrFE) 0–3 nanocomposite films, *Composites Part a: Applied Science and Manufacturing* 30 pp (163-167)

Zhang, Q.Q, Chan, H. L. W., Ploss, B., Zhou, Q. F., and Choy, C. L., (1999).*J. Non. Cryst. Solids*, 254 pp (118-125)

Zhang, Q. Q, Chan, H. L. W., Ploss, B., and Choy, C. L. (2001). PCLT/P(VDF-TrFE) nanocomposite pyroelectric sensors, *IEEE Trans. Ultra., Ferro., and Freq. Cont.*, 48(1) pp(154-160)

Zhang, Q.Q, Ploss, B., Chan, H. L. W., and Choy, C. L. (2000). *Sensors and Actuators A*, 86 216

Zhang, Q. Q, Ploss, B. Chan, H. L. W., and Choy, C. L. (2000). Integrated pyroelectric arrays based on PCLT/P (VDF-TrFE) composite, *Sensors and Actuators A* 86 pp (216-219)

Zook, J. D., and Liu, S. T. (1976). *Ferroelectrics*. 11 371

13

Designing Bio-Inspired Composite Materials for Medical Applications

Oana Craciunescu and Lucia Moldovan
Department of Cellular and Molecular Biology, National Institute R&D for Biological Sciences, Bucharest, Romania

1. Introduction

Composite materials are multi-phased combinations of two or several components, which acquire new characteristic properties that the individual constituents, by themselves, cannot obtain. A composite material typically consists of a certain matrix containing one or more fillers which can be made up of particles, sheets or fibers. When at least one of these phases has dimensions less than 100 nm, the material is named a nanocomposite and offers in addition a higher surface to volume ratio. There are natural composite materials, like wood and plant leaves, in the vegetal kingdom and bird feathers, silky threads spun by the spider and shells, in the animal kingdom. Also, connective tissues from animal and human body are natural composite materials due to their composition and structure. Connective tissues are the major supporting tissues of the body. They are named after their main function, i.e. packing and binding other structures together, and also providing a framework for the body. Connective tissues are generally soft tissues (e.g., skin, cartilage, cornea, etc), excepting bone which is a dense connective tissue.

Similar to other natural composite materials, bone consists of an organic part that forms the matrix and an inorganic part representing the filler. Bone matrix is a framework mainly composed of collagen fibers which together with small quantities of other non-collagenous proteins, proteoglycans, lipids, peptides and water form a hydrogel (~ 30 % from bone dry weight). The filler, that reinforces bone matrix, is formed of nano-sized crystals of carbonated calcium phosphate apatite (~ 70 % from bone dry weight). This natural nanocomposite material has superior strength and toughness than its individual components. Bone matrix is a source for nourishing bone cells, such as osteoclasts, osteoblasts and osteocytes, which grow inside it. It also increases cell biological activities like adhesion, proliferation and differentiation. Bone has key functions as skeletal tissue of the body, including support of softer tissues, mechanical protection for many internal organs and storage of minerals. In the development phase, bone attains the most suited structure to resist the forces acting upon it. During life time, bone is subjected to various diseases that are inherited (*osteogenesis imperfecta*) or caused by metabolism disturbances, such as osteoporosis, osteosarcoma, osteoarthritis. Bone is also affected by traumas, i.e. fractures, micro-fractures. Each year, millions of people are treated in hospitals for fractures presenting risk of developing into delayed union or nonunion. Also, progressive aging of population and the related pathologies lead to loss of variable quantities of bone that have to be replaced. The demand for bone substitutes is extremely large in orthopedic clinics from all over the world. That is why a growing interest in different aspects of creation, characterization, testing and application of composite materials for biomedical applications is registered.

Composite materials for medical application are developed for pathologies of osseous tissues from different parts of the body, like long bones, vertebrae, cartilage and teeth. Two-dimensional and three-dimensional structures are fabricated and commercialized as composite materials or combined with therapeutic organic substances (drugs, growth factors, etc) (table 1). There are several treatments for bone repairing: patching, replacing the missing tissue using allografts or xenografts, or self-healing initiated by materials containing signal molecules for tissue remodelling. The existing methods and techniques for treatment of large bone defects, as a result of trauma or tumor, do not satisfactory restore bone tissue. The classic technique for bone repair consists in autologous bone implantation, but is limited by the availability of transplanted material, the morbidity of the donor, difficulties in harvesting, longer hospitalization period and higher treatment costs. The last decade, initiated the utilization of resorbable materials, tailored with structures having controlled porosity, as medical devices for *in vivo* tissue regeneration (Silva et al., 2005; Patterson et al., 2008). The structure of the composite material has a role in the transport of nutrients, metabolites and regulator molecules towards and from the cells. New rapid prototyping techniques, like 3-D printing, selective laser sintering, stereolithography, allow the development of desired structures, similar to natural bone, having reproducible, well-defined shapes and controlled pore morphology and density. Composite material has a microporosity referring to the free spaces remaining between ceramic material particles bound on the polymer and a macroporosity meaning the pores larger than 100 μm from its structure. The morphological characteristics have a direct impact on the uniform distribution of cells within the porous material. An optimal pore size and interconnectivity facilitate cell colonization into the construct and influence the geometry of the new developed tissue. As for material properties, there is a need to improve its mechanical characteristics, in order to obtain a controlled biodegradability and good biocompatibility.

The main constituent of bone matrix – collagen, can be prepared using standardized techniques and purified at high levels (>90 %, w/w). For better mimicking natural bone composition, chondroitin sulfate is added into the collagen matrix. Chondroitin sulfate represents the glycanic part of the small proteoglycan, named biglycan, found in bone. It is a glycosaminoglycan having a linear chain of repeating units of anionic, acidic sugars. Chondroitin sulfate plays an important role in the regulation of mineralization process and in repair of bone defects in animals (Douglas et al., 2008). Also, chondroitin sulfate carries negative charges which are known to enhance osseous cell proliferation (Ohgaki et al., 2001). The other main component of composite materials intended for bone substitution is one of the several constituents of calcium phosphate minerals class (Barrere et al, 2006). Calcium hydroxyapatite, ($Ca_{10}(PO_4)_6(OH)_2$) and tricalcium phosphate (TCP), ($Ca_3(PO_4)_2$) are the most commonly investigated ceramics for biomedical applications because they possess the ability to improve new bone formation, showing osteoconductive properties (Laurencin et al, 2006). When their crystal size is similar to the nanometer size of the apatite from the natural bone, an increase in protein adsorption and osteoblast adhesion is expected (Webster et al., 1999). Both ceramics are thoroughly used in bone substitutes, but it is demonstrated a better conductivity, osteocompatibility and resorption rate for tricalcium phosphate than for highly crystalline, sintered hydroxyapatite (Vaccaro, 2002). The latter is limited in use because of its brittleness and difficult processing (Fujita et al, 2003). Bone apatite also contains various trace elements, such as magnesium, copper, zinc, silicate, fluoride, which are showed to have an effect on bone quality. Magnesium ion is the most abundant in the development phase of cartilage and bone tissues and sharply decreases when the bone is mature. Magnesium depletion alters bone and mineral metabolism which results in bone loss and is a risk factor for osteoporosis (Rude et al., 2009).

Product	Conditioning form	Composition	Delivered substance	Target tissue	Reference
MinerOss	Powder	Bone allograft	-	Periodontal	Gapski et al., 2008
Cortoss	Injectable	Resin with glass ceramic particles	-	Spinal	Bae et al., 2010
rh-BMP-2 (development)	Sponge	Collagen and titanium mesh	rhBMP-2	Spinal	Mulconrey et al., 2008
Healos	Sponge	Collagen type I with hydroxyapatite coating	rhGDF-5, gentamicin, marrow aspirate	Bone	Magit et al., 2006 Carter et al., 2009 Furstenberg et al., 2010
ProOsteon	Sponge	Sea coral with hydroxyapatite	-	Bone	Jensen et al., 2007
Immix	Microsphere	Polylactic acid/glycolic acid	-	Bone	Chenite & Chaput, 2010
Carticel	Hydrogel	Synthetic and natural polymers	growth factors, cells	Cartilage	De Bie, 2007

(rh-BMP-2 is recombinant human bone morphogenetic protein-2, rhGDF-5 is recombinant human growth and differentiation factor-5)

Table 1. Several composite products for bone repair

A series of collagen-calcium phosphate composite materials are tailored and used as temporary scaffolds in studies on animals and humans for tissue regeneration (Wang et al., 2004; Chen et al., 2009). The addition of collagen to a ceramic material provides many advantages for medical applications: shape control, spatial adaptation and ability for clot formation (Scabbia & Trombelli, 2004). Collagen could also serve as efficient bonding agent for ceramic particles. Conversely, the addition of calcium phosphates to collagen scaffolds improves the osteoconductive properties of the material (Takahashi et al., 2005; Kretlow et al., 2007). Composite properties are strongly dependent on synthesis conditions, like the calcium phosphate/collagen ratio, temperature and pH. The cohesion between the two materials is based on the interaction of calcium ions from ceramic material and the carboxyl groups from collagen (Zou et al., 2005).

The main advantage of a composite material tailored from collagen and calcium phosphate is the excellent biocompatibility property, due to collagen and its ability to allow bone cell attachment and differentiation. The crystals of synthetic calcium phosphate remain undistorted, for a long period after material implantation, sustaining the formation of new tissue.

The main difficulty in using these devices is that they are easily degraded and reabsorbed by the body. After hydration, they don't possess strength and their mechanical properties are relatively low in comparison to bone (Matsuno et al., 2006). The problem to be solved is to obtain a controlled degradation of the composite material so that to ensure as long as possible a scaffold where cells could deposit new bone tissue. Cross-linking could be used in order to control composite biodegradation rate and its mechanical characteristics, but it might compromise the biocompatibility.

Several bone regeneration therapies use a combination of collagen-ceramic composite materials, cell population and signal substance delivery to initiate more rapidly the healing process (Arinzeh et al., 2005; Guo et al., 2006). It were tested various growth factors, like transforming growth factor-beta, basic fibroblast growth factor or bone morphogenetic proteins (McKay et al., 2007; Evans, 2010).

This chapter describes the design of bio-inspired composite materials, as bone substitutes, choosing the most appropriate composition and structure to fulfill the unique morphological characteristics and biological properties of natural bone. Several aspects regarding the fabrication of a composite material that mimic porous bone structure are discussed in this chapter. A key aspect is composite cross-linking and a discussion on carbodiimide advantages is presented. *In vitro* experimental models on cell cultures are used for composite material biocompatibility evaluation. A bioactive implant consisting of osteoblast cells injected into the composite material and cultivated *in vitro* is analyzed for osteogenic properties by cell adhesion assay and osteoblast-specific marker expression.

2. Preparative methods

At present, there are known 29 types of collagen, having various structural and functional properties, depending on the connective tissue where they are found. Therefore, collagen extraction from different tissues it is not a standard procedure. Collagen can be obtained in its insoluble form, acid soluble form, neutral salt soluble form or its denatured form - gelatin using three types of extraction methods with neutral salt solutions, dilute acid solvents, chemical agents (acids or bases) with proteolytic enzymes. The neutral salt and dilute acid extraction methods are efficiently applied only to extract collagen from young animal tissues. For mature tissues, chemical reagents and proteolytic enzymes are used together to yield triple-helical molecules of collagen. Collagen type I can be extracted from animal tissues like skin, tendon and cornea and is commercialized in its insoluble form as an acidic solution. Tendon contains a high quantity of collagen (86 % from dry weight mass) which is made up of 97 % collagen type I. There are technologies applied for collagen extraction that aim to obtain soluble, but non-denatured collagen molecules with an intact triple-helix conformation. The chemical and enzymatic processes used in these technologies remove the non-helical polypeptidic ends (telopeptides) from the collagen molecule and break up the intermolecular cross-links. The enzymatic reaction does not succeed in completely cleavage of these chemical bonds present in all the three-dimensional structure of collagen. Therefore, a collagenous extract containing more than 70 % intact atelocollagen macromolecules is obtained. When natural conditions are induced (temperature 37 °C, pH 7.4), these intact triple-helical macromolecules are able to spontaneously aggregate to form fibers.

2.1 Tissue processing

Bovine tendons were obtained from the local abattoir after animal slaughtering. They were immediately rinsed in cooled water (4 °C) or phosphate-buffered saline (PBS) (pH 7.4). After transportation to the lab, in a cooler box (4 °C), they were peeled from adherent tissues with a scalpel and washed in cold tap water. The tendons were minced in 1-2 mm^3 pieces, and kept at -18 °C until processing.

2.2 Collagen extraction

Small pieces of tissue were put in a one-liter Berzelius glass and 0.5 M acetic acid containing pepsin (E.C. 3.4.23.1, Sigma), in a weight ratio of 1:10 (w/w) enzyme:dry tissue was added. The extraction process was conducted at 4 °C, with gentle stirring, for 24 h. The obtained gel was filtered and the remaining tissue was again extracted as above. The two viscous solutions were combined in the same glass and a precipitation-step was achieved by slowly adding 0.7 M NaCl in the gel and leaving the mixture at 4 °C, for 20 h. The precipitate was separated by centrifugation at 4000 rpm, for 20 min and it was dissolved in acetic acid 0.5 M by homogenizing on a magnetic stirrer at 500 rpm, for 2 h. The purified collagen type I solution was dialyzed against distilled water using cellulose tubes (molecular mass cut-off 12,400) for one week, renewing the outer solution three times a day. All extraction steps were performed at 4 °C in order to prevent denaturation of collagen.

The obtained collagen solution was characterized by analytical techniques and the results indicated 9.98 % hydroxyproline content, 84.94 % collagen, 88.40 % total protein content, 10.80 % hexosamines and pH 6.0. The value of its average molecular weight, determined by viscosimetry (Turkovski et al., 2008) was 308 kDa, comparable to that of tropocollagen (300 kDa). This observation indicates that the used enzymatic extraction is a non-denaturing method, which preserves the native triple helix structure of collagen. At the same time, the method eliminates collagen telopeptides to yield a non-immunogenic polymer. The purity analysis, conducted by SDS-polyacrylamide gel electrophoresis (Miller & Rhodes, 1982), revealed the presence of five distinct bands, corresponding to α constituent chains of collagen type I (α_1 and α_2), two β dimers and a γ trimer, having identical mobilities to the control collagen (Biocolor, UK) (fig. 1). The ratio between the specific α_1(I) and α_2(I) chains was very close to the natural value of 2:1, that confirmed the native structure of extracted collagen type I, $[\alpha_1(I)]_2\alpha_2(I)$. It was showed by electron microscopy that collagen obtained by this method and used to prepare composite scaffolds is mostly organized as fibril aggregates and a few fibers having the 67-nm characteristic banding pattern (Zarnescu et al., 2010).

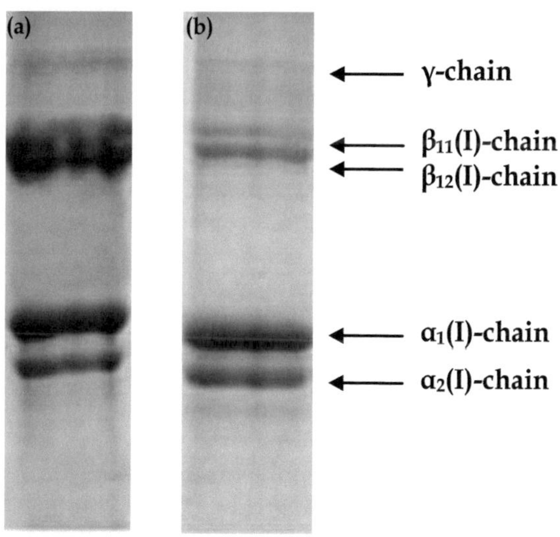

Fig. 1. SDS-polyacrylamide gel electrophoresis of collagen type I from control (a) and extracted from bovine tendon (b) showing the presence of two constituent α-chains, α1(I) and α2(I), small quantities of dimer (β) and trimer (γ) chains, clearly separated

2.3 Collagen/n-β-TCP composite material preparation

Several ceramics were investigated for their role in bone regeneration, but they lack structural stability and it is difficult to maintain them at the defect site. Therefore, TCP must be included into a polymeric scaffold of collagen.

Nanopowder of β-TCP (n-β-TCP) was obtained from the Ceramic lab of INCDIE ICPE-CA Bucharest, Romania. The particles had unit cell parameters similar to ASTM data and their diameter was lower than 84 nm. Nano-sized β-TCP powder presented a good *in vitro* biocompatibility in cell culture (Tardei et al., 2010). The dimension of ceramic crystals is an important factor involved in the first phase of cell-biomaterial interactions.

A nanocomposite material, collagen/n-β-TCP, consisting of the two main components of natural bone was prepared from a 0.8 % (w/w) collagen type I solution and β-TCP nanopowder mixed in a ratio of 50:50 (w/w) and homogenized with a manual speed-stirrer (Xenox, Germany) at 6000 rpm, at room temperature. In the next step, the mixtures were poured into glass molds (15 mm diameter) and were frozen at – 20 °C, overnight.

The polymeric matrix of natural collagen was not only used to achieve a stable composite, but to prevent rapid release of calcium ions in surrounding medium and to improve the interaction with osteoblast cells. Both components, collagen type I and calcium phosphate stimulate osteoblastic differentiation in cell cultures (Xie et al., 2004) and together, in composite materials, accelerate osteogenesis and allow the achievement of mechanical and biological properties, superior to their individual ones (Wahl and Czernuszka, 2006). Fang et al. (2009) showed that nanocomposites containing biomimetic HA deposited from simulated body fluid facilitate adhesion and spreading of human mesenchymal stem cells. The majority of temporary bone substitutes developed in the last decade are resorbable composite materials consisting of fibrils of collagen type I and calcium phosphate crystals, mimicking the composition and tissue structure (Yamauchi et al., 2004).

3. Composite material design

Designing an artificial bone substituent involves a process of optimizing its composition and structure that influence the osteoconductive properties and interaction with cells. Recent scientific progress in material science and engineering evolved in a biomimetic approach for bone substitute fabrication. A biomimetic composite material can be any artificial material designed to mimic one or several features of the natural one. Natural bone biomimetism implies using collagen type I and apatite in the form of nanoparticles. The osteoinductive property of collagen combined with the bioactivity and osteoconductive property of calcium phosphates give a high biocompatibility to the composite material and favor cell growth (John et al., 2001).

A collagen sponge prepared by freeze-drying (fig. 2A) is similar in structure to trabecular bone, where mineralized fibrils are arranged in a network of trabeculae and voids which are filled by bone marrow *in vivo* (Fantner et al., 2006). Attachment of calcium phosphate particles doesn't significantly modify collagenous network porosity as they are tightly bond to the collagen fibrils, wrapping the skeleton with a fine layer (fig. 2B).

Collagen-calcium phosphate composites can be conditioned in various shapes and forms. Sheets are obtained by mixture drying at room temperature or electrospinning, a technique that yields nanostructures. Cells could adhere to these two-dimensional materials, but proliferation is restricted in comparison to three-dimensional constructs. Composites conditioned as hydrogels are preferred as injectable form for tissue repair in order to avoid complicated operation; they are three-dimensional structures, but they lack mechanical strength (Hunt & Grover, 2010).

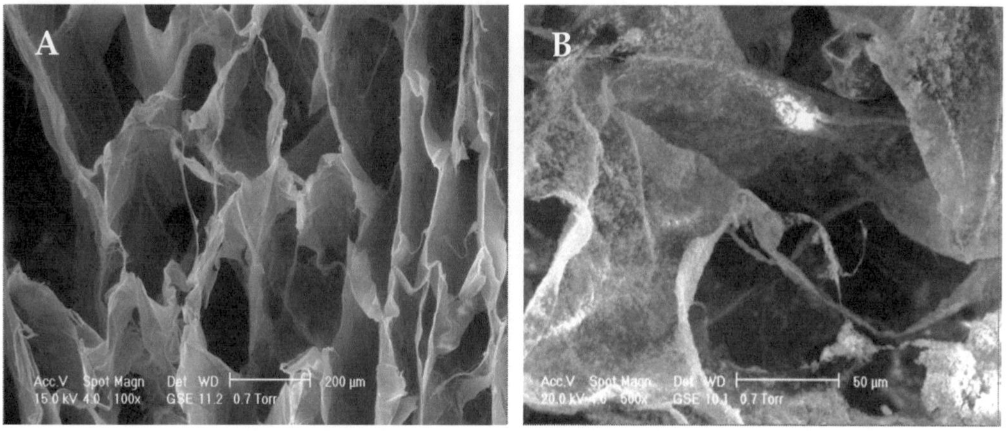

Fig. 2. Scanning electron micrograph of a freeze-dried collagen material presenting a network of fibrils and voids (A) and a detail of a freeze-dried collagen/TCP 50/50 (w/w) composite material showing the collagenous skeleton wrapped with a fine layer of tricalcium phosphate particles (B)

Collagen-calcium phosphate composites can be conditioned in various shapes and forms. Sheets are obtained by mixture drying at room temperature or electrospinning, a technique that yields nanostructures. Cells could adhere to these two-dimensional materials, but proliferation is restricted in comparison to three-dimensional constructs. Composites conditioned as hydrogels are preferred as injectable form for tissue repair in order to avoid complicated operation; they are three-dimensional structures, but they lack mechanical strength (Hunt & Grover, 2010).

Selection of bone substituent composition indicates an optimum ratio between collagen and calcium phosphate of 50:50 (w/w). A higher calcium phosphate concentration in a composite with 50:100 (w/w) ratio between the two components led to an intense loaded collagenous network having less strength, a lower value of porosity and decreased biocompatibility (Moldovan et al., 2009).

3.1 Freeze-drying of composite material

Freeze-drying technique is based on two processes: first, a solution is frozen and then, the solvent, which is usually water, is removed under vacuum, at low temperatures by sublimation. The solvent crystals formed during freezing have the same size and morphology as the pores of the material after drying. Therefore, parameters like the rate and temperature of freezing, concentration and pH of the solution, and the presence or absence of other macromolecules influence pore morphology and size in the final product.

The mixture, consisting of collagen type I and n-β-TCP particles, was subjected to freeze-drying using a freezing temperature of -35 °C. The programme of the freeze-dryer (Christ, Germany) continued with a 0 °C-step, at 0.26 mbar, for 17 h and drying at + 30 °C. This process yielded a nanocomposite material, conditioned as porous scaffold, which was sealed in a plastic bag and exposed to UV-radiation, for 8 h, in a sterilization cabinet (Scie-Plas, UK). A collagen solution was identically processed and used as control material.

All the operations for collagen/n-β-TCP composite material fabrication are summarized in fig. 3.

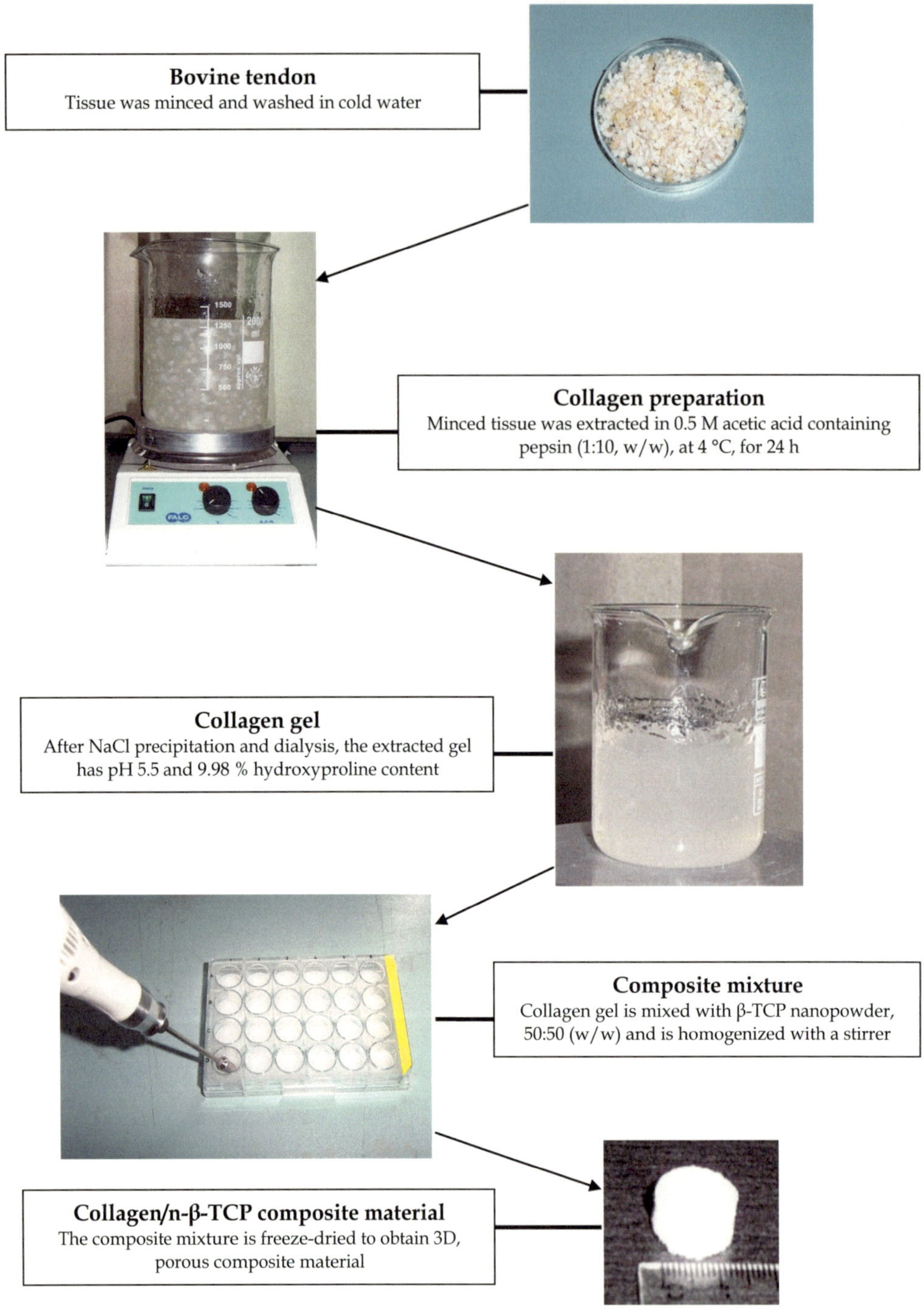

Fig. 3. Schematic diagram showing the experimental procedures used to prepare collagen/n-β-TCP nanocomposite material from collagen type I solution and β-TCP nanopowder by freeze-drying technique

Freeze drying is a technique used for the fabrication of porous materials (Schoof et al., 2001). The final porosity of three-dimensional composites can be controlled by varying the freezing temperature at -20 °C, -78 °C or -196 °C, respectively; namely, the lower the temperature, the smaller the pore size (Karageorgiou & Kaplan, 2005). Still, a temperature of -196 °C gives too small pores for a medical implant and its mechanical properties are altered. Porous materials having pores in the range of 50-1500 μm can be obtained by freeze-drying (Li, 2000).
The materials having high porosity are preferred because they have a high void volume within cells can grow and form new tissue. A network formed from well defined and interconnected pores is necessary in view of viable implant development. The porosity value for a composite material used in bone tissue engineering must be above 70 % in order to allow cell growth and proliferation (Boland et al., 2004). The size of the pores from bone substitutes used as cell scaffolds must have at least 100 μm because osteoblast cells have sizes in the range of 10-60 μm, depending on the species and cell line (Xu & Simon, 2004).

3.2 Cross-linking of composite material

Collagen-based composite materials used in bone repair mimic the ultrastructure of native extracellular matrix, but possess high sensitivity to enzymatic degradation. Therefore, new covalent bonds must be introduced in collagen structure in order to register less biodegradability. Cross-linking of collagen-based composite materials must be a compulsory step in their fabrication in order to control collagen biodegradation rate and their mechanical characteristics. However, depending on the used reagent, composite material biocompatibility might be compromised.
Chemical cross-linking is a technique that involves the formation of covalent bonds between two different or identical protein molecules. It uses bifunctional reagents, containing reactive groups that react with functional groups present on the side chains of amino acid residues, such as the amino group of lysine, arginine, glutamine and asparagine, or sulfhydryl from cysteine. The commonly used cross-linking agents for collagen-based materials, including glutaraldehyde, formaldehyde and epoxy compounds can be used by directly mixing with the protein or in a vapour chamber. They are cytotoxic owing to reactive moieties covalently coupled between neighbour collagen fibrils (Badylak, 2002). A neutralization of the cytotoxic residues after cross-linking is achieved using 10 mM sodium borohydride, at 4 °C, for 24h prior to implantation. Glutaraldehyde is also involved in development of calcification that occurs subsequently to implantation (Schoen & Levy, 2005). An efficient cross-linking method uses the heterobifunctional carbodiimide, 1-ethyl-3-(3-dimethylaminopropyl) carbodiimide hydrochloride (EDC), so-called zero-length agent because it does not incorporate itself into the polymer macromolecules, thus improving the biocompatibility of the material. This is an important advantage over other chemical cross-linking agents. EDC reagent is suited for use with collagen type I materials as showed Pieper et al. (1999) and electrospun collagen type II materials as showed Barnes et al. (2007). At present, EDC is used to cross-link composites like collagen-glycosaminoglycan skin substitutes (Powell & Boyce, 2006), collagen-elastin-glycosaminoglycan vascular scaffolds (Daamen et al., 2008) or gelatin-hydroxyapatite for bone repair (Chang & Douglas, 2007). Polyethylene glycol is used in biomedicine as dispersing agents, solvents, ointment, and suppository bases and is currently tested as cross-linking agent (Popescu et al., 2009).
Physical methods of cross-linking, such as ultraviolet irradiation (Lee et al., 2001), photo-oxidation (Turek & Cwalina, 2010) and dehydrothermal treatment (Haugh et al., 2009) have

a mild effect on collagenous materials, but they obtain an improved enzymatic resistance and biocompatibility.

Recently, natural plant polyphenolic compounds, like tannic acid (Isenburg et al., 2005), genipin (Bi et al., 2011), proanthocyanidins (Chen et al., 2008), catechin (Madhan et al., 2005), and riboflavin (Ashwin & McDonnell, 2010) have been shown to stabilize collagen structure through hydrogen bonding and hydrophobic interactions, while preserving its cytocompatibility.

Enzymatic cross-linking using transglutaminase (EC 2.3.2.13) is applied to denatured collagen-based composite materials, conditioned as films or gels, leading to formation of covalent amidic bonds between carboxyl group of glutamine residues and ε-amino group of lysine residues (Chen et al., 2003). The process results in irreversible network junctions, similar to that formed by chemical agents. Microbial transglutaminase, a calcium-independent enzyme, has a higher specific activity that stimulated new applications, especially in food industry (Colligan et al., 2002; Garcia et al., 2007).

Three cross-linking protocols using different agents (EDC, glutaraldehyde and catechin) were comparatively evaluated in terms of efficiency on collagen porous material. First, the cross-linking process was carried out in ethanolic solution of EDC, at pH 5.5, by slowly shaking on a platform, at room temperature, for 18 h (Pieper et al., 2002). Alternatively, collagen sample was treated with EDC/N-hydroxysuccinimide (NHS) solution, in the same conditions and was shaken for 4 h. After cross-linking, several washing steps of the samples were carried out in order to eliminate any unreacted intermediates. The samples were washed in solutions of 0.1 M sodium phosphate (pH 9.1), 1 M and 2 M NaCl. After the final washing in distilled water, samples were once again lyophilized.

The second cross-linking process used glutaraldehyde and took place in a special chamber in which the solution does not contact the material sample. A solution of 3 % (w/w) glutaraldehyde was put at the bottom of the chamber and collagen material was exposed to the vapors, at room temperature, for 18 h. The sample was then washed in distilled water, renewed every 1 h.

The third method of cross-linking used catechin, a natural polyphenol from green tea and was carried out by simply immersion of collagen sample in 10 mM catechin solution and shaking at 300 rpm, at room temperature, for 18 h.

The cross-linking degree can be assayed using physical or chemical methods for the determination of shrinkage and denaturation temperature, content in amino free groups, *in vitro* enzymatic digestion or mechanical properties. In order to compare the cross-linking degree of collagen materials, it was calculated the percentage of free amino groups lost during the process by spectrophotometric assay using 2,4,6-trinitrobenzene sulfonic acid (TNBS) (Barnes et al., 2007). The degree of cross-linking was expressed as percentage loss in free amino groups after cross-linking and was calculated as follows:

$$\%\ \text{Cross-linking degree} = 1 - \left(\text{ABS}_{\text{CL}} / \text{MASS}_{\text{CL}}\right) \ / \ \left(\text{ABS}_{\text{NCL}} / \text{MASS}_{\text{NCL}}\right) \quad (1)$$

where ABS is absorbance at 346 nm, MASS is sample weight, CL is the cross-linked sample and NCL is the non-cross-linked sample.

The results showed that there is a similar cross-linking degree of collagen materials treated with 30 mM EDC and EDC/NHS, compared to 3 % glutaraldehyde vapors and 10 mM catechin (fig. 4). The cross-linking of collagen porous material was achieved at a rate above 50 % for all used agents.

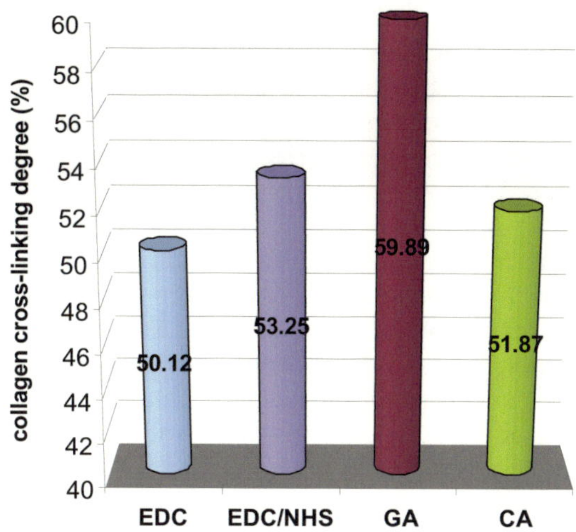

Fig. 4. Variation of cross-linking degree for porous collagen material treated with EDC, EDC/NHS, glutaraldehyde (GA) and catechin (CA)

EDC cross-linker has a mechanism of action consisting of reaction with carboxyl groups converting them into O-acylisourea intermediates that further react with lysine residues or other available primary amines. The amines exert a nucleophilic attack on the intermediate, resulting in a stable amide bond (see reaction on Thermo Fisher Scientific). EDC is released as a soluble urea derivative and can be removed by several washings (Pieper et al., 1999, 2002).

NHS is often included in EDC coupling protocols to improve efficiency. The reaction of NHS with carboxyl groups is mediated by EDC and results in formation of an NHS ester intermediate. The advantage of NHS is the higher stability of its intermediate, compared to EDC intermediate. This property allows a more efficient conjugation to the primary amines. The amide bond is formed in shorter periods of time, compared to EDC cross-linking and NHS is released.

EDC cross-linking is an efficient process when takes place in acidic conditions (pH 5.5) and a suitable buffer for carbodiimide reaction is 4-morpholinoethanesulfonic acid buffer. A higher pH (up to 7.2) could lower the efficiency of the reaction but can be compensated by increasing the amount of EDC. Still, a high quantity of EDC could prevent cross-linking reaction. The optimum molar ratio between protein and EDC reactive groups is 1:1, as earlier established in studies on collagenous materials (Olde Damink et al., 1996). NHS cross-linking reaction is optimal at physiologic or slightly alkaline pH (up to 8.5); thus, phosphate, bicarbonate or carbonate, HEPES or borate buffers are commonly used. The optimal temperature for EDC reaction is room temperature and higher than 40 °C for NHS reaction. The final washings are important to eliminate intermediary products formed during the reaction and to avoid cytotoxicity.

In the particular case of collagen cross-linking, there are aspartic and glutamic acid available residues within the α chains that interact with lysine and hydroxylysine residues from the same chain, the neighbor chain or from other collagen molecule/fibril, forming intra- or intermolecular covalent cross-links. When cross-linking is applied to porous collagenous structure, ethanol is recommended to prevent pore morphological changes. Ethanol reduces dipolar forces and allows bond changes due to its lower dielectric constant of 35, compared to 81 for water (Barnes et al., 2007). Also, ethanol molecules could act as proton donors in reaction with EDC, improving cross-linking yield.

The first study on applying EDC and EDC/NHS cross-linking methods to collagen/n-β-TCP 50/50 (w/w) composite material was performed to compare their efficiency. The protocols were identical to those applied to collagen material (see above). The results showed that the cross-linking process resulted in a decrease of the free amine group content relative to non-cross-linked composite material. The value for the composite cross-linking degree was 25.64 % using EDC. A stronger capability of cross-linking was registered for EDC/NHS agent, 28.01 % (fig. 5). An increase in TCP quantity (50/100 ratio, w/w) resulted in a decrease of composite cross-linking degree to values of 20.76 % for EDC and 24.34 % for EDC/NHS. According to TNBS assay, the cross-linking degree for collagen material was significantly higher than for composite material, taking into account that the same quantity of collagen was used to prepare simple and composite materials. This result might indicate that the presence of β-TCP nanoparticles on collagen fibril surface partially hindered lysine interactions with aspartic and glutamic acid residues within the three α-chains of collagen fibrils. A similar action was reported for 1,4-butanediol diglycidyl ether used as cross-linking agent, namely the process was more evident for collagen without mineral phase than for the composite (Tampieri et al., 2008).
Composite material cross-linking influences its structure, biodegradability, calcium release and biocompatibility.

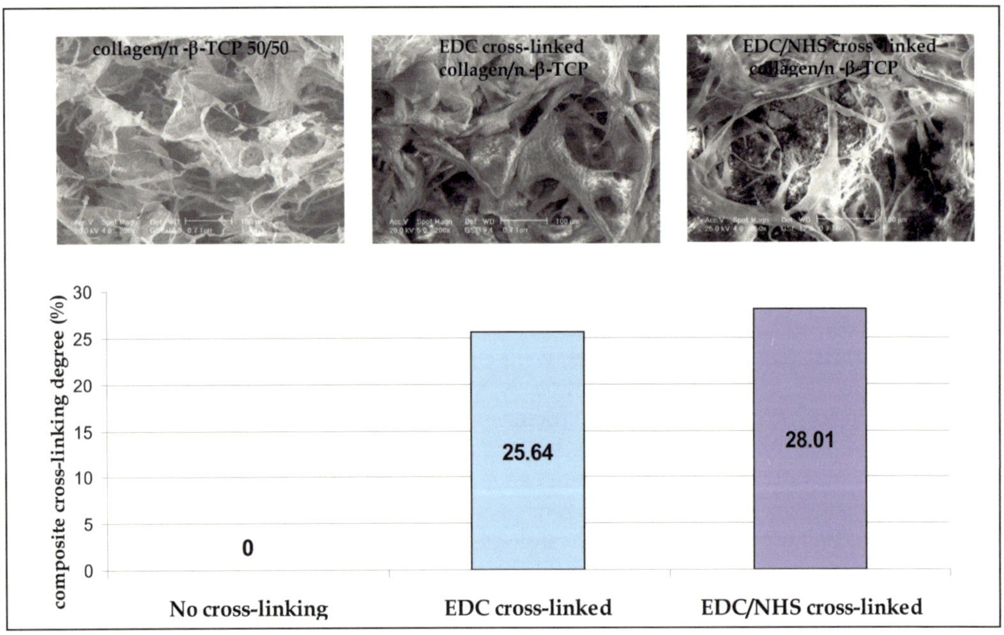

Fig. 5. Micrographs of scanning electron microscopy of collagen/n-β-TCP 50/50 (w/w) composite material before and after cross-linking with EDC and EDC/NHS and their percentage of cross-linking degree assayed with TNBS

4. Morphological characterization of cross-linked composite material

4.1 Scanning electron microscopy of cross-linked composite material

The microstructure of the obtained composite material before and after cross-linking was examined by environmental scanning electron microscopy. The sample was mounted on carbon pads attached to aluminium stubs and visualized at an ESEM apparatus (Quanta 400, FEI, Philips, Holland) using the low vacuum mode. Micrographs showed white deposits of β-TCP nanoparticles disposed on the surface of collagen fibrils (fig. 5). Images of non-cross-linked composite variant showed a typical structure of lyophilized collagenous materials, with regular pores, which favor a good biological behavior. Pore morphology was affected by the cross-linking process. The ordered structure with interconnected pores of non-cross-linked composite material was replaced by a structure with unevenly sized pores, ranging from 20 to 200 μm in the EDC and EDC/NHS cross-linked materials. Sample rehydration and the additional lyophilization process that occur during the cross-linking treatment could induce a slightly collapse of pore network and led to a dense material.

4.2 Porosity of cross-linked composite material

The porosity (ε) of composite materials was measured by water displacement method (Zhang et al., 2003). The following equation was used to calculate the porosity value:

$$\varepsilon = \left(v_1 - v_3\right) \ / \ \left(v_2 - v_3\right) \times 100 \qquad (2)$$

where v_1 is a known volume of water in a graded test tube, v_2 is the total volume of water plus the water-impregnated composite material sample, after 3h of incubation to allow water to penetrate and fill the pores, v_3 is the residual volume of water after removing the water-impregnated composite material from the test tube.

In comparison to the non-cross-linked composite material, having a porosity of 94.83 %, the values for cross-linked composite materials decreased according to the used method of cross-linking and the degree of cross-linking. Thus, the porosity value for the EDC-treated material was 83.76 % and 78.25 % for the EDC/NHS-treated material. Composite material porosity decrease, registered after EDC or EDC/NHS cross-linking could be due to the newly formed cross-links. Results showed that the used freeze-drying process, with freezing temperatures of -35 °C, led to composites with a porosity of at least 78 %, a value that allows a good infiltration of cells.

5. Biochemical and biological properties of cross-linked composite material

5.1 Collagenase degradation of cross-linked composite material

Bone regeneration takes place over duration of several months. It is important for a composite material used for tissue repair to degrade in a controlled fashion while new tissue is formed. An *in vitro* experimental model using bacterial collagenase mimics the enzymatic attack on the collagenous composite material implanted *in vivo*. This enzyme acts specific on the amino bond of glycine from the peptidic repetitive sequence, -X-Gly-Pro-, from the helical region of collagen. This model shows if the covalent cross-links introduced in the collagen molecule by carbodiimide treatment could hinder and protect the cleavage site, block bacterial collagenase action and reduce material degradability.

To quantify collagen/n-β-TCP composite degradation, each sample of material was weighed and pre-incubated in TES buffer, pH 7.4, containing 50 mM $CaCl_2$, at 37 °C, for 30 min. In the next step, 100 μl bacterial collagenase type IA (Sigma-Aldrich) in TES buffer were added and the degradation took place in a water bath, at 37 °C, for different periods of time (6h, 12h, 18h and 24 h). At the end of each incubation period, the reaction was stopped with EDTA, at 0 °C and the protein content of the supernatant was assayed by ninhydrin method. The percentage of biodegradation was calculated using the equation:

$$\% \text{ biodegradation} = (\mu\text{M aminoacids}_{CL} / w_{CL}) \; / \; (\mu\text{M aminoacids}_{NCL} / w_{NCL}) \times 100 \quad (3)$$

where w is the material weight in grams, CL is the cross-linked sample and NCL is the non-cross-linked sample which is completely degraded after 6h of incubation (control).
Fig. 6 compares the biodegradability of collagen/n-β-TCP composite before and after the cross-linking treatment. The non-cross-linked composite material had been thoroughly degraded after incubation in collagenase solution for only 6 h (100 % biodegradability). After cross-linking treatment, the biostability of the material was enhanced according to the cross-linking agent. The EDC-cross-linked material was only 58.17 % degraded in 24 h. The EDC/NHS-cross-linked material had a better ability to resist collagenase degradation (max. 43.41 %) due to its higher cross-linking degree. These results reveal that both cross-linking methods improve collagen/n-β-TCP material biostability, but EDC/NHS treatment is faster and more efficient. The cross-linking treatment lowered the biodegradability of the composites which were less susceptible towards collagenase attack. It was observed a good correlation between the cross-linking degree and the biodegradability of each sample. Composites with a higher degree of cross-linking yielded a smaller quantity of degraded collagen. These results indicated a better stability of EDC/NHS cross-linked composite material over EDC cross-linked one and both values over untreated sample.

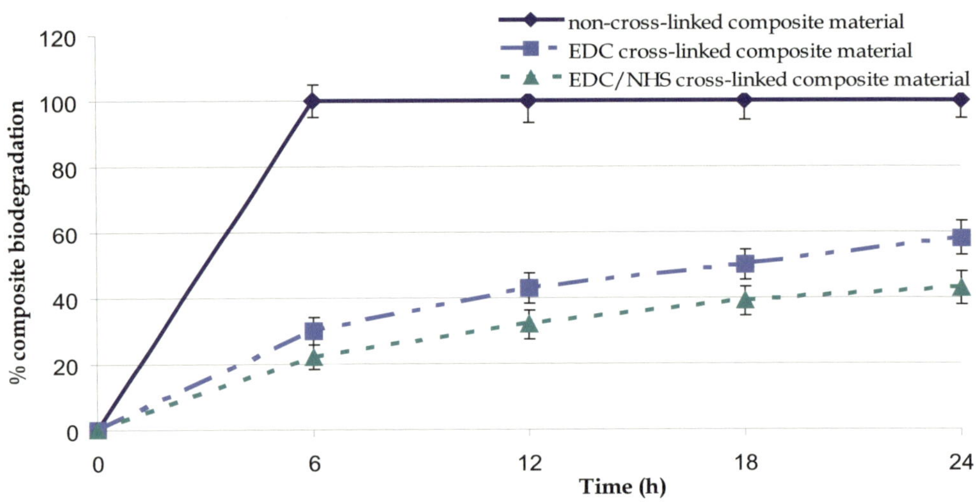

Fig. 6. The biodegradation degree of collagen/n-β-TCP 50/50 (w/w) composite material before and after cross-linking with EDC and EDC/NHS, after incubation with bacterial collagenase, for 24 h

5.2 *In vitro* calcium release from cross-linked composite material

In vivo dissolution of ceramic particles takes place by a decrease in crystal size and increase in macroporosity and microporosity (LeGeros et al., 2003). When ceramics are soaked in buffer solution, a dissolution reaction leads to increasing calcium and phosphate ion concentrations in the solution. A decrease of calcium concentration in the medium is registered when the reprecipitation reaction occurs (Wang et al., 2004).
The dissolution of β-TCP nanoparticles attached to collagen fibrils in composite material was analyzed by assessment of calcium ions released in solution, in physiological

conditions. Samples of material were incubated in phosphate buffer saline, pH 7.4, for 5 days. At each 24 h, 5 μl of supernatant were transferred to a 96-well plate and the same volume of fresh PBS was added to the reaction tube. The calcium content of the supernatant was determined using the QuantiChrom Calcium Assay kit (BioAssay Systems, USA), according to the instructions. After 3 min of incubation with reaction reagent, the optical density was read at 612 nm using a plate-reader (Tecan, Austria). The concentration of calcium was calculated using a standard curve in the range 0-200 μg/ml. The dynamic of the dissolution behavior of n-β-TCP particles from non-cross-linked and cross-linked material variants is shown in fig. 7. The calcium quantity released from the non-cross-linked material increased in the first 48 h and was higher than the value for the cross-linked ones after 5 days. The cross-linked composite materials had a similar pattern for calcium release, regardless of the cross-linking method, having a maximum value of calcium ions after 24 h of incubation. It was concluded that the cross-linking process is beneficial for collagen/n-β-TCP composite materials because the calcium release takes place in a controlled-fashion, in comparison to the non-treated composite materials.

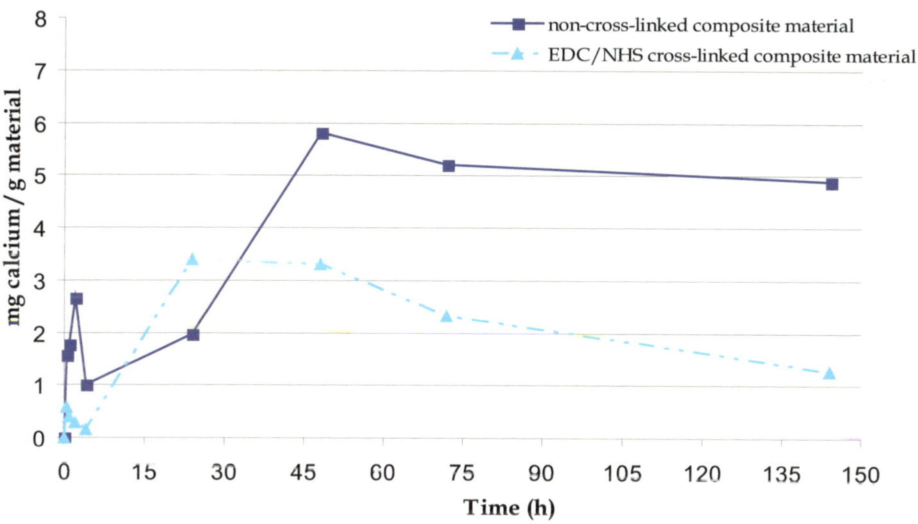

Fig. 7. Variation of calcium ion concentration released from non-cross-linked and EDC/NHS cross-linked collagen/n-β-TCP composite materials in 0.1 M PBS, pH 7.4

All the above tested properties and the obtained results lead us to conclude that EDC/NHS cross-linking protocol allows to fabricate a stable collagen/n-β-TCP composite material, with a controlled release of calcium and a porous microstructure adequate for cell infiltration and proliferation.

5.3 Preliminary *in vitro* biological testing of cross-linked composite material

In the last decade, porous composite materials prepared from synthetic and/or natural polymers combined with a ceramic component are tailored and tested for their efficacy in regeneration of wounded tissue (Ge et al., 2008). These composites serve as scaffolds for cell cultivation *in vitro* or as temporary bone substitutes *in vivo* guiding cell proliferation and new extracellular matrix formation (Leong et al., 2003). Composite materials designed for medical applications must be first tested for *in vitro* cytotoxicity on cell cultures, before *in vivo* pre-clinical and clinical trials. According to the International Standard ISO 10993-5, there are several specific methods to analyze the cytotoxicity of medical devices, which involves different aspects of cell function, like viability and proliferation, loss of membrane integrity, decrease in cell adhesion, cell morphology. The viability of cells cultured in the presence of a medical device could be assayed using MTT assay (Mossman, 1983). This is a colorimetric assay based on the reduction of yellow soluble salts of 3-(4,5-dimethylthiazol-2-yl)-2,5-diphenyl tetrazolium bromide (MTT) to dark purple insoluble formazan crystals by dehydrogenases from cell mitochondria. Since conversion of MTT takes place only in metabolically active cells, the level of enzymatic activity measured as optical density is directly proportional to the viability of the cells.

The standard extract protocol consisted in sterile samples of cross-linked and non-cross-linked collagen/n-β-TCP composite material immersed in the culture medium DMEM supplemented with 10 % fetal calf serum (FCS) and incubated in a humidified atmosphere of 5 % CO_2 and 95 % air, for 24 h. The used ratio between the surface area of the composite and the volume of culture medium was 1 cm^2/ml, ranging between 0.5-6.0 cm^2/ml, the values from ISO standard. The conditioned medium, named extract, was used for MTT assay.

Fibroblasts from NCTC cell line (ECACC) were seeded in the wells of a 24-well culture plate, at a density of $5x10^4$ cells/ml and cultivated in DMEM containing 10 % FCS and 1% antibiotic mixture. The plate was placed into an incubator, with 5 % CO_2 atmosphere, at 37 °C, for 24h, to allow cell adhesion. The culture medium was, then, replaced with the same volume of extract and the plates were incubated in humidified atmosphere with 5 % CO_2, at 37 °C, for 48h.

In order to assay the viable cells after *in vitro* culture with the extracts, the medium was removed and fresh medium with MTT solution, in a 10:1 (v/v) ratio was added. The plates were incubated at 37 °C, for 3h. The medium was removed and 500 μl isopropanol were added to each well and the plate was gently shaken on a platform, for 3 h, to dissolve the formazan crystals. The colored solution was transferred to another 96-well plate and the optical density was read at 570 nm, with reference settled at 630 nm, using a microplate reader (Sunrise Tecan, Austria). The cells cultured with complete culture medium were negative control (nontoxic) and cells cultured in the presence of hydrogen peroxide were positive control (toxic). The experiment was performed with three samples per each group (n=3). The results were calculated as mean values ± standard deviation for cells cultured with the control or the material extract, respectively and expressed as percentage from the negative control, considered to be 100 % viable cells.

The extract method allows the evaluation of possible toxic compounds released from the material in the medium and that could modulate the cellular activity. As shown in fig. 8, the viability of cells cultured with composite material extracts was superior approx. 1.2-fold to the negative control, after 48h of cultivation. The extract of cross-linked material induced a higher viability to NCTC fibroblasts in comparison to the non-cross-linked material extract. These values indicated that cross-linked composite was a bioactive material, able to control and stimulate the cellular activity, better than the non-cross-linked variant. Other collagen-based composite materials were shown to modulate fibroblast activity in culture (Zhang et al., 2003; Jantova et al., 2009).

Other 24-well plate was seeded with NCTC fibroblasts and cultivated in the presence of composite extract using the same protocol as for the extract method (described above). After 48 h of incubation, cells cultured on polystyrene plate were fixed in methanol and Giemsa stained. The morphology of cells grown in extract medium was observed by light microscopy. The micrographs showed that cells maintained their normal phenotype, presenting euchromatic nuclei with 1-2 nucleoli and a clear cytoplasm (fig. 8, right).

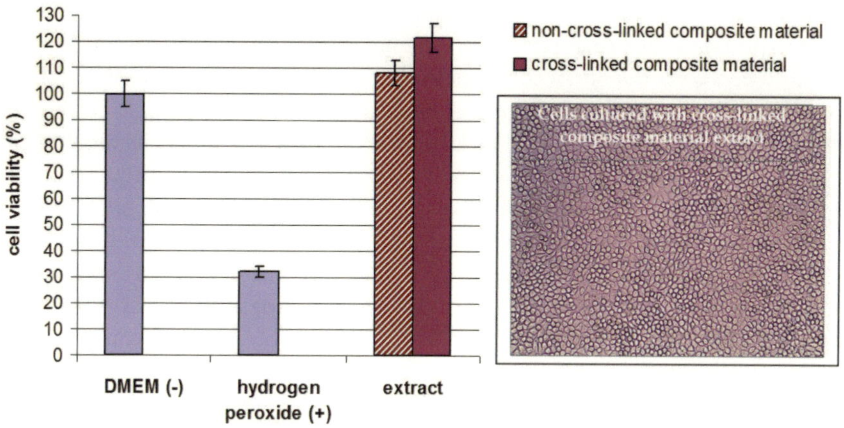

Fig. 8. Cell viability measured by MTT assay after 48h in culture with DMEM (negative control), DMEM containing hydrogen peroxide (positive control) and collagen/n-β-TCP composite extract. On the right a light micrograph showing fibroblast cell morphology after 48h of cultivation in extract of collagen/n-β-TCP cross-linked composite material.

The culture media removed from the MTT test was analyzed by lactate dehydrogenase (LDH) assay. LDH assay is based on the reduction of nicotinamide adenine dinucleotide (NAD) by LDH. The resulting NADH is used to convert a tetrazolium salt to a colored formazan, which strongly absorbs in the visible range 490-520 nm. The quantified LDH activity is an indicator of cell viability as only lysed cells are able to release this enzyme from the cytosol into the medium. The NCTC fibroblast membrane integrity was assayed in this experiment as a function of the amount of cytoplasmic LDH released into the culture medium, according to the instructions of the kit (Cayman Chemical Co., USA). After an incubation of the centrifuged culture medium with reagent mixture with gentle shaking, at room temperature, for 30 min, the optical density was measured at 490 nm using a microplate reader (Sunrise Tecan, Austria). LDH activity (mU/ml) was calculated from the standard curve plotted with standard LDH in the range 0-1 mU. Results were reported in arbitrary units, the negative control being considered equal to 1.

No increase in LDH leakage was observed for cells cultivated with composite material extracts for 48 h (non-cross-linked-0.95; cross-linked-1) when compared to the negative control. This indicates that composites have no cytotoxicity related to NCTC fibroblasts. The LDH analysis in the culture medium from the same experiment with cells analyzed by MTT allows meaningfully comparison of the results. Both experiments confirmed a high *in vitro* biocompatibility of collagen/n-β-TCP that is an important property for a polymeric composite material of medical utility, intended to support tissue repair.

5.4 Osteogenic properties of cross-linked composite material

Many cell types need an appropriate adhesion surface in order to maintain their proliferation ability and specific or differentiated functions. Cell adhesion is an important factor influenced by the surface characteristics of the material. Material efficacy in fullfilling these requirements depends mainly on the chemical characteristics of the surface, that determines cell-substrate interaction, but also on cell morphology and the relation between the cells and the material.

It was demonstrated that this organizatoric role can be successfully played by porous composite materials using an *in vitro* experimental model, i.e. rat osteoblasts from a primary culture cultivated into the three-dimensional, porous collagen/n-β-TCP 50/50 composite material. The cell-composite construct, cultured for different periods of time, yield a bioactive implant that can be used in tissue repair. In order to assess its medical utility, cell proliferation and adhesion was evaluated by total DNA quantification. The primary cell culture of rat osteoblasts was established from parietal and frontal bones by enzymatic method, as described by Gu et al. (2002), according to the international guidelines for care and use of laboratory animals. Cells were cultivated in DMEM supplemented with 0.05 mM ascorbic acid (Sigma) and 10 mM β-glycerophosphate (Sigma), for 21 days and alkaline phosphatase activity and calcium phosphate deposits were histochemically detected, to confirm cell osteoblast phenotype (Oprita et al., 2008).

Sterile samples of collagen/n-β-TCP cross-linked composite material (5x5x5 mm^3) were placed into the wells of a 24-well culture plate. Rat osteoblasts in 200 μl DMEM supplemented with 10 % FCS, at a density of $4x10^6$ cells/cm^3, were injected into the samples and the plates were incubated in humidified atmosphere, with 5 % CO_2, at 37 °C. After 4h, 0.5 ml of the same medium were added into each well to cover the cell-composite construct and they were incubated at 37 °C, for 6 days. Cell culture medium was renewed twice a week.

Total DNA content was fluorimetrically assayed in cell lysate. After 1 day and 6 days of cultivation, respectively, each cell-composite construct was washed three times in phosphate-buffered solution and was frozen at -80 °C until analysis. After thawing, the constructs were cut in very small pieces and the fragments were incubated with saline-sodium citrate buffer, pH 8.5 containing 0.02 % SDS, at 37 °C, with occasional stirring, for 1 h. After centrifugation at 10000 g, an aliquot of 10 μl of cell lysate was transferred to a test tube to determine the DNA content, with Quant-iT dsDNA HS Assay kit (InVitrogen, USA) on a Qubit fluorometer (InVitrogen, USA). Results were reported as cell number, assuming a standard quantity of 8 pg DNA per cell (Ahlfors & Billiar, 2007).

The number of adhered cells after 1 day of cultivation was approx. 54 % related to the seeded cell number. An approx. 1.4-fold increase in cell number was observed from $2.7x10^5$ cells in day 1 to $3.8x10^5$ cells in day 6. The cross-linked composite material allowed osteoblast adhesion and proliferation. As Anselme (2000) reported, rat osteoblast adhesion does not occur preferentially to ceramic crystals or to collagen fibrils and it is independent of the roughness of the material surface. Other authors demonstrate that collagen-based composites, organized as three-dimensional scaffolds, enhance the contact guidance process of osteoblasts inoculated onto the material surface (Rodrigues et al, 2003). It is known that collagen has binding sites that promote cell attachment through focal contacts and adhesion plaques, providing an increased cell adhesion to composites designed for tissue regeneration (Douglas et al., 2008).

In another experiment, osteoblast-composite constructs were cultivated in normal medium (DMEM) and osteogenic medium (DMEM supplemented with 0.05 mM ascorbic acid, 10 mM β-glycerophosphate and 100 nM dexamethasone), for 21 days. After 7, 14 and 21 days of cultivation, the constructs were washed twice in PBS and frozen at -80 °C. After thawing, cell lysis was conducted as described above. The alkaline phosphatase activity was analyzed using an artificial substrate, p-nitrophenylphosphate reagent, pH 9.8, freshly prepared. The color developed after incubation of the lysate with substrate reagent was read at 410 nm, using a microplate reader (Sunrise Tecan, Austria). The standard curve was constructed using p-nitrophenol in the range of 0-1.0 mM concentration, as the resulting product from the enzymatic reaction. The results were expressed as micromoles of p-nitrophenol per min reported to the total protein content.

Rat osteoblasts injected into composite materials expressed alkaline phosphatase during the 3 weeks of cultivation. Temporal expression of alkaline phosphatase showed a gradually increase in the first 14 days of culture, peaked around day 14 and then, a decrease to day 21 (fig. 9). Comparing the culture conditions, it were registered higher values of alkaline phosphatase activity for constructs cultured in DMEM than in osteogenic medium.

Expression of alkaline phosphatase activity demonstrated maintenance of the osteoblastic phenotype after cell cultivation into collagen/n-β-TCP cross-linked composite material and its osteogenic properties. The increase in alkaline phosphatase activity indicates the presence of mature osteoblasts and its decrease corresponds to their differentiation into osteocytes (Heinemann et al., 2008).

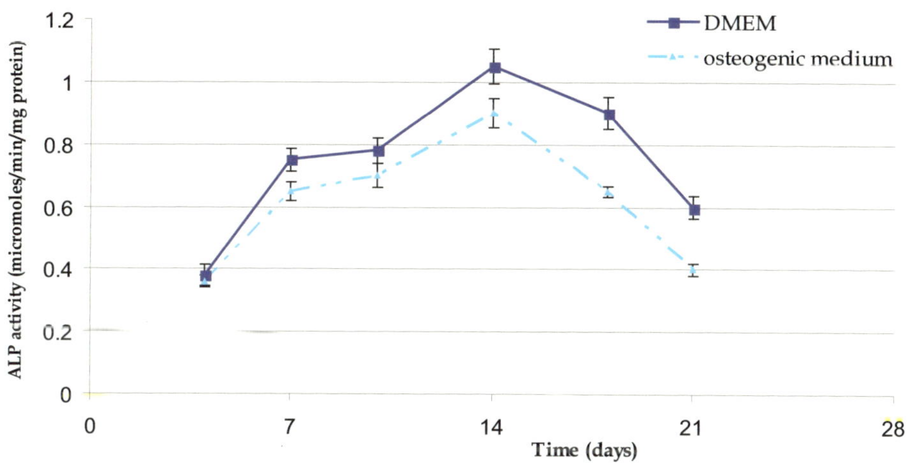

Fig. 9. Temporal expression of alkaline phosphatase (ALP) activity by rat osteoblasts injected into collagen/n-β-TCP cross-linked composite material and cultured in DMEM (straight line) and osteogenic medium (dot line), for 21 days

6. Applications

Bone is the second most transplanted tissue in the body after blood transfusions. Autologous bone transplantation is limited by the transplantable quantity and the damages produced in healthy body parts after harvesting. The cell-composite constructs cultured *in vitro* for different periods of time yield bioactive implants that can be used in tissue-engineered bone transplantation, a new medical technology used for the regeneration of bone and joints. Therefore, cooperation between a human or veterinary clinic and a research institution with experienced cell culture laboratory and biomaterial engineers must be established. A flow of clinical application research was verified as showed in fig. 10.

The bone tissue of new born rats was transferred from a veterinary clinic to the clean room of the cell culture facility where was processed in culture flasks using cell culture medium to obtain a primary osteoblast culture that multiplied. These cells were seeded into a three-dimensional, porous collagen/n-β-TCP cross-linked composite material, biomimetic to bone and cell biocompatible. The construct was cultured for three weeks in cell culture medium to achieve a bioactive implant. The cells maintained their osteogenic phenotype during *in vitro* culture into the composite material, as indicated by osteoblast specific marker identification. This tissue-engineered artificial bone has to be transferred back to the clinic for implantation in adult rats to test its inflammatory response. Implant ability to induce new bone formation is verified in animal defect experimental models and comparison to existent treatments is necessary.

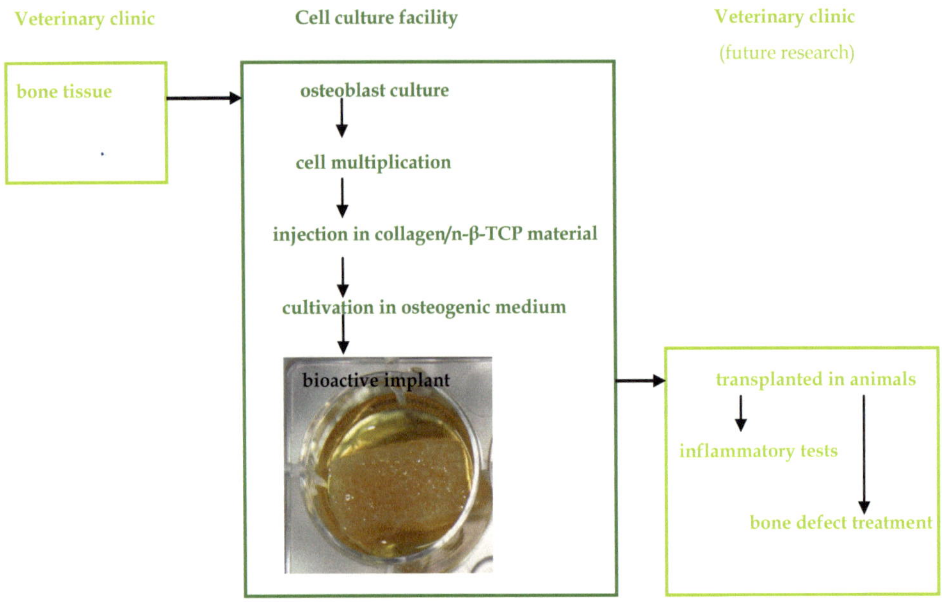

Fig. 10. Biotechnological research of a cell culture facility in collaboration with a veterinary clinic for achievement of a bioactive implant after osteoblast cell seeding into collagen/n-β-TCP cross-linked composite material and *in vitro* cultivation for three weeks

7. Conclusions and future research

A composite material mimetic to trabecular bone was prepared by freeze-drying a mixture of β-TCP nanopowder and collagen type I solution. It was reported a fast and efficient chemical cross-linking method using EDC/NHS. The cross-linked composite material had a porosity near 80 %. It was 50 % more stable than the non-cross-linked variant in the presence of bacterial collagenase and released calcium ions in a controlled manner. The cross-linked composite material was tested in a cell line of NCTC fibroblasts and showed a good biocompatibility after 48h of cultivation. An *in vitro* experimental model using rat osteoblasts from a primary cell culture showed that collagen/n-β-TCP cross-linked composite material allowed cell adhesion to its walls and cell proliferation. It was also observed that osteoblasts from the three-dimensional cell-composite construct maintained their phenotype after 21 days in culture. All these results demonstrate that collagen/n-β-TCP cross-linked composite material designed as a three-dimensional porous scaffold is useful in cell culture studies or bioactive implant development for bone tissue engineering.

These observations supply a basis for future studies regarding osteoblast activity and differentiation when cultivated in collagen/n-β-TCP composite material to achieve bioactive implants for bone repair. The evaluation is associated with the synthesis of extracellular matrix components (collagen type I, osteocalcin, etc). In order to examine collagen/n-β-TCP cross-linked composite material ability to induce bone repair, *in vivo* experimental models are necessary.

8. Acknowledgement

This work was supported by Projects No. 61012 and No. 62059 from Romanian Research Programme 4 - *Partnership in Strategic Domains.*

9. References

Ahlfors J.E.W. & Billiar K.L. (2007). Biomechanical and biochemical characteristics of a human fibroblast-produced and remodeled matrix, *Biomaterials*, Vol. 28, pp. 2183-2191

Anselme K. (2000). Osteoblast adhesion on biomaterials, *Biomaterials*, Vol. 21, pp. 667-681

Arinzeh T.L., Tran T., Mcalary J. & Daculsi G. (2005). A comparative study of biphasic calcium phosphate ceramics for human mesenchymal stem-cell-induced bone formation, *Biomaterials*, Vol. 26, pp. 3631-3638

Ashwin P.T. & McDonnell P.J. (2010). Collagen cross-linkage: a comprehensive review and directions for future research, *British Journal of Ophthalmology*, Vol. 94, pp. 965-970

Badylak S.F. (2002). The extracellular matrix as a scaffold for tissue reconstruction, *Seminars in Cell and Developmental Biology*, Vol. 13, pp. 377-383

Bae H., Shen M., Maurer P., Peppelman W., Beutler W., Linovitz R., Westerlund E., Peppers T., Lieberman I., Kim C. & Girardi F. (2010). Clinical experience using Cortoss for treating vertebral compression fractures with vertebroplasty and kyphoplasty: twenty four-month follow-up, *Spine*, Vol. 35, pp. E1030-E1036

Barnes C.P., Pemble IV C.W., Brand D.D., Simpson D.G. & Bowlin G.L. (2007). Cross-linking electrospun type II collagen tissue engineering scaffolds with carbodiimide in ethanol, *Tissue Engineering Part A*, Vol. 13, pp. 1593-1605

Barrere F., van Blitterswijk C.A. & de Groot K. (2006). Bone regeneration: molecular and cellular interactions with calcium phosphate ceramics, *International Journal of Nanomedicine*, Vol. 1, pp. 317-332

Bi L., Cao Z., Hu Y., Song Y., Yu L., Yang B., Mu J., Huang Z. & Han Y. (2011). Effects of different cross-linking conditions on the properties of genipin-cross-linked chitosan/collagen scaffolds for cartilage tissue engineering, *Journal of Materials Science: Materials in Medicine*, Vol. 22, pp. 51-62

Boland F.D., Espy P.G. & Bowlin G.L. (2004). Tissue engineering scaffolds, In: *Encyclopedia of Biomaterials and Biomedical Engineering*, G.L. Bowlin and G. Wnek (Eds.), pp. 2828-2838, Marcel Dekker Inc., New York, USA

Carter J.D., Swearingen A.B., Chaput C.D. & Rahm M.D. (2009). Clinical and radiographic assessment of transforaminal lumbar interbody fusion using HEALOS collagen-hydroxyapatite sponge with autologous bone marrow aspirate, *The Spine Journal*, Vol. 9, pp. 434-438

Chang M.C. & Douglas W.H. (2007). Cross-linkage of hydroxyapatite/gelatin nanocomposite using imide-based zero-length cross-linker, *Journal of Materials Science: Materials in Medicine*, Vol. 18, pp. 2045-2051

Chen T., Embree H.D., Brown E.M. & Taylor M.M. (2003). Enzyme-catalyzed gel formation of gelatin and chitosan: Potential for *in situ* applications, *Biomaterials*, Vol. 24, pp. 2831-2841

Chen K.Y., Shyu P.C., Chen Y.S. & Yao C.H. (2008). Novel bone substitute composed of oligomeric proanthocyanidins-crosslinked gelatin and tricalcium phosphate, *Macromolecular Bioscience*, Vol. 8, pp. 942-950

Chen Y.S., Kuo S.M. & Yao C.H. (2009). A review for gelatin used for artificial nerve and bone implants, *BioMedical Engineering-Applications Basis Communications*, Vol. 21, pp. 233-238

Chenite A. & Chaput C. (2010). Injectable *in situ* self-forming mineral-polymer hybrid composition and uses thereof, *Patent US 20100021545*

Collighan R., Cortez J. & Griffin M. (2002). The biotechnological applications of transglutaminases, *Minerva Biotechnologies*, Vol. 14, pp. 143-148

Daamen W.F., Nillesen S.T., Wismans R.G., Reinhardt D.P., Hafmans T., Veerkamp J.H. & van Kuppevelt T.H. (2008). A biomaterial composed of collagen and solubilized elastin enhances angiogenesis and elastic fiber formation without calcification, *Tissue Engineering Part A*, Vol. 14, pp. 349-360

De Bie C. (2007). Genzyme: 15 years of cell and gene therapy research, *Regenerative Medicine*, Vol. 2, pp. 95-97

Douglas T., Hemper U, Mietrach C., Viola M., Vigetti D., Heinemann S., Bierbaum S. & Worch H. (2008). Influence of collagen-fibril-based coatings containing decorin and biglycan on osteoblast behavior, *Journal of Biomedical Materials Research*, Vol. 84, pp. 805-816

Evans C.H. (2010). Gene therapy for bone healing, *Expert Reviews in Molecular Medicine*, Vol. 12, pp. e18-e26

Fang B., Wan Y.Z., Tang T.T., Gao C. & Dai K.R. (2009). Proliferation and osteoblastic differentiation of human bone marrow stromal cells on hydroxyapatite/bacterial cellulose nanocomposite scaffolds, *Tissue Engineering Part A*, Vol. 15, pp. 1091-1098

Fantner G.E., Rabinovych O., Schitter G., Thurner P., Kindt J.H., Finch M.M., Weaver J.C., Golde L.S., Morse D.E., Lipman E.A., Rangelow I.W. & Hansma P.K. (2006). Hierarchical interconnections in the nano-composite material bone: fibrillar cross-links resist fracture on several length scales, *Composites Science and Technology*, Vol. 66, pp. 1202-1208

Fujita R., Yokoyama A., Nodosaka Y., Kohgo T. & Kawasaki T. (2003). Ultrastructure of ceramic-bone interface using HA and beta-tricalcium phosphate ceramics and replacement mechanism of beta-tricalcium phosphate in bone, *Tissue & Cell*, Vol. 35, pp. 427–440

Furstenberg C.H., Wiedenhofer B., Putz C., Burckhardt I., Gantz S., Kleinschmidt K. & Schroder K. (2010). Collagen hydroxyapatite (Healos) saturated with gentamicin or levofloxacin. *In vitro* antimicrobial effectiveness - a pilot study, *Orthopade*, Vol. 39, pp. 437-443

Gapski R., Misch C., Stapleton D., Mullins S., Cobb C., Vansanthan A. & Reissner M. (2008). Histological, histomorphometric and radiographic evaluation of a sinus augmentation with a new bone allograft: a clinical case report, *Implant Dentistry*, Vol. 17, pp. 430-438

Garcia Y., Collighan R., Griffin M. & Pandit, A. (2007). Assessment of cell viability in a three-dimensional enzymatically cross-linked collagen scaffold, *Journal of Materials Science: Materials in Medicine*, Vol. 18, pp. 1991-2001

Ge Z., Jin Z. & Cao T. (2008). Manufacture of degradable polymeric scaffolds for bone regeneration, *Biomedical Materials*, Vol. 3, pp. 022001-022011

Gu Q., Zhu H.M. & Zhang X.J. (2002). Apoptosis of rat osteoblasts in process of calcification *in vitro*, *Acta Pharmacologica Sinnica*, Vol. 23, pp. 808-812

Guo X., Zheng Q., Kulbatski I., Yuan Q., Yang S., Shao Z., Wang H., Xiao B., Pan Z. & Tang S. (2006). Bone regeneration with active angiogenesis by basic fibroblast growth factor gene transfected mesenchymal stem cells seeded on porous beta-TCP ceramic scaffolds, *Biomedical Materials*, Vol. 1, pp. 93-99

Haugh M.G., Jaasma M.J. & O'Brien F.J. (2009). The effect of dehydrothermal treatment on the mechanical and structural properties of collagen-GAG scaffolds, *Journal of Biomedical Materials Research*, Vol. 89, pp. 363-369

Heinemann C., Heinemann S., Bernhardt A., Worch H. & Hanke T. (2008). Novel textile chitosan scaffolds promote spreading, proliferation and differentiation of osteoblasts, *Biomacromolecules*, Vol. 9, pp. 2913-2920

Hunt N.C. & Grover L.M. (2010). Cell encapsulation using biopolymer gels for regenerative medicine, *Biotechnological Letters*, Vol. 32, pp. 733-742

Isenburg, J.C., Simionescu, D.T. & Vyavahare, N.R. (2005). Tannic acid treatment enhances biostability and reduces calcification of glutaraldehyde fixed aortic wall, *Biomaterials*, Vol. 26, pp. 1237-1245

Jantova S., Letasiova S., Theiszova M. & Palou M. (2009). Comparison of murine fibroblast cell response to fluor-hydroxyapatite composite, fluorapatite and hydroxyapatite by eluate assay, *Acta Biologica Hungarica*, Vol. 60, pp. 89-107

Jensen T.B., Overgaard S., Lind M., Rahbek O., Bunger C. & Soballe K. (2007). Osteogenic protein-1 increases the fixation of implants grafted with morcellised bone allograft and ProOsteon bone substitute: an experimental study in dogs, *Journal of Bone and Joint Surgery*, Vol. 89, pp. 121-126

John A., Hong L., Ikada Y. & Tabata Y. (2001). A trial to prepare biodegradable collagen-hydroxyapatite composites for bone repair, *Journal of Biomaterials Science, Polymer Edition*, Vol. 12, pp. 689-705

Karageorgiou V. & Kaplan D. (2005). Porosity of 3D biomaterial scaffolds and osteogenesis, *Biomaterials*, Vol. 26, pp. 5474-5491

Kretlow J.D., Klouda L. & Mikos A.G. (2007). Injectable matrices and scaffolds for drug delivery in tissue engineering, *Advanced Drug Delivery Reviews*, Vol. 59, pp. 263-273

Laurencin C., Khan Y. & El-Amin S.F. (2006). Bone graft substitutes, *Expert Review of Medical Devices*, Vol. 3, pp. 49 57

Lee J.E., Park J.C., Hwang Y.S., Kim J.K., Kim J.G. & Suh H. (2001). Characterization of UV-irradiated dense/porous collagen membranes: morphology, enzymatic degradation and mechanical properties, *Yonsei Medical Journal*, Vol. 42, pp. 172-179

LeGeros R.Z., Lin S., Rohanizadeh R., Mijares D. & LeGeros, J.P. (2003). Biphasic calcium phosphate bioceramics: preparation, properties and applications, *Journal of Materials Science: Materials in Medicine*, Vol. 14, pp. 201-209

Leong K.F., Cheah C.M. & Chua C.K. (2003). Solid freeform fabrication of three-dimensional scaffolds for engineering replacement tissues and organs, *Biomaterials*, Vol. 24, pp. 2363-2378

Li S.T. (2000). Biologic biomaterials: tissue-derived biomaterials, In: *The biomedical engineering handbook*, 2nd ed., Vol. I, J.D. Bonzino (Ed.), pp. 42.1-42.20, CRC Press, Springer-Verlag, Boca Raton, Florida, USA

Madhan B., Subramanian V., Rao J.R., Nair B.U. & Ramasami T. (2005). Stabilization of collagen using plant polyphenol: Role of catechin, *International Journal of Biological Macromolecules*, Vol. 37, pp. 47-53

Magit D.P., Maak T., Trioano N., Raphael B., Hamouria Q., Polzhofer G., Drespe I., Albert T.J. & Grauer J.N. (2006). Healos/recombinant human growth and differentiation factor-5 induces posterolateral lumbar fusion in a New Zealand white rabbit model, *Spine,* Vol. 31, pp. 2180-2188

Matsuno T., Nakamura T., Kuremoto K., Notazawa S., Nakahara T., Hashimoto Y., Satoh T. & Shimizu Y. (2006). Development of beta-tricalcium phosphate/collagen sponge composite for bone regeneration, *Dental Materials Journal*, Vol. 25, pp. 138-144

McKay W.F., Peckham S.M. & Badura J.M. (2007). A comprehensive clinical review of recombinant human bone morphogenetic protein-2, *International Orthopaedics*, Vol. 31, pp. 729-734

Miller E.J. & Rhodes R.K. (1982). Preparation and characterization of the different types of collagen, In: *Methods in Enzymology*, L.W. Cunninghaam & D.W. Frederiksen (eds.), pp. 33-64, Academic Press, New York, USA

Moldovan L., Craciunescu O., Oprita E.I., Balan M., Zarnescu O. (2009). Collagen-chondroitin sulfate-hydroxyapatite porous composites: preparation, characterization and *in vitro* biocompatibility testing, *Romanian Biotechnological Letters*, Vol. 14, pp. 4459-4466

Mossman T. (1983). Rapid colorimetric assay for cellular growth and survival: application to proliferation and cytotoxic assays, *Journal of Immunological Methods*, Vol. 65, pp. 55-63

Mulconrey D.S., Bridwell K.H., Flynn J., Cronen G.A. & Rose P.S. (2008). Bone morphogenetic protein (rhBMP-2) as a substitute for iliac crest bone graft in multilevel adult spinal deformity surgery: minimum two-year evaluation of fusion, *Spine*, Vol. 33, pp. 2153-2159

Ohgaki M., Kizuki T., Katsura M. & Yamashita K. (2001). Manipulation of selective cell adhesion and growth by surface charges of electrically polarized hydroxyapatite, *Journal of Biomedical Materials Research*, Vol. 57, pp. 366-373

Olde Damink L.H.H., Dijkstra P.J., van Luyn M.J.A., van Wachem P.B., Nieuwenhuis P. & Feijen J. (1996). Cross-linking of dermal sheep collagen using a water-soluble carbodiimide, *Biomaterials*, Vol. 17, pp. 765-773

Oprita E.I., Moldovan L., Craciunescu O. & Zarnescu O. (2008). *In vitro* behaviour of osteoblast cells seeded into a COL/beta-TCP composite scaffold, *Central European Journal of Biology*, Vol. 3, pp. 31-37

Patterson T.E., Kumagai K., Griffith L. & Muschler G.F. (2008). Cellular strategies for enhancement of fracture repair, *The Journal of Bone and Joint Surgery*, Vol. 90, pp. 111-119

Pieper J.S., Oosterhof A., Dijkstra P.J., Veerkamp J.H. & van Kuppevelt T.H. (1999). Preparation and characterization of porous cross-linked collagenous matrices containing bioavailable chondroitin sulphate, *Biomaterials*, Vol. 20, pp. 847-858

Pieper J.S., van der Kraan P.M., Hafmans T., Kamp J., Buma P., van Susante J.L.C., van den Berg W.B., Veerkamp J.H. & van Kuppevelt T.H. (2002). Crosslinked type II collagen matrices: preparation, characterization and potential for cartilage engineering, *Biomaterials*, Vol. 23, pp. 3183-3192

Popescu R.G., Morega A., Iordachescu D. & Demetrescu I. (2009). Ternary biopolymeric films based on collagen gel, polyethylene glycol and hydroxyapatite, *Materiale Plastice*, Vol. 46, pp. 140-143

Powell H.M. & Boyce S.T. (2006). EDC cross-linking improves skin substitute strength and stability, *Biomaterials*, Vol. 27, pp. 5821-5827

Rodrigues C.V.M., Serricella P., Linhares A.B.R., Guerdes R.M., Borojevic R., Rossi M.A., Duarte M.E.L. & Farina M. (2003). Characterization of a bovine collagen-hydroxyapatite composite scaffold for bone tissue engineering, *Biomaterials,* Vol. 24, pp. 4987-4997

Rude R.K., Singer F.R. & Gruber H.E. (2009). Skeletal and hormonal effects of magnesium deficiency, *Journal of the American College of Nutrition*, Vol. 28, pp. 131-141

Scabbia, A. & Trombelli, L. (2004). A comparative study on the use of a HA/collagen/chondroitin sulphate biomaterial (Biostite) and a bovine-derived HA xenograft (Bio-Oss) in the treatment of deep intra-osseous defects, *Journal of Clinical Periodontology*, Vol. 31, pp. 348-355

Schoen F.J. & Levy R.J. (2005). Calcification of tissue heart valve substitutes: progress toward understanding and prevention, *The Annals of Thoracic Surgery*, Vol. 79, pp. 1072-1080

Schoof H., Apel J., Heschel I. & Rau G. (2001). Control of pore structure and size in freeze-dried collagen sponges, *Journal of Biomedical Materials Research*, Vol. 58, pp. 352-357

Silva G.A., Marques A.P., Gomes M.E., Coutinho O.P. & Reis R.L. (2005). Biodegradable Systems in *Tissue Engineering and Regenerative Medicine*, CRC Press, Reis, R.L., San Roman, J. (eds), pp. 339-361, Boca Raton, Florida, USA

Takahashi Y., Yamamoto M. & Tabata Y. (2005). Enhanced osteoinduction by controlled release of bone morphogenetic protein-2 from biodegradable sponge composed of gelatin and beta-tricalcium phosphate, *Biomaterials*, Vol. 26, pp. 4856–4865

Tampieri A., Sandri M., Landi E., Pressato D., Francioli S., Quarto R. & Martin I. (2008). Design of graded biomimetic osteochondral composite scaffolds, *Biomaterials*, Vol. 29, pp. 3539-3546

Tardei C., Moldovan L. & Craciunescu O. (2010). Hybrid bioresorbable composites with high bioactivity properties, *Romanian Journal of Materials*, Vol. 40, pp. 41-49

Thermo Fisher Scientific, www.piercenet.com/products/browse.cfm?fldID=CE4D6C5C-5946-4814-9904-C46E01232683

Turek A. & Cwalina B. (2010). Some irradiation-influenced features of pericardial tissues engineered for biomaterials in *New Developments in Biomedical Engineering*, InTech, pp. 543-563, Vienna, Austria

Turkovski I.I., Paramonov B.A., Antonov S.F., Kozlov D.N., Klimova O.A. & Pomorski K.P. (2008). Comparative evaluation of the depth of collagen and hyaluronic acid hydrolysis *in vitro* by collagenase and hyaluronidase preparations, *Bulletin of Experimental Biology and Medicine*, Vol. 146, pp. 81-82

Vaccaro A.R. (2002). The Role of the Osteoconductive Scaffold in Synthetic Bone Graft, *Orthopedics*, Vol. 25, pp. 571–578

Wahl D.A. & Czernuszka J.T. (2006). Collagen-hydroxyapatite composites for hard tissue repair, *European Cells & Materials*, Vol. 11, pp. 43-56

Wang X., Grogan S.P., Rieser F., Winkelmann V., Maquet V., Berge M. L. & Mainil-Varlet P. (2004). Tissue engineering of biphasic cartilage constructs using various biodegradable scaffolds: an *in vitro* study, *Biomaterials*, Vol. 25, pp. 3681-3688

Webster T.J., Siegel R.W. & Bizios R. (1999). Osteoblast adhesion on nanophase ceramics, *Biomaterials*, Vol. 20, pp. 1221-1227

Xie J., Baumann M.J. & McCabe L.R. (2004). Osteoblasts respond to hydroxyapatite surfaces with immediate changes in gene expression, *Journal of Biomedical Materials Research*, Vol. 71, pp. 108-117

Xu H.H. & Simon C.G. (2004). Self-hardening calcium phosphate cement-mesh composite: reinforcement, macropores and cell response, *Journal of Biomedical Materials Research*, Vol. 69, pp. 267-278

Yamauchi K., Goda T., Takeuchi N., Einaga H. & Tanabe T. (2004). Preparation of collagen/calcium phosphate multilayer sheet using enzymatic mineralization, *Biomaterials*, Vol. 25, pp. 5481-5489

Zarnescu O., Craciunescu O. & Moldovan L. (2010). Collagen-chondroitin sulphate-hydroxyapatite porous composites: a histochemical and electron microscopy approach, *Microscopy & Microanalysis*, Vol. 16, pp. 137-142

Zhang S.M., Cui F.Z., Liao S.S., Zhu Y. & Han L. (2003). Synthesis and biocompatibility of porous nano-HA/collagen/alginate composite, *Journal of Materials Science: Materials in Medicine*, Vol. 14, pp. 641-645

Zou C., Weng W., Deng X., Cheng K., Liu X., Du P., Shen G. & Han G. (2005). Preparation and characterization of porous beta-tricalcium phosphate/collagen composites with an integrated structure, *Biomaterials*, Vol. 26, pp. 5276-5284

14

Silicon Based Composite Anode for Lithium Ion Battery

Angathevar Veluchamy[1] and Chil-Hoon Doh[2]
[1]*Central Electrochemical Research Institute,*
[2]*Korea Electrotechnology Research Institute,*
[1]*India*
[2]*Republic of Korea*

1. Introduction

The invention of Voltaic pile during 1800 by the Italian Physicist, Alessandro Volta brought new light and energy into the world through Direct Current (DC) producing device known as battery which performed operations in domestic, industrial and transport equipments. These batteries are divided broadly into two categories one the primary battery and the other rechargeable battery. Few examples of primary batteries are the non-aqueous lithium batteries, and the zinc and magnesium based batteries. The prominent and well established rechargeable batteries are lithium-ion, lead-acid, nickel-metal hydride and nickel- cadmium, silver-zinc systems (Balasubramanian et al., 1994, 1995; Jose Benedict et al., 1998; Renuka et al. 1992; Veluchamy et al., 2001, April 2009).

Recently, the cost escalation of petroleum fuel has turned the attention of policy makers and researchers toward battery powers systems in order to partly/completely replace the petroleum fuels for automotive applications. Among the battery systems considered for electric vehicle (EV) applications, the lithium ion batteries show promise of meeting both energy and power requirements. The possibility of developing high energy and safe lithium ion battery has been reinforced further by the statement of (Panasonic, 2010) which reports a silicon-carbon nanocomposite anode for the lithium ion battery, with 30% higher capacity than the graphite based lithium ion cells, has been intended to provide power to laptop computers during the fiscal 2012. To make economic viability and practical implementation, especially for EV application still more work has to be done on fronts such as improvements on energy/power capability and safety. To meet this requirement work has to be pursued on battery materials development with a special emphasis on cost and environments. For transport applications attempts have to be made to replace combustible organic electrolytes, oxygen releasing cathodes and reactive lithium anode. This chapter presents development of high specific capacity anode materials especially the silicon composite anode material for lithium ion batteries. The resource material for this chapter has been gathered from recent research publications on lithium anode materials development.

2. Anode development for lithium batteries

The early development on non-aqueous lithium battery began with lithium (Li) metal anode. Its use has been restricted to primary batteries and application to rechargeable systems was beset with dendrite growth over the Li anode. The dendrite appeared profusely with cycling, in most cases punctured the separator, caused internal short between anode and cathode leading to rise of cell temperature which in the presence of organic electrolyte resulted in thermal runaway and in extreme cases cell explosion.

The discovery by (Yazami & Touzain, 1982, 1983) during 1980 that lithium could be inserted and de-inserted between the graphite layers and the report by (Mizushima et al., 1980) that lithium in 'lithium cobalt oxide' could be intercalated and de-intercalated made Sony Energytec in 1991 to roll the first lithium ion batteries into commercial market. Later, newer versions of lithium ion batteries with different cathodes such as lithium manganese oxide, lithium nickel oxide and lithium iron phosphate have been developed to meet different electrical requirement of the battery market. However, no break through was made on the development of anode materials to replace the graphite anode (LiC_6). LiC_6 has a theoretical capacity ~ 372 mAh g^{-1}, which is only 10% of the lithium metal anode capacity (3800 mAh g^{-1}) and its practical capacity lied between 300 and 350 mAh g^{-1}. In addition to lower specific capacity it has other drawbacks such as first cycle irreversible capacity, capacity degradation with cycling and safety issues (Brummer et al., 1980; Huggins, 2002). In the mean time the announcement by Fujifilm that a new amorphous metal oxide has been introduced as negative electrodes for its lithium ion battery has turned the researchers still more vibrant especially to bring out newer high capacity anode materials for lithium ion batteries (Fujifilm, internet, 1996).

In the search for high capacity anode materials, silicon and tin apparently qualified as anode materials due to their high theoretical energy densities ~4190 and ~990 mAh g^{-1} corresponding to the formation of binary alloys $Li_{22}Si_5$ and $Li_{22}Sn_5$. These anodes could not find immediate application because of their large volume variation ~328% for 'Si' anode and ~258% for 'Sn'. Among the two anode materials, much attention was paid on silicon anode. The reaction of lithium with elemental silicon was known for a long time. Prior to the findings that Li-Si alloy for Li-ion batteries it was exploited as a negative electrode for molten salt electrolyte batteries which operate at 400 °C (Singh et al., 2004).

The alloy forming reaction between 'Si' and 'Li' takes place when 4.4 atoms of Li combine with one atom of Si which is represented by the equation (1)

$$4.4Li + Si \rightarrow Li_{4.4}Si \tag{1}$$

The volume increase of the product is 3.7 times that of the reactants. Such large crystallographic volume changes between the charged (alloyed) and discharged (de-alloyed) state cause stress in the electrode leading to internal cracks in the electrode. During cycling the electrode particles are pulverized into micro/nano particles. Such morphological changes results in loss of electrical contact, increased impedance leading to thermal run away and cell failure (Benedek & Thakeray, 2002).

3. Capacity degradation

The capacity degradation with cycling in a battery is attributed to volume variation of the electro-active masses. The magnitude of volume variations during lithium insertion and de-insertion differs for most anode/cathode materials which cause particle fracture. Every material has a characteristic terminal particle size beyond which the particle will not fracture or disintegrate. Particle with smaller sizes do not continue to fracture and it is recommended to employ smaller starting alloy particle in the electrode for better cycling behavior (Yang et al., 2007). Such finding encouraged research work on the synthesis and use of composites containing nano-particles.

The particle fracture causes electrical disconnection between the particles and current collector. The cause for such decrepitation/crumpling of electrode particles has been described through a simple one dimensional model which states that when two phase formation occurs simultaneously with volume mismatch the result will be the development of fracture in the material. If the two phases occurs with no change in the volume, the particle will not undergo fracture. Further the critical particle size could be bigger for a particle having greater toughness (Huggins & Nix. 2000).
It is also suggested micro-crack does not happen below a critical grain size. The predicted critical grain size is less than the unit cell size for a majority of single-phase materials. This suggests that it may not be practically possible to reach the particle size to solve the mechanical instability problem associated with Li-alloys (Wolfenstine, 1999). The following passage brings out published research work on silicon based anode which includes description of mechanism toward minimum volume changes of the electrode for better cycle characteristics.

4. Silicon composite anode material - state of the art

In most of the research papers on silicon composite anode the insito generated Cr-Si, $NiSi_2$, NiSi, $FeSi_2$, TiN, SiC, TiC, SiO_2 active nano– silicon, the externally added metal particles Fe and Cu, also the graphite and carbon blacks all serve to minimize the volume changes that arise during lithiation/de-lithiation of Si in the electrodes.
For both binary Li-Si and ternary Li-Si-Cr composites containing dendritic copper powder (Weydanz et al, 1999) obtained reversible capacities of about ~ 500 and 800 mAh g^{-1} respectively. The higher capacity of 'Li-Si-Cr' composite is attributed to insito generated Cr-Si phase which acts as an inactive conductive matrix along with active Li-Si phase.
Mg_2Si/acetylene black composite reported by (Roberts et al, 2002) delivers an initial discharge capacity ~ 830 mAh g^{-1} where the Li insertion into Mg_2Si forms Li_2MgSi which then tends to decompose into binary alloys, Li-Si and Li-Mg at low potential regions causing capacity degradation with cycling. (Cairns & Reimer, 2002) have also reported on magnesium silicide anode. (Chen et al., 2006) has reported on the preparation of nano structured Si/C composites.
(Shi et al, 2001) has shown the better reversibility of Li-Mg anode alloys prepared through Kinetically Controlled Vapor Deposition (KCVD) method is attributed to high diffusion coefficient of lithium atoms in the Li-Mg alloy compared to Li-Mg alloy prepared by reacting Mg and Li in molten states in a Glove box .
(Wolfenstine, 2003) reported two composite anodes, one made out of sieved $CaSi_2$ and the other ball milled $CaSi_2$ with carbon conductive material and found both materials showing not so promising results.
Through EDAX (Kim et al, 2000) reported a composite made of amorphous Si and nanosized TiN which gave a reversible capacity of 300 mAh g^{-1} where TiN acted as a buffer matrix. In the nano-composite consisting of amorphous silicon and nano-crystalline TiC, the electrochemically inactive TiC served as a buffer matrix and the electrode delivered a reversible capacity of 380 mAh g^{-1}(Guo, 2005).
(Park, 2005, 2006) obtained an alloy through arc-melting followed by high energy mechanical milling (HEMM) which consisted inactive phases $NiSi_2$, NiSi and an active Si phases which delivered a reversible capacity of ~ 800 mAh g^{-1}. (Wang et al 2000) showed that the NiSi/C nano-composite to perform better than Fe-Si nano-composite and found both electrodes exhibited capacity decline with cycling.

Composite with low Si content gave better cycle capacity ~ 600 mAh g^{-1} even at 20th cycle where $FeSi_2$ served as a volume buffer. The irreversible capacity usually observed between 0.9 and 0.7V is caused by electrolyte decomposition that contribute to passive film formation (Lee & Lee, 2002).

(Kim et al., 2005) achieved acceptable anode capacity for the multilayer films consisting of active (Si) and inactive (Fe) metal elements. (Dong, 2004) showed for a Fe-Si composite ($FeSi_2$ + Si)/C composed of a sandwich structure with the alloy particles in the middle core and the graphite layer at outer shells has exhibited high initial capacity ~ 680 mAh g^{-1} with a reversible capacity ~ 500 mAh g^{-1} even after 15 cycles. The good performance was attributed to the effective buffering of the volumetric changes of the Fe-Si particles by the graphite shell.

In (Zuo et al., 2006) the Si-Mn composites after 60 h ball milling showed best performance, especially when annealed at 300 °C for 2 h, which had a reversible capacity of 455 mAh g^{-1}. Mn acted as an inactive buffer. The decline in cycle life was attributed to agglomeration of nano-sized Si-Mn particles. In another report the authors show that a Si-Mn-C composite annealed at 200 °C exhibited an initial reversible capacity of 463 mAh·g^{-1} retaining 387 mAh g^{-1} even after 40 cycles.

In (Wang et al, 2007) a composite prepared by incorporating silicon powder into an inverse emulsion polymerised resorcinol-formaldehyde (RF) followed by carbonization in an inert atmosphere gave 910 mAh g^{-1}. The addition of Cu further improved the cycling performance of the composite.

(Dimov et al., 2003) proposed trapping of lithium ions in the form of Li-Si alloy into the electrode takes place during cycling, making progressive decrease in the availability of lithium for extraction. Such decrease in capacity with cycling is responsible for the capacity fading with cycling in the carbon coated silicon powder. (Wilson et al., 1997) have pyrolysed 50 different silicon containing polymers including polysilanes, polysiloxane, pitch silane blends and showed through a diagram the extent of formation of SiC, SiO_2 and C is due to pyrolization.

In (Lee et al., 2005) a carbon coated $Ni_{20}Si_{80}$ alloy/graphite composite is presented with an improved cycle performance and reduced initial irreversible capacity loss attributed to buffering and conductive actions of the carbon coated graphite present in the composite. Around 200 compositions of combinatorial materials comprising Si-M (M =Cr + Ni, Fe, Mn) has been reported by (Fleischauer, 2005) where the anode capacity strongly depended upon Si content where the capacity varied from over 3000 mAh g^{-1} for nearly pure silicon to effectively zero for ~50-60 at% of silicon.

Multiple component alloy composite (Dong et al., 2003) with Graphite/Ba-Fe-Si alloy has been observed to provide 500 mAh g^{-1} at 15th cycle.

Si-AB5 composite alloy (Zhang, 2007), where Fe is one of the components, exhibited initial and maximum reversible capacity ~ 370 and 385 mAh g^{-1} with high capacity retention even after 50 cycles. The Si-AB5 composite containing 20 wt% Si provides discharge capacity ~ 420 mAh/g with good capacity retention (72%) even after 50 cycles where the inactive AB5 alloy is said to accommodate large volume changes of Si nano-particles distributed on the surface of Si-AB5 composites during cycling.

(Zhang et al., 2004) reported, for the silicon-based thin film Si/TiN nano-composites formed by pulsed laser deposition method, better capacity attributed to dispersal of nanometer range amorphous Si particles in the presence of an inactive matrix of TiN.

Similarly the role of copper in Cu_5Si-Si/C anode has been explored and the report states that the anode delivered ~ 612 mAh g^{-1} capacity with a capacity fade during initial cycles. The capacity fade formed was attributed to SEI film formation (NuLi, 2006).

(Kim, J. H. 2005) for a carbon coated Si–Cu_3Si–Cu composite the copper silicide formed by pyrolysis is not an electro-active material and the improved performance of the composite was attributed to better electrical contact of metal silicide and pyrolyzed carbon.
A novel silicon/graphite material prepared by depositing a gelatinous silicon precursor over a porous natural graphite gave a stable 100 cycles at ~ 840 mAh g^{-1} ascribed to good embedding of the silicon in the graphite matrix (Fuchsbichler, 2011).
(Dimove et al., 2003) have deposited carbon over silicon through thermal vapour deposition using benzene or toluene as carbon source and nitrogen as a carrier gas. Among the three binary alloys of the type, Ti_x-Si_y with variable x and y values only the carbon coated Ti-Si alloy exhibited good cycle performance, especially when the Si content in Ti-Si alloy is kept below the eutectic value of (Si + $TiSi_2$) in the composite (Lee et al, December, 2006).
This part of the discussion has made clear that any one composite that is prepared with carbon coating, insito generated innert phases with or without nano-active silicon particle produced during charge/discharge process deliver better electrochemical performances.
In the following sub headings the research contributions of the authors on the topic 'Silicon composite anode for lithium ion batteries' has been presented (Doh et al., Jan 2008, Feb 2008, April 2008, June 2008, May 2010; Veluchamy et al., Feb 2009, April 2009)

5. Experimental

5.1 Fe-Cu-Si/C composite anode

5.1.1 Composite powder

Appropriate quantities of Cu (<10µm, 99% pure, Sigma–Aldrich), Fe (<53µm, 99.9% pure, High Purity Chemical Research Company, Japan) and Si (1–5µm, SI-100, >99 purity, AEE, NJ) with atomic ratio, Fe:Cu:Si = 1:1:2.5 were ball milled using a stainless steel (SS) grinding vial along with SS balls. The weight ratio of SS ball to the material was kept at 10:1 and the millings were carried out at 350 rpm. The vial was filled with argon gas and tightly closed with a gasket to prevent ingress of atmospheric oxygen. The material ball milled for 3h was then mixed with graphite in equal proportions in weight ratio, 50:50 and again ball milled for 24h. The composites ball milled for 3, 10, and 24h are hereinafter referred as 3HBM, 10HBM, and 24HBM, respectively. The composite thus arrived with the ratio Fe: Cu: Si = 1:1:2.5/graphite is represented as Fe–Cu–Si/C.

5.1.2 Composite electrode

The composite powder from 5.1.1 was then blended with 10 wt% polyvinylidene difluoride (PVDF) dissolved in 1-methyl-2-pyrrolidinone in an agitator for 10 min to obtain a slurry mass. The slurry mass was coated onto a copper foil current collector and dried in a hot air oven at 110 °C for 2 h and pressed using a SS roller in order to reduce the thickness to ~ 75%. The active material coated copper foils were annealed at 110, 150, 200, and 250 °C in vacuum oven for 12 h.

5.1.3 Cell construction

Annealed composite electrode was cut in the form of a circular disc of diameter 1.4 cm and coupled with lithium foil counter electrode separated by Celgard-2700 separator. The electrolyte from Techno Semichem. Ltd., Korea is 1M $LiPF_6$ dissolved in a co-solvent consisting of ethylene carbonate (EC) and ethyl methyl carbonate (EMC) in 1:1 (v/v) ratio with 2 wt% vinylene carbonate (VC). The coin cells assembled in a dry room maintained at ~21°C with dew point temperature between -65 and -70 °C were subjected to life cycle test using charge-discharge analyzer, Toyo System, Ltd., Japan between 0 and 2V versus Li+/Li at a constant current of 0.253 mA cm^{-2}.

5.1.4 XRD and SEM investigations

The ball milled composites were examined using Philips 1830 X-ray diffractometer with nickel-filtered Cu-Kα radiation at a scan rate of 0.04° s^{-1} over 2θ range between 10–80°. The surface morphology of the active material coated copper foil was scanned using Hitachi S-4800 scanning electron microscope (SEM). In Fig. 1, the XRD pattern for Si, Cu, Fe, graphite powder(C), 3HBM-Fe-Cu-Si, 10HBM-Fe-Cu-Si, and 24HBM-{3HBM-Fe-Cu-Si:graphite = 50:50(w/w)} are presented.

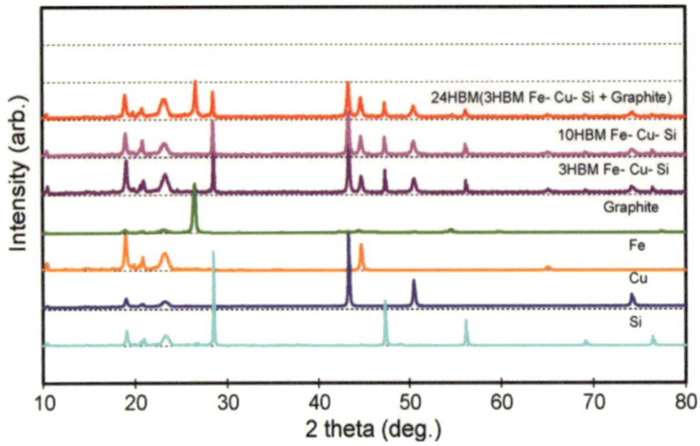

Fig. 1. XRD pattern of Fe, Cu, Si, C, 3HBM- Fe-Cu-Si, 10HBM- Fe-Cu-Si, and 24HBM-{3HBM- Fe-Cu-Si: C = 50:50(w/w}.

All these suggest ball mill duration does not change the elemental particles, rather reduces them into smaller particulates as evident from the minimum change in the intensity of Braggs peak in the XRD patterns.

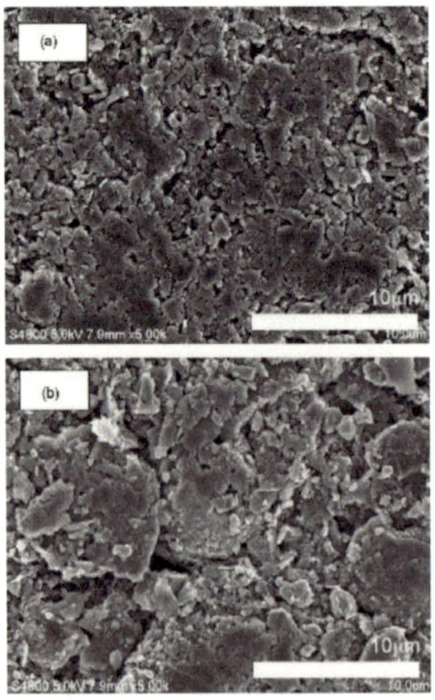

Fig. 2. SEM pictures of Fe-Cu-Si / C composite electrode annealed. Electrode --'a' at 110 °C and electrode ---'b' at 200 °C.

The SEM pictures of Fe-Cu-Si/C composite with PVDF binder annealed at 110 °C was compared with the best performing electrode annealed at 200 °C and presented in Fig. 2 as 'a' and 'b', respectively. The pictures show the cathode particles in 'b' remain in agglomerated state with boundries and distinct interspaces compared to that of the particles in 'a' with particle sizes varying from < ~1 to ~10 µm. Such interfaces and voids formed in the composite electrode annealed at high temperature is expected to provide better ionic pathway for the ions thereby providing better conductivity and also the voids are expected to minimise the volume change of the electrode during charge/discharge cycles.

The superimposition of Braggs peaks appearing in the XRD pattern of individual materials such as Si, Cu, Fe over the XRD spectrum of Si, Cu, Fe & graphite composite shows that the superimposed XRD spectrum coincide exactly with spectrums obtained for 3HBM- Fe-Cu-Si, 10HBM- Fe-Cu-Si, and 24HBM-{3HBM-Fe-Cu-Si: C (graphite) = 50:50(w/w)} respectively.

5.1.5 Charge/discharge characteristics

The typical discharge-charge profiles for the electrode Fe-Cu-Si/C annealed at 200 °C presented in Fig. 3 shows the initial lithiation and delithiation capacities respectively are 809 and 464 mAh g^{-1}. The capacity difference between the two processes is 345 mAh, nearly 42% of the initial lithiation capacity. This implies that during the first cycle ~ 42% of Li is retained within the electrode as irreversible capacity. This irreversible capacity is due to the reaction of graphite, metal particles and silicon with the chemically bonded or adsorbed oxides present within the electrode. During the initial discharge, inserted lithium reacts irreversibly at first with these oxides and water impurities forming Li_2O, and then forms Li-Si and LiC_6. The amount of Li_2O formed is proportional to irreversible capacity during cycling. Thus formed Li_2O along with Fe, Cu and Graphite acts as a buffer, absorbs the volume expansion/contraction during discharge/charge process. Also other products formed due to oxidation of electrolyte and binder along with Li_2O are Li_2CO_3, LiF and alkyl carbonates. All these products forms part of SEI films of silicon and graphite.

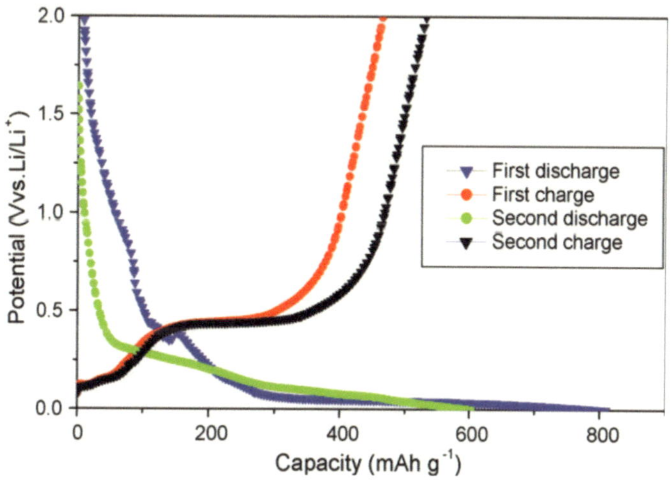

Fig. 3. Discharge-charge profiles for Fe-Cu-Si/C composite electrode annealed at 200 °C.

For the present system we propose the following reactions that are responsible for the reversible reaction (2) and irreversible reactions such as lithium oxide formation reaction (3), organic solvent decomposition reaction (4), & electrolyte decomposition reaction (5).

$$(x+y)Li^{+} + (Si, Fe, Cu, C) + (x+y)e^{-} \leftrightarrow Li_xSi + Li_yC_6 + (Fe, Cu) \tag{2}$$

$$2Li^{+} + -O- + 2e^{-} \rightarrow Li_2O \tag{3}$$

$$Li^{+} + (EC + EMC\ldots\ldots) + e^{-} \rightarrow SEI\ (Li_2CO_3,\ ROCO_2Li,\ \ldots.) \tag{4}$$

$$LiPF_6 + 2xLi^{+} + 2xe^{-} \rightarrow Li_xPF_{5-x} + (x+1)LiF \tag{5}$$

In addition to the reactions (3), (4) and (5) other contributing factors responsible for capacity degradation with cycle life are particle fracture followed by loss of electrical contact between the electro-active species and also between electro-active species & current collector (Aurbach & Schechter,2004). In the second cycle the discharge and charge capacities respectively are 607 and 531mAh g^{-1} which shows a considerable reduction in irreversible capacity equal to ~12%. The authors in (NuLi, 2006; Wang, 2007; & Aurbach, 2004&2005) describe that the irreversible capacity in the graphite electrode is due to the reaction of intercalated Li with the adsorbed oxygen and solvent molecules leading to the formation of SEI film over the graphite particles which comprises mainly LiF, Li_2CO_3 and $ROCO_2Li$ etc.

5.1.6 Capacity with cycle number at different electrode annealing temperatures

The discharge and charge capacities of the electrodes annealed at different temperatures are depicted in Fig. 4 and Fig. 5 respectively.

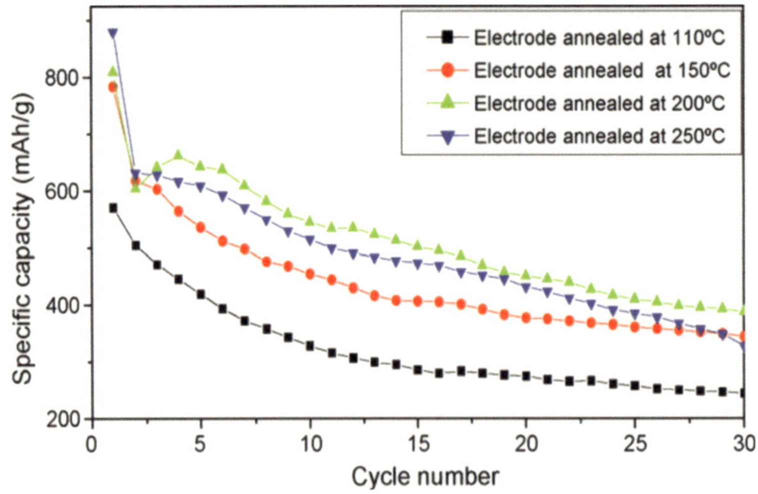

Fig. 4. Discharge capacity as a function of cycling with electrode annealing temperature.

The Fig. 4 shows an improvement of discharge capacity with annealing temperature. The curve shows discharge capacity 387 and 327 mAh g^{-1} for the electrode annealed at 200 and 110°C respectively at 30th cycle. In Fig. 5 the initial charge capacity values for charging process remains at a low value, after few cycles the capacity raises and then stabilizes in the subsequent cycles. The gain in charge/discharge capacity after the first cycle in all four temperature except at 110°C (Fig.4 & 5) is attributed to realignment/redistribution of constituents of the SEI film. The higher capacity obtainrd for the electrode annealed at 200 °C could be attributed to enhanced electrical conductivity and easy ionic diffusion of the electrode.

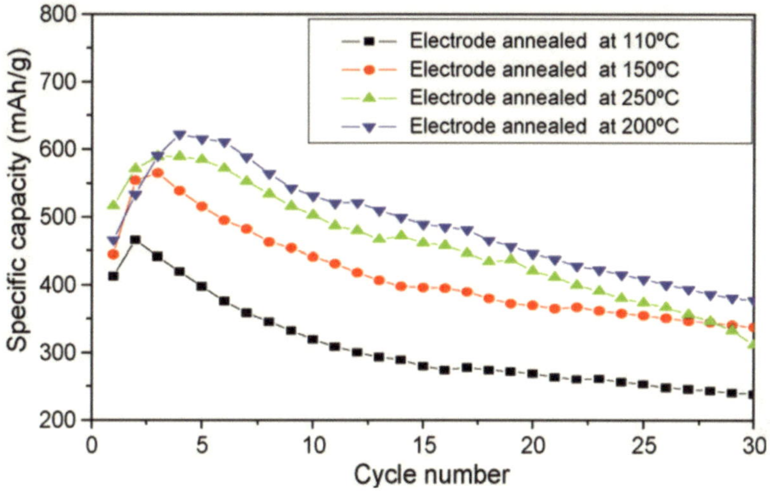

Fig. 5. Charge capacity as a function of cycling with electrode annealing temperature

Further support for the improved capacity of the annealed electrodes could be understood from the SEM picture which shows well separated electrode particulates with interspaces between the electrode particles which is expected to favour ionic diffusivity and electrochemical behaviour. Similar reports have been presented by (Zuo et al., 2006) for SiMn/C anode which exhibited improved reversible capacity (426 mAh g^{-1}) for the electrode annealed at 200 °C.

5.1.7 Irreversible capacity loss

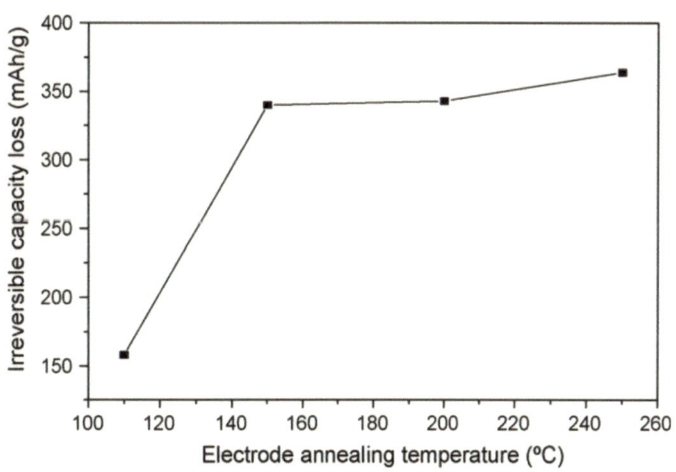

Fig. 6. Irreversible capacity loss as a function of electrode annealing temperature.

The Fig.6 illustrates irreversible capacity loss by Fe-Cu-Si/C electrode during first cycle (lithiation process) for the electrodes annealed at different temperatures. The curve shows higher irreversible capacity for the electrodes annealed above 150 °C. The high temperature annealing causes better particle segregation which makes easy diffusion of lithium ions and effective conversion of all adsorbed oxides present in the electrode into Li_2O (reaction 3). The high charge consumption for SEI film formation in Cu-Si alloy (NuLi, 2006) anode has been explained as due to increase of surface area through longer ball milling duration.

5.2 SiO/C composite anode

5.2.1 Composite preparation

Equal proportions of SiO (Aldrich, −325 mesh) and graphite (Sodiff New Materials Co. Ltd., Korea, −400 mesh) powders were placed together in a 200 ml stainless steel (SS) vial. The weight ratio of the SS ball to the material was maintained at 10:1 and the vial was filled with argon gas. The contents in the vial were milled for 12, 18, 24 and 30 h by means of high-energy ball milling (HEBM) at 350 rpm and four different samples were collected. The samples were subjected to physical and electrochemical investigation following the procedures as in 5.1.4 and 5.1.5.

5.2.2 XRD and SEM Investigation

The XRD patterns for the starting materials SiO, graphite(C) and milled SiO/C composites are presented in Fig. 7. The XRD pattern shows that the graphite considered for ball milling is crystalline whereas SiO is amorphous. The patterns for milled SiO/C samples reveal that the peak corresponding to graphite decreases with increase in ball mill duration and also no new peaks observed. This suggests that ball milling causes only particle size reduction of the graphite and does not produce any new compound.

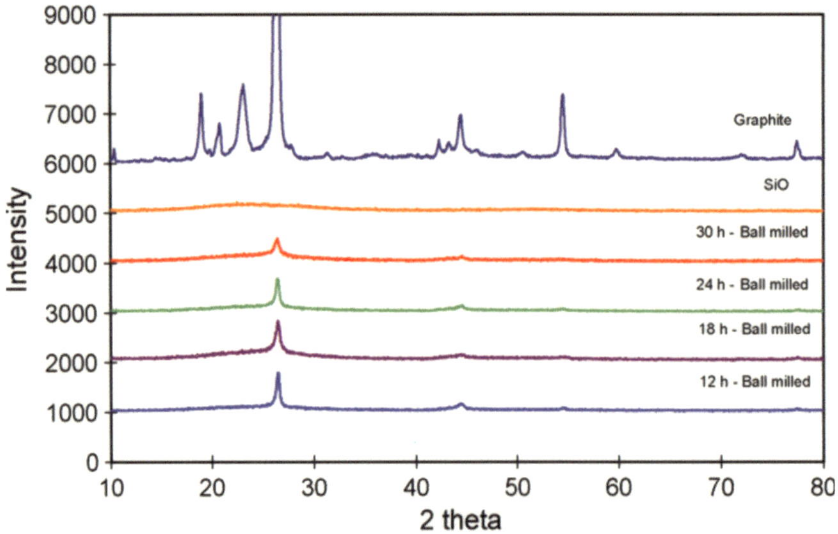

Fig. 7. XRD patterns of SiO, graphite (C), and samples (SiO and graphite) ball-milled for different durations.

The scanning electron micrographs (SEM) of samples ball-milled for different durations are given in Fig. 8. The SEM pictures show a gradual decrease in the crystallinity of the particulates of the samples with ball mill duration which is in accordance with the XRD patterns. The particles in the sample milled for 12 h exist as discrete crystalline particles whereas the 30 h-milled sample lacks definite particle shape and shows continuity in the particle distribution. In some domains, the particle distribution in the 24 h ball-milled sample remains between 5 and 20 µm whereas in other regions the continuity is maintained.

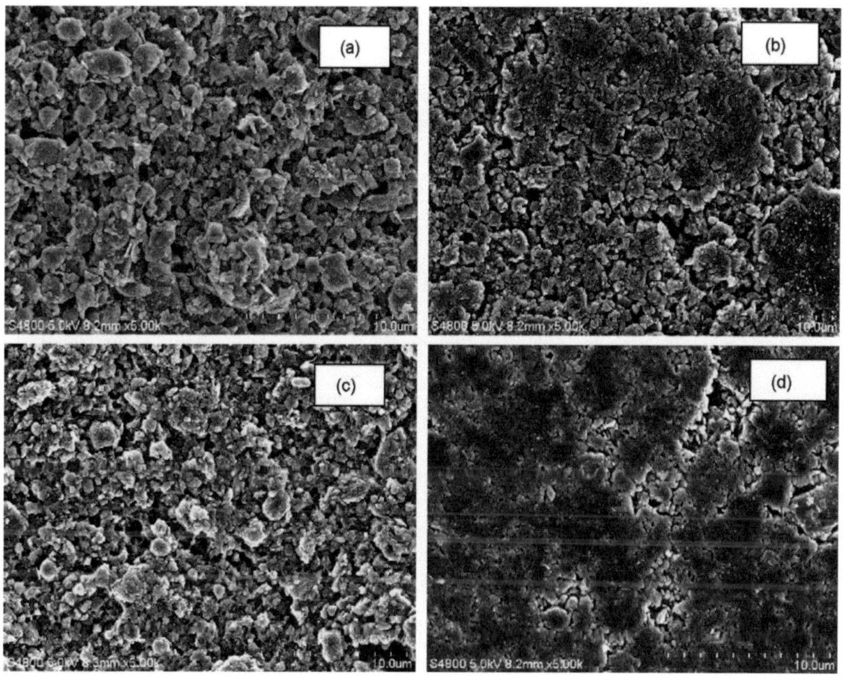

Fig. 8. SEM micrographs of samples (SiO and graphite) ball-milled for different durations: (a) 12 h; (b) 18 h; (c) 24 h; (d) 30 h.

5.2.3 Charge/discharge characteristics

The composite powder from the section 5.2. 1 has been coated over copper foil following the procedures described in 5.1.2 and cell was fabricated as described in 5.1.3 to get SiO/C anode for electrochemical characterization study. A typical discharge charge profile of the SiO/C composite electrode is given in Fig. 9. The difference between the initial discharge (1556 mAh g^{-1}) and charge (693 mAh g^{-1}) capacities equal to 55% of capacity is known as irreversible capacity. In the second cycle, the irreversible capacity is reduced to 14%. It is well known that during the first lithiation process, lithium reacts with SiO and forms nano-silicon and Li_2O. The nano-silicon then reacts with Li and forms Li-Si alloy (Lee & Lee, 2004). Contrary to this explanation (Miyachi et al., 2006) showed through O 1s spectra analysis that during the first lithiation process a direct absorption of Li by SiO takes place giving rise to the formation of Li_4SiO_4 and Li_2O.

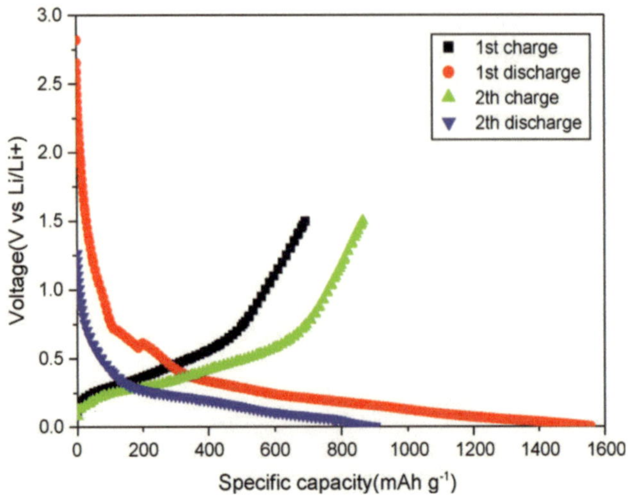

Fig. 9. Voltage - specific capacity profile of the SiO/C composite electrode

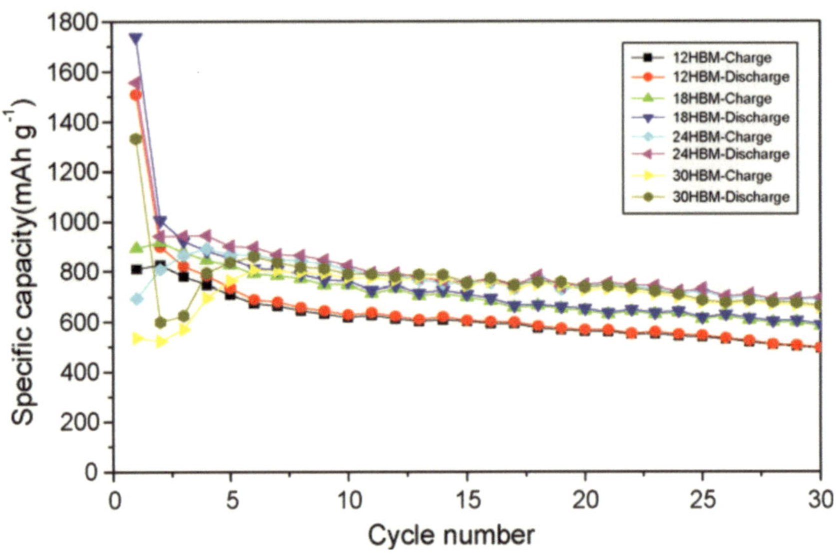

Fig. 10. Specific capacity - cycle number for the electrodes with ball mill durations

Among the composites prepared, the 24 h ball-milled sample exhibits higher reversible capacity. This might be attributed to optimum particle size distribution at 24 h ball milling which provided better lithium ion diffusion compared with other compositions. The loss of capacity shown by the sample with cycling could be attributed to trapping of Li^+ ions within the electrode particles and decomposition of the organic solvent toward formation of SEI film. The other factors responsible for capacity degradation with cycling are particle fracture and loss of electrical contact between electro-active species and current-collector.

The discharge and charge capacity values for the initial and 30th cycles obtained from Fig. 10 are listed in Table 1. The total irreversible capacity is due to irreversible reaction between Li reacting with, 1) SiO into forming Li_2O through the reaction $SiO + 2Li^+ \rightarrow Si + Li_2O$, 2) oxygen impurities present in the graphite, and 3) electrolyte and binders during initial charging. These reaction products form the SEI film which constitute organic and inorganic carbonates, Fluorides, oxides etc., (Datta & Kumta, 2007; Aurbach, 2002 & 2005).

	1st cycle			30th cycle	
Ball mill duration (h)	*Discharge capacity* (mAh/g)	*Charge capacity* (mAh/g)	*Irreversible capacity* (%)	*Discharge capacity* (mAh/g)	*Charge capacity* (mAh/g)
12	1508	809	46	495	493
18	1738	892	48	589	580
24	1556	693	55	696	688
30	1330	535	59	663	656

Table 1. The capacity values arrived at 1st and 30th cycle at different ball mill duration.

The reversible and irreversible capacity values appear to be interdependent as their values show simultaneous increase until the 24h ball-milled sample. This interdependency also explains the beneficial role played by a higher percentage of Li_2O which buffers the volume change during the alloying and de-alloying processes.

5.3 Carbon coated SiO/C composite anode

5.3.1 Composite preparation

This part compares performance of two composites. One composite prepared by ball milling equal proportion of 'SiO' and 'C' for 15 h, named as SiO/C composite and denoted as 'A' by following the same procedure described 5.2.1. Another composite is prepared by mixing 2.5 g of the composite 'A' in 5 ml of the solution made of propylene carbonate (PC) and acetone (AC) in the ratio 1:1 (v/v) in a beaker using a magnetic stirrer. During stirring the volume of the slurry will be reduced due to the evaporation of acetone. The slurry was poured into an alumina boat and heated in a furnace at a rate of 5 °C under an argon atmosphere and maintained at 750 °C for 3 h and then spontaneously cooled down to reach room temperature. Thus obtained composite is named as composite 'B'. The thermal decomposition of the organic molecules (Ng et al., 2007) under argon atmosphere in the presence of SiO/C composite may be represented by the equation (6)

$$\text{Propylene carbonates}(C_3H_6CO_3) + \text{trace / trapped acetone}(CH_3COCH_3), + \\ -O- + -H- \rightarrow C_x + CO_2 + H_2O + \text{energy} \tag{6}$$

The components -O- and -H- come from trace of adsorbed water molecules and also from physically/chemically attached -O- from SiO and graphite. In C_x, the x value of carbon varies from 1 to 3. The material obtained after thermal treatment of the slurry appeared as agglomerated clusters presumably due to wrapping up of C_x carbon particles which upon crushing in a mortar converted into a powder named as composite 'B'. Similar approach has been followed by (Ng et al,2007) for carbon coating of silicon.

5.3.2 XRD and SEM investigation

The XRD pattern obtained for SiO, graphite (C) and the composite 'A', and the composite 'B' are presented in Fig. 11. The pattern for composite 'A' shows that the graphite considered for ball milling is crystalline one and SiO is an amorphous variety. Thermally treated organic slurry composite powder shows a slight enhancement of the amplitude of the peak at 26° showing enhanced crystalline nature of the graphite. Further this XRD pattern suggests ball milling and the heat treatment with organic solutions have not produced the formation of any new compound.

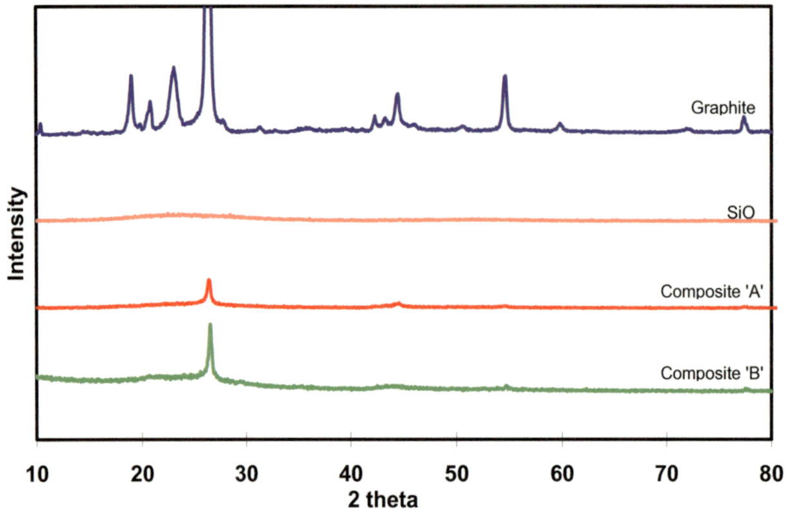

Fig. 11. XRD pattern of SiO; Graphite(C); Ball milled SiO/C Composite → 'A' ; Blend made of (SiO/C + PC + AC) and heated to 750 °C for 3 h → 'B'.

The SEM pictures of the electrode-A and electrode-B are shown in Fig. 12 It is evident from the SEM pictures that the electrode-B has larger particles with voids and interspaces compared to the particle arrangement in the electrode-A. The partial removal of oxygen and trace -H- present in the composite could have effected bonding rearrangement within the particles invoking reinforcement and compactness to the resultant structure in concomitant with trapping/wrapping by formed Cx particles.

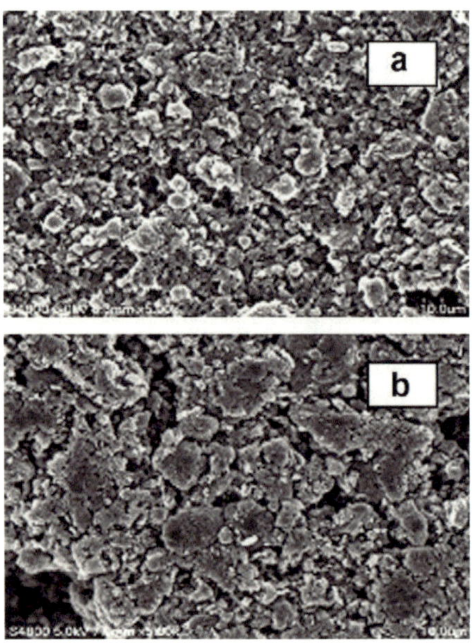

Fig. 12. SEM micrographs of samples, a- Electrode –A; b- Electode- B

The comparative curves showing the values of specific capacity with cycle number for the electrode-A and electrode-B are given in Fig. 13. The figure illustrates that the electrode-A exhibits higher charge/discharge capacity values until 53rd cycle and then the capacity falls below the value of the electrode-B. The initial irreversible capacities (863 mAh g^{-1} for the electrode-A and 733 mAh g^{-1} for electrode-B) denote the utilization of these capacities for the formation of SEI film Li_2O. SEI film protects the anode and Li_2O acts as a buffer component for improving the cycle behavior. The difference in irreversible capacities (863-733= 130 mAh g^{-1}) suggests lowering of 130 mAh g^{-1} capacity in the composite -B compared to the electrode-A. From this we can infer that 130 mAh g^{-1} equivalent of oxygen/water molecule are chemically/physically present in the composite 'A' which is absent in the composite 'B'. During carbon coating these oxygen/water molecules could have liberated as CO_2.. This also suggests that Li_2O in composite-B is lower by 130 mAh g^{-1} capacity equivalent compared to that in composite-A. Even though the electrode-B has lower buffer component than in electrode-A, it shows better cycle behavior, attributed possibly to structural rearrangement during partial removal of oxygen/water and wrapping of the electrode particle by C_x to form a well reinforced but a flexible structure to allow Li^+ ion diffusion for making the electrode-B to withstand the volume changes during cycling. It may be noted from 1st to 4th cycle for the electrode-A, the charge capacity value increases from 693 to 890 mAh g_{-1} and for the electrode-B from 613 to 652 mAh g_{-1}. The increase in the charge capacity values calculated as 197 and 39 mAh g_{-1} may be termed as the charge capacity recovery values for the respective electrode-A and electrode-B. The attributes of the electrode-B such as lower charge capacity recovery and lower capacity degradation with cycling further support the favourable structural rearrangement that occurred during thermal treatment with organic solution.

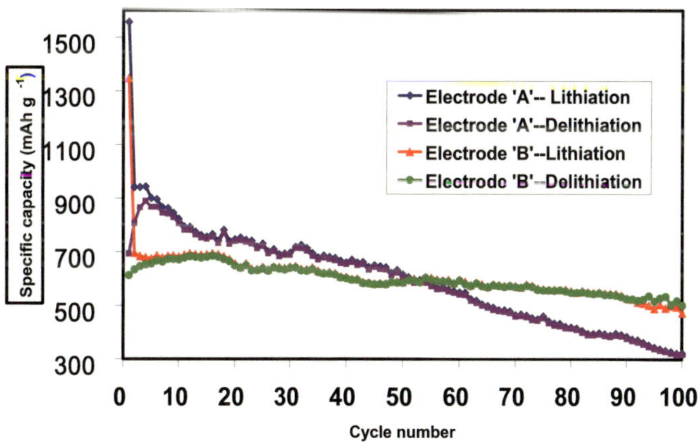

Fig. 13. Specific capacity as a function of cycle number for Electrode-A and Electrode-B

The cyclic voltammograms of the electrode-A and electrode-B scanned between 0 and 1.5 V at a scan rate of 0.1 mV s^{-1} is shown in Fig. 14. The delithiation process displayed in the 5th cycle is considered for discussion. The delithiation process of the electrode-A is associated with an increase of current with potential whereas the electrode-B shows an increase of voltage by 0.3 V but exhibits nearly a stable current between 0.3 and 0.6 V. This shows the over-potential dependent current flow through the electrode-A which is possible for an electrode with a loosely bound dispersed structure. The electrode-B nearly does not show increase of current as the potential is scanned from 0.3 to 0.6 V, illustrating that the electrode-B behaves apparently like a resistor over this narrow potential range. The resistor like behaviour illustrates the compact nature of the structure which could stream line the current flow with a tolerable volume changes during charge/discharge process. This is well supported by the SEM micrographs which show the presence of large interspaces and voids among the agglomerated particles. These voids and interspaces could accommodate enough electrolytes for easy lithium ion diffusion. Since the capacity degradation is less with cycles it may be presumed that the electrode has a flexible structure with sufficient pores to buffer the volume changes during charge/discharge cycles.

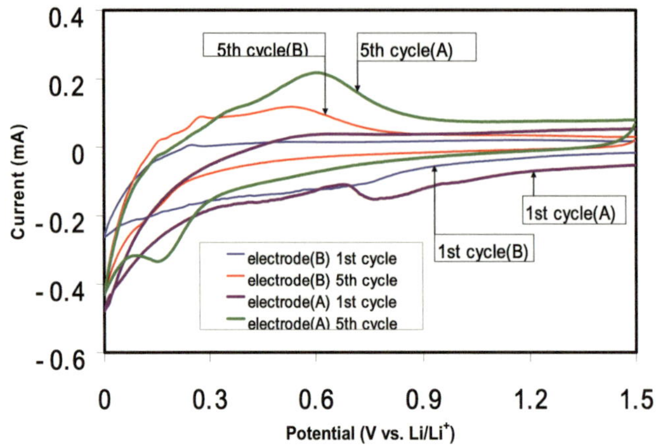

Fig. 14. Cyclicvoltammogram for Electrode-A and Electrode-B at 0.1mVs^{-1}.

5.4 Thermochemically generated Li_4SiO_4 phase in the composite

5.4.1 Composite preparation

Equimolecular quantities of lithium hydroxide monohydrate $LiOH{\cdot}H_2O$ (Aldrich) and SiO (−325 mesh, LTS Chemicals Inc.) were mixed in a mortar and then heated to 550 °C for 3 h in an argon atmosphere. Graphite (C) was then added to maintain the weight ratio of C and SiO at 1:1 and ball-milled for 15 h. The composite was named as 'A'. In order to understand the behaviour of the new composite 'A' another composite 'B' was prepared by just ball milling SiO and C in a weight ratio of 1:1 for 15 h. The samples were subjected to physical and electrochemical investigation following the procedures described in sections 5.1.4 and 5.1.5 in the following section.

5.4.2 X-ray diffraction analysis

The XRD patterns of graphite (C), SiO, LiOH, heat-treated (SiO + LiOH) and ball-milled material made of heat-treated (SiO + LiOH)/C presented in Fig. 15 shows graphite(C) and LiOH are crystallites, and SiO is amorphous. The constituents C, SiO and LiOH are noted in figure as (a) (b) and (c). The precursor material obtained by heating (SiO + LiOH) at 550 °C does not show peaks corresponding to LiOH but exhibits new peaks. Examination with the JCPDS file shows that the new peaks are related to the compound Li_4SiO_4. Also the base line above the horizontal line between 2θ values 20° and 30° suggests the presence of SiO in the heat-treated materials. In the sub-section XR-2, the material (d) = (b) + (c) heated to 550 °C exhibits a peak at 2θ =32° which could be ascribed to Li_4SiO_4 and not to Li_2O as there exists no other peak that can be attributed to Li_2O.

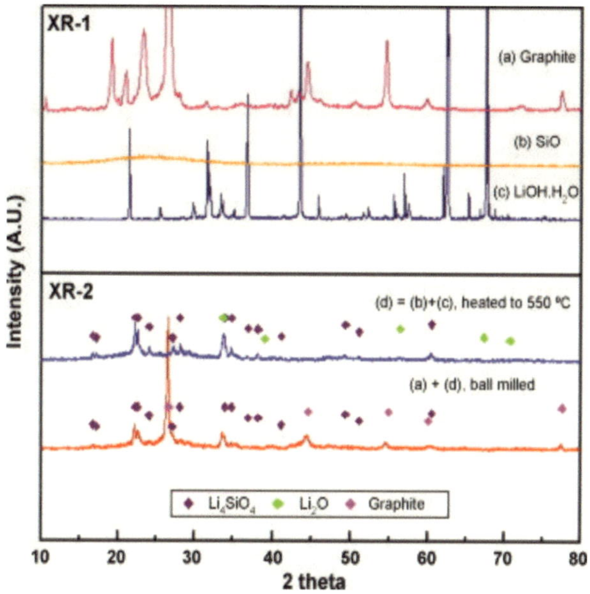

Fig. 15. XRD pattern of materials: (a) graphite; (b) SiO; (c) LiOH; (d) = (b) + (c) heated at 550 °C in argon atmosphere; (a) + (d) ball-milled composite material.

This observation leads to the conclusion that all the Li_2O has reacted with SiO at 550 °C to form Li_4SiO_4. There are also several minor peaks for Li_4SiO_4. Hence, it may be proposed that before the melting point (from ~450 to 470 °C) the lithium hydroxide is dehydrated to form Li_2O as represented by reaction (7). The reaction of Li_2O with SiO may be represented by reaction (8)

$$2LiOH{\cdot}H_2O \rightarrow Li_2O + 3H_2O \tag{7}$$

$$4SiO + 2Li_2O \rightarrow Li_4SiO_4 + 2SiO + Si \tag{8}$$

Reaction (8) indicates the product contains nano-silicon, Li_4SiO_4 in addition to SiO in the precursor material (d). The compound Li_4SiO_4 has a high negative free energy change ($\Delta G°298K$ = −2366 kJ/mole) (International, 2004) and is considered to be an electrochemically inactive species in the composite (yang, 2007). The XRD pattern of the ball milled materials (d) with graphite (a) shows a reduction in the peak heights of both graphite and Li_4SiO_4 which implies these particles have undergone size reduction during ball milling.

5.4.3 Cycle/capacity behaviour

The variation of specific capacity with cycle number for electrodes 'A' and 'B' is presented in Fig. 16. Data obtained from Fig. 16 is presented in Table 2. The irreversible capacity 387 mAh g^{-1} (44%) exhibited by electrode 'A' is considerably lower compared to ~ 863 mAh g^{-1} (55%) as shown by electrode 'B'.

The electrode 'A' also exhibits a lower initial lithiation capacity (862 mAh g^{-1}) than electrode 'B'. The reduction in initial capacity and also the irreversible capacity are attributed to the reduction of ¼ part of SiO in the composite 'A', which has been converted into Li_4SiO_4 during precursor formation. The amount of SiO present in the composite has a direct relationship with irreversible capacity and initial specific capacity, which is in agreement with an early report (Doh et al., May 2008). The slightly higher charge capacity shown by electrode 'A', even at the 100th cycle could be attributed to the buffer action provided by Li_4SiO_4. The capacity values such as initial lithiation, initial delithiation, irreversible capacity and charge capacity values obtained for the electrode 'A' and electrode 'B' is presented in the Table. 2 which illustrate the superior quality of thermochemically treated composite over the untreated.

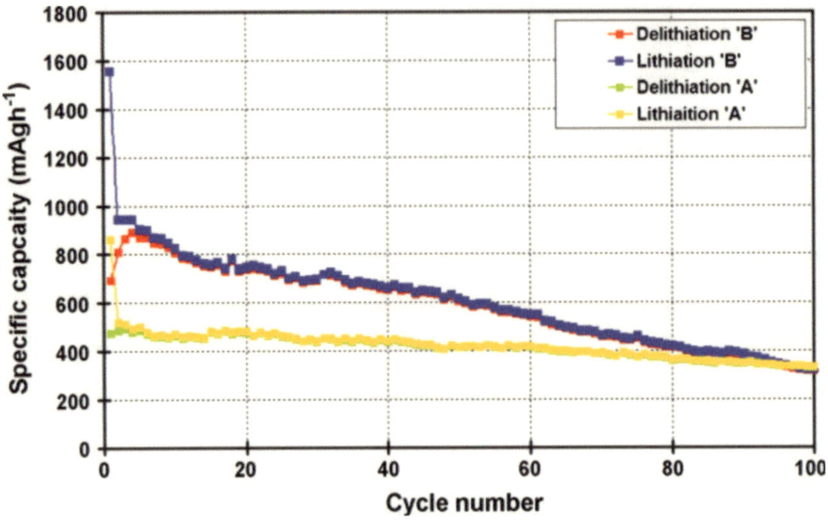

Fig. 16. Profile of specific capacity versus cycle number: delithiation 'B' and lithiation 'B' represent charge and discharge curves, respectively, for SiO/C electrode 'B'; delithiation 'A' and lithiation 'A' represent charge and discharge curves for heat treated (SiO + LiOH)/C electrode 'A'.

Composition of the electrodes	Initial lithiation capacity (mAh g^{-1})	Initial delithiation capacity (mAh g^{-1})	Irreversible capacity (mAh g^{-1})	Charge capacity at 100th cycle (mAh g^{-1})
Heat-treated (SiO + LiOH) with C electrode 'A'	862	475	387 (44%)	333
SiO with C electrode 'B'	556	693	863 (55%)	318

Table. 2 Capacity values obtained from Fig. 16 for two different composite electrodes.

5.5 Comparative performance

5.5.1 Composites with SiO, Si and graphite as constituents

The Charge/discharge and cycle capability for four different composites each containing 4g of C (Graphite or graphite with carbon) along with constituents Si and SiO are compared in the Fig. 17 and the data from the figure is presented in the Table 3.

The table shows maximum delithiation capacity values for the composite (a) is 1400 mAh g^{-1} at 4th cycle for Si:SiO:C = 3:1:4, for (b) is 1450 mAh g^{-1} at 8th cycle for Si:SiO:C = 2:2:4, for (c) is 870 mAh g^{-1} at 4th cycle for SiO:C = 1:1, and for (d) the carbon coated composite SiO:C = 1:1 is 700 mAh g^{-1} at 16th cycle. The same composites show delithiation capacities at 50th cycle for the composite (a) 480, for (b) 750, for (c) 610, and for (d) 580 mAh g^{-1}. Therefore, the capacity retentions at the fiftieth cycle compared to maximum specific capacity values are

34%, 52%, 70%, and 83% for electrodes (a), (b), (c), and (d), respectively. Fig. 17 and Table-3 show electrode with lower silicon content exhibit better electrochemical performance.

Electroactive material in the electrode (ratio)	1st LC	1st DLC	1st IRC	DLC_{max}@cycle no.	50th DLC	capacity retention (%)	Si:C (gram ratio)
	(mAh/g)						
(a) Si:SiO:C = 3:1:4	1839	933	906	1400@4	480	34	3.63:4
(b) Si:SiO:C = 2:2:4	986	523	463	1450@8	750	52	3.27:4
(c) SiO:C = 1:1	1556	693	863	870@4	610	70	2.52:4
(d) SiO:C = 1:1 (carbon coating)	1346	606	740	700@16	580	83	2.50:4

1st LC = first lithiation capacity; 1st DLC = first delithiation capacity; IRC = initial irreversible capacity; DLC_{max} = the maximum delithiation capacity; 50th DLC = 50th delithiation capacity; capacity retention = capacity retention at 50th cycle compared to DLC_{max}; Si:C denotes ratio of silicon and C (graphite with or without carbon coating).

Table 3. The different capacity values, capacity retention and the silicon content present in the composites.

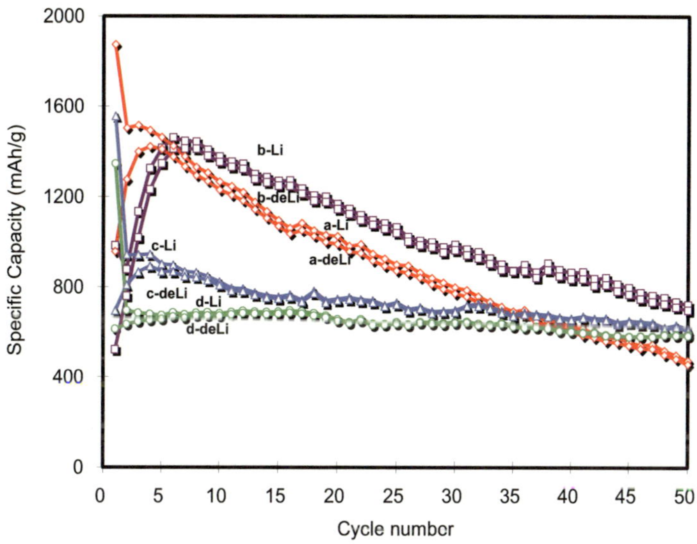

Fig. 17. Profile of capacity retention with cycling: a-Li & a-deLi, b-Li & b-deLi, c-Li & c-deLi, d-Li & d-deLi are lithiation and delithiation curves for the electrodes (a) Si:SiO:C = 3:1:4; (b) Si:SiO:C = 2:2:4; (c) SiO:C = 1:1; (d) SiO:C = 1:1 (carbon coated).

Low silicon content in the anode composite reduces the electrode strain during the lithiation and delithiation processes. The order of the capacity retention for the composite electrodes was (d) > (c) > (b) > (a). Even though the amount of carbon coating was small, its contribution to the capacity retention of the electrode was considerable; this is attributed to increase in conductivity of the electrode because of the carbon coating.

Usually delithiation capacities are lower than that for lithiation. The differences in lithiation and delithiation capacities are high when there is a high rate of decrease in capacity retention. Thus lithiation and delithiation capacities are correlated with capacity retention property.

5.5.2 Cyclic voltammograms of the composites (SiO/Si/C)

The lithiation and delithiation reactions that take place in the electrode may be represented in the form of equation as (9), (10) and 11) .

$$C_6(\text{graphite}) + xLi^+ + xe^- \rightleftharpoons Li_xC_6 \quad (9)$$

$$xLi^+ + Si + xe^- \rightleftharpoons Li_xSi \quad (10)$$

$$2SiO(Si + SiO_2) + xLi^+ + xe^- \rightleftharpoons 2\,Li_xSiO(Li_xSi + SiO_2) \quad (11)$$

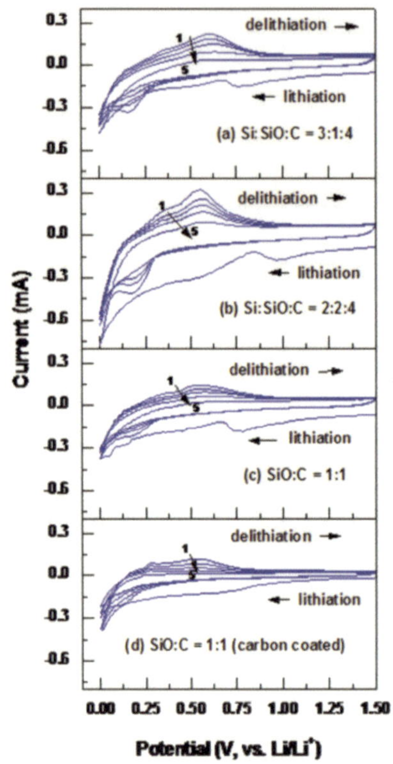

Fig. 18. Cyclic voltammograms of (a) Si:SiO:C = 3:1:4; (b) Si:SiO:C = 2:2:4; (c) SiO:C = 1:1; (d) SiO:C = 1:1 (carbon coated) with scan rate of 0.1 mV s^{-1}.

Both forward reactions toward the formation Li_xC_6 (x = 0 to 1) and Li_xSi (x = 0 to 4.4) or the corresponding reverse reactions(delithiation) appear to be pseudo-homogeneous phase reactions as understood from lithiation and delithiation process exhibited by the voltammograms (Fig. 18) which shows a continuous change of potential with x values. The lithiation and delithiation in graphite, silicon, and silicon monoxide are expected to occur simultaneously. The cyclic voltammogarm shows higher delithiation peak currents for the composites containing silicon. The composite SiO/C devoid of silicon content and the same composite with carbon coated shows different pattern. The carbon coated one has flat peak current peak current value with potential change indicating the stability of the composite. (Schulmeister and Madar, 2003) stated that the commercial silicon monoxide is a two-phase material with regions rich in silicon and others rich in oxygen. The silicon-rich phase contains ~1000 to 2000 atoms, with the phase-separated regions having diameters of 3 - 4 nm. The silicon-rich phase alloys and de-alloys resulting in volume changes of up to ~328%.

During initial lithiation, the oxygen-rich region is converted into Li_2O following reaction (3) and remains as an inert buffer material in the subsequent cycles. Other physically bonded -O-present in the composite also forms Li_2O and serves as a volume buffer during the cycling process. Lower silicon content should provide less volume change and better capacity retention to the electrodes. Furthermore, the electrolyte undergoing decomposition into forming SEI film during the initial and subsequent cycles may be represented by the reactions (3), (5) and (12) ,

$$2Li^+ + -O- + 2e^- \rightarrow Li_2O \tag{3}$$

$$LiPF_6 + 2xLi^+ + 2xe^- \rightarrow Li_xPF_{5-x} + (x+1)LiF \tag{5}$$

$$Li^+ + (EC + EMC...) + e^- \rightarrow SEI(Li_2CO_3, ROCO_2Li) \tag{12}$$

5.5.3 Comparative delithiation behaviour of composites based on Fe, Cu, Si & C

Since the delithiation properties are directly related to the amount of lithium available for lithiation process in any practical lithium ion battery, the comparative delithiation behavior of three compositions are presented in Fig. 18. The curve 'a' obtained for the composite Fe(1)Cu(1)Si(3.5)/SPB containing high silicon content does not provide good cycle life attributed to large volume expansion and crumbling of the active material. However, the same composite mixed with graphite as Fe (1) Cu (1) Si (3.5)/Graphite shown as 'b' was able to prolong the cycle life with very low capacity which is not useful. The figure clearly shows that the composition Fe(1)Cu(1)Si(2.5)/Graphite shown as 'c' depicts better capacity and high cycle life. The high performance of the Fe(1)Cu(1)Si(2.5)/Graphite is attributed to the effective buffer action of internally generated Li_2O during first lithiation process. These materials are actually double phase material as both silicon and graphite are active towards Li^+ in the same potential window.

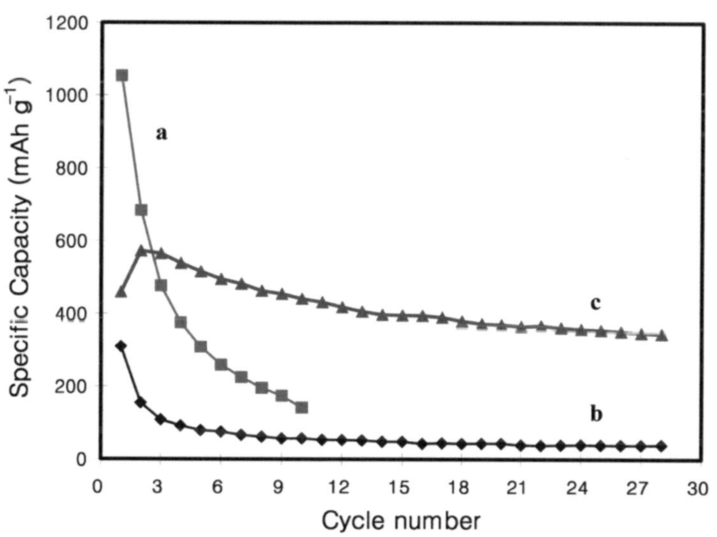

Fig. 19. Comparative delithiation behavior of a) Fe(1)Cu(1)Si(3.5) composite with Super 'P' black; b) Fe(1)Cu(1)Si(3.5)/Graphite ; c)Fe(1)Cu(1)Si(2.5)/Graphite.

6. Conclusion

The chapter on ' Silicon based Composite anode for Lithium ion battery' presented the state of the art on the silicon anode development and anode preparation using ball milling techniques, carbon coatings of the SiO/C anode using simple organic liqiuds, thermochemical treatment of the SiO with lithium hydroxide followed by ball milling with graphite and characterisation of prepared silicon anode composite through physical and electrochemical techniques.

In the composite Fe–Cu–Si/C, the metal elements act as buffer component. Both silicon and graphite participate in lithiation and de-lithiation processes. In addition the graphite has the tendency to serve as a buffer and is able to absorb volume expansion of silicon during charge/discharge process. The electrode annealed at 200 °C exhibits high initial discharge and charge capacity of 809 and 464 mAh g^{-1}, respectively with a sustainable reversible capacity of~385 mAh g^{-1} at 30th cycle.

Comparative data on two different composites (1) Cu:Fe:Si = 1:1:2.5 and (2) Cu:Fe:Si = 1:1:3.5 shows that higher silicon content is deletrious to cycle performance.

The high reversible capacity ~ 690 mAh g^{-1} at the 30th cycle by SiO/C (50:50) composite is unique in the sense that it does not involve any additional process or incorporation of a third element to act as buffers. This investigation also shows the interdependency between irreversible capacity and reversible capacity of the SiO-based system.

The charge capacity value for carbon coated carbon coated SiO/C composite anode is 500 mAh g^{-1} where as the uncoated composite anodes shows only 318 mAh g^{-1} at 100th cycle. The higher cycle stability in the carbon coated composite is attributed to the reinforcement of the structure by Cx particles. The cyclic voltammetry study shows constant peak current value from the voltage range 0.3 to 0.6 V during delithiation which explains the compactness of the electrode structure and resistor like behaviour of the composite..

The sustained cycle capacity follow the order Si:SiO:C = 3:1:4 < Si:SiO:C = 2:2:4 < SiO:C = 1:1 < SiO:C = 1:1 (carbon coated). A comparison of the capacity and cycle stability for the composites containing silicon shows smaller silicon content favours stability of the composite electrodes.

Even though the carbon-coated composite delivered moderate capacity during cycling compared to other composites investigated, its low capacity degradation made the anode to exhibit better performance.

The heat treat treatment of (SiO + LiOH) followed by graphite mixing resulted a composition with Li_4SiO_4, SiO & nano-silicon particles named as (SiO + LiOH)/C gave a charge capacity ~ 333 mAh g^{-1} at the 100th cycle with a low-capacity fade on cycling. The presence of Li_4SiO_4 is considered to provide better charge/discharge behavior to the composite.

The dicussion in the chapter shows there is ample scope replacing the graphite anode of lithium ion battery with silicon composite anode. The reason being it is environmentally benign, has higher specific capacity and prolonged cycle life. Cost effective and viable manufacturing process is possible as evident from the discussion in the chapter. Hence, the most promising anode candidate for the next generation Li-ion battery that will power automotive and other electronic gadgets will be the silicon based composite anode.

7. Acknowledgment

One of the authors, Angathevar Veluchamy wishes to thank Korean Federation of Science and Technology, Korea for awarding Brain Pool Fellowship for two years from October 2006 to September 2008 for the research work on anode materials for lithium ion batteries especially on 'Silicon composite anode' at Korea Electrotechnology Research Institute, Changwon, the master degree students Dong-Hun Kim and Hye-Min Shin for their support during my research period, and also the Central Electrochemical Research Institute/Council of Scientific and Industrial Research, India for granting extraordinary leave during this period.

8. References

Aurbach, D (2005). A review on new solutions, new measurements procedures and new materials for rechargeable Li batteries, *Journal of Power Sources, Vol. 146, (August 2005) 71-78*

Aurbach, D. & Schechter, W. A. (2004) in: Nazri, G. A.; Pistoia, G. (Eds.). Lithium batteries Science and Technology: Advanced Liquid Electrolyte Solution, *Kluwer Academic Publishers*, Boston, (2004) 1110

Aurbach, D. (2002) in: Schechter, W. A.; Scrosati, B (Eds.). Advances in Lithium Ion Batteries. *Kluwer Academic*, New York, (2002) 79-101

Balasubramanian, R.; A.Veluchamy, A.; Venkatakrishnan, N & Gangadharan, R.(1995) Electrochemical characterization of magnesium silver chloride battery. *Journal of Power Sources*, 96(1995) 197-199

Balasubramanian, R.; A.Veluchamy, A. & Venkatakrishnan, N(1994). Gasometric corrosion rate studies of magnesium alloy in magnesium batteries. *Journal of Power Sources, Vol. 52(1994) 305-308.*

Benedek, R & Thakeray, M. M. (2002). Lithium reactions with intermetallic compound electrodes. *Journal of Power Sources*, Vol. 110(August, 2002) 406-411

Brummer, S. B. (1980) in: Yeager, E.B.; Schumm, B.; Blomgren, G.; Blankenship, D.R.; Leger. V & Akridge. J, (Eds.). Lithium Non-aqueous Battery Electrochemistry. *The Electrochemical Society Proceedings Series*, Princeton, NJ, USA, 1980, Proc. vol. 80–7, pp. 130–142

Cairns, E. J. & Reimer, J. A. (2002). Magnesium silicide as a negative electrode material for lithium-ion batteries. *Journal of Power Sources*, Vol. 110, (2002) 424-429.

Chen, L, Xie, X.; Wang, B.; Wang, k & Xie, J (2006) Spherical nanostructured Si/C composite prepared by spray drying technique for lithium ion batteries anode. *Material Science and Engineering*. Vol.131(July 2006) 186-190.

Datta, M. K. & Kumta, P. N. (2007). Silicon, graphite and resin based hard carbon nanocomposite anodes for lithium ion batteries. *Journal of Power Sources*, Vol. 165, (2007) 368-378.

Dimove, N.; Kugino, S. & Yoshio, M. (2003). Carbon-coated silicon as anode material for lithium ion batteries: advantages and limitations. *Electrochimica Acta*, Vol. 48, (2003) 1579-1587.

Dimove, N.; Fukuda,K.; Umeno, T.; Kugino, S & Yoshio, M. (2003). Characterization of carbon-coated silicon: Structural evolution and possible limitations. Journaal of Power sources. *Vol. 114, (February 2003) 88-95.*

Doh, C.H.; Shin, H.M.; Kim, D.H.; Jeong, Y.D.; Moon, S.I.; Jin, B.S.; Kim, H.S.; Kim, K.W.; Oh, D.H & Veluchamy, A.(2008). A. new composite anode, Fe-Cu-Si/C for lithium ion battery. *Journal of alloys and compounds*, Vol. 461, (June 2008) 321-325.

Doh, C.H.; Park, C.W.; Shin, H.M.; Kim, D.H.; Chung, Y.D.; Moon, S.I.; Jin, B.S.; Kim, H.S & Veluchamy, A.(2008). A new SiO/C anode composition for lithium -ion battery. *Journal of Power Sources*, Vol. 179 (April, 2008) 367-370.

Doh, C.H.; Shin, H.M.; Kim, D.H.; Ha, Y.C.; Jin, B.S.; Kim, H.S.; Moon, S.I & Veluchamy, A.(2008). Improved anode performance of thermally treated SiO/C composite with an organic solution mixture. *Electrochemistry Communications* Vol. 10 (Feb, 2008) 233-237.

Doh, C.H.; Min, H.M.; Kim, D.H.; Chung, Y.D.; Moon, S.I.; Jin, B.S.; Kim, H.S.; Kim, K.W.; Oh, D.H & Veluchamy, A.(2008). Effect of silicon content over Fe-Cu-Si/C based composite anode for lithium ion battery. *Bull. Korean Chem. Soc.,* Vol. 29 (Feb, 2008) 309-312.

Doh, C.H.; Kim, D.H.; Lee, J.H.; Ha. K.H.; Shin, H.M.; Jin, B.S.; Kim, H.S.; Park, C.W.; Moon, S.I & Veluchamy, A.(2010). A comparative study on the performance of anodes based on SiO, Si and graphite for lithium ion battery, *Bull. Korean Chem. Soc.*, Vol. 31, No. 5, (May 2010) 1257-1261

Doh, C. H.; Kim, D. H.; Kim, H. S.; Shin, H. M.; Jeong, Y. D.; Moon, S. I.; Jin, B. S.; Eom, S. W Kim, H. S.; Kim, K. W.; Oh, D. H.; Veluchamy, A. (January, 2008). Thermal and electrochemical behaviour of C/LixCoO2 cell during safety test. *Journal of Power Sources* 175 (Jan, 2008) 881–885

Dong, H.; Feng, R. X.; Ai, X. P.; Cao, Y. L. & Yang, H. X. (2004). Structural and electrochemical characterization of Fe-Si/C composite anode for Li-ion batteries synthesized by mechanical alloying. *Electrochimica Acta*, Vol. 49, (2004) 5217-5222.

Dong, H.; Ai, X. P. & Yang, H. X. (2003). Carbon/Ba-Fe-Si alloy composite as high capacity anode, materials for Li-ion batteries. *Electrochemistry Communications*, Vol. 5, (2003) 952-957.

Fleischauer, M. D.; Topple, J. M. & Dahn, J. R. (2005). Combinatorial investigations of Si-M (M = Cr + Ni, Fe, Mn) thin film negative electrode materials. *Electrochemical and solid state letters*. Vol. 8, (2005) A137.

Fuchsbichler, B.; Stangl, C. ; Kren, H. ; Uhlig, F. & Koller, S.(March, 2011). High capacity graphite–silicon composite anode material for lithium-ion batteries. *Journal of Power Sources, Vol. 196(March 2011) 2889-2892*

Fujifilm, Internet(1997), http://www.fujifilm.co.jp/eng/news_e/nr079.html (1996).

Guo, Z. P.; Zhao, Z. W; Liu, H. K. & Dou, S. X. (2005). Lithium insertion in Si–TiC nanocomposite materials produced by high-energy mechanical milling. *Journal of Power Sources*, Vol. 146, (2005) 190-194.

Huggins, R. A. & Nix & W. D. (2000). Decrepitation model for capacity loss during cycling of alloys in rechargeable electrochemical system, *Solid State Ionics*, Vol. 6, (2000) 57-63.

Huggins, R. A. (2002). Alternative materials for negative electrodes in lithium systems. *Solid State Ionics*, Vol. 152–153, (2002) 61-68

International Application published under the patent corporation treaty (PCT)(2004). International Publication Number WO 2004/093223 A2 dated 28.10.2004, PCT/US2004/011350.

Jose Benedict, T.; Banumathi, S.; Veluchamy. A.; Gangadharan, R.; Zulfihar Ahemed, A. &Rajendran, S(1998). Characterization of plasticized soild polymer electrolyte by XRD and AC impedance methods. *Journal of Power Sources*, Vol. 75(1998) 171-174.

Kim, J. B.; Jun B. S. & Lee. S. M. (2005). Improvement of capacity and cycleability of Fe/Si multilayer thin film anodes for lithium rechargeable batteries. *Electrochimica Acta,* Vol. 50, (2005) 3390-3394.

Kim, I. S.; Kumta, P. N. & Blomgren, E. (2000). Si/TiN Nanocomposites Novel anode Materals for Li-Ion Batteries, *Electrochemical and solid state letters.* Vol. 3, (2000) 493-496

Kim, J. H.; Kim. H & Sohn, H. J (2005). Addition of Cu for carbon coated Si-based omposites as anode materials for lithium-ion batteries. *Electrochemistry Communications. Vol. 7(May 2005) 557-561*

Lee, Y. S.; Lee, J. H.; Kim, Y. W.; Sun, Y. K. & Lee, S. M. (2006). Rapidly solidified Ti-Si alloy/carbon composites as anode for Li-ion batteries. *Electrochimica Acta*, Vol. 52, (2006) 1523-1526

Lee, H. Y.; Kim, Y. L.; Hong, M. K. & Lee, S. M. (2005). Carbon-coated $Ni_{20}Si_{80}$ alloy–graphite composite as an anode material for lithium-ion batteries. *Journal of Power Sources*, Vol. 141, (2005) 159-162.

Lee, H. Y. & Lee, S. M. (2002). Graphite–FeSi alloy composites as anode materials for rechargeable lithium batteries, *Journal of Power Sources*, Vol. 112, (2002) 649–654.

Lee, H. Y. & Lee, S. M. (2004). Carbon-coated nano-Si dispersed oxides/graphite composites as anode material for lithium ion batteries. *Electrochemistry Communications*, Vol. 6, (2004) 465–469

Miyachi, M.; Yamamoto, H.; Kawai, H.; Ohta, T. & Shirakata, M. (2006). Analysis of SiO Anodes for Lithium-Ion Batteries. *Journal of the electrochemical society*, Vol. 152, (2005) A2089-A2091.

Mizushima, K.; Jones, P. C.; Wiseman, P. J. & Goodenough, J. B. (1980). Li_xCoO_2 ($0<x<-1$): A new cathode material for batteries of high energy density. *Materials Research Bulletin*, Vol. 15, (1980) 783-789.

Ng, S. H.; Wang, J.; Konstantinnov, K.; Wexler, D.; Chew, S. Y.; Guo, Z. P. & Liu, H. (2007). Spray-pyrolyzed silicon/disordered carbon nanocomposites for lithium-ion battery anodes, *Journal of Power Sources*,. Vol. 174, (2007) 823-827.

NuLi, Y.; Wang, B.; Yang, J.; Yuan, X. & Ma, Z. (2006). Cu_5Si–Si/C composites for lithium-ion battery anodes. *Journal of Power Sources*, Vol. 153, (2006) 371-374

Panasonic's New Li-Ion Batteries Use Si Anode for 30% Higher Capacity (2010), http://techon.nikkeibp.co.jp/article/HONSHI/20100223/180545/

Park, M. S.; Rajendran, S.; Kang, Y. M.; Han, K. S.; Han, Y. S. & Lee, J. Y. (2006). Si–Ni alloy–graphite composite synthesized by arc-melting and high-energy mechanical milling for use as an anode in lithium ion batteries. *Journal of Power Sources*, Vol. 158, (2006) 650-653.

Park, M. S.; Lee, Y. J.; Rajendran, S.; Song, M. S.; Kim, H. S. & Lee, J. Y. (2005). Electrochemical properties of Si/Ni alloy–graphite composite as an anode material for Li-ion batteries. *Electrochimica Acta*, Vol. 50, (2005) 5561-5567.

Renuka, A.; Veluchamy, A.; Venkatakrishnan, N.; Nathirabegam, S.; Chidambaram, V. R.; Sabapathi, R(1992). Improved cycle life performance of Zz/NiOOH cells using stabilized zinc electrode. *Journal of Applied electrochemistry*, Vol.*22(1992) 182-184.*

Roberts, G. A.; Cairns, E. J. & Reimer, J. A. (2002). Magnesium silicide as a negative electrode material for lithium-ion batteries, *Journal of Power Sources*, Vol. 110, (2002) 424-429

Schulmeister, K & Mader, W. (2003). TEM investigation on the structure of amorphous silicon monoxide, *Journal of Non-Crystalline Solids* 320 (2003) 143–150

Shi, Z.; Liu, M.; Naik, D. & Gole, J. L. (2001). Electrochemical properties of alloy electrodes for lithium batteries, *Journal of Power Sources*, Vol. 92, (2001) 70-80

Singh, P.; Guidotti, R. A & Reisner, D. (2004). AC impedance measure ments of molten salt thermal batteries, *Journal of Power Sources*, Vol. 138, (2004) 323-326.

Veluchamy, A.; Doh, C. H.; Kim, D. H.; Lee, J. H.; Lee, D. J,; Ha, K.H.; Shin, H.M.; Jin, B.S.; Kim, H.S.; Moon, S.I & Park, C.W.(2009). Improvement of cycle behavior of SiO/C anode composite by thermochemically generated Li_4SiO_4 inert phase for lithium battery. *Journal of Power Sources*, Vol. 188, No. 2, (Feb 2009) 574-577.

Veluchamy, A.; Doh, C.H.; Kim, D. H.; Lee, J. H.; Shin, H. M.; Jin, B. S.; Kim, H. S.& Moon, S. I (2009). Thermal analysis of LixCoO2 cathode material of lithium ion battery *Journal of Power Sources*, Vol. 189(April 2009) 855-858.

Veluchamy, A.; Ikuta, H & Wakihara, M (2001). Boron-substituted manganese spinel oxide cathode for lithium ion battery. Solid state Ionics. *Vol. 143,(June, 2001)161-171*

Wang, G. X.; Sun, L.; Bradhurst, D. H.; Zhong, S.; Dou, S. X. & Liu, H. K. (2000). Innovative nanosize lithium storage alloys with silica as active centre. *Journal of Power Sources*, Vol. 88(2), (2000) 278-281

Wang, K.; He, X.; Wang, L.; Ren, J.; Jiang, C. & Wan. C. (2007). Si, Si/Cu core in carbon shell composite as anode material in lithium ion batteries. *Solid State Ionics*, Vol. 178, (2007) 115-118.

Weydanz, W. J.; Wohlfahrt-Mehrens, M. & Huggins, R. A. (1999). A room temperature study of the binary lithium–silicon and the ternary lithium–chromium–silicon system for use in rechargeable lithium batteries. *Journal of Power Sources*, Vol. 81-82, (1999) 237-242

Wilson, A. m.; Zank, G.; Eguchi, K.; Xing, W. & Dahn, J. R. (1997). Pyrolysed silicon-containing polymers as high capacity anodes for lithium- ion batteries. *Journal of Power Sources*, Vol.68 (1997) 195-200.

Wolfenstine, J. (1999). Critical grain size for micro-cracking during lithium insertion. *Journal of Power Sources*, Vol. 79, (1999) 111-113

Wolfenstine, J. (2003) $CaSi_2$ as an anode for lithium-ion battery, *Journal of Power Sources*, Vol. 124, (2003) 241-245.

Yang, X.; Wen, Z.; Xu, X.; Lin, B.; Huang, S.(2007). Nanosized silicon-based composite derived by in situ mechanochemical reduction for lithium ion batteries. *Journal of Power sources, Vol.164,(Feb 2007) 880-884.*

Yazami, R. & Touzain, Ph. (1982). International Meeting on Lithium Batteries. Rome, April 27–29, 1982, C.L.U.P. Ed. Milan, Abstract # 23.

Yazami, R. & Touzain, Ph. (1983). A reversible graphite-lithium negative electrode for electrochemical generators. *Journal of Power Sources*, Vol. 9, (1983) 365-371.

Zhang, Y.; Fu, Z. W & Qin, Qi. Z.(2004). Microstructure and Li alloy formation of nano-structured amorphous Si and Si/TiN composite thin film electrodes. *Electrochemistry communications, Vol. (May 2004) 484-491*

Zhang, X. N.; Huang, P. X.; Li, G. R.; Yan, T. Y.; Pan, G. L. & Gao, X. P. (2007). Si–AB5 composites as anode materials for lithium ion batteries, *Electrochemistry Communications*. Vol. 9, (2007) 713-717

Zuo, P.; Yin, G. & Tong, Y. (2006). SiMn–Graphite composites as anodes for lithium ion batteries, *Solid State Ionics*, Vol. 177, (2006) 3297-3301.

Zuo, P & Yin, G. (2006). Si–Mn composite anodes for lithium ion batteries, *Journal of alloys and compounds*, Vol. 414, (2006) 265-268.

15

Nano-Structured Calcium Silicate Phase Change Materials for Packaging Temperature Sensitive Products

James H. Johnston
School of Chemical & Physical Sciences and The MacDiarmid, Institute for Advanced Materials and Nanotechnology, Victoria University of Wellington, New Zealand

1. Introduction

The safe transportation together with en route and temporary storage of temperature sensitive and perishable materials, particularly chilled food and medical products, often require the use of packaging that provides thermal buffering against unwanted transient temperature rises and falls as a result of rapid changes in the temperature of the immediate environment of the package. Effective insulation can alleviate unwanted temperature rises or falls to some extent for a short time. However, this is often insufficient when the packages are exposed to higher or lower temperatures for several hours without effective temperature control (Dodds, 2009; Johnston et al., 2007).

New Zealand and other food producing countries export considerable quantities of fresh food to international markets in response to the increasing global demand for such produce. Much of this food cannot be frozen as the quality and nutritional value are adversely affected, but has to be chilled to maintain freshness. After harvesting and packing, chilling is usually carried out in a coolstore. A major issue is the unwanted warming of the food when the packages are exposed to warm temperatures on airport tarmacs and in temporary (un-refrigerated) storage in transit during air transportation, which can adversely affect the food quality and export value. This is due to the limited thermal insulation and poor thermal buffering capacity of the paperboard containers that are typically used. Polystyrene containers with better insulating properties can be used but there is now a move away from them due to environmental issues associated with their production and disposal. The warm temperature fluctuations often encountered are for durations typically up to 1 hour or so, but can be up to several hours, which is sufficient time to cause spoilage of the perishable food in the package. Hence there is a need for an effective material that provides insulation and thermal buffering properties and can be accommodated in readily re-useable liners for inserting into these packages to insulate against and also absorb the transient heat from the external environment, thereby protecting the valuable and perishable products in the package (Amcor Kiwi Packaging Ltd, 2011; Carter Holt Harvey Packaging, 2011; Cool pack, 2011; Johnston and Dodds, 2011; Johnston et al., 2007; Plastics New Zealand, 2011; Powell and Mathews, 1987).

A similar application is in the transportation of temperature sensitive medical products, often of a high value, which must be kept at a requisite low temperature, but not frozen. Also, comparable problems arise with the transport of products which require maintaining temperatures above ambient. Such products include the delivery of hot fast foods to homes and functions where the desired serving temperature of the food has to be maintained. Similarly in specialist applications such as the transportation of tropical fish, the water has to be maintained within a small temperature range above ambient temperature.
Phase change materials (PCMs) such as alkanes (paraffin oils and waxes) can provide passive thermal buffering properties due to the relatively large latent heat content (~140-210 kJ kg^{-1}) associated with their reversible solid-liquid phase transition. They are non-toxic and the operating temperature (melting point) can be tuned by altering the hydrocarbon chain length (Farid et al., 2004; Rubitherm Technologies GmBH, 2011; Feldman et al., 1986, Zalba et al., 2003). Hence they can absorb heat energy from the immediate environment on melting and release it back on solidifying or freezing. Farid et al., (2004) and Zalba et al., (2003) have reviewed the properties and use of phase change materials in energy storage applications. Khudhair and Farid (2004) have similarly reviewed energy conservation approaches in buildings using phase change materials to capture heat energy from the sun during the day and release it at night. In packaging applications, if the alkane PCM is in thermal contact with a temperature sensitive material or is contained in the surrounding packaging material, the PCM has the effect of buffering or minimising the unwanted sudden temperature changes due to heating and cooling transients in the immediate environment. Here the heat energy is absorbed or released accordingly to counteract the unwanted change in temperature (Farid et al., 2004; Feldman et al., 1986; Hawlader et al, 2003; Johnston and Dodds, 2011; Rubitherm Technologies GmBH, 2011; Schossig et al., 2003; Tyagi and Buddhi, 2007; Zalba et al., 2003; Zhang et al., 2007). However in the practical application of this concept, it is necessary to contain the liquid phase of the PCM upon melting in order to prevent loss of the PCM and the contamination of the material it is thermally protecting. This poses a major problem as the PCM has to be accommodated in robust leak proof and preferably flexible containers around or within the package. This has been addressed by encapsulating the PCM in a microcapsule and then incorporating these microcapsules into the final product such as building materials (Hawlader et al., 2003; Schossig et al., 2003). It is possible these microcapsules can rupture or become dislodged whereupon the PCM thermal buffering capacity is reduced. An alternative approach is to contain the PCM in a highly porous matrix as is discussed here. For this, a composite proprietary nano-structured calcium silicate (NCS) material with a high pore volume has been used to accommodate the alkane PCM to provide a novel nano-structured calcium silicate - phase change material (PCM) in the form of a dry powder (NCS-PCM) (Johnston et al., 2006) that can be contained in flexible plastic packaging liners made of polythene or bubblewrap as detailed below.

2. Preparation of nano-structured calcium silicate – phase change materials

Nano-structured calcium silicate is a proprietary new material comprising nano-size platelets stacked together in a unique open framework structure forming discrete particles of about 1-5 microns in size Fig. 1 (Johnston et al., 2006). It can be prepared by the reaction of Ca^{2+} with the silicate anion, typically $H_3SiO_4^-$ at room temperature under alkaline conditions at a pH typically about pH=11, according to the following general equation:

$$Ca^{2+} + H_3SiO_4^{-} + OH^{-} \rightarrow CaSiO_x(OH)_y + H_2O \qquad \text{Where } x \approx 2 \text{ and } y \approx 2$$

Fig. 1. Nano-structured Calcium Silicate showing the framework of nano-size platelets stacked together to form discrete particles. Some agglomerates are visible.

The $H_3SiO_4^-$ species can be provided by sodium silicate or another source of dissolved silica such as geothermal water. The Ca^{2+} and OH^- can be provided separately from a soluble calcium salt and sodium hydroxide, or provided collectively by a partially neutralised slurry of $Ca(OH)_2$ to remove 1 mole of OH^- ions. Upon mixing the reactants, the NCS precipitates immediately and is then aged for up to several hours to develop the nano-structure shown in Fig. 1. During ageing, the pore volume or oil absorption capacity and the surface area increase with time (Fig. 2) to provide a material with a high pore volume with a consequent high liquid absorbency of up to about 500-600 g oil $100g^{-1}$ silicate (ASTM Oil Absorption test D281), and a high surface area of up to about 500 m^2 g^{-1} (Johnston et al., 2006).

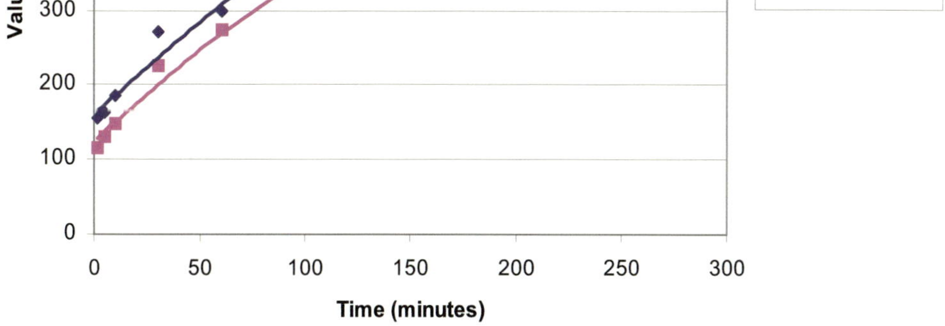

Fig. 2. The increase in oil absorption and surface area of nano-structured calcium silicate as the framework and pore structure develop with ageing.

Alkane PCMs are commercially available from a number of suppliers. For example *Rubitherm GmbH* in Germany currently offers a number of such PCMs with melting points ranging from -4 °C to 100 °C (Rubitherm Technologies GmbH, 2011). The practical problem mentioned above of containing the liquid phase on melting, has been overcome here by incorporating the alkane PCM into the highly porous NCS matrix. It is a relatively simple process to mix and work the liquid PCM into the readily accessible pores of the NCS with a spatula for small test samples, or with a food mixer or other such mixing equipment for larger scale preparations. In the transportation and temporary storage of chilled food where it is desirable to maintain a temperature of about 7-10 °C to retain freshness, a NCS-PCM composite containing either *Rubitherm* RT2 or RT6 alkane phase change materials can be used to provide thermal buffering in this temperature range. The RT2 has an approximate melting point of 6 °C, a congealing point of 2 °C and a heat storage capacity of 214 kJ kg^{-1}. RT6 has an approximate melting point of 8 °C, a congealing point of 6 °C and a heat storage capacity of 174 kJ kg^{-1} (Rubitherm Technologies GmbH, 2011).

Levels of 100–400 wt % RT6 PCM in the NCS have been tested to determine the highest loading achievable whilst still retaining the free-flowing powder characteristics. An optimum loading level of PCM in the NCS was found to be about 300 wt % PCM on a NCS basis (i.e. 1 kg of NCS-PCM composite comprises 250 g NCS and 750 g PCM). The resulting composite NCS-300PCM remains a powdery solid above the alkane melting point and can be readily incorporated into flexible packaging materials to provide heat buffering of about 100-120 kJ kg^{-1} of composite. In addition, the open framework structure of the NCS itself provides some insulation properties. This thermal buffering packaging can be a simple robust plastic bag containing the NCS-PCM or a common bubble wrap lined plastic bag containing the NCS-PCM which provides both thermal insulation and thermal buffering properties. These can be placed as thermal buffering liners inside conventional paperboard packages. Alkane PCMs are commercially available from a number of suppliers.

Hence by accommodating a PCM of a particular melting point in the NCS pores, it is possible to produce a NCS-PCM composite powdered material and contain it in a conventional plastic bag or bubble wrap plastic bag to form a flexible liner with effective thermal buffering properties at selected temperatures in the range from -4 °C to 100 °C (Fig. 3). This can therefore provide thermal buffering capacity to negate or minimise the deleterious effect of short term, high or low temperature fluctuations on temperature sensitive perishable produce and materials.

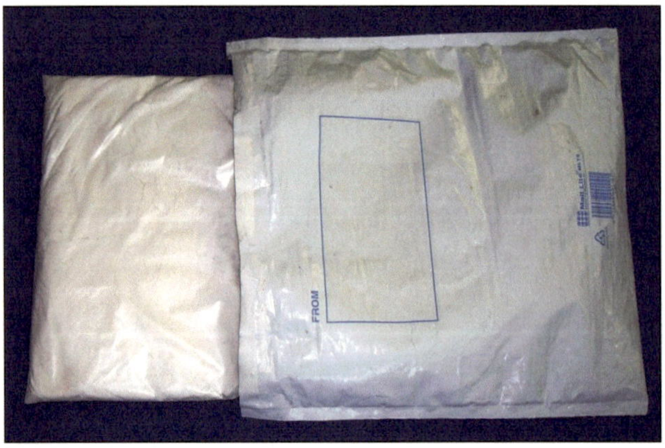

Fig. 3. The polythene bag containing a sample of NCS-PCM (left) being inserted into a bubblewrap bag (right).

3. Thermal properties of nano-structured calcium silicate – phase change materials

The thermal buffering properties of the NCS-PCM composites with 300 wt% and 400 wt% PCM loadings of different PCMs have each been measured by Differential Scanning Calorimetry (DSC) using a Shimadzu DSC-60 Differential Scanning Calorimeter. The results showed that 300 wt% PCM could readily be accommodated into the NCS without unwanted weeping out of the PCM in repeated heating and cooling cycles and at the same time provide useful thermal buffering properties. The DSC curves of the NCS-300PCM composite material with *Rubitherm* RT6 showed the typical reversible endotherms and exotherms across the solid-liquid and liquid-solid phase transitions respectively with a measured thermal buffering capacity of about 101 kJ kg^{-1} (Fig. 4). Some small hysteresis is observable.

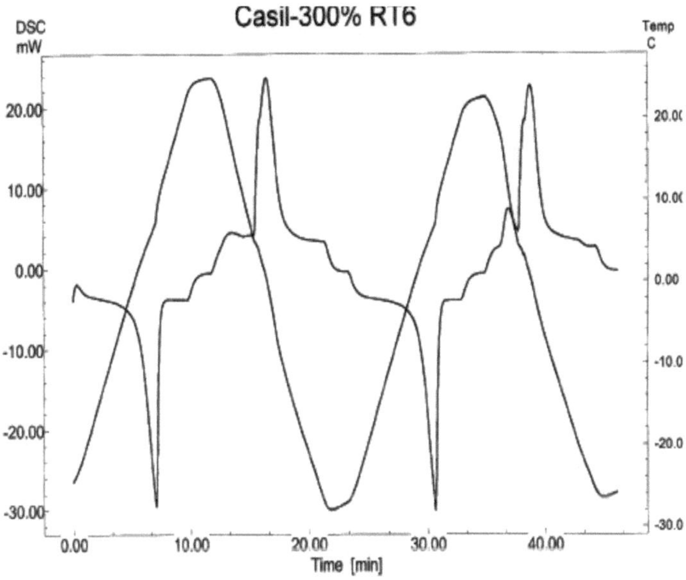

Fig. 4. The Differential Scanning Calorimetry results for the NCS-300PCM composite material for two heating and cooling cycles from -25 °C to 25 °C. The triangular curve is the temperature sweep. The DSC curve shows the energy absorbed (endotherm) on melting and released (exotherm) on freezing.

When considering the usefulness of a composite phase change material it is important to also consider the thermal insulation properties by measuring their thermal resistance and thermal conductivity and comparing them with conventional insulation materials. Hence measurements of thermal resistance and thermal conductivity of NCS-300PCM composites containing *Rubitherm* RT2 and RT6 respectively in polythene and bubblewrap bags were carried out with a Togmeter (Fig. 5) (Johnston and Dodds, 2011; Dodds, 2009). This is a device used to measure the temperature difference and the heat flow across a flat plate-like sample from which the thermal resistance and thermal buffering properties can be determined. It comprises a 30 cm diameter thermostatically controlled bottom plate which can be heated above ambient temperature and a 30 cm diameter top plate which can be held at or below ambient temperature by refrigeration. Amounts of 429 g of NCS-RT2, 394 g of NCS-RT6 and 151 g of NCS by itself were placed in a 773 cm^2 polythene bags respectively to form essentially flat sheet like liners (Fig. 3) for which heat transfer and thermal buffering properties could be measured across the liner using the Togmeter. The liners were also placed inside 280 mm x 325 mm bubblewrap bags and similarly measured to show the additional effect of a layer of bubblewrap insulation surrounding the liners (Fig. 3). Overall the density of NCS-PCM in the liner was about 0.5 g cm^{-2} (Dodds, 2009).

The sample being measured is sandwiched between the plates under an appropriate pressure and uniform thickness to ensure good contact between the plates and the sample (Fig. 5 (left). The respective temperatures of the bottom plate (*T1*), the sample lower surface (*T2*), the sample upper surface (*T3*) and ambient air temperature (*T4*) are measured by thermocouples as shown in Fig. 5 (right). Measurements were carried out in accordance with the international standard ISO 5085-1.

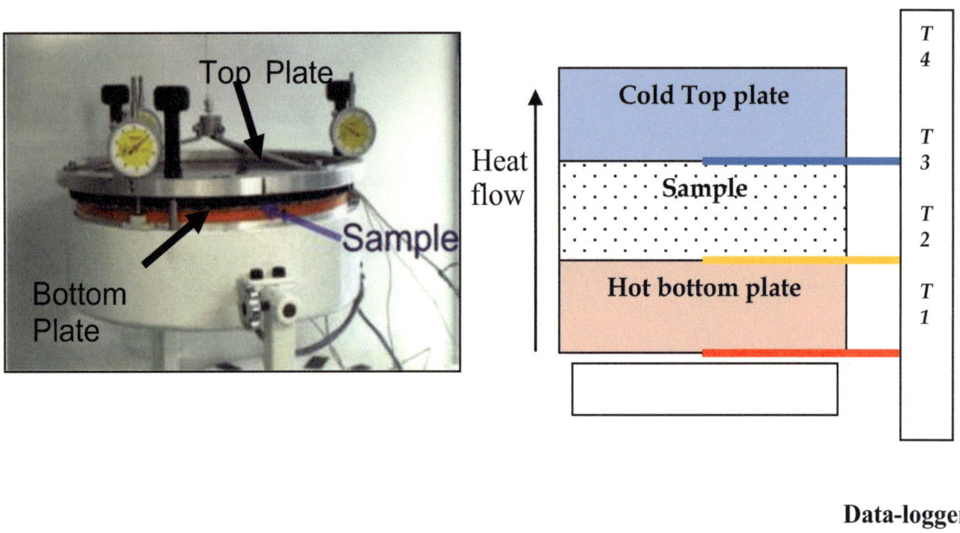

Fig. 5. (left) The Togmeter showing the heated bottom plate, cooled top plate and sample positioned between these plates. The thickness gauges ensure uniform sample thickness and contact between the plates; (right) the placement of thermocouples *T1-T4*. *T1* measures the temperature of the hot plate, *T2* the lower surface of the sample, *T3* the upper surface of the sample, and *T4* the ambient temperature (Dodds, 2009).

The measurement of thermal resistance *Rs* is made by measuring the temperatures *T1, T2* and *T3* across the sample under the condition of steady state heat transfer between the hot bottom plate and cooler top plate (Fig. 5).

$$Rs = Rst - Rc\,(T2-T3)/(T1-T2) \qquad \left(\mathrm{m^2 K\, W^{-1}}\right)$$

Where:
Rs = thermal resistance
Rc = contact resistance (known for instrument)
Rst = resistance of calibration standard
The Thermal Conductivity *k* is calculated from the thermal resistance and the sample thickness *d*

$$k = d / Rs \qquad \left(\mathrm{W\,m^{-1}K^{-1}}\right)$$

For thermal resistance and thermal conductivity measurements the sample is placed between the plates, the bottom plate temperature set and the temperatures *T1-T3* recorded on a data logger until they reach constant values, confirming a steady state heat flow across the sample is reached. These temperature values were used to calculate thermal resistance and thermal conductivity for the sample (Dodds. 2009).
The thermal resistance and thermal conductivity values for NCS and NCS-PCM in the polythene bags and bubblewrap bags are presented in Table 1. The values measured here for corrugated paperboard and expanded polystyrene (EPS) are also provided as a comparison.

Sample ID	*d* mm	*Rs* m^2KW^{-1} *±0.002*	*k* $Wm^{-1}K^{-1}$ *±0.002*
NCS in polythene bag	12.5	0.427	0.030
NCS in polythene bag in bubble wrap bag	19.5	0.520	0.038
NCS-RT2 in polythene bag	12.5	0.337	0.037
NCS-RT2 in polythene bag in bubble wrap bag	19.5	0.367	0.053
NCS-RT6 in polythene bag	12.5	0.183	0.068
NCS-RT6 in polythene bag in bubble wrap bag	19.5	0.264	0.069
Bubble wrap bag	5.05	0.139	0.036
Corrugated paperboard	11.9	0.243	0.049
10mm Expanded polystyrene	10.3	.372	0.032

Table 1. Thermal resistance and thermal conductivity values (Dodds, 2009).

The results (Table 1) show that the thermal conductivity *k* of NCS contained in the polythene bag is similar to that for 10 mm expanded polystyrene which is a widely used insulating and packaging material. The NCS thermal conductivity is slightly better than the bubblewrap bag and much better than that for corrugated paperboard. The excellent insulation properties of NCS result from its open framework and low density structure due to its high pore volume and consequent large number of air voids. When a PCM is incorporated into the NCS pores, the thermal conductivity of the NCS-PCM composite increases as the air is displaced by the PCM which has a higher thermal conductivity. This is shown for the NCS-PCM with RT2 and RT6 in the polythene and bubblewrap bags. Even though these thermal conductivity values have increased with the incorporation of the PCM into the NCS pores, they are substantially better than the value of 0.2 $Wm^{-1}K^{-1}$ quoted for RT2 and RT6 (Rubitherm Technologies GmbH, 2011). This clearly shows the advantage of incorporating these PCMs into the NCS material to provide the composite NCS-PCM products developed here.

4. Thermally buffered paperboard packages

The thermal buffering effectiveness of the NCS-PCM composite for chilled packaged food was tested using a paperboard package containing asparagus as a typical food that requires chilled temperatures of about 7-10 ^{o}C during storage and transportation to maintain optimum freshness. The package was a rectangular container (box) with dimensions 270 mm long x 200 mm wide x 180 mm high, constructed from 4 mm fluted paperboard and filled with 2 kg of freshly picked asparagus sealed in a polythene bag (Fig. 6). A series of six bubblewrap bags of the same dimensions as the walls, base and top of the package respectively, were prepared as removable and re-useable liners (Figs. 3 and 6). These were filled with the NCS-300PCM containing the *Rubitherm* RT6 alkane as the PCM to provide thermal buffering around 7-10 °C. Three different total amounts of NCS-300PCM collectively in the six liners of 200g, 400g and 600g being 10 wt%, 20 wt% and 33 wt% of the asparagus respectively were used. The results showed that the thermal buffering properties depended linearly on the amount of NCS-PCM as expected and hence only those detailed results for the 400 g quantity of NCS-PCM in the liners surrounding the paperboard package are presented here.

Fig. 6. Paperboard package with bubble wrap bag liners containing 400 g NCS-300PCM composite with RT6, and 2 kg of asparagus sealed in a polythene bag. The positions of the *Dallas Semiconductor* I-buttons used to measure the temperature / time profiles are shown by the arrows.

The thermal buffering properties of the paperboard package by itself, with the bubblewrap liners containing the NCS-300PCM, and also with the 2 kg of asparagus were each measured by recording the temperature / time profiles for the particular package configuration as it was cooled down in refrigerator to about 0 °C, then warmed up in the ambient environment to room temperature. The temperature / time profiles were recorded using *Dallas Semiconductor* I-buttons placed outside the package, inside the package on the inner side of the NCS-300PCM liner and at the centre of the package. When the package was filled with asparagus, I-buttons were placed throughout the package amongst the asparagus (Fig. 6). The temperature was recorded every 4 minutes. These temperature/time profiles are shown in Figs. 7-10.

The temperature / time cooling and heating profiles for the empty package without the liners or asparagus, measured on the outside, inside wall and centre of the package show the paperboard package walls provide no effective insulation and no thermal buffering capacity (Fig. 7). The inside of the package cools down and warms up at essentially the same rate as the outside temperature / time profile. There is a very slight lag in the heating up profile between the inside and outside of the package but this is negligible.

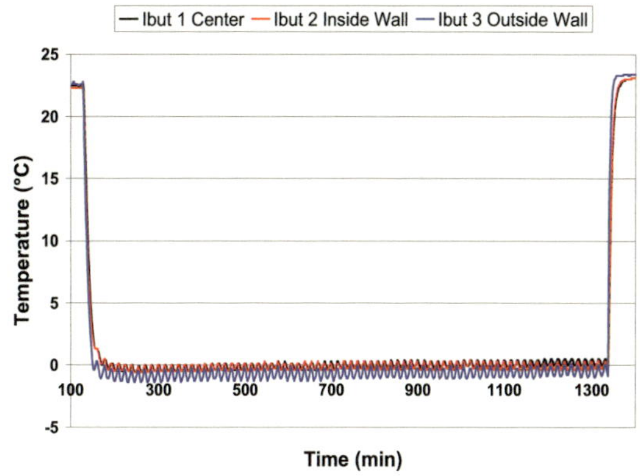

Fig. 7. The temperature / time cooling and heating profiles for the empty paperboard package.

The temperature / time profiles for the package with the empty bubblewrap liners (Fig. 8) show a small lag in the cooling and heating profiles for the inside wall and centre of the package when compared with the outside temperature, demonstrating a relatively small insulating effect of the bubblewrap liners. For comparison purposes a temperature of 10 °C which is at the higher end for chilled foods and a temperature of 7 °C which is more typical were chosen. These of course vary for particular foods but serve as a general guideline here to evaluate the thermal buffering performance of the paperboard package and the NCS-300PCM composite material. Using this approach, the heating up profiles show that it took about 8 minutes longer and 6 minutes longer for the inside temperature of the package to warm to 10 °C and 7 °C respectively, than for the time taken for the outside temperature to similarly increase. This shows that the bubblewrap liners do serve to increase the thermal insulation properties of the paperboard package, but not markedly so.

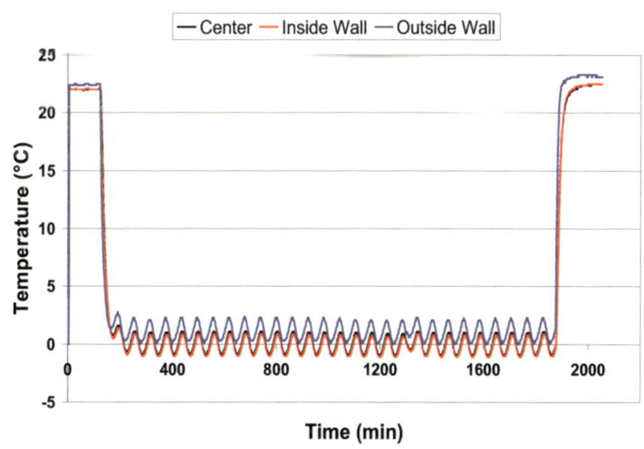

Fig. 8. The temperature / time cooling and heating profiles for the paperboard package with empty bubblewrap liners placed on the inside walls, base and top of the package.

When the bubblewrap liners collectively containing 400 g of NCS-300PCM composite were placed inside the package there was a distinct lag in the cooling down and heating up temperature / time profiles between the inside and outside of the package (Fig. 9). The cooling profiles for inside and outside the package are similar down to about 7 °C which is around the freezing point of the RT6 PCM. The freezing process takes additional (phase change) heat energy from the immediate environment and hence the cooling rate is slowed significantly. Below 7 °C the cooling profile inside the package plateaus somewhat for about 100 minutes during which time the RT6 solidifies progressively and the cooling then continues at a slower rate until it reaches the outside temperature. On warming, the heating profiles for inside and outside the package are similar until a temperature of about 5 °C is reached whereupon the RT6 starts to melt and this continues progressively up to about 10 °C as the additional heat energy (phase change) required to melt the PCM must come form the immediate outside environment (Fig. 9). For comparison, the time lag for the inside temperature of the package with the bubblewrap liners containing the NCS-300PCM to reach 10 °C is 97 minutes and to reach 7 °C is 67 minutes. These are much longer periods and show that the 400 g of NCS-300PCM contained in the bubblewrap liners provides effective thermal buffering to the inside of the package for about 1-1.5 hours. The temperature / time profiles on the inside wall and the centre of the package are very similar showing the heat spreads quickly and uniformly through the inside of the box. Also, the small differences between the temperature / time profiles inside and outside the package between about 10 °C and ambient temperature reflect the thermal insulation properties of the bubblewrap as noted above in Fig. 8.

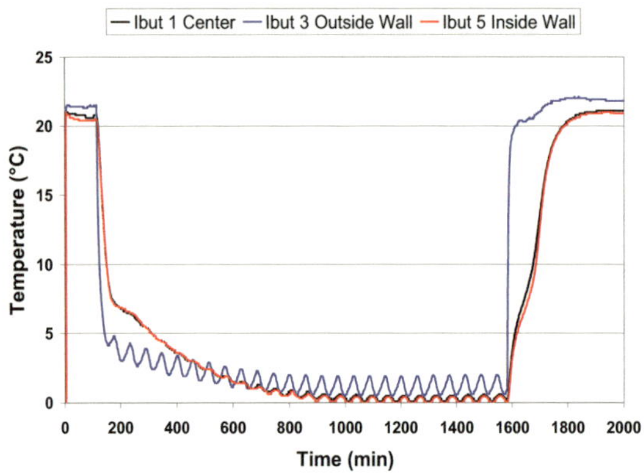

Fig. 9. The temperature / time cooling and heating profiles for the paperboard package with bubblewrap liners collectively containing 400 g of the NCS-300PCM composite and placed on the inside walls, base and top of the package.

Fig. 10 shows the temperature/time profiles for the paperboard package lined with bubblewrap liners collectively containing 400 g of the NCS-300PCM composite and placed on the inside walls, base and top of the package which is also filled with 2 kg of asparagus in a sealed plastic bag to ensure moisture retention. There is a noticeable difference between the profiles measured on the inside wall of the bubblewrap liner and the centre of the package resulting from the progressive absorption of the heat energy by the asparagus which is close to the wall of the package and the progressive flow of this heat energy through the asparagus to the centre of the package. This also shows the inherent thermal capacity for the asparagus itself. At 10 °C the time lag in the heating up profile between the asparagus at the centre of the package and the outside environment is an impressive 550 minutes or over 9 hours. The time lag between the inside wall of the liner and the outside environment is 260 minutes or over 4 hours which is also very significant. At 7 °C these time lags are 426 minutes and 171 minutes respectively. These large time lags collectively demonstrate considerable thermal buffering at such temperatures.

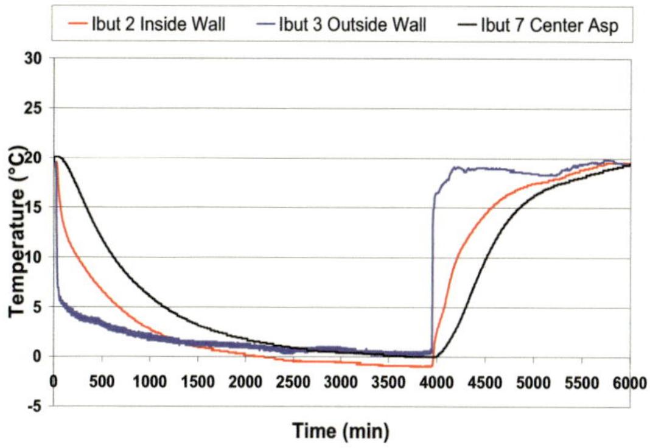

Fig. 10. The temperature / time cooling and heating profiles for the paperboard package with bubblewrap liners collectively containing 400 g of the NCS-300PCM composite and placed on the inside walls, base and top of the package and filled with 2 kg of asparagus.

The results show that this model system clearly demonstrates the thermal buffering capacity and effectiveness of our novel NCS-300PCM thermal buffering composite material when contained in bubblewrap liners. These flexible liners can easily be inserted inside a paperboard package containing temperature sensitive perishable food to maintain the temperature of the food at a suitably chilled level for about 4-9 hours even though the outside of the package has warmed up to higher temperatures during this time. This thermal buffering capacity is ideally suited to protect chilled and temperature sensitive food and other products against unwanted higher outside temperatures which are often encountered during the transportation and temporary en route storage from the coolstore to the international market place. The PCM which is incorporated in the high pore volume nano-structured calcium silicate host is selected such that the melting temperature is close to the optimum chilled temperature required for maintaining the freshness of the food. Hence, the use of flexible bubblewrap liners for containing the NCS-PCM is an ideal way to provide such required thermal buffering capacity to paperboard packages containing the food. These liners are readily recyclable and re-useable. The food, NCS-PCM liners and package are all cooled to just below the PCM melting point for a long enough time to ensure thermal equilibrium is reached and all the PCM is in the solid form. The package is then usually transported in a chilled container to the airport where may sit on the tarmac or in uncontrolled temporary storage for a few hours before being loaded onto an aircraft for international export. The NCS-PCM will prevent the perishable food contents of the package from warming up and spoiling as the PCM will slowly melt and absorb the unwanted ambient heat before it reaches the perishable food in the package. Once in further chilled storage, the PCM will solidify again ready to provide thermal buffering against the next unwanted transient temperature rise. The same system can be used to provide thermal buffering for other temperature sensitive materials such as medical supplies, where again the PCM is chosen to give the required temperature range.

5. Thermally buffered water containers

The NCS-PCM composite can also be used in maintaining the temperature of the contents of a package above that of ambient and to temporarily prevent or buffer the cooling rate of the contents of the package. This is useful for slowing the cooling of hot fast foods during delivery and in boutique applications such as maintaining the required water temperature for transporting tropical fish. In this application, a PCM with a suitably higher melting point is selected. The PCM is a solid up to the required buffering temperature range but has to be heated above the buffering temperature to ensure it is in the liquid phase initially. Upon cooling, the liquid to solid phase change releases heat energy to the inside of the package which buffers or slows down the cooling process, thereby enabling the above ambient temperature to be maintained in the package for a longer period of time. An example of the use of the NCS-PCM in the transportation of tropical fish where it is desirable to keep the water temperature slightly above about 18 - 20 °C is presented here. *Rubitherm* RT20 PCM which has a melting point and hence thermal buffering range of about 20 °C was chosen. This was heated to a liquid and worked into the pores of the NCS to a level of 300 wt %. The resulting NCS - 300PCM was similarly placed in bubblewrap liners. An 11 kg quantity of water was placed in a rectangular plastic container and heated to about 26 °C. I-buttons were placed in the centre of the water volume and also outside (beside) the container to record temperature / time profiles. The container was then placed in a room where the temperature was about 13 °C and the temperature / time cooling profiles were recorded over 2000 minutes as the container cooled down from about 26 °C to about 13 °C (Fig. 11). The temperature profile is characteristic of a typical exponential cooling curve.

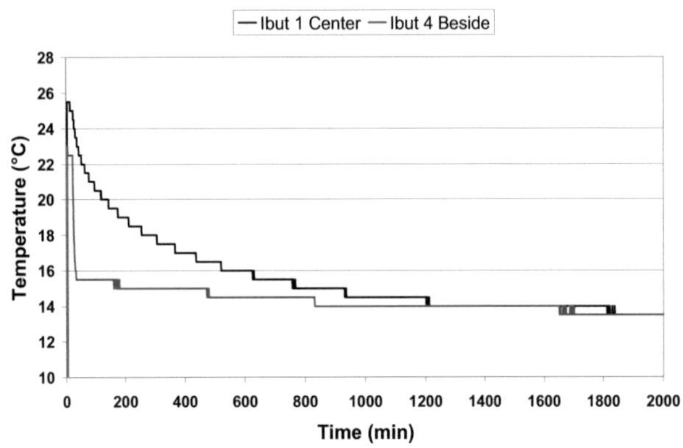

Fig. 11. The temperature / time cooling profile for an 11 kg volume of water in a rectangular plastic container.

The plastic container of water was then surrounded by 6 bubblewrap liners containing a total of 600 g of the NCS - 300PCM composite with RT20 to provide thermal buffering at about 20 °C. The water was again heated to about 26 °C together with the liners to make sure all the PCM in the NCS-PCM composite was in the liquid phase. The liners were then placed around the container and the temperature / time cooling profiles recorded (Fig. 12). A comparison of the cooling profile for the water with and without the NCS-PCM liners clearly shows that the rate of cooling when the container is surrounded by the NCS-PCM composite bubblewrap liners is significantly slower. (Figs. 11 and 12). This is particularly noticeable in the region from about 26 - 18 °C which spans the freezing point of the PCM. Here the PCM gives out heat on freezing which buffers the cooling rate of the water. The results show that the times taken for the water to cool to 18 °C and to 14 °C without the PCM liners are 265 minutes and 1220 minutes respectively. With the PCM liners these cooling times are much longer being 690 minutes and 1770 minutes respectively. The slower cooling rate achieved with the use of the NCS-PCM bubblewrap liners makes it possible to maintain water above a particular temperature for a longer period of time without additional external heating, as is required in the transport of hot food and tropical fish.

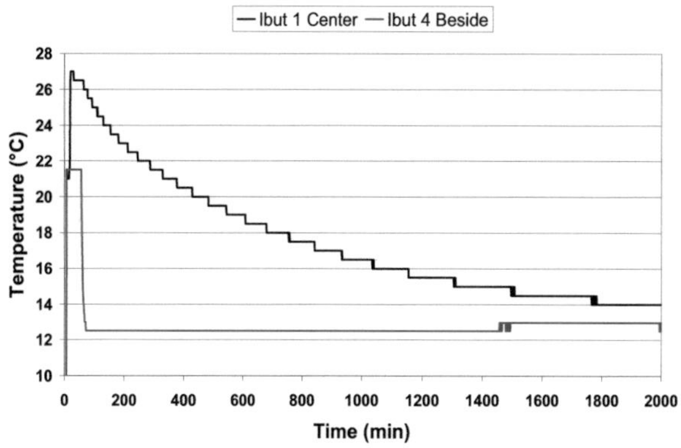

Fig. 12. The temperature / time cooling profile for an 11 kg volume of water in a rectangular plastic container surrounded by bubblewrap liners containing 600 g of RT20 NCS-PCM.

6. Conclusion

The research has shown that the proprietary nano-structured calcium silicate material can accommodate alkane phase change materials to form NCS-PCM composites which effectively contain the liquid phase PCM in the large NCS pore volume when the composite is at temperatures above the PCM melting point. Below the melting point the NCS-PCM contains solid alkane PCM in the solid NCS host and above the melting point the NCS-PCM contains liquid PCM in the solid NCS host, but the overall composite is still a dry free flowing powder. The optimum level of PCM was found to about 300 wt% PCM on a NCS basis. The NCS-PCM composite can readily be contained in bubblewrap plastic bags as liners inside paperboard packages to provide thermal buffering for the contents. It has been successfully shown that a NCS-PCM composite containing 300 wt% of *Rubitherm* RT6 alkane PCM in bubblewrap liners inserted into a paperboard package around the inside walls, can provide effective thermal buffering for temperature sensitive chilled food to prevent unwanted warming and hence food spoilage, during temporary storage and transportation enroute from the supplier to the international market place. For a package containing 2 kg of chilled asparagus and lined with bubblewrap liners containing 400 g of NCS-PCM using RT6 PCM, the inside temperature of the asparagus can be maintained below about 10 ^{o}C for some 4-9 hours even though the ambient temperature had warmed well above this. In a similar way the cooling rate of temperature sensitive contents such as hot food or tropical fish can be reduced by using these NCS-PCM liners in which the PCM has a higher melting point consistent with the preferred temperature for the contents of the package. The NCS-PCM composite material developed here and contained in easily useable and re-useable bubblewrap liners therefore has wide range of potential applications in the packaging industry for the safe transport of temperature sensitive and perishable produce and materials, below and above ambient temperature.

7. Acknowledgement

Financial support received from the NZ FRST RFI Grant C04X0401 as well as the work of Margaret Dodds and James Grindrod are gratefully acknowledged.

8. References

Amcor Kiwi Packaging Ltd. (2011) <www.amcor.com>

Carter Holt Harvey Packaging. (2011) <www.chhpackaging.co.nz>

Cool Pack Cold Chain. (2011) <coolpack.com/learning_center>

Dodds, M.M. (2009) The Development of thermally responsive Packaging Materials for Transporting Perishable Foods. *MSc Thesis* Victoria University of Wellington.

Farid, Mohammed M.; Khudhair, Amar M.; Razack, Siddique Ali K.; & Al-Hallaj Said (2004) A Review on Phase Change Energy Storage: Materials and Applications. *Energy Conversion and Management*, 45 (9-10) pp 1597-1615

Feldman, D.; Shapiro, M.M.; & Banu, D. (1986) Organic Phase Change Materials for Thermal Energy Storage. *Solar Energy Materials*, 13, pp 1-10

Hawlader, M.N.A.; Uddin, M.S.; & Khin, Mya Mya. (2003) Microencapsulated PCM Thermal-energy Storage System. *Applied Energy*, 74, Issues 1-2, pp 195-202

Johnston, J. H.; Borrmann, T.; & Mcfarlane, A. J. (2006) Nano-structured Silicate, Functionalised Forms Thereof, Preparation and Uses. *NZ Patent Specification No. 537747 ; International PCT Application PCT/NZ2006/000003*

Johnston, James H.; & Dodds, Margaret. (2011) The Development of a Flexible Re-useable Thermal Buffering and Insulating Liner for Packaging Temperature Sensitive Products. *Appita*, 64(2), pp 153-157

Johnston, J.H.; Grindrod, J.E. & Dodds, M.M. (2007) Nano-structured Calcium Silicate Phase Change Material: A New Product for Thermal Buffering In Packaging. *Proceedings of the 61st Appita Conference,* pp 366-369, Brisbane, May 2007

Khudhair, Amar M.; & Farid, Mohammed M. (2004) A Review on Energy Conservation in Building Applications with Thermal Storage by Latent Heat Using Phase Change Materials. *Energy Conversion and Management*, 45 (2), pp 263-275

Plastics New Zealand (2011) <www.plastics.org.nz>

Powell, F.J.; & Matthews, S.L. (1987) Thermal Insulation: Materials and Systems, ASTM Committee C-16 on Thermal Insulation. *Published by ASTM International, 1987 ISBN 0803104936, 9780803104938,* 755 pages

Rubitherm Technologies GmbH. (2011) <www. Rubitherm.com>

Schossig, P.; Henning, H.-M.; Gschwander, S.; & Haussmann, T. (2005) Micro-encapsulated Phase-change Materials. Integrated into Construction Materials *Solar Energy Materials and Solar Cells*, 89, pp 297-306

Tyagi, Vineet Veer; & Buddhi, D. (2007) PCM Thermal Storage in Buildings: A State of Art. *Renewable and Sustainable Energy Reviews*, 11, pp 1146-1166

Zalba, Belén; Marín, José; Cabeza, Luisa F.; & Mehling, Harald (2003) Review On Thermal Energy Storage With Phase Change: Materials, Heat Transfer Analysis and Applications. *Applied Thermal Engineering,* 23, pp 251–283

Zhang, Yinping; Zhou, Guobing; Kunping, Qunli; Zhang, Lin; & Di, Hongfa. (2007) Application of Latent Heat Thermal Energy Storage in Buildings: State-of-the-art and Outlook. *Building and Environment*, 42, pp 2197-2209

16

Matrix Active Micro- and Nanocomposites Based on the Polymer, Semiconductive and Ferropiezoceramic Materials

Mahmud Kerimov[1], Mirza Kurbanov[2], Azad Bayramov[2] and Ali Mamedov[2]
[1]Presidium of Azerbaijan National Academy of Sciences,
[2]Institute of Physics Azerbaijan National Academy of Sciences,
Azerbaijan

1. Introduction

Active composite materials are a wide class of material used in radio engineering, electronics, optoelectronics, pyro- and piezotechnic. The possibility of application of created electrets, piezo- and pyroelectrics, posistors, piezoresistors, varistors, photoresistors, photovoltaic elements and photoelectrets based on the active composites is multiform beginning from various acoustoelectric transducers, acoustic generators, sensors of heat flow, image transducers, photoelements of information recording and storing, until using for radiation protection of spaceships, revealing of underwater objects, seismic and geophysics exploration, also for solving many problems of creating of alternative energy sources (for example, photovoltaic) and medical-biological problems.

At present, there are mainly four directions in area of creating abovementioned elements based on the solid dielectrics and semiconductors. First of them is connected with creating new ferropiezoelectric and semiconductive materials mainly based on the barium titanate, plumbum-zirconium-titanate (PZT), $A^{II}B^{VI}$, $A^{III}B^{V}$, $A^{III}B^{VI}$, $A^{IV}B^{VI}$, also semiconductors doped by rare-earth elements. Second direction is based on the synthesis of new polymer dielectrics revealing electret, piezo-, pyroelectric and photoelectric properties. Third direction is connected with creating various composite materials based on the polymer (matrix) dispersed by inorganic ferropiezoelectric or semiconductive fillers. Forth direction discovers new possibilities connected with creation various nanocomposite active materials including hybrid materials consisting of matrix nano- and micropiezoelectric composites.

The establishment of interconnection between characteristics of composite materials and properties of polymer, inorganic, ferroelectric and semiconductive phases is a key task on the way of solving of the very important problem of creating micro- and nanoelectric material. Numerous investigations of ferro- and piezoelectric materials have shown that the usage multicomponent piezomaterials based on PZT as inorganic phase is most perspective for obtaining high effective pyro-, piezoelectrics and electrics.

It should be noted that at present by variation of composition and number of components in PZT family it is impossible further to improve properties of piezo-, pyroelectric and electret materials. It is caused by that, the increase of piezo- and pyroelectric coefficients due to regulation (increase) of ferrosoftness and electromechanical characteristics is directly accompanied by a decreasing of pyro- and piezoelectric responses owing to increase of dielectric constant. But the increase of ferrostiffness with aim of decrease of a dielectric constant is accompanied by a decreasing of piezo- and pyrosensitivity due to decrease piezo- and pyroelectric constants. So, an appreciable gain of characteristics of the piezo-, pyroelectric and electric ceramic materials isn't provided by application abovementioned technological methods.

The wide and various application of piezo- and pyroelectric dielectrics in electroacoustics, pyro- and piezotechnics in many cases demands to create large-size flexible transducers of various configuration. These demands aren't practically satisfied by application of inorganic ferropiezoelectrics and semiconductors. As it was noted, developments of composite materials with organic and inorganic components discover new possibilities in material science. At present, the numerous kinds of polymer materials with various physical-mechanical, electrophysical, thermophysical and photoelectrical characteristics are used as organic matrix [Furukawa T. et al., 1979; Hiroyoshi Ueda et al., 1986; Jamada T. et al., 1982;; Furukawa T. et al., 1976; Chinchurreta F.J. & Montero de Espinosa F., 1988; Topolov V.Yu. et al., 2006; Kerimov M.K. et al., 2005, 2007, 2009, 2011; Kurbanov M.A. et al., 2009, 2011; Ploss B. et al., 2005]. Along with this, it should noted that also numerous semiconductive and ferropiezoelectric materials with various properties are as inorganic phase of composites. Such combination of composite components properties allow to create both microactive and nanoactive dielectrics. Created such kind the micro-, nanocomposites and their hybrid have properties not inherent in organic and inorganic phases separately.

Really, polymer piezo- and pyroelements have low piezo- and pyrocoefficients owing to low value of stabile dipole orientational polarization as in case of ferropiezoelectrics. They have low dielectric constants 2÷15 and their low piezo- and pyrocoefficients don't give a large gain in sensitivity to external actions. The composites of polymer-ferropiezoelectric type have positive properties of ferropiezoelectrics and polymers, and can have large a piezo- and pyrosensitivity. Ferropiezoelectrics have more thermic stability than polymer materials, but their parameters are low owing to high electric permeability. The thermal stability of polymer piezo-, pyroelectric and electric materials is low, despite there are traps with enough high concentrations, stable remanent orientational polarization, higher value of electret potential difference and better dielectric properties in these materials.

It is known, that polymer electrets, piezo- and pyroelements with relatively high electrophysical and electromechanical characteristics are produced mainly from fluorine-containing high molecular compounds [Kerimov M.K., 2009; Sessler G.M., 1980]. But, simple and more technological thermoplastic polymers of polyolefine family, for example, polyethylene and polypropylene, have very low electret, piezo- and pyroelectric characteristics. So, they aren't used widely in practice, though also they have high electric and mechanic properties.

One of main way of development of the stabile piezo-, pyroelectric and electret materials is a creation of phycotechnological basis of composite materials based on the polyolefines with various organic and inorganic fillers.

The perspective direction using composite polymer dielectrics is a development on their basis flexible nonlinear elements - posistors, varistors, piezoresistors and thermoresistros. It is known, that for obtaining nonlinear elements, particularly varistors and posistors, from polycrystal materials the below conditions are needed:

- the element must consist of minimum two phases - crystalline and amorphous;
- the potential barrier must form on the boundary of these phases;
- the element must has bipolar conductivity and symmetric voltage-current characteristic;

- the energy gap width of dielectric and semiconductive phases must differ as far as possible for aim of obtaining high potential barrier on the interface.

In composite materials of polymer-inorganic phase abovementioned demands are satisfied efficiently. In addition to these demands in composite materials the potential barrier with large width is formed due to noticeable difference of components density on the interface. Finally, this allows more effectively regulating the parameters of potential barrier and composite nonlinear element.

As it was noted, one of perspective directions using composite active materials is a performance on their basis the film flexible posistors that is the resistors with a positive temperature resistance coefficient. As known, the posistors are mainly created on the basis of ferrosemiconductive materials, particularly, barium titanate the specific resistance of which is decreased by doping of impurities of rare-earth elements of lanthanum, cerium etc. The formation of posistor effect in this ferroceramic is connected with the tetragonal-cubic phase transition when it is occurs an increasing of barrier width on the boundary of ceramic grains. It is accompanied by sharp increasing of electric resistance and decreasing of dielectric constant.

By synthesis of ferrosemiconductive posistors on the basis of barium titanate we need to accept a quite a number of disadvantages: a high temperature sintering, a narrow concentration interval of doped admixtures, complexity of technology due to using superpurity raw material. Posistor effect in composites based on the crystallizable polymers and ferropiezoelectrics is caused by presence of barrier layer on the boundary of ferropiezoelectric particles the height of which is defined by dielectric, electronic, electrophysical and thermophysical parameters of components, and also by supramolecular structure of polymer matrix after its dispersion.

Dielectric and thermophysical characteristics of posistor composites are defined by both properties of inorganic dispersing agents and polymer matrix, and also by temperature-time regime of crystallization of the polymer composites in whole. It gives a possibility to regulate the formation of the posistor effect in composites in wide temperature interval, also to develop the posistors with stepless control of temperature of the posistor maximum occurrence.

It should be noted, that the operation of composite transducers in various conditions and regimes has revealed theirs advantages and disadvantages, and targeted at their improving. The reliability and stability of developed transducers is much depended from right solution of this problem. For expansion of range and conditions of composite transducers applications we need to define mechanical and electrical strength properties of their active dielectrics. It is one of the most important operating indexes of composite transducers in every range of their application. For example, when a shock wave passes through electret, pyro- and piezoelectric transducers, the mechanical stress and electric field are appeared in these materials, which promote an accelerated damage of these elements. It is known also, that when electrothermopolarization the composite material is hold long time in electric field at various temperatures, and in some cases under the simultaneous action of electric, mechanical and temperature fields.

2. Obtaining active composites and methods of their properties investigation

2.1 Choice components, obtaining homogeneous mixtures and composites on their basis

The process of obtaining some composites includes at least three stages: 1) preparation and choice initial components; 2) mixing components; 3) obtaining composite material from components of homogeneous mixture. A choice component of the composite and method for obtaining a composite active material is dictated by the set of properties that the final material should possess. In connection with this, when development of composites the polymer phases should meet next demands: easy processing, high plasticity and mechanical strength, heightened dielectric constant and electric strength, minimal dielectric loss.

The assessment of specific demands of occurrence processes of the internal stress and relaxation of structures when obtaining and processing composites based on the selected matrix is not less important. The melting temperature must be enough high for that the structure and charge state of obtaining composites aren't damaged by electrothermoprocessing. The glass transition temperature must be low for providing stable relaxing properties defined the temperature interval of reproducibility of the composite parameters.

The analysis of thermophysical, electrical and physical-mechanical properties of polymers showed that the thermoplastic nonpolar polymers of polyolefine series and the fluorocarbon polar polymers (high density polyethylene HDPE, low density polyethylene LDPE, polypropylene PP, polyvinylidene fluoride PVDF, copolymer vinylidene fluoride with tetrafluorethylene, polyvinylchloride PVC and polytrifluoethylene) are suitable under abovementioned demands. At chose of polymer matrix, in point of view of the pyro- and piezoelectric composites, the determinative factor is high values of the dielectric constant and specific volume resistance; but in point of view of electret composites the matrix must possess deep centers of charges capture. Besides, it should note, that already an industry developed production of thermoplastic polymers of polyolefine series and the fluorocarbon polymers. It is a key factor in economic point of view for development of transducers based on the polymer composites. Pointed polymers are distinguishing their physical-mechanical properties and structures. For example, polyethylene and polypropylene are nonpolar polymers with low values of dielectric constant and high volume resistance. But fluorocarbon polymers are polar polymers with high values of dielectric constants, but low volume resistance. Fluorine-containing polymers possess not bad piezo- and pyroelectric, electret and light-sensitive properties. It should note that by changing of isothermal regimes of moulding and temperature-time conditions of crystallization we easily can vary of supramolecular structure of pointed polymers [Sessler G.M., 1980; Marikhin V.A. & Myasnikova L.R., 1977].

The demands to active elements are differentiated by effects waited from composite materials. As active fillers the most effective ferropiezoelectric and ceramic semiconductors with various structures were used for electret, piezo- and pyroelectric, posistor and varistor composites.

The investigations of piezoceramic materials brought into the obtaining of effective piezoelectrics and pyroelectrics based on the PZT. Then, on the basis of PZT the more effective piezoceramic materials were produced by doping of ternary oxides [Fesenko E. G., 1983; Lines M.E. & Glass A.M., 1977; Burfoot J.C. & Taylor G.W., 1979; Rez I.S. and Poplavko Yu.M., 1989]. In connecting of this, as fillers (inorganic dispersing agents) the multicomponent piezoceramic materials of PZT family were used. As will be shown below, obtained on the basis of various polymer dielectrics and multicomponent ferropiezoelectrics the composite materials are more effective and possess high electret, piezo- and pyroelectric properties in comparison with composites having double component PZT as dispersing agent. Used multicomponent ferropiezoelectric materials had rhombohedral (Re), tetragonal (T) and compound (heterogeneous Re+T) structures. At this a morphotropy range is considered as important reference point when choice an active phase (dispersing agents) dividing system on rhombohedral and tetragonal phases. So, used main piezoceramic materials are solid ferroactive solutions of multicomponent systems of type:

$$PbTiO_3 - PbZrO_3 - \sum_n PbB'_{1-\alpha}B''_{\alpha}O_3, \quad n=2,3,...,$$

which possess various symmetry, piezo-, pyroelectric characteristics and dielectric properties in dependence of position relative to morphotropy range.

By development of piezoresistive, posistor, varistor, photoresistor, photovoltaic, photoelectret and nanostructured systems, the simple semiconductors of Ge, Si, Se, GaSe, CdS, ZnS, ferrocene, monochalkogenide and one-and-a-half chalcogenide of rare-earth elements were used. A choosing of their as active phase is connected with that some of them were recommended themselves as effective materials for sensors of various purposes.

Next technological operation is a mechanical mixing of prepared powder-like thermoplastic polymers with powder-like inorganic dispersing agents. The technology of such mixing is known well. It includes a stage of preliminary dry mixing of powder-like components (polymer and active phase) and a stage of homogenizing in a powder-like state and then in a melt. It should be noted, when doping filler, a viscosity of polymer melt increases, and this complicates its processing in a product by high effective methods of extrusion and press molding. The limiting level of filling is determined by two indications: permissible changing of properties of the material and its rheological characteristics.

As far as, when development of active elements it is needed to obtain composites with volume contents of particles of the dispersing agents till 80%, an application of modern methods of extrusion and press molding was became practically impossible. Really, in according with published articles, and also carried out experiments, mixing granulated equipment, extruders for forming of product are used only for composites contented 25-35% vol. particles of the dispersing agents. But even in case of less content of filler, an extrusion and calendering don't provide obtaining nondefect on particles distribution composites. In this, the method of hot pressing of preliminarily carefully homogenized mixture of composite component is most acceptable. This method becomes irreplaceable due to possibility of carrying out thermoprocessing and even polarization in single technological cycle. In experimental part it will be shown, that such processing appreciable increases an efficiency of obtained active materials.

So, the method of obtaining of a composite material is consisted in the following: 1) a homogeneous mixture is obtained from powder polymers and dispersing agents (active phase) in a mechanical mixer (a laboratory ball mill); 2) pellets are obtained from mixture by cold pressing; 3) during some time, pellets are hold at the melt temperature under a small pressure; 4) at this temperature, a pressure is slowly increased till pressing pressure and then sample is hold over time of pressing; 5) then a pressure is removed and a sample is cooled in water with ice, or in liquid nitrogen, or slowly cooled under a pressure.

Taking into account, that generally, thermoplastic polymers are used as polymer matrix of composites, a pressure and temperature of pressing of their composite are little differed, and change into 15÷20 MPa и 443÷550 K intervals, respectively. When choosing a pressure and a temperature of composite pressing it is need to take into account the temperature of decomposition of a polymer matrix and mechanical destruction particles of the dispersing agents under the mechanical stress arising during pressing.

2.2 Investigation method of charge state of composites-thermoactivated spectroscopy

The methods of thermoactivated current spectroscopy are widely applied for investigations of electrophysical properties and charge state of semiconductive, dielectric and ferropiezoelectric materials. The basis of this method is that when a linear changing of temperature an object transits from non-equilibrium state to new thermodynamic equilibrium one. On of the methods of thermoactivated current spectroscopy is a method of thermostimulated depolarization (TSD). It consists of next: studied dielectric is preliminary polarized under the action of intense electric field in a combination of other actions (heat, visible or UV lighting, radiation, mechanical stress). At this, polarization can occur on next ways: orientation of polar molecules or domains along of field; a displacement (migration) of free charge carriers on a macrodistance with following locking on the traps; an injection of charge carriers from electrodes and following their capture by traps; a displacement of free charge carriers on a microdistance with their capture by traps and an accumulation of charges on the interphase boundary in polycrystalline and composite materials.
Polarized dielectric is cooled under the field till less temperature, and then it is heated on the certain law in short-circuited state. As at this, charge carriers, dipoles and domains, change theirs positions in space, then, in common case, there is a depolarized current in an external circuit. There is a changing the value of induced charge on electrodes. A density of depolarized current can be describe by formula [Sessler G.M., 1980]

$$j(t)=q\cdot\mu\cdot n(t,x)\cdot F(t,x)+\lambda(t)\cdot F(t,x)+\frac{\partial}{\partial t}[\varepsilon\varepsilon_0\cdot F(t,x)+P(t)]$$

where, q, μ, n are a charge, mobility and concentration of free polarized charge, F is an intensity of internal electric field, $\lambda(t)$ is a equilibrium conductivity of sample, ε is a dielectric constant of sample, P(t) is a relaxion polarization of sample.
There is in the expression, the first term describes nonequilibrium conductivity current; the second one describes a equilibrium conductivity current; the third one describes a displacement current due to relaxation of macroscopic heterogeneous and homogeneous polarization. Electrodes with the help of special cable are connected to electrometer for registration of depolarized current. The control of sample's temperature on a linear law is carried out with the help of the special regulator of temperature.
The method of thermoactivated current spectroscopy is effective for investigation of electret, piezo- and pyroelectric properties of composites as far as it allows to define the value of stabilized charge and domain-orientational polarization formulated in the processes of electrothermopolarization. For investigation of TSD of composites in temperature range of 77÷520K, the special set, regulator of temperature, was developed. It allows to investigate TSD in the regime of factional heat, and also under the simultaneous action of a mechanical stress on the composite.
On the TSD curves the parameters of dipoles, electron and ionic centers of captures of the charge carriers can be determined. These parameters are: activation energy, frequency factor and concentration. The most simply these parameters are determined by TSD curves received in the regime of linearly heat of composites, contented one kind of electrical active centers. There are suggested many various methods of determine of activation energy of charge relaxion on TSD current curves. Particularly, it can be determined on temperature position of maximums of TSD current, by the method of a heat rate variation, on the character points of TSD curve [Sessler G.M., 1980].

Pizoelectrical and electricmechanical properties of composites are measured by quasi-static method of resonance and antiresonance, and in dynamic regime using acoustic equipments of Bruel & Kjaer firm.
Also, the remanent reorientation polarization P_r of the composites was determined from their thermally stimulated depolarization curves. The same technique was used to determine the pyroelectric coefficient $\gamma(\Delta P_r/\Delta T)$, which is a measure of the pyroelectric activity of the composites. The current gradually decreases in sequential heating-cooling cycles and is reproduced in magnitude beginning with the seventh to eleventh cycle. The reproducible current was approximately ten times lower than the current measured in the first heating cycle. Therefore, the thermally stimulated current contains the reproducible and irreproducible components. The irreproducible component is associated with the release of charge carriers from shallow traps at the sample surface and in the polymer phase. The reproducible current component is determined by the change in the remanent reorientation polarization of the pyroelectric ceramic phase as a function of temperature. It should be noted that a symmetric dependence of the pyroelectric current in a heating-cooling cycle is observed only alter the reproducible thermally stimulated current is attained. This is one of the indications of the pyroelectric effect in materials, in particular, polymer-pyroelectric ceramic composites. In this case, the pyroelectric current reversed direction, whereas the current magnitude remains identical to that observed upon heating of the sample. Upon change in sign, the current of thermally stimulated depolarization undercooling initially passes through zero due to the sharp increase in the relaxation time and then rapidly increases. The symmetric dependence of the pyroelectric current in the heating-cooling cycle is ensured by the quantity dP/dT. After the symmetric dependence was attained, the total thermally stimulated depolarization curves were measured and the remanent reorientation polarization in the composites was determined from the areas under these curves.
Electret potential difference and electret charge were determined by using TSD methods and registration of induction current.

3. Composite matrix piezo- and pyroelectric materials

The composites are widely applied almost in all fields of technology and constitute the background of the most natural materials. In this connection a lot of works are devoted to their fabrication technology and studying of their macroscopic characteristics [Sessler G.M., 1980; Kerimov M.K., 2005, 2007, 2009; Kurbanov M.A., 2009, 2011; Topolov V.Yu. et al. 2006; Smay J.E., 2002; Zhang Q.M. & Geng, 1994; Mamedov G. et al, 2010], significantly, of the practical directionality. As far as the structures and features of composites are extremely various, the general approach for explanation of their properties the formation practically is absent.
The most common representation of composites feature formation is based on account of characteristics of the constituents. In this case the geometry (size, shape) and their relative position are taken into account. The interaction between the phases is considered negligible, i.e. contributions of phase properties to integral characteristics of composites are taken additive. By this approach there is no necessity in separation of composite material physics. It follows also from adopted definition of composites as the heterogeneous system, obtained from two or more components with conservation of individual properties each of them. However, one cannot neglect the interaction between the phases formed the composites as far as it becomes decisive in formation of their properties for many cases. The obvious example of the impossibility of determination of the integral features of composites starting from additivity of the contributions of the phases is a system consisted of sequentially connected phases between which *p-n* junction is formed.

Below we shall make an attempt to show the necessity of accounting of the interface interaction, the electron states of phase interface and charge stabilizations init formation of the composite properties on example of disordered polymer-piezoelectric system.

High density and low density polyethylene (HDPE and LDPE), polypropylene (PP), polyvinylidene fluoride (PVDF) are used as a matrix of composites. Multicomponent piezoceramics of lead-zirconate-titanate family of PCR-piezoceramics Rostov type are used as active filler. The USA analogues of used PCR piezoceramics are presented in table 1.

Used PCR-piezoceramics Rostov type and its analogous in brackets	PCR-1 (PZT-5A)	PCR-3M (PZT-7A)	PCR -5 (PZT-75)	PCR -7M (PZT-5H)	PCR -8 (PZT-8)
$^2d_{33}/\varepsilon_0\varepsilon$, [Vm/N]	$33.0 \cdot 10^{-3}$	$24.0 \cdot 10^{-3}$	$29.0 \cdot 10^{-3}$	$15.8 \cdot 10^{-3}$	$23.2 \cdot 10^{-3}$

Table 1. The comparison of the piezoelectric properties of the composites based on the same polymer with different piezoelectric fillers.

Samples are obtained based on the homogeneous mixture polymer-piezoparticles by hot pressing method. Diameter of the piezoparticles was changed from 63 to 100 mkm. Thickness of samples was varied from 80 to 200 mkm. Piezocoefficient (d_{ij}) of composites was measured in the quasistatic condition with error not more 8%. Volume fraction of piezophase was changed from 40 to 70%. Polarization of composites was brought off in simultaneously action of constant electric field and temperature. Magnitudes of electric intensity (E_p) and temperature (T_p) of polarization were limited by electric strength of breakdown and melting temperature of samples.

Let's consider the role of polarization processes taking place in a polymer matrix and on polymer-piezoelectric interfaces on the basis of following experimentally stated facts:

1. The dependences of piezocoefficient of HDPE+50%vol.PCR-1 composites vs polarization electric fields E_p, and temperature T_p are shown in fig.1a,b. With increasing of E_p and T_p the composite piezocoefficient (d_{33}) is increased, passes via maximum and then is decreased. As far as the temperatures and electric field intensities of polarization is smaller than Curie temperature of piezophase and composites electric breakdown intensity correspondingly, then these dependencies mustn't have a maximum.

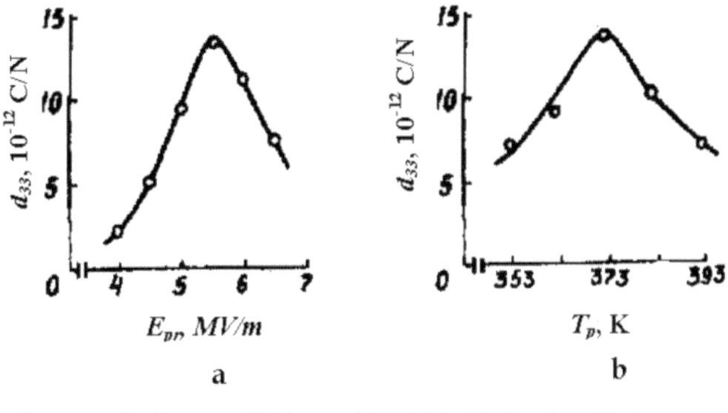

Fig. 1. The dependences of piezocoefficient of HDPE+50%vol.PCR-1 composites vs a) polarization electric fields E_p at T_p=373K; and b) vs temperature T_p at E_p=5,5 MV/m. The sizes of piezoparticles are 63 mkm.

2. A comparison of d_{33} values of the composites on the base of the low and high-density polyethylene and polypropylene with the same piezofiller (fig.2) indicates that their piezoelectric coefficient corresponding to the optimum polarization conditions are significantly distinguished.

A theoretical consideration of two phase system based on a polymer dielectric dispersed by piezoelectric particles uniformly distributed in volume, gives [Furukawa T. et al., 1976, 1979; Jamada T. et al., 1982; Wede H. Et al., 1986]

$$d_{33} = F \cdot \frac{3\varepsilon_1}{2\varepsilon_1 + \varepsilon_2 + F(\varepsilon_1 - \varepsilon_2)} \cdot \frac{5C_2^2 d_{33}}{3C_1 + 2C_2 - 3F(C_1 - C_2)}, \tag{1}$$

where ε_1 and ε_2 are dielectric constants of the polymer matrix (phase I) and piezoceramics (phase II), accordingly, C_1 and C_2 are appropriate flexibleity modules. F is the volume fraction of piezofiller; ${}^2d_{33}$ is a piezocoefficient of piezofiller.

With taking into account $\varepsilon_1 >> \varepsilon_2$ and $C_2 >> C_1$ one obtains

$$d_{33} = \frac{15F}{(1-F)(2+3F)} \cdot \frac{{}^2d_{33}}{\varepsilon_1} \varepsilon_1 \tag{2}$$

It is seen, that the piezocoefficient of composite is proportional to ε_1 of polymer phase and piezoceramics sensitivity (${}^2d_{33}/\varepsilon_2$). Thus, this experimental fact has no explanation within the scope of theory, since ε for the polymers of the polyolefin chain is the same.

3. The comparison of the piezoelectric properties of the composites based on the same polymer with different piezoelectric fillers (table 1) also shows that the piezocoefficients of the composites (fig.3) are not proportional to the ratio ${}^2d_{33}/\varepsilon_2$ as it is predicted by theory.

4. It is seen from preceding results that for the optimal polarization the composites have significant piezocoefficient that points to high polarization of piezophase. Nevertheless the effective field E acted on isolated spherical piezoparticles in composite is defined according to Frelich theory with taking into account $\varepsilon_1 << \varepsilon_2$ by relation

$$E = \frac{3(1+2F)}{1-F} \cdot \frac{\varepsilon_1}{\varepsilon_2} \cdot E_p \tag{3}$$

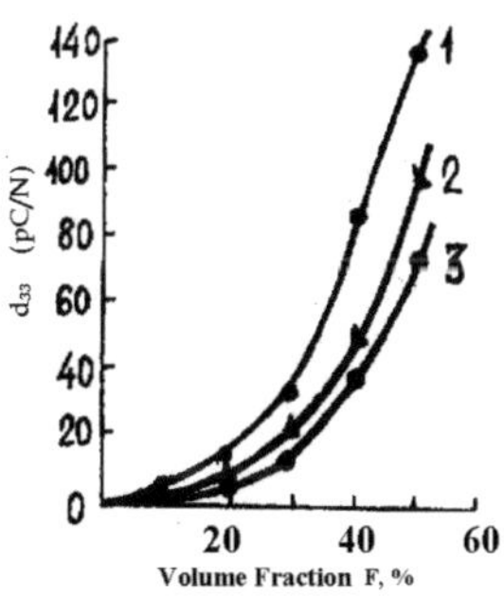

Fig. 2. The composite piezocoefficient values on the base of the low and high-density polyethylene and polypropylene with the same piezofiller: 1-PP+PCR-3M; 2-LDPE+PCR-3M; HDPE+PCR-3M. The sizes of piezoparticles are 63 mkm.

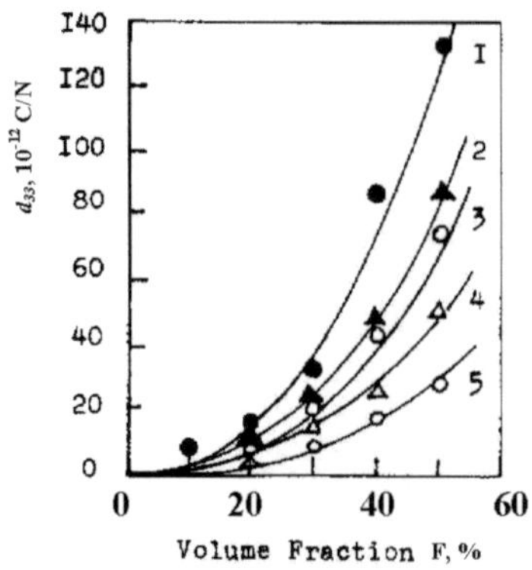

Fig. 3. Piezocoefficient d_{33} as a function of piezoceramic volume fraction for composites on the basic of polypropylene and piezoceramics with different ${}^2d_{33}/\varepsilon_2$: 1) PP/PCR-3M, ${}^2d_{33}/\varepsilon_2$=0.024 for PCR-3M; 2) PP/PCR-5, ${}^2d_{33}/\varepsilon_2$=0.029 for PCR-5; 3) PP/PCR-7M, ${}^2d_{33}/\varepsilon_2$=0.0158 for PCR-7M; 4) PP/PCR-1, ${}^2d_{33}/\varepsilon_2$=0.033 for PCR-1; 5) PP/PCR-8, ${}^2d_{33}/\varepsilon_2$=0.24 for PCR-8. The sizes of piezoparticles are 63 mkm.

It is seen, that the field intensity per of piezoparticles during thermopolarization much smaller than the applied field of polarization E_p and coercive field of piezoceramic and consequently the piezoparticles must not be efficiently polarized.

The attempts were made to explain the significant distinctions between the experimental data and theoretical calculations with taking into account the third phase supposedly arising as a result of the interaction of the piezoparticles and polymer [Hiroyoshi Ueda et. al., 1986; Furukawa T. Et al., 1976, 1979; Wede H. Et.al. 1986; Jamada T. Et al. 1982]. However, the use of such model leads to no reasonable values for parameters of the intermediate third phase, considered as fitted one for the agreement of the theory and experiment.

Numerous experimental results [Topolov V.Yu. et al., 2006; Zhang Q.M. & Geng X., 1994; Kerimov M.K. et al. 2005, 2007, 2009, 2011; Kurbanov M.A. et al., 2009, 2011] for the composites based on piezoceramics and polymers with the various structures polarities verify the fact listed above and point to necessity for different approach in explanation of the piezoelectric properties formation of composites. In particular, all these facts may be explained if one takes into account an interconnection of charge accumulation processes in polymer matrix and piezophase polarization of composite.

A charge state of the composites has been investigated by the thermostimulated depolarization method (TSD). In fig.4 the typical TSD spectra of composites based on HDPE and PZT-19 are shows. These are two maximums on spectra. The first, low temperature one is coincides with the TSD maximum of pure polymer. Activation energy of the first TSD maximum of the composite is close in a magnitude to the activation energy of TSD maximum of a base polymer.

One can come to conclusion that the first maximum of TSD spectrum is caused by liberation of charge carriers from the traps in polymer. The second maximum of TSD current of a composite is connected with a relaxation of charge captured by traps with high activation energy. One can suppose that these traps are located in polymer phase the structure of which is formed under the influence of piezoparticles surface, that is, in boundary layer of polymer near the particles of the piezoceramic.

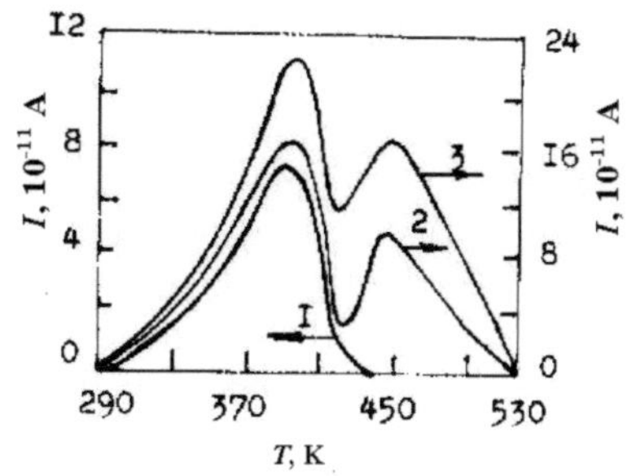

Fig. 4. TSD spectra of composites based on HDPE and PZT-19. 1-HDPE; 2-HDPE+30%vol.PZT-19; 3-HDPE+50%vol.PZT-19. E_p=5.5 MV/m; T_p=373K. Polarization time - 0.5 hour.

A role of the stabilized charge on the polymer-piezoceramic interface in formation of the piezoelectric properties of the composites is verified by following experimental facts:

1. The dependencies of the total charge Q, charges Q_1 and Q_2 corresponding to the first and second TSD maximum vs E_p, and T_p (fig.5) for PP+50%PZT-19 and content of the piezofiller (fig.6) have the same form as the dependencies d_{33} vs these parameters (fig.1,2,3).

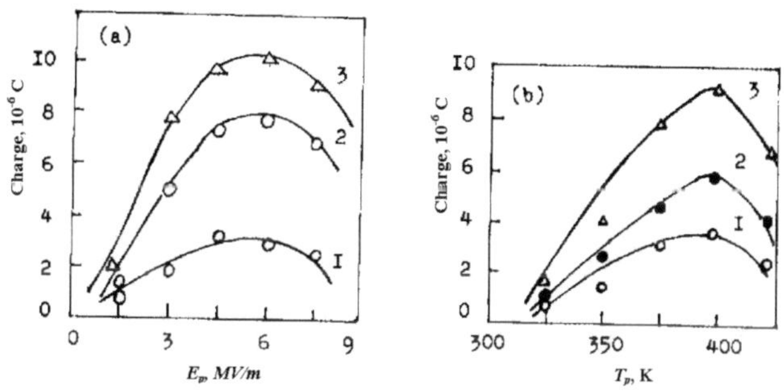

Fig. 5. The dependencies of the total charge Q, charges Q_1 and Q_2 corresponding to the first and second TSD maximum vs E_p and T_p for PP+50%PZT-19. a) 1-Q=$f(E_p)$; 2- Q_1=$f(E_p)$; 3- Q_2=$f(E_p)$. b) 1-Q=$f(T_p)$; 2- Q_1=$f(T_p)$; 3- Q_2=$f(T_p)$.

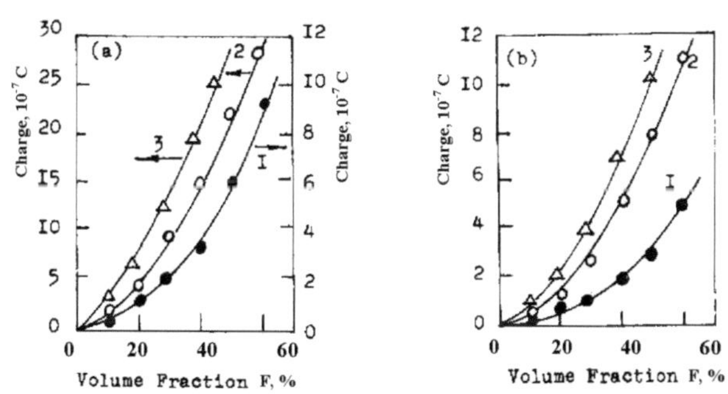

Fig. 6. a) Total charge Q and b) charge corresponding to the second maximum Q_2 of TSD spectra as function of piezoceramic volume fraction. 1) HDPE/PZT-19; 2) PP/PZT-19; 3) LDPE/PZT-19.

2. For determination of this bond we use dependences of d_{33}/Q, d_{33}/Q_2 , and d_{33} on the Q и Q_2 (fig. 7). It is shown, for all contents of piezophase d_{33}/Q_2 is approximately equal. But with increasing F, d_{33}/Q and d_{33}/Q_1 are changed by complicated manner (fig. 7a). Received for various contents of piezophase the dependence of Q_2 on d_{33} is direct proportional, but the dependence of Q on d_{33} has not direct behavior (fig. 7b). Analogous results are received for another composites (fig. 8 a,b): PP-50% vol.PZT-19. From this we can conclude, that a value of piezocoefficient of investigated composites is well correlated with value of charge corresponded to second maximum of TSD spectrum. Such dependencies is not observed for the charges Q, Q_1 and d_{33} (fig. 7,8).

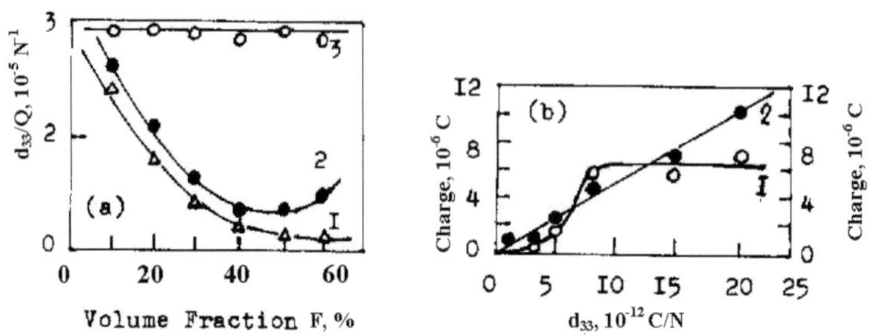

Fig. 7. a) Dependences of d_{33}/Q (1), d_{33}/Q_1 (2) and d_{33}/Q_2 (3) of HDPE-PZT-19 on F. Q-full charge, Q_1 - charge of the first maximum of TSD spectrum, Q_2 - charge of the second maximum of TSD spectrum. b) Dependences Q (1) and Q_2 (2) on d_{33} of HDPE-PZT-19 composite.

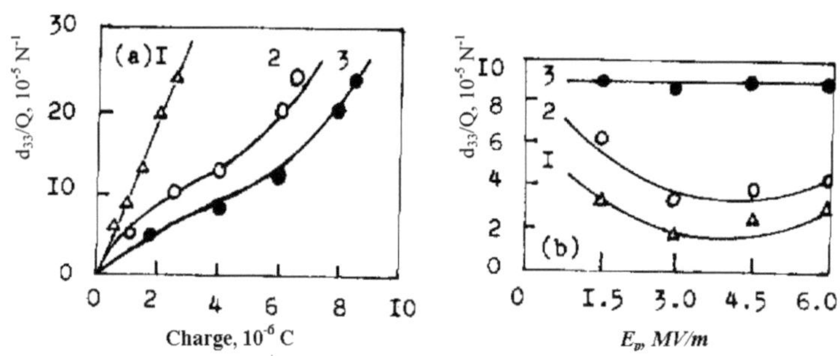

Fig. 8. a) Dependence d_{33} on Q_2 (1), Q_1 (2) and Q (3) for various E_p and T_p=273K. b) Dependence d_{33}/Q (1), d_{33}/Q_1 (2) and d_{33}/Q_2 (3) on E_p of PP-50% vol.PZT-19 composite.

3. The values of $Q_2/Q`_2$ and $d_{33}/d`_{33}$ practically is constant (fig.9), but $Q/Q`$ and $Q_1/Q`_1$ is entirely relaxed (fig.9 and 10) with the heating till temperature of TSD spectrum minimum. Here, $Q`$, $Q`_1$, $Q`_2$ and $d`_{33}$ are value of pointed parameters after first heat of polarized composite. It is shown from fig.10, after the first heat, the first maximum of TSD spectrum disappears and second maximum decreases slightly. After first heat, also a piezocoefficient of composite decreases slightly too.
4. The simultaneous relaxation Q_2 and d_{33} is observed at temperatures beyond second TSD maximum temperature (fig.11).

So that above listed experimental results indicate that the tight binding between the accumulated charges on the phase interface in polymer matrix and piezophase polarization is formed. Such interconnection may be explained on the basis of a simplified model shoved in fig. 12.

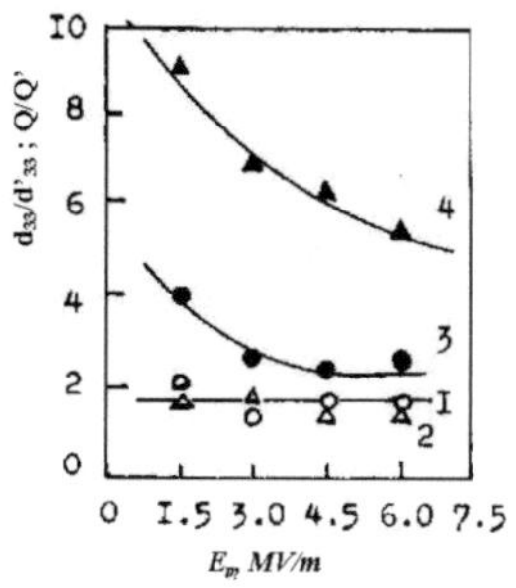

Fig. 9. The dependences of values of $d_{33}/d`_{33}$ (1), $Q_2/Q`_2$ (2), Q/ Q` (3) and $Q_1/Q`_1$ (4) vs E_p for PP-PZT-19 composite F-5-%; T_p=393K.

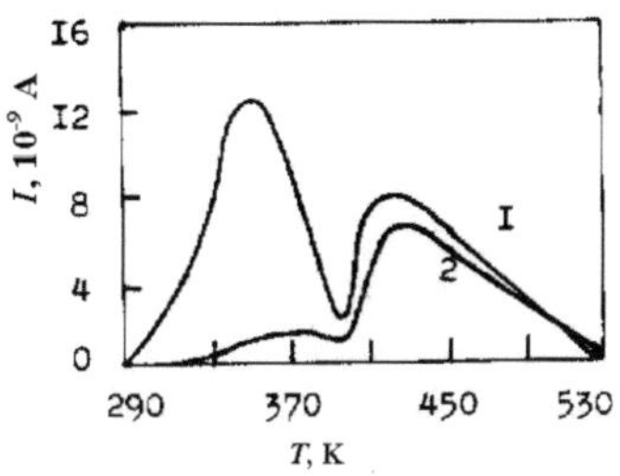

Fig. 10. TSD curves of LDPE-30%vol. PZT-19. 1-at first heating till 413K and cooling till room temperature; 2-at second heating.

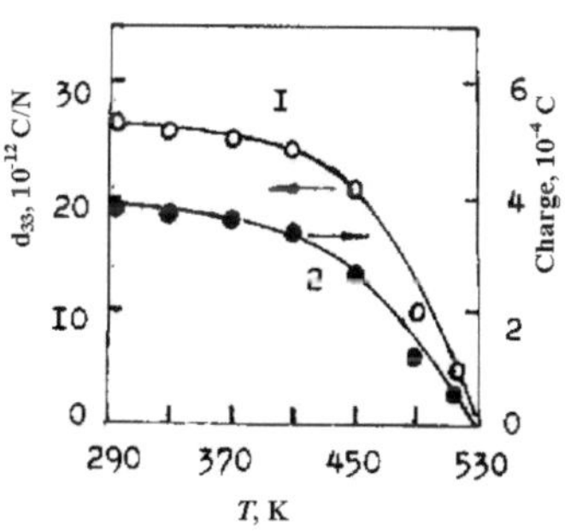

Fig. 11. The dependences of Q_2 and d_{33} vs temperature LDPE-30% vol.PZT-19 composite.

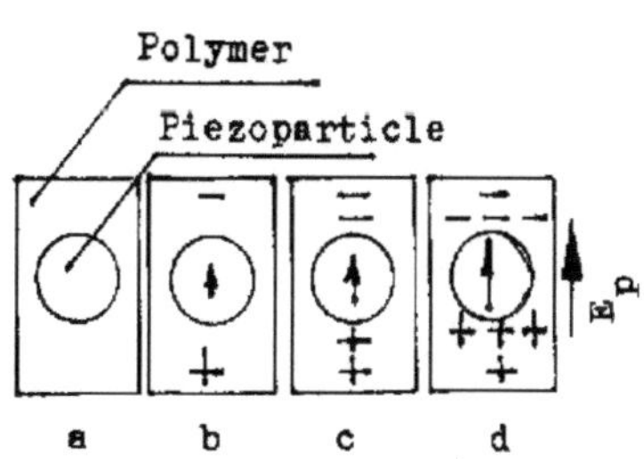

Fig. 12. A simplified model of the accumulated charges on the phase interface in polymer matrix

According to this model the polarization process goes on by following way. At $E_p = 0$ a total polarization of piezoparticles is equal to zero (fig.12a). During electrothermopolarization in the initial moment the charges are injected from electrode to the composite and at the same time insignificant polarization of piezoparticles takes place (fig.12b). This insignificant polarization compensates a fraction of per particle external field in accordance to the dielectric constants and phase conductivities for choosed T_p and E_p. Then, the injected changes drift towards the piezoparticles and are localized on the traps near the phase interface. This leads to the amplification of the local field on a piezoparticles and increase is polarization (fig.12,c).

A formation of quasi-neutral complexes injected charges-oriented domain favors the further injection and drift of the charges towards the piezoparticles and their appropriate to their polarization (fig.12,d). This explains an effectivity of composite polarization. It would seem that process must be continued till a total polarization of the particles. However, in practice a number of facts may prevent it, for instance a deficiency of amount of the deep traps on the interface, a screening of the piezoparticles by accumulated charge and etc.

The efficiency of the piezocomposites polarization becomes dependent on the piezoceramic structure. In fig.13 for example the dependencies of d_{33} and Q on a volume fraction of piezophase of various structures are shown. It is seen, that the magnitude of stabilized charge and the piezocoefficient value are greater in the case of using of the piezoceramic of rhombohedral (R_h) (PCR-3M) structure as a piezofiller compared with the tetragonal (T) structure (PCR-7M) piezoceramic. At the same time the base ceramic have reverse relation of the piecoefficient.

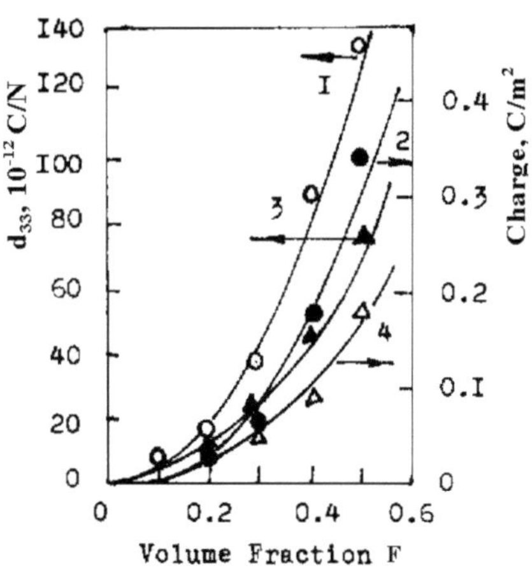

Fig. 13. d_{33} (1,3) and Q_2 (2.4) of composites as a function of piezoceramic volume fraction of different structure. Composites are obtained on the basis of polypropylene. (1,2) for PP-PCR-3M composite; (3.4) for PP-PCR-7M composite. E_p=3 MV/m, T_p=410K, t_p=0.5 hour.

In [Fesenko E.G. et al., 1983] a high value of the piezocoefficient of tetragonal piezoceramics is associated with a large amount of a dielectric constant ε and the known relation is used:

$$d_{ij} = 2Q_s\varepsilon_0\varepsilon P_r \tag{4}$$

where, Q_s is the electrostriction coefficient, P_r is an amount of the reorientational polarization of domain different from 180^0.

Dielectric constant in composites is significantly smaller than in the piezoceramics because of a low value of ε at a polymer phase (table 2). Therefore P_r plays a main role and it is greater in the rhombohedral ceramics. On the other hand in the rhombohedral ceramics a mobility of the domain walls is greater, that in turn leads to increasing of a charge moving to the interphase boundaries. That is a cause of a great magnitude Q_2.

Thus, the above model interactively explains both the polarization curves and the dependencies d_{33} = f(E_p,T_p), and dependence of the piezocoefficient on the piezofiller structure. An additional verification of a workability of the proposed model one may obtain from calculation of d_{ij} of the composites using the expression (4). Taking into account a major role of the interface charges in piezoparticles polarization it should be used Q_2 instead of P_r in formula (4). A magnitude of Q_s for composites is taken to be equal to its value for a base piezoceramic. Such approach is justified by the fact, that the polarization is concentrated in the piezophase. The calculated quantities d_{33} are illustrated in table 2. It is seen, that a good agreement takes place between the calculated and experimental values of d_{33}. Thus, the experimental data and suggested model demonstrate a significant role of one kind of the interface interaction in the formation of the composites properties. This interaction leads to seemed anomalous dependencies of composite properties on a composition.

Composition	ε composition	Q_r, C/m^2	d_{33}, 10^{-12} C/N (experiment)	d_{33}, 10^{-12} C/N (calculation)
PVDF+7M; F=50%	95	0.22	65	69.9
PVDF+7M; F=50%	70	0.40	80	92.4
PVDF+7M; F=50%	62	0.56	87	118.4
PVDF+7M; F=50%	102	0.25	135	133.0
PP+7M; F=50%	70	0.02	45	4.6
PP+PZT-2; F=50%	75	0.03	55	6.8
PP+PCR-3M; F=50%	80	0.40	120	100.0
PVDF+PZT-2; F=50%	86	0.35	120	111.0
PVDF+PCR-3M; F=50%	110	0.45	160	161.5
LDPE+PCR-3M; F=50%	100	0.42	130	139.9
HDPE+PCR-3M; F=50%	70	0.14	76	30.0

Table 2. The calculated quantities d_{33} for various composites.

The amount of the composite piezocoefficients for the certain compositions becomes greater than d_{ij} of both components. If the interface interactions would not play a significant role in the formation of piezoelectric effect, then such maximum must be absent.

In the case, when the interaction between the charges localized in a polymer phase and particle polarization as taking into account, this maximum points to the best polarization of piezophase in a composite compared to a base ceramic. In general case the appearance of maximum on diagrams composition-composite properties shows the significant role of the interaction on the interface. In particular, the above pictured interaction on the interface, that is lead to the formation of the system of oriented domain-localized boundary charge, explains the piezoelectric and electret effects in composites.

In fig. 14 pyroelectric current curves for a composite PVDF-PCR-3M is shown. It should be noted, that the pyroelectric effect in composites becomes greater than in base components.

This is in agreement with the above pictured model, which explains an occurrence of the large reorientation polarization responsible for piezoelectric effect [Lines M.E. & Glass A.M., 1977; Kerimov M.K., 2011]. With the existence of coupling charge-domain system, the occurrence of the stable electret state in the polymer-piezoelectric composites for appropriate proportioning their composition is also explained.

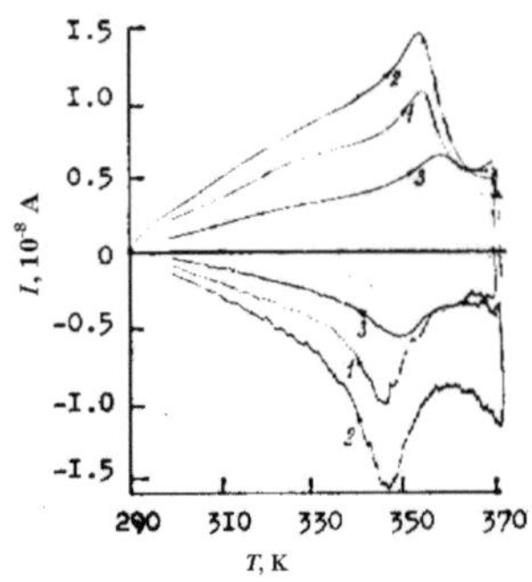

Fig. 14. Curves of pyroelectric current of PVDF-PCR-3M composite: F=40%, t_p=0.5 hour, E_p=3 MV/m, (1), T_p=353K, (2) T_p=373K, (3) T_p=353K.

A stabilization of charges on the phase interface and their relaxation significantly depends on a structure of polymer on the phase interface. In its turn, this structure is defined by interaction between the polymer chains and piezoparticles surface. In particular, in fig.15 the temperature dependencies of the piezocoefficient and thermal expansion coefficient (α) of composites based on the single matrix and piezoceramic of rhombohedral and tetragonal structures are shown.

It is seen that a temperature decline of the piezocoefficient and a noticeable increase of the thermal expansion coefficient is higher in the case of a piezofiller of the rhombohedral structure.

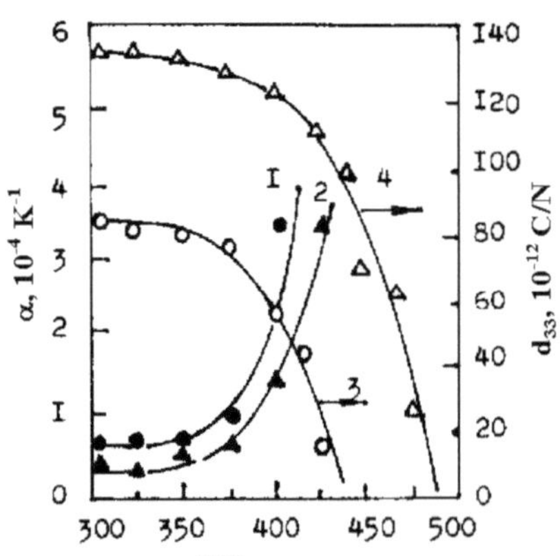

Fig. 15. The temperature dependencies of the piezocoefficient and thermal expansion coefficient (α) of composites based on the single matrix and piezoceramic of rhombohedral and tetragonal structures.

It should be noted, that the piezocoefficient decline is associated with the boundary charges relaxation which, in its turn, is defined by mobility of the polymer chains. It shows that the

interaction of the polymer chains with the piezoparticles surface of rhombohedral structure is stronger. By the same way the dependencies of α vs T are explained also.
A structure and property of near-surface layer affect on a posistor effect in a composite. It is known, that the posistor effect in the piezoceramic is associated with a transition from a ferroelectric phase to paraelectric one [Burfoot J.C. &Taylor G.W., 1979; Lines M.E. &Glass A.M., 1977; Rez I.S. & Poplavko Yu.M., 1989]. In polymers this effect practically is absent. In the composites a significant posistor effect is observed (fig.16). It can not be associated in composites with phase transition in piezoparticles, as far as a temperature of sharp increase of composite resistance is remarkable lower than the piezophase Curie temperature.
The posistor effect in composites is associated with a sharp changing of the potential barrier on the phase interface in melting of ordered phase of a polymer matrix. Such effect is specific only for the composites and may be a basis of a new structure-sensitive method of their investigation. A decrease of mobility of the polymer change both the occurrence temperature and half-width of the posistor peak (fig.16).
It should be noted, that numerous types of interactions are on an interface. The contributions of different interactions naturally will be depending on considered properties. One of principal problems of a compositional material physics must be revelation of major types of interactions on the phase interface and their correct accounting in formation of the corresponding effects. Physics of composite materials must be based on consistent account of contributions of separate type of interactions between phases in the formation of specifically properties of composites.

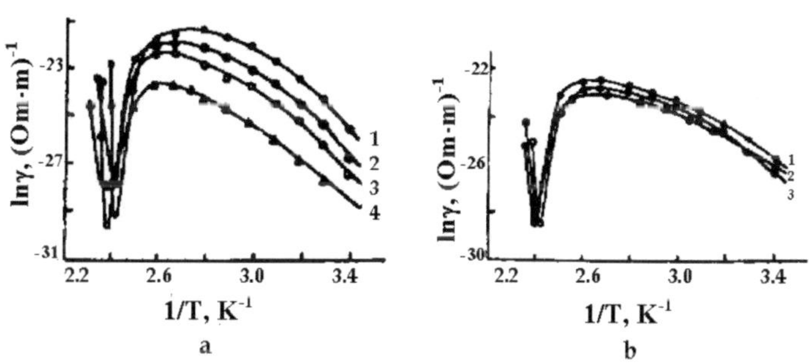

Fig. 16. Observed significant posistor effect in the composites.

Charge states and various interface interactions can be varied by regulation of structure and electrical negativity of components (polymer and ferroelectrical phase), by activation of surface of polymer particles, by doping 3-rd component which increases an adhesion of polymer to ferrophase. Electronic states of polymer-ferropiezoceramic boundary, concentration and activation energy of local levels in quasi-forbidden band of polymer phase are of importance.
We propose a new technological method for varying the structural heterogeneity of polymers and their composites [Kerimov M.K., 2005]. The method is based on the air-discharge plasma-assisted crystallization (electric discharge-enhanced thermal crystallization), which begins at the melting point. We also studied the effect of the electric discharge-enhanced thermal crystallization on the pyroelectric properties of polymer composites.
The electric discharge-enhanced thermal crystallization was carried out as follows. After briefly compressing the composite at the melting point, the upper die was removed from the

surface of the specimen and a special discharge cell was attached to the die. The discharge cell enabled one to activate electric discharges of different intensity in air. An electric discharge was generated in a 4-mm-thick air gap between the teflon plate of the cell and a composite specimen: therefore, a partial discharge was induced [Bagirov M.A., 1972; Kerimov M.K., 2005; Kurbanov M.A., 2009]. A 15-kV sinusoidal electric voltage (50 Hz) was applied to the cell. Using the partial discharge for the electric discharge-enhanced thermal crystallization made it possible to treat the composite surface uniformly without causing noticeable mechanical damage.

Beginning from the melting point of the polymer phase of the composite, the temperature of the electric charge-enhanced plasma-assisted crystallization was changed stepwise (by 10 K over 0.5 h) at the first stage of cooling and then at a rate of 0.25-2 K/min down to the polarization temperature (~373 K) or room temperature. The changes in the structure were studied using the infrared spectra of the polymer phase of a composite. The following polymers were used as a polymer phase: high-density polyethylene (HDPE), polypropylene (PP), and polyvinylidene fluoride (PVDF). As the pyroelectric phase, we used the following pyroelectric ceramics with rhombohedral or tetragonal structure and different Curie temperatures T_c and pyroelectric coefficients γ: PCR-7M (T_c = 448 K, γ= 5.6 x 10^{-4} C/m^{2}K). PCR-3M (T_c = 533 K, γ= 5 x 10^{-4} C/m^{2}K), and PCR-57 (T_c = 463 K, γ= 6 x 10^{-4} C/m^{2}K).

Figure 17 shows the simultaneously recorded spatial (electron-optic converted image) and time (oscillogram) diagrams of the electric-discharge evolution in the air gap between the teflon plate and a composite. As is seen from Fig. 17, the electric discharge in the air gap between the composite and the dielectric is accompanied by separate series of discrete microdischarges. Each voltage pulse on the oscillogram (oscillogram *4* in Fig. 17) corresponds to one series of microdischarges. Each series, in turn, consists of a number of microdischarges generated in different regions of the sample surface. This character of the evolution of partial discharges (microdischarges) enables one to treat the composite surface uniformly and thus carry out effective crystallization of a composite under conditions of a discharge plasma in air with electronegative components.

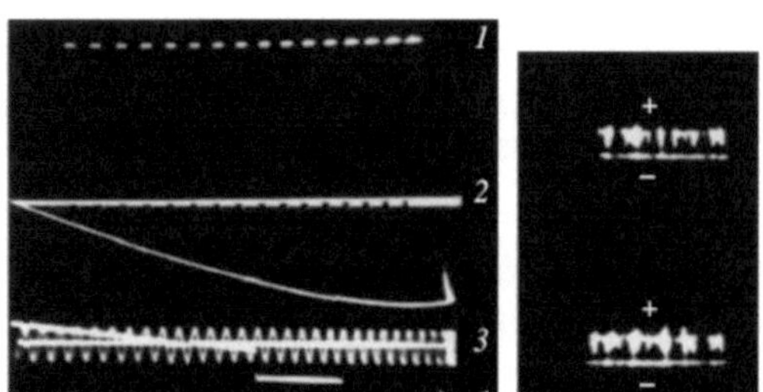

Fig. 17. Simultaneously recorded (a) electrical and (b) optical patterns of the discharge evolution in the air gap between dielectrics. In the oscillogram: (1) gate pulses of the electron-optic converter, (2) voltage change in the test cell, (3) calibrating voltage (T = 200 mks), and (4) voltage pulses corresponding to separate series of microdischarges.

Figure 18 shows the changes in the optical density of the bands corresponding to C-O-C, C=0, and OH that appeared in the infrared spectrum of the polymer matrix of the HDPE-PCR-57 composite during its crystallization under the simultaneous action of temperature, electric discharge, and discharge radiation. The results obtained by us show that the structure of the polymer matrix undergoes substantial changes in the course of electric-discharge crystallization. Indeed, the appearance of C-O-C, C=0, and OH groups in the infrared spectrum indicates changes in the chemical structure of the macromolecules, an enhancement of intermolecular interactions due to the polarity of these groups, and the formation of C-O-C bridges between macromolecules. Clearly, these chemical changes are accompanied by changes in the crystallization conditions and. therefore, in the physical structure of the polymer phase of the composite.

Figure 19 shows the temperature dependence of the pyroelectric current for several successive heating and cooling cycles of the PP-50 vol.% PCR-57 composite. Curve *1* corresponds to the first heating-cooling cycle and represents the nonreproducible part of the pyro-electric current. Curves *2-4* correspond to the reproducible part of the pyroelectric current. The heating (cooling) rate of the given pyroelectric element is 6 K/min. The pyroelectric current reverses direction under switching from heating to cooling, but its value remains the same; in other words, the heating-cooling cycle is characterized by a symmetric temperature-dependence plot of the reproducible part of the pyroelectric current.

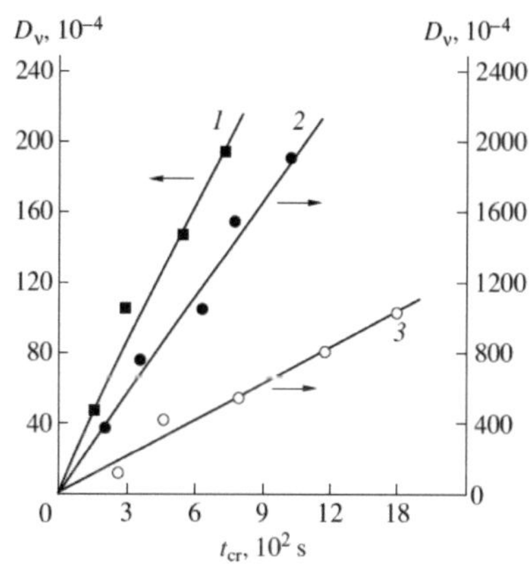

Fig. 18. Variations in the optical density of the bands that arise in the infrared spectrum of the polymer phase of the HDPE + 50 vol % PCR-57 composite during electric discharge-enhanced thermal crystallization. The cooling rate is 2 K/min, U_p=11.8 kV, and ΔW= 1.4x10^{-6} J; *(1)* ν=3380 cm^{-1}, (2) ν = 1735 cm^{-1} (C=0), and (3) ν = 1280 cm^{-1}.

Figure 20 shows the dependence of the pyroelectric coefficient of the PVDF-PCR-57 composite on the volume fraction *F* of the pyroelectric-ceramic phase. It can be seen that, under the same crystallization conditions, γ increases faster than linearly with *F*. However, the pyroelectric coefficient of electrothermally crystallized composites is much higher than γ of thermally crystallized pyroelectric elements, all other polarization conditions (E_p, T_p) and the temperature of measurements *T* being equal.

We also studied the pyroelectric characteristics of a PVDF-based composite filled with particles of a PCR-3M pyroelectric ceramic, which has rhombohedral structure and a relatively high Curie temperature. The pyroelectric ceramics PCR-3M, as well as PCR-57, is characterized by a high pyroelectric coefficient, which, in combination with the low dielectric constant, ensures the high quality factor of this pyroelectric detector, M_r = γ/e = 1.4 x 10^{-6} C/m^2K [Fesenko E. G.., 1983]. The fact that its pyroelectric coefficient is high at room temperature (Fig. 21) makes this composite promising for pyroelectric elements.

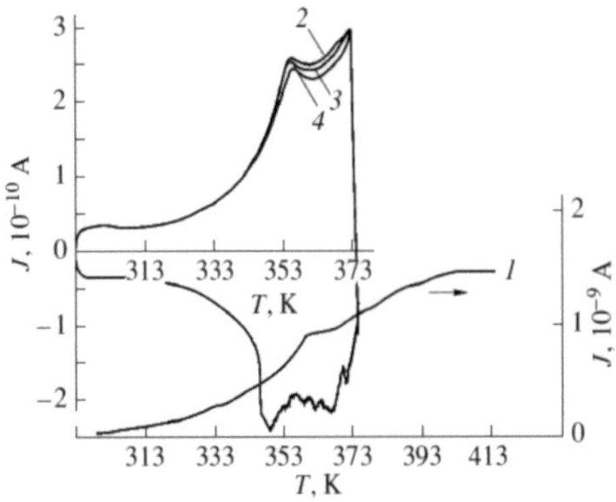

Fig. 19. Dependence of the pyroelectric current on the measuring temperature for the PP - 50 vol % PCR-57 composite. E_p = 8 MV/m, T_p = 373 K, and t_p = 1 h; *(1)* the nonreproducible part of pyroelectric current and *(2-4)* the reproducible pyroelectric current.

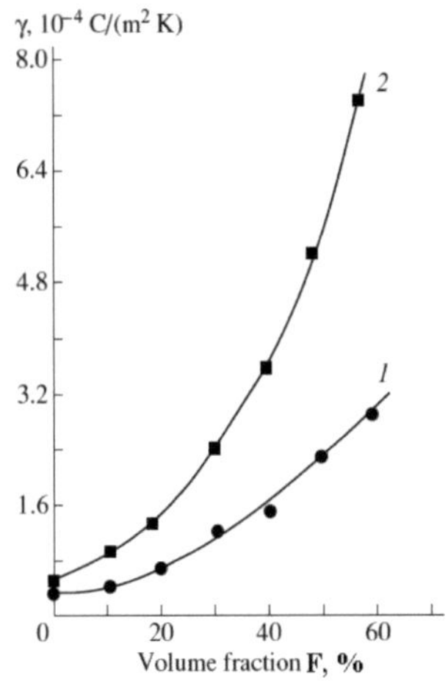

Fig. 20. Dependence of the pyroelectric coefficient γ on the volume fraction of the pyroelectric phase *F* in the PVDF -50 vol. % PCR-57 composite. T_p = 373 K, E_p = 4 MV/m. t_p = 1 h, and *T*= 358 K: (1) after thermal crystallization and (2) after electric discharge-enhanced thermal crystallization.

It can be seen in Fig.21 that the electric discharge-enhanced thermal crystallization of PVDF-PCR-3M composites results in a significant increase in their pyroelectric coefficients throughout the entire range of measurement temperatures *T*, including room temperature. Furthermore, the electric discharge-enhanced thermal crystallization significantly shifts the temperature of the onset of an abrupt increase in the value of the pyroelectric coefficient and. therefore, of the pyroelectric current. This experimental result shows that the electric charge-enhanced thermal crystallization increases the interphase interactions, which restrict the mobility of macromolecules of the polymer phase, and causes relaxation of charges on the polymer-pyroelectric-ceramic interface. The measured pyroelectric properties matrix composites that were produced using electric discharge-enhanced plasma-assisted crystallization are shown in the Table 3.

Composites	F, %vol.	T, K	Polarization conditions		γ, 10^{-4} C/m²K	
			E_p, MV/m	T_p, K	after thermal crystallization	after electric discharge-enhanced thermal crystallization
PVDF+PCR-7M	50	353	6	353	1.1	4.8
HDPE+PCR-7M	50	353	6	353	0.55	2.7
HDPE+PCR+57	50	353	6	353	0.82	4.2

Table 3. Pyi'oelectiic coefficients of different composites after thermal crystallization and after electric discharge-enhanced thermal crystallization

It is of great interest to consider the kinetics of charge (electron) accumulation at the various stages of crystallization under discharge conditions. Figure 22a shows the space charge generated in an HDPE-PCR-57 composite as a function of the duration of electric discharge-enhanced thermal crystallization. The charge value was found from the depolarization current spectra immediately after electric discharge-enhanced thermal crystallization. The low values of the charge arising at the initial stage of crystallization can be due to electron traps being destroyed under the action of temperature, discharge, and voltage pulses generated during each series of microdischarges and also due to electrons being captured by free radicals of the polymer phase [Vannikov A.V. & Grishina A.D., 1984]. It should be noted that, in ac electric fields, a fraction of the injected charges reappears at the electrode when the polarity is changed. The charges localized in deeper traps can enhance the electric-field intensity in the polymer composite at each air-gap breakdown during the electric discharge-enhanced thermal crystallization. The electron bombardment and the strong local field also initiate the formation of radicals [Bagirov M.A., 1972].
The results obtained allow us to suggest the following mechanism for changing the pyroelectric properties of composites during electric discharge-enhanced thermal crystallization. We believe that, in the course of electrical loading of the composite-gaseous medium-composite system, partial discharges occur and conditions are created for rather complex physicochemical processes in the bulk of the polymer phase caused by electron-ion bombardment, radiation, local heating of the contact between the microdischarge channel and the polymer (Fig. 17b), and active gas-discharge products (O_3, O, OH, NO). Together, these components of a partial discharge predetermine the specific features of the formation, transformation, and accumulation of free radicals, which account for all structural and energy changes in the polymer phase of the composite. It should be noted that the structural rearrangements that accompany the transition of the polymer from the fluid state to a highly elastic and finally to a glassy state in the course of the crystallization result in an abrupt change in the amplitude of the rotational segmental motion. Furthermore, under the action of temperature and electric discharge-induced plasma, these rearrangements of the structure and the changes in the amplitude of the rotational segmental motion will be significantly different due to electronic, ionic, destructive, and oxidizing processes. For this reason, the conditions of injection, generation, and transfer of charge carriers and their interactions with the macromolecules are changed in the course of electric discharge-enhanced crystallization. Under the action of electron-ion bombardment and discharge radiation, free radicals will form during crystallization, especially in the temperature ranges corresponding to the fluid and highly elastic states. The most probable mechanism consistent with the partial discharge conditions is the mechanism based on discharge radiation ionization of macromolecules

followed by decomposition of the molecular cation *(M~)* into a free radical *(R[)* and a cation fragment *(R'₂)*: $M \rightarrow M^{+} + e$; $M^{+} \rightarrow R^{*}_{1} + R^{+}_{2}$

In ionized macromolecules, the chemical bonds are excited. Therefore, they break relatively easily, and oxidizing processes begin, resulting in the appearance of new local energy levels in the quasigap of the polymer and. therefore, in the formation of additional carrier traps, which is experimentally confirmed by the results shown in Fig. 22. The effective accumulation of earners at the polymer-pyroelectric-ceramic interfaces and the strong orientational domain polarization of the pyroelectric phase (curves *3, 4* in Fig. 22b) significantly increase the pyroelectric coefficient (curves *L,2* in Fig. 22b) of the composites produced through electric discharge-enhanced plasma-assisted crystallization.

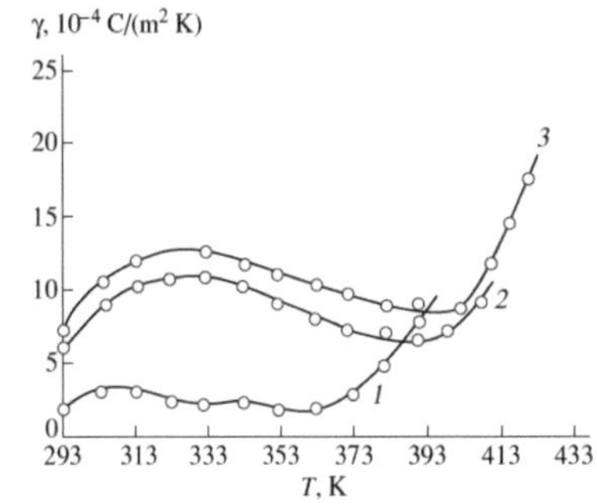

Fig. 21. Dependence of the pyroelectric coefficient γ on the measuring temperature T for the PVDF + 50 vol % PCR-3M composite. $T_{„}$ = 373 K and t_p = 1 h; (1) after thermal crystallization (E_p = 1.5 MV/m) and (2, 3) after electric discharge-enhanced thermal crystallization at E_p = 1.5 and 3 MV/m, respectively.

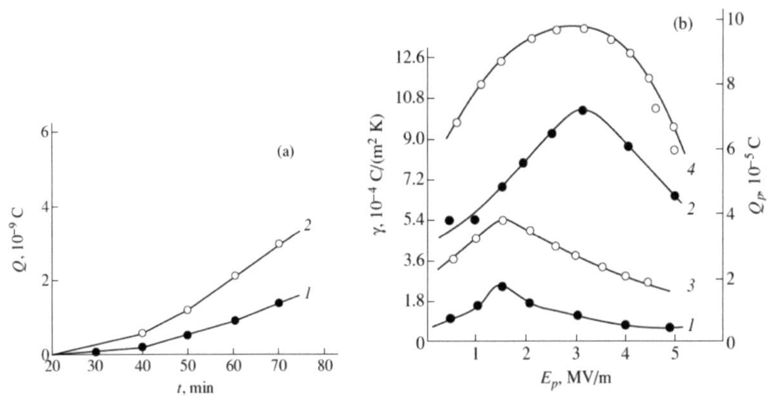

Fig. 22. (a) Dependence of the stabilized charge Q on the cooling duration in the course of electric discharge-enhanced thermal crystallization of the HDPE - 40 vol. % PCR-57 composite. The cooling rate is 1.5 K/min; the treatment voltage U is *(1)* 16 and (2) I8 kV. (b) Dependence of (1,2) the pyroelectric coefficient γ and (3, 4) the charge Q_p accumulated during polarization on the electric-field intensity E_p of the PVDF - PCR-3M composite. T_p = 373 K, t_p =1 h, and T= 373 K; (1, 3) after thermal crystallization and (2, 4) after electric discharge-enhanced thermal crystallization.

The results obtained by us show that the method of electric discharge-enhanced thermal crystallization should be promising for the development of highly efficient active composites. However, these interesting results currently cannot be exhaustively explained, because these studies are at the initial stage. It may be said that the pyroelectric properties of polymer-pyroelectric-ceramic composites produced through electric discharge-enhanced plasma-assisted crystallization are significantly improved and that the concentration of charge carriers accumulated in them during electrothermal polarization increases.

The results obtained enable us to make the important conclusion that, in order to design highly efficient pyroelectric composites based oil polymers and pyro-electric ceramics, it is necessary to increase the density of localized states in the quasigap of the polymer phase and at the interphase boundary, which can be achieved using electric discharge-enhanced thermal crystallization of the polymer phase of composites.

4. New class of active materials based on the hybrid of matrix nano- and microferropiezoelectric composites

At present, the matrix system, consisting of polar and nonpolar polymers, dispersed piezoceramic particles, is the most effective electret, pyro-and piezoelectric composites [Vannikov A.V. & Grishina A.D., 1984; Topolov V.Yu. et al. 2006; Kerimov M.K. et al., 2005, 2007, 2009, 2011; Kurbanov M.A. et al., 2009, 2011]. For various purposes the power-intensive and low-power acoustoelectric and electroacoustic transducers are developed based on these composites. Piezoelectric transducers have some advantages and some disadvantages. For example, polymer piezoelectric composites in the mode of receiving acoustic waves (mechanical perturbations) have high piezosensitivity, defined as $g_{ij}=d_{ij}/\varepsilon\varepsilon_0$, where d_{ij} is a piezoelectric modulus, ε is a dielectric constant of pyezocomposit element, ε_0 is a electrical constant. However, piezoelectric polymer composites in the mode of radiation elastic waves are less effective than the piezoceramic elements because of the relatively low values of electromechanical coupling coefficient K_{ij}, specific acoustic power $(d_{ij}Y)^2$ (Y is Young's modulus), mechanical Q_M and piezoelectric quality factor $K_{ij}^2 \cdot Q_M$, electromechanical efficiency ($K_{ij}^2 / tg\delta$) and high mechanical compliance S_{11}^E [Rez I.S. & Poplavko Yu.M., 1989; Fesenko E. G. et al. 1983]. In addition to these properties, the frequency sensitivity range of amplitude-frequency characteristics of pyezocomposit converters is not very wide, and the value of the output signal decreases appreciably when $f \geq 7$ kHz. The application of pyezocomposit elements as powerful sources of radiation elastic waves is very important to create a receiving-transmitting acoustic antenna for detection of submarine objects. However, the low values of electromechanical parameters $(d_{ij}^2 Y_{11}^2, K_{ij}^2 / tg\delta, K_{ij}^2 \cdot Q_M, K_{ij}, Q_M)$ of pyezocomposit elements reduce the efficiency of acoustic antennas in the radiation mode.

Physical-mechanical and, consequently, electro-mechanical properties of matrix composites depend on the mobility of the macromolecules of the polymer phase and the interfacial interaction. Therefore, by varying these factors, we can specifically regulate the piezoelectric and electro-mechanical characteristics of composites, making them effective in the manufacture of energy-intensive sources of acoustic waves. Nanotechnology advances help solve this problem by creating a new class of piezoelectric materials by the combination of polymer composites with nano dimensional dielectric and micro dimensional piezoceramic particles.

We propose a technology for creating piezoelectric composites, combining nano- and micro composites integrated with polymer matrix (a hybrid system). The principle of the development of hybrid nano- and micro composites based on the difference in the temperatures of dissolution and melting regions of the polymer composite phase, whose structure is formed under the influence and not under the influence of the surface of micropiezoelectric particles. Progress in nanotechnology allows us to create power-intensive and low-power piezoelectric transducers. At presents, due to lack of nanoscale piezoceramic particles of different structures, determining the piezoelectric properties of the hybrid matrix of nano- and piezocomposites the main component are microscale piezoelectric particles. From the literature it is known that the piezoceramic material of PZT family have a stable structure and are the most effective piezoelectric material to develop on their basis both power-intensive and low-power converters, in particular, receiving and transmitting antennas, and acoustic sensors [Marikhin V.A. & Myasnikova L.R., 1977; Sessler G.M., 1980; Kerimov M.K. et al. 2009].

We proposed polymeric nano - and microhybrid materials combine the phases responsible for the creation of low-power and power-intensive transducers. Nano particles of barium titanate ($BaTiO_3$) and quartz of various structures have low stability and the Curie temperature. Therefore, they are ineffective materials to create piezoelectric composites. We offer a hybrid piezoelectric composites using nano-materials, which combine the positive qualities of PZT piezoceramic materials of different structures (tetragonal and rhombohedral) and properties (high Curie temperature, piezoelectric modulus, mechanical quality factor, Young's modulus and high coefficient of electromechanical coupling and electromechanical characteristics).

Tumerous experimental results show that depending on the size of particles of piezo-phase the piezoelectric coefficient and Young's modulus of the composites vary not symbatically. Piezo-coefficient of the composite increases monotonically, and Young's modulus decreases with increasing size of piezoceramic particles. This is the reason for the low value of the specific acoustic power $(d_{ij}Y)^2$, electromechanical coupling coefficient K_{ij}, piezoelectric $K_{ij}^2 \cdot Q_M$ and mechanical Q_M quality factor and the high value of mechanical compliance.

Experiments on the influence of dispersion of piezoceramic fillers on the physical, mechanical and piezoelectric properties of piezocomposites showed that for high piezo-module (d_{ij}) the average particle size of piezoceramic must be within 100-250 microns, and for high Young's modulus and low-compliance - under 50 microns (Fig. 23).

Let us primarily consider the technology of the creation of active centers of clusterization of nanoparticles in the polymer phase of the composites by the performance of their crystallization under the effect of the plasma of an electrical discharge [Kerimov M.K. et al. 2011]. The crystallization of the composites under the plasma of an electric discharge involves heating of the material up to the melting point and keeping it at that temperature for 5—30 min (and, not relieving the effect of the discharge) with further cooling down to the temperature of the crystallization of the polymer phase with a rate from 0.5 to 4 K/min with the aim of the directed changing of the chemical (oxidation) and physical (supramolecular) structures of the polymer matrix. The melting point of the composite was determined by the temperature of the third peak of the spectrum of the thermodepolarization current (Fig. 24, curve *2)*. The duration of the crystallization t_{kr} was limited to the saturation of the optical density of the oxygen-containing groups (for example, OH and C=0), which evolved in the IR spectrum of the polymer—piezoceramic composite as a result of the action of the electrical discharge, and changed in the range from 5 to 30 minutes.

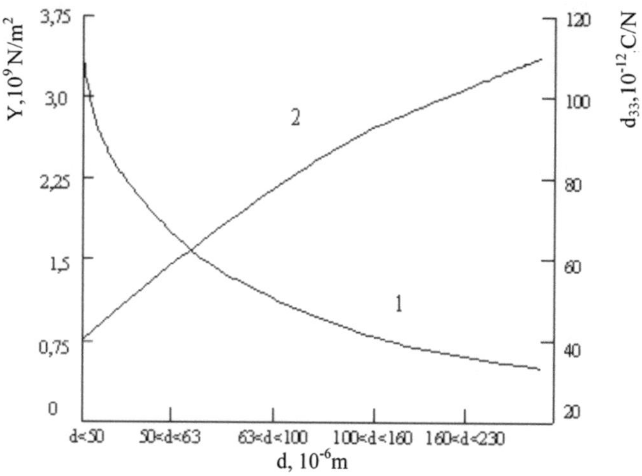

Fig. 23. Change of Young's modulus (curve 1) and piezo-electric coefficient (curve 2) of high-density polyethylene +50% vol. piezoceramic-PZT composite, depending on the size of the piezo-electric particles (piezo-phase).

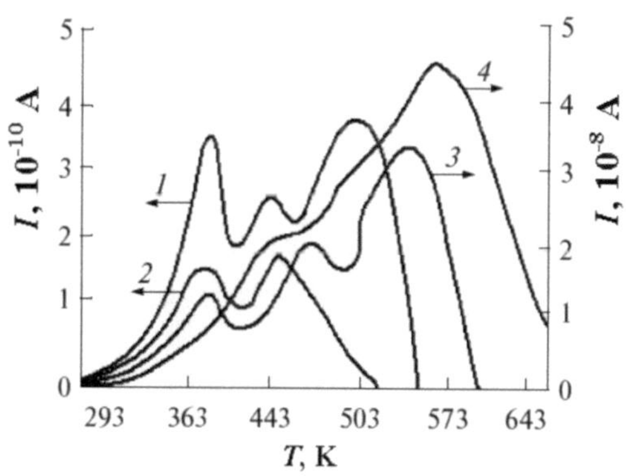

Fig. 24. Spectra of the thermostimulated depolarization current PVDF and the composites on its basis. (1) PVDF; (2) PVDF+50 vol % PZT 5A; (3) PVDF+1 vol % $BaTi0_3$ + 49.9 vol % PZT-5A, *W*= 26.3 x $I0^{-6}$ J; and *(4)* PVDF + 1 vol % $BaTi0_3$ + 49 vol % PZT-5A, *W*= 335 x $I0^{-6}$ J. The conditions of the polarization: the electric field of the polarization E_p= 3x10^6 V/m, the temperature is T_p = 413 K, and the time of the polarization is t_p= 0.5 h.

The experimental results demonstrate that, in the IR-spectrum of the polymer matrix (PVDF) of the composite, after its crystallization under the effect of the plasma of the electrical discharge, new highly polar oxygen-containing groups, C=0 and OH, are formed. In Fig. 25, the change in the optical density *D* of the IR spectrum is given for the mentioned groups in the polymer matrix that crystallized under the conditions of the plasma of an electrical discharge. It is seen that the change in the optical density *D* within the time of the crystallization up to 15 min has a linear character. The centers of oxidation can be the seeds for the clusterization of nanoparticles. The next stage of the immobilization is the dissolution of the polymer in toluene. After that, the $BaTiO_3$ nanoparticles (<50 nm) are introduced into the solution in the range from 0.5 to 3.0 vol %.

The structure of the discharges in the gas phase between the dielectric and the polymer solution with $BaTiO_3$ nanoparticles was recorded by an electron-optical transducer (Fig. 26). As is seen, the discharge forms at various sites of the dielectric surface and the polymer solution.

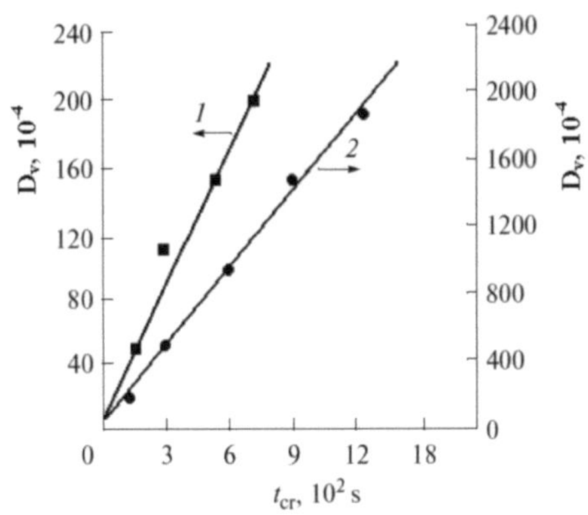

Fig. 25. Chance in the optical density of the bands evolved in the IR spectrum of the polymer phase of the composite PVDF+50 vol % PZT-5A during the electrothermodischarges of the crystallization. The cooling rate is 2K/min. U_p= 11.8 kV, *W*= 1.4 x IO-6J. (1) ν=3380(OH)cm-1, (2) ν=1735(C-O) cm-1 and (3) ν=1280(C-O-C)cm-1.

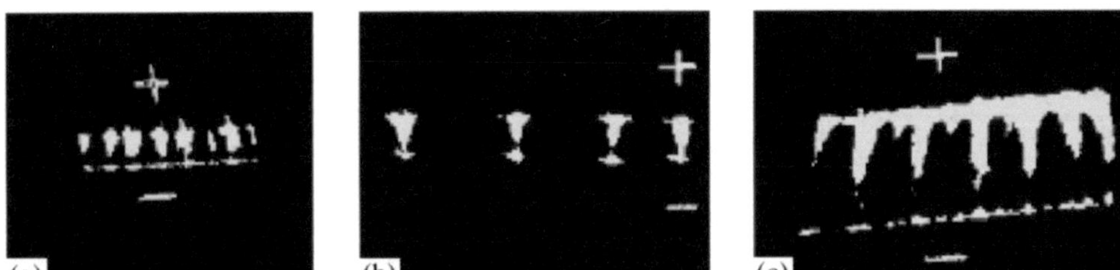

Fig. 26. Optical image of the development of microdischarges in the air medium between the dielectric (polytetrafluoroethylene with a thickness *d* = 2 mm) and the surface of the polymer solution. *(a) d* = 4 mm, *(b) d* = 6 mm, and *(c) d*= 10 mm; *U* = 35 x IO3 V.

Let us consider the technology of the preparation of the hybrid composed of nano- and micropiezoelectric composites. We used for the first time the effect of nanostructuring of polymers for the creation of a novel class of piezoelectric composites. A general model and a scheme of the individual steps of the technology suggested are given in Fig. 27. The technology involves the following main stages: the preparation of the piezo-composite composed of micropiezoceramic particles, a polymer interface layer, and a near-surface polymer region (Fig. 27, a); the dissolution of the near-surface polymer phase (area) and the preparation of the piezo-composite structure (substrate) composed of piezoceramic particles and the polymer interface layer having direct contact with the piezoparticles (Fig. 27, b); the nanostructuring of the dissolved polymer phase (Fig. 27, a, element *3)* by the dielectric $BaliO_3$ nano-particles; the chemical precipitation of the nanostructured dissolved polymer on the surface of the piezo-composite substrate; and the preparation of the hybrid (Fig. 27, c) from the polymer—nanoparticles $BaTiO_3$ and polymer—micropiezoceramic particle matrix systems.

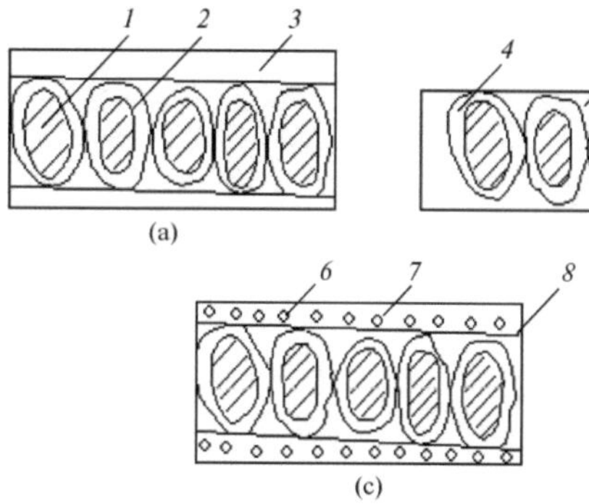

Fig. 27. Scheme of the structure formation of the hybrid of nano- and micropiezoelectric composites PVDF—nanoparticles and $BaTiO_3$—piezoceramics PZT-5A. (a) micropiezoelectric composite PVDF— PZT-5A, (b)the micropiezoelectric composite in the case of the dissolved subsurface polymer layer, and (c) the hybrid nano- and micropiezocomposite PVDF— $BaTiO_3$— PZT-5A: (1) piezoparticles, *(2)* interface polymer layer, *(3)* subsurface polymer range (phase), *(4)* interface layer in the case of the absence of the subsurface polymer range, (5) piezocomposite structure (substrate) (b. 5), *(6)* nanoparticles $BaTiO_3$, (7) nanostructure polymer subsurface layer, and *(8)* nano- and piezohybrid composite.

5. Conclusion

So, the wide investigations were carried out in range of physics and technology of active dielectrics - from synthesis of new polymers, ferropiezoelectrics, semiconductors, nano- and microcomposites till creation of hybrid composites, in particulars, hybrids on the basis of matrix nano- and micropiezoelectrics. For creation functional ensembles including micro- and nanopiezoelectric composites (synergetic system) we need to know how these structures can joint each with other, to know possible methods of nano- and micro composites interconnection into ensembles.

It is possible to synthesis nanoscale particles with regulated sizes and compounds, and then to immobilize these particles in polymer phase possessing unique properties In chapter of this book this problem was solved by next way: 1) Firstly, the experimental results on determination of the role of interface electron-ion and polarized processes for formation of piezo- and pyroelectric properties of composites are considered. 2) The role of electronic properties of interface boundaries are revealed; it is shown that a formation of quasi neutral system on the interface, injected charge in process of polarization and an oriented domain are key factors for formation of high piezo-, pyroelectric and electret properties. 3) Local energetic levels of high concentration and activation energy are created by crystallizing of polymer under the conditions of plasma of electric discharge; the presence of such deep traps on the interface increases essentially electrophysical properties of composites owing to increasing value of re-orientation polarization of piezophase. 4) It is revealed that intensive oxidation of polymer phase under the action of plasma of electric discharge reduces essentially a mobility of macromolecules on the interface owing to growing interface interactions and promotes increasing of electromechanical properties of composites. 5) By investigation of influence of piezoelectric particles sizes on electromechanical, mechanical and electrophysical properties it is determined that decreasing of particle sizes is accompanied by increasing of mechanical characteristics of composites and decreasing of piezo-, pyro- and electret properties. 6) By investigation of physical-mechanical properties of composites, using nanosized CO_2 and $BaTiO_3$ particles as inorganic phase, it is determined that the nanodispersion of polymer phase leads to noticeable increasing of mechanical and electromechanical parameters of composites. 7) The noticeable difference of molecular moving and supramolecular structures of polymer matrix on the interface and in the volume, where the influence of piezoelectric particles is absent, promoted a development of hybrid active composites based on the matrix nano- and micropiezoelectric composites.

6. Acknowledgment

The work was implemented under the Science Development Foundation under the President of the Republic of Azerbaijan, grant EİF-2010-1(1)-40/04-M-3.

7. References

Bagirov M.A., Nuraliev N.A., Kurbanov M.A.. (1972). Investigation of discharge in air gap and method of dtermination of the impulse number. Sov. Phys. Tech. Phys. 42(3),, pp.629-634. ISSN 54

Burfoot J.C., Taylor G.W. (1979). *Polar dielectrics and their application.* The Macmillan Press LTD, London.

Chinchurreta F.J., Montero de Espinosa F. (1988). Guides to characterize piezoelectric materials from disk shaped samples. *Ferroelectrics*, v.81, pp.229-232, ISSN: 0015-0193

Fesenko E. G., Dantsiger A.Ya., Razumovskaya O.N. (1983). *New Piezoelectric Materials.* Rostov. Gos. Univ., Rostov-on-Don. [in Russian].

Furukawa T., Ishida K., Fukuda E. (1979) Piezoelectric properties in the composite systems of polymers and PZT ceramics. *J.Appl.Phys.* 50(7), pp.4904-4912. ISSN 0021-8979.

Furukawa T., Fujino K., Fukada E. (1976) Electromechanical properties in the composites of epoxy resin and PZT ceramics. *Japan J.Appl.Physics*, v.15, pp.2119-2129. ISSN 0021-8979.

Hiroyoshi Ueda, Fukada Eichi, Karasz Frank E. (1986) Piezoelectricity in three-phase systems: Effect of the boundary phase. *J.Appl.Phys.* 60(8), pp.2672-2677. ISSN 0021-8979.

Jamada T., Ueda T., Kitayama T. (1982). Piezoelectricity of a high-content lead zirconate titanate/polymer composite. *J.Appl.Phys.*, 53(6), pp.4328-4332. ISSN 0021-8979.

Kerimov M.K., Kerimov E.A., Musaeva S.N., Panich A.E., Kurbanov M.A. (2007). The Influence of Structural and Electrical Parameters of the Pyroelectric Phase on the

Pyroelectric Properties of a Polymer–Pyroelectric Ceramic Composite. *Physics of the Solid State,* 49(5), pp. 925–929. ISSN 1063-7834,

Kerimov M.K., Kurbanov M.A., Tatardar F.N., Mekhtili A.A., Sultankhamedova I.S., Aliev G.G., Yakh'yaev F.F., Yusifova U.V. (2011). A New Technology of the Immobilization of Nanoparticles in Polymers and the Development of Piezoelectrics Based on a Hybrid Matrix of Nano- and Micropiezoceramic Composites. *Surface Engineering and Applied Electrochemistry,* 47(1), pp.76-83. ISSN 1068-3755.

Kerimov M.K., Kurbanov M.A., Agaev F.G., Musaeva S.N., Kerimov E.A. (2005) Pyroelectrical effect in composites crystallized in conditions of electrical discharge plasma acting. *Physics of the Solid State,* 47(4), pp.710-714. ISSN 1063-7834.

Kerimov M.K., Kurbanov M.A., Musaeva S.N., Geidarov G.M., Aliev G.G. (2009). New Technology of Polymer–Piezoceramic Composites with High Piezoelectric and Electromechanical Properties. *Technical Physics Letters,* 35(2), pp. 166–169. ISSN 1063-7850

Kurbanov M.A., Sultanahmedova I.S., Aliev G.G., Geydarov G.M. (2009) Piezoelectric properties of polymer-piezoelectric ceramic crystallized under the action of electric discharge plasma. *Journal of Technical Physics,* 79(7), p.63-69. ISSN 0044-4642

Kurbanov M.A., Sultanakhmedova I.S., Kerimov E.A., Aliev Kh.S., Alieva G.G., Geidarov G.M. (2009) Plasma Crystallization of Polymer–Ferroelectric/Piezoelectric Ceramic Composites and Their Piezoelectric Properties *Physics of the Solid State,* 51(6), pp.1223–1230. ISSN 1063_7834

Lines M.E., Glass A.M. (1977). Principles and application of ferroelectrics and related materials. Clarendo Press-Oxford, London.

Mamedov G., Kurbanov M.A., Ramazanova I.S., Mekhtili A.A., AlievX.S., Aliev G.G., Tatardar F.N., Orudjev I.N., Kulieva G.X., Xudayarov B.G. (2010) A new technology of the immobilization of nanoparticles in polymer matrix of hybrid micropiezoceramic Ccmposites. Proc. of the International conference "Nano-technology and its use in technique", Baku, pp.14-18

Marikhin V.A. and Myasnikova L.R. (1977) *Permolecular Structure of Polymers.* Khimiya. Leningrad. [in Russian].

Ploss B., Wong Y.W., Shin F.G. (2005). Pyroelectric Ceramic/Polymer Composite with Electrically Conducting Matrix Material *Ferroelectrics,* 325:165–169. ISSN: 0015-0193

Rez I.S., Poplavko Yu.M. (1989). Dielectrics: Main properties and applications in electronics. Radio I Svyaz, Moscow, [in Russian]

Topics in Applied Physics Founded by Helmut K.V.Lotsch (1980). vol.33 *Electrets.* Ed. by G.M. Sessler. Springer. Berlin-Heidebberg-New York.

Topolov V.Yu., Glushanin S.V., Bowen C.R. (2006). Piezoelectric response of porous ceramic and composite materials based on $Pb(Zr,Ti)O_3$: experiment and modeling. *Advances in applied Ceramics,* 105(1), pp.1-5. ISSN AAC0561.3d

Smay J.E., Cesareno J.I., Tuttle A.B., Lewis J.A. (2002). Piezoelectric properties of 3-X periodic $Pb(Zr_xTi1-_x)O_3$–polymer composites. *J. Applied Physics.* 92(10), p. 6119-6128, ISSN 0021-8979.

Vannikov A.V. and Grishina A.D. (1984) *Photochemistry of Polymeric Donor-Acceptor Complexes* Nauka. Moscow. [in Russian].

Topolov V.Yu., Panich A.E., Kurbanov M.A. (2006). Peculiarities of piezoelectric properties of 0-3 composites on the basis of PKR ceramic. *Nano- and microchip technics*. № 1, pp. 34-38, ISBN 621. 38.H 254.

Wede H., Fukada E., Karasg Frank E. (1986). Piezoelectricity in three phase systems: Effect of the boundary phase. *J.Appl.Phys.* 60(8), pp.2672-2677. ISSN 0021-8979

Zhang Q.M., Geng X. (1994). Dynamic modeling of piezoceramic polymer composite with 2-2 connectivity. *J. Applied Physics*. 76(10), pp. 6014-6016. ISSN 0021-8979